THE PRINCIPLES OF CHEMICAL EQUILIBRIUM

WITH APPLICATIONS IN CHEMISTRY AND CHEMICAL ENGINEERING

BY

KENNETH DENBIGH, F.R.S.,

Principal of Queen Elizabeth College,
University of London,
formerly Courtaulds Professor of Chemical
Engineering at Imperial College, London

THIRD EDITION

CAMBRIDGE UNIVERSITY PRESS

CAMBRIDGE

LONDON · NEW YORK · MELBOURNE

Published by the Syndics of the Cambridge University Press
The Pitt Building, Trumpington Street, Cambridge CB2 1RP
Bentley House, 200 Euston Road, London NW1 2DB
32 East 57th Street, New York, NY 10022, USA
296 Beaconsfield Parade, Middle Park, Melbourne 3206, Australia

© Cambridge University Press 1966, 1971

Library of Congress catalogue card number: 74–152683

ISBN 0 521 08151 3 hard covers
ISBN 0 521 09655 3 paperback

First published 1955
Reprinted 1957 1961 1964
Second edition 1966
Third edition 1971
Reprinted 1973 1978

Printed in Great Britain at the
University Press, Cambridge

PREFACE TO THE FIRST EDITION

My aim has been to write a book on the general theory of chemical equilibrium, including its statistical development, and displaying its numerous practical applications, in the laboratory and industry, by means of problems. It is hoped that the book may be equally useful to students in their final years of either a chemistry or a chemical engineering degree.

Thermodynamics is a subject which needs to be studied not once but several times over at advancing levels. In the first round, usually taken in the first or second year of the degree, a good deal of attention is given to calorimetry, before going forward to the second law. In the second or third rounds—such as I am concerned with in this book —it is assumed that the student is already very familiar with the concepts of temperature and heat, but it is useful once again to go over the basis of the first and second laws, this time in a more logical sequence.

The student's confidence, and his ability to apply thermodynamics in novel situations, can be greatly developed if he works a considerable number of problems which are both theoretical and numerical in character. Thermodynamics is a quantitative subject and it can be mastered, not by the memorizing of proofs, but only by detailed and quantitative application to specific problems. The student is therefore advised not to aim at committing anything to memory. The *three or four* basic equations which embody the 'laws', together with a few defining relations, soon become familiar, and all the remainder can be obtained from these as required.

The problems at the end of each chapter have been graded from the very easy to those to which the student may need to return several times before the method of solution occurs to him. At the end of the book some notes are given on the more difficult problems, together with numerical answers.

Questions marked C.U.C.E. are from the qualifying and final examinations for the Cambridge University Chemical Engineering degree, and publication is by permission. The symbols which occur in these questions are not always quite the same as in the text, but their meaning is made clear.

In order to keep the size of the book within bounds, the thermodynamics of interfaces has not been included. The discussion of galvanic cells and the activity coefficients of electrolytes is also rather brief.

Part I contains the basis of thermodynamics developed on traditional lines, involving the Carnot cycle. Part II contains the main development in the field of chemical equilibria, and the methods adopted here have been much influenced by Guggenheim's books, to which I am greatly indebted. Part III contains a short introduction to statistical mechanics along the lines of the Gibbs ensemble and the methods used by R. C. Tolman in his *Principles of Statistical Mechanics*.

It is a great pleasure to acknowledge my gratitude to a number of friends. In particular, my best thanks are due to Dr Peter Gray, Professor N. R. Amundson, Dr J. F. Davidson and Dr R. G. H. Watson, for helpful criticism and suggestions, and to Professor T. R. C. Fox, for stimulating and friendly discussions on thermodynamics over several years. Finally I wish to express my appreciation of the good work of the Cambridge University Press, and my thanks to Messrs Jonathan and Philip Denbigh, for help with the proof correcting, and to my wife for help in many other ways.

K. G. D.

CAMBRIDGE
October 1954

PREFACE TO THE THIRD EDITION

Since this book first appeared a number of correspondents have sent very helpful comments. I am particularly indebted to Professor Peter Gray, Professor E. A. Guggenheim and Professor J. S. Rowlinson, most of whose suggested improvements have been incorporated in the text.

The revision for the third edition has included the following:

(1) Complete adoption of the notation recommended in the 1969 I.U.P.A.C. publication *Manual of Symbols and Terminology for Physicochemical Quantities and Units* (Butterworths, 1970). This has involved thousands of small changes in the text; if readers find symbols which have escaped alteration, or any other sorts of error, I should be most grateful if they would kindly let me know.

(2) Adoption of the I.U.P.A.C. recommendation that work should be counted as positive when done *on* the system of interest. This is a rational proposal in so far as it gives work the same sign as an increment of internal energy, but it does have the unfortunate effect of making the important notion of *maximum work* rather less easy to explain. This will be seen from equations such as (2·16) of the new edition where I have felt obliged to use a double negative in order to make the meaning clear.

(3) Partial adoption of the SI units†. Most of the examples concerning physical processes have been converted to kg, m, J, Pa, etc., but not the examples of Chapters 4 and 10. In these I have retained the use of the calorie ($= 4.184$ J) for enthalpies and free energies of reaction and C_p power series, because this is the unit in which the only extensive compilations of these quantities are at present available. For the same reason I have continued in places using the atmosphere as a unit of pressure (1 atm $=$ 101.325 kPa $= 101\ 325$ Nm^{-2}). Indeed the atmosphere will be difficult to displace for as long as 1 atm pressure continues to be used as a standard state. Otherwise a tiresome numerical factor must be introduced into the equation $\Delta G_T^0 = -RT \ln K_p$.

The present edition is thus not a ' completely SI edition '. I hope nevertheless that it will provide students of chemistry and chemical engineering with many exercises in the use of SI and that they will thereby be in a better position to adopt the system fully as soon as the necessary compilations of basic data become available.

Additional changes to the text are the inclusion of ξ, the extent of reaction, certain improvements in the presentation of reaction equilibrium and the correction of a false impression concerning ergodity in Chapter 11.

September 1970   K. G. D.

NOTE ON THE 1978 REPRINTING

I am greatly indebted to various correspondents who have so kindly taken the trouble to inform me about errors and misprints. Professor Thomas W. Weber gave me a list of forty items and several other correspondents have also contributed very helpfully to the improved accuracy of this reprinting.

December 1977   K. G. D.

---

† SI units are briefly described on p. xxi.

CONTENTS

PART I: THE PRINCIPLES OF THERMODYNAMICS

Chapter 1: First and Second Laws

PART II: REACTION AND PHASE
EQUILIBRIA

Chapter 4: Equilibria of Reactions Involving Gases

Chapter 5: Phase Rule

Chapter 6: Phase Equilibria in Single Component Systems

Chapter 7: General Properties of Solutions and the Gibbs–Duhem Equation

Chapter 8: Ideal Solutions

Chapter 9: Non-Ideal Solutions

PART III: THERMODYNAMICS IN RELATION TO THE EXISTENCE OF MOLECULES

Chapter 11: Statistical Analogues of Entropy and Free Energy

Chapter 12: Partition Function of a Perfect Gas

Chapter 13: Perfect Crystals and the Third Law

Chapter 14: Regular Solutions and Adsorption

Chapter 15: Chemical Equilibrium in Relation to Chemical Kinetics.

LIST OF SYMBOLS

		Definition	
		Equation	Page
a_i	Activity of ith species	—	287
A	Helmholtz free energy of a system	2·1	63
c_i	Molarity of ith species	9·11	275
C	Compressibility factor of a gas	3·51	124
C	Number of independent components of a system	—	171, 184, 187
c_p	Molar heat capacity at constant pressure	2·87	96
C_p	Heat capacity of system at constant pressure	2·86	95
c_V	Molar heat capacity at constant volume	2·87	96
C	Heat capacity of system at constant volume	2·86	95
e	Symbol for an electron	—	—
E	Any extensive property of a system	—	8
E	Electromotive force	—	75, 164
E	Total energy of a system	—	17
E_i	Energy of the ith quantum state of a macroscopic system	—	342
f	Fugacity	3·45 and 3·56	122, 125
f	Molecular partition function	12·9	364
F	The Faraday	—	75
F	Degrees of freedom of system	—	185
G	Gibbs free energy of system	2·3	63
$\Delta_f G_T^0$	Standard free energy of formation from the elements at temperature T	—	148
ΔG_T^0	Standard free energy change in reaction at temperature T	4·17	142
G^E	Excess free energy	—	285
h	Planck constant	—	—
h	Enthalpy per mole of a pure substance	—	100
H	Enthalpy of system	2·1	63

List of Symbols

		Defin	ition
		Equation	Page
s	Entropy per mole of pure substance	2·99	98
S	Entropy of system	1·13	32
S_i	Partial molar entropy of ith species	2·104	99
S'	$S' \equiv -k\Sigma P_i \ln P_i$	11·14	343
S''	$S'' \equiv k \ln \Omega$	11·15	343·
T	Temperature on thermodynamic scale	1·12	31
u	Internal energy per mole of pure substance	2·99	98
U	Internal energy of system	—	17
U_i	Partial molar internal energy of ith species	2·104	99
v	Volume per mole of pure substance	2·99	98
V	Volume of system	—	—
V_i	Partial molar volume of ith species	2·104	99
w	Work done on system	—	14
w'	Work done on system, not including that part which is due to volume change	—	66
w_{ij}	Potential energy of a pair of molecules of types i and j	—	243, 430
x_i	Mole fraction of ith species in condensed phase	—	—
y_i	Mole fraction of ith species in vapour phase	—	—
z_+, z_-	Charges of positive and negative ions respectively in units of the proton charge	—	73, 163, 300
α	Coefficient of thermal expansivity	2·88	94
β	A statistical parameter	11·10	342
γ_i	Activity coefficient of ith species	9·2, 9·3 and 9·16	269, 274
γ	Surface tension	—	—
Δ	Sign indicating excess of final over initial value	—	—
ϵ_i	Energy of the ith quantum state of a molecule	—	—
θ	Temperature on any scale	—	11
κ	Compressibility coefficient	2·89	94
μ_i	Chemical potential of ith species	2·39 and 2·41	76, 77
μ_i^0	Gibbs free energy per mole of pure substance at unit pressure and at the same temperature as that of the mixture under discussion	—	Ch. 3

		Definition	
		Equation	Page
μ_i^*	Gibbs free energy per mole of pure substance at the same temperature and pressure as that of mixture under discussion	—	Chs. 8 and 9
$\mu_i^\square$	Chemical potential of ith species in a hypothetical ideal solution of unit molality at the same temperature and pressure as solution under discussion	—	276
μ	Joule–Thomson coefficient	3·42	120
ν_i	Stoichiometric coefficient for ith species in a reaction	—	134
ν_+, ν_-	Numbers of positive and negative ions respectively formed on dissociation of one molecule of electrolyte	—	302
ν	$\nu \equiv \nu_+ + \nu_-$	10·34	305
ξ	Extent of reaction	4·2	135
Π	Continued product operator sign	—	141
Π	Osmotic pressure	8·54	263
ρ	Density	—	—
σ	Created entropy	1·15	39
Σ	Summation operator sign	—	—
ϕ	Potential energy	—	17, 87
χ_i	Fugacity coefficient of ith species	3·54	125
ω_i	Degeneracy of the ith molecular energy level	—	361, 367
Ω	Number of accessible quantum states of a macroscopic system of constant energy and volume	—	335, 338
Ω_i	Ditto as applied to the particular energy state E_i	—	353, 367
$\approx$	Denotes an approximate equality	—	—
$\doteqdot$	Denotes a very close approximation	—	—
$\equiv$	Used where it is desired to emphasize that the relation is an identity, or a definition.	—	—

VALUES OF PHYSICAL CONSTANTS

Ice-point temperature	$T_{ice} = 273.15$ K
Boltzmann constant	$k = 1.380\ 54 \times 10^{-23}$ J K^{-1}
Planck constant	$h = 6.625\ 6 \times 10^{-34}$ J s
Avogadro constant	$L = 6.022\ 52 \times 10^{23}$ mol^{-1}
Faraday constant	$F = 96\ 487$ C mol^{-1}
Charge of proton	$e = 1.602\ 10 \times 10^{-19}$ C
Gas constant	$R = 8.314\ 3$ J K^{-1} mol^{-1}
	$= 1.987\ 2$ cal K^{-1} mol^{-1}
	$= 82.06$ cm^3 atm K^{-1} mol^{-1}

THE SI UNITS

The basic SI units are the metre (m), kilogramme (kg), second (s), kelvin (K; *not* °K), mole (mol), ampere (A) and candela (cd). The mole is the unit of 'amount of substance' and is defined as that amount of substance which contains as many elementary entities as there are atoms in 0.012 kg of carbon-12. It is thus precisely the same amount of substance as, in the c.g.s. system, had been called the 'gramme-molecule'. Some of the SI derived units which are important in the present volume, together with their symbols, are as follows:

for energy . . . the joule (J); kg m^2 s^{-2}
for force . . . the newton (N); kg m s^{-2} = J m^{-1}
for pressure . . . the pascal (Pa); kg m^{-1} s^{-2} = N m^{-2}
for electric charge . . . the coulomb (C); A s
for electric potential difference . . . the volt (V);
$$\text{kg m}^2\,\text{s}^{-3}\,\text{A}^{-1} = \text{J A}^{-1}\,\text{s}^{-1}$$

In terms of SI units two 'old-style' units which are also used in this book are:

the thermochemical calorie (cal) = 4.184 J
the atmosphere (atm)   = 101.325 kPa = 101 325 N m^{-2}

PART I

THE PRINCIPLES OF
THERMODYNAMICS

CHAPTER 1

FIRST AND SECOND LAWS

1·1. Introduction

One reason why the study of thermodynamics is so valuable to students of chemistry and chemical engineering is that it is a theory which can be developed in its entirety, without gaps in the argument, on the basis of only a moderate knowledge of mathematics. It is therefore a self-contained logical structure, and much benefit—and incidentally much pleasure—may be obtained from its study. Another reason is that it is one of the few branches of physics or chemistry which is largely independent of any assumptions concerning the nature of the fundamental particles. It does not depend on 'mechanisms', such as are used in theories of molecular structure and kinetics, and therefore it can often be used as a check on such theories.

Thermodynamics is also a subject of immense practical value. The kind of results which may be obtained may perhaps be summarized very briefly as follows:

(a) On the basis of the first law, relations may be established between quantities of heat and work, and these relations are not restricted to systems at equilibrium.

(b) On the basis of the first and second laws together, predictions may be made concerning the effect of changes of pressure, temperature and composition on a great variety of physico-chemical systems. These applications are limited to systems at equilibrium. Let χ be a quantity characteristic of an equilibrium, such as the vapour pressure of a liquid, the solubility of a solid, or the equilibrium constant of a reaction. Then some of the most useful results of thermodynamics are of the form

$$\left(\frac{\partial \ln \chi}{\partial T}\right)_p = \frac{(\text{A characteristic energy})}{RT^2},$$

$$\left(\frac{\partial \ln \chi}{\partial p}\right)_t = \frac{(\text{A characteristic volume})}{RT}.$$

The present volume is mainly concerned with the type of results of (b) above. However, in any actual problem of chemistry, or the chemical industry, it must always be decided, in the first place, whether the essential features of the problem are concerned with *equilibria* or with *rates*. This point may be illustrated by reference to two well-known chemical reactions.

In the synthesis of ammonia, under industrial conditions, the reaction normally comes sufficiently close to equilibrium for the applications of thermodynamics to prove of immense value.† Thus it will predict the influence of changes of pressure, temperature and composition on the maximum attainable yield. By contrast in the catalytic oxidation of ammonia the yield of nitric oxide is determined, not by the opposition of forward and backward reactions, as in ammonia synthesis, but by the relative speeds of two independent processes which compete with each other for the available ammonia. These are the reactions producing nitric oxide and nitrogen respectively, the latter being an undesired and wasteful product. The useful yield of nitric oxide is thus determined by the *relative speeds* of these two reactions on the surface of the catalyst. It is therefore a problem of rates and not of equilibria.

The theory of equilibria, based on thermodynamics, is much simpler, and also more precise, than any theory of rates which has yet been devised. For example, the equilibrium constant of a reaction in a perfect gas can be calculated exactly from a knowledge only of certain macroscopic properties of the pure reactants and products. The rate cannot be so predicted with any degree of accuracy for it depends on the details of molecular structure and can only be calculated, in any precise sense, by the immensely laborious process of solving the Schrödinger wave equation. Thermodynamics, on the other hand, is independent of the fine structure of matter,‡ and its peculiar simplicity arises from a certain condition which must be satisfied in any state of equilibrium, according to the second law.

The foundations of thermodynamics are three facts of ordinary experience. These may be expressed very roughly as follows:

(1) bodies are at equilibrium with each other only when they have the same degree of 'hotness';

(2) the impossibility of perpetual motion;

(3) the impossibility of reversing any natural process in its entirety.

In the present chapter we shall be concerned with expressing these facts more precisely, both in words and in the language of mathematics. It will be shown that (1), (2) and (3) above each gives rise to the definition of a certain function, namely, temperature, internal energy and entropy respectively. These have the property of being entirely determined by the state of a body and therefore they form exact differentials. This leads to the following equations which contain the *whole* of the fundamental theory:

† The error in using thermodynamic predictions, as a function of the extent to which the particular process falls short of equilibrium, is discussed by Rastogi and Denbigh, *Chem. Eng. Science*, 7 (1958), 261.

‡ In making this statement we are regarding thermodynamics as having its own secure empirical basis. On the other hand, the laws of thermodynamics may themselves be interpreted in terms of the fine structure of matter, by the methods of statistical mechanics (Part III).

$$\mathrm{d}U = \mathrm{d}q + \mathrm{d}w,$$

$$\mathrm{d}S = \mathrm{d}q/T, \quad \text{for a reversible change,}$$

$$\mathrm{d}S \geqslant 0, \quad \text{for a change in an isolated system,}$$

$$\mathrm{d}U = T\mathrm{d}S - p\mathrm{d}V + \Sigma\mu_i\mathrm{d}n_i \quad \text{for each homogeneous part}$$
of a system.

Subsequent chapters of Parts I and II will be concerned with the elaboration and application of these results. The student is advised that there is no need to commit any equations to memory; the four above, together with a few definitions of auxiliary quantities such as free energy, soon become familiar, and almost any problem can be solved by using them.

In conclusion to this introduction it may be remarked that a new branch of thermodynamics has developed during the past few decades which is not limited in its applications to systems at equilibrium. This is based on the use of the principle of microscopic reversibility as an auxiliary to the information contained in the laws of classical thermodynamics. It gives useful and interesting results when applied to non-equilibrium systems in which there are coupled transport processes, as in the thermo-electric effect and in thermal diffusion. It does not have significant applications in the study of chemical reaction or phase change and for this reason is not included in the present volume.†

1·2. Thermodynamic systems

These may be classified as follows:

Isolated systems are those which are entirely uninfluenced by changes in their environment. In particular, there is no possibility of the transfer either of energy or of matter across the boundaries of the system.

Closed systems are those in which there is the possibility of energy exchange with the environment, but there is no transfer of matter across the boundaries. This does not exclude the possibility of a change of internal composition due to chemical reaction.

Open systems are those which can exchange both energy and matter with their environment. An open system is thus not defined in terms

† For an elementary account of the theory see the author's *Thermodynamics of the Steady State* (London, Methuen, 1951), or de Groot's more comprehensive monograph *Thermodynamics of Irreversible Processes* (Amsterdam, North-Holland Publishing Co., 1951). Also Prigogine's *Introduction to the Thermodynamics of Irreversible Processes* (Wiley, 1962), Callens' *Thermodynamics* (Wiley, 1960), Fitt's *Non-Equilibrium Thermodynamics* (McGraw-Hill, 1962), van Rysselberghe's *Thermodynamics of Irreversible Processes* (Hermann, 1963) and de Groot and Mazur's *Non-Equilibrium Thermodynamics* (North Holland Publishing Co., 1962).

of a given piece of material but rather as a region of space with geometrically defined boundaries across which there is the possibility of transfer of energy and matter.

Where the word *body* is used below it refers either to the isolated or the closed system. The preliminary theorems of thermodynamics all refer to bodies, and many of the results which are valid for them are not directly applicable to open systems.

The application of thermodynamics is simplest when the system under discussion consists of one or more parts, each of which is spatially uniform in its properties and is called a *phase*. For example, a system composed of a liquid and its vapour consists almost entirely of two homogeneous phases. It is true that between the liquid and the vapour there is a layer, two or three molecules thick, in which there is a gradation of density, and other properties, in the direction normal to the interface. However, the effect of this layer on the thermodynamic properties of the overall system can usually be neglected. This is because the work involved in changes of interfacial area, of the magnitudes which occur in practice, is small compared to the work of volume change of the bulk phases. On the other hand, if it were desired to make a thermodynamic analysis of the phenomena of surface tension it would be necessary to concentrate attention on the properties of this layer.

Thermodynamic discussion of real systems usually involves certain approximations which are made for the sake of convenience and are not always stated explicitly. For example, in dealing with vapour-liquid equilibrium, in addition to neglecting the interfacial layer, it is customary to assume that each phase is uniform throughout its depth, despite the incipient separation of the components due to the gravitational field. However, the latter effect can itself be treated thermodynamically, whenever it is of interest.

Approximations such as the above are to be distinguished from certain *idealizations* which affect the validity of the fundamental theory. The notion of isolation is an idealization, since it is never possible to separate a system completely from its environment. All insulating materials have a non-zero thermal conductivity and allow also the passage of cosmic rays and the influence of external fields. If a system were completely isolated it would be unobservable.

1·3. Thermodynamic variables

Thermodynamics is concerned only with the macroscopic properties of a body and not with its atomic properties, such as the distance between the atoms in a particular crystal. These macroscopic properties form a large class and include the volume, pressure, surface tension, viscosity, etc., and also the 'hotness'. They may be divided into two groups as follows:

The *extensive* properties, such as volume and mass, are those which are additive, in the sense that the value of the property for the whole of a body is the sum of the values for all of its constituent parts.

The *intensive* properties, such as pressure, density, etc., are those whose values can be specified at each point in a system and which may vary from point to point, when there is an absence of equilibrium. Such properties are not additive and do not require any specification of the quantity of the sample to which they refer.

Consider the latter class and let it be supposed that the system under discussion is closed and consists of a single phase which is not undergoing a chemical reaction. For such a system it is usually found that the specification of any two of the intensive variables will determine the values of the rest. For example, if $I_1, I_2, ..., I_j, ..., I_n$, are the intensive properties then the fixing of, say, I_1 and I_2 will give the values of all the others. Thus†

$$I_j = f(I_1, I_2) \quad (j = 3, 4, ..., n). \tag{1·1}$$

For example, if the viscosity of a sample of water is chosen as 0.506×10^{-3} N s m^{-2} and its refractive index as 1.3289, then its density is 0.0881 g cm^{-3}, its 'hotness' is 50 °C, etc. In the next section, instead of choosing viscosity and refractive index, we shall take as our reference variables the pressure and density, which are a more convenient choice. On this basis we shall discuss what is meant by 'hotness' or 'temperature' (which it is part of the business of thermodynamics to define), and thereafter we shall take pressure and temperature as the independent intensive variables, as is always done in practice.

What has just been said, to the effect that two intensive properties of a phase usually determine the values of the rest, applies to mixtures as well as to pure substances. Thus a given mixture of alcohol and water has definite properties at a chosen pressure and density. On the other hand, in order to specify which particular mixture is under discussion it is necessary to choose an extra set of variables, namely, those describing the *chemical composition* of the system. These variables depend on the notion of the *pure substance*, namely, a substance which cannot be separated into fractions of different properties by means of the *same processes* as those to which we intend to apply our

---

† The equation means that I_3, I_4, etc., are all functions of I_1 and I_2. Thus $I_3 = f(I_1, I_2)$ might stand for $I_3 = I_1^2 I_2^4$, $I_3 = I_1/I_2^2$, etc. A simple relation of this kind is $T = $ constant p/ρ for the temperature of a gas as a function of its pressure and density. However, for most substances and properties the precise form of the functional relationship is unknown.

thermodynamic discussion; for example, the simple physical processes such as vaporization, passage through a semi-permeable membrane, etc. If there are q pure substances present, the composition may be expressed by means of $q - 1$ of the mole fractions, denoted $x_1, x_2, \ldots,$ x_{q-1}. Thus in place of the previous relation we have

$$I_j = f(I_1, I_2, x_1, x_2, \ldots, x_{q-1}). \tag{1·2}$$

These considerations apply to each phase of the system.

Turning now to the *extensive* properties it is evident that the choice of only two of these is insufficient to determine the state of a system, even if it is a pure substance. Thus if we fix both the volume and mass of a quantity of hydrogen, it is still possible to make simultaneous changes of pressure and of 'hotness'. An extensive property of a pure phase is usually determined by the choice of *three* of its properties, one of which may be conveniently chosen as the mass (thereby determining the quantity of the pure phase in question) and the other two as intensive properties. For example, if $E_1, E_2, \ldots, E_r$ are extensive properties, then any one of them will usually be determined by the same two intensive variables, I_1 and I_2, as chosen previously, together with the total mass m. Thus

$$E_i = m \times f(I_1, I_2) \quad (i = 1, 2, \ldots, r). \tag{1·3}$$

This equation expresses also that E_i is proportional to m, since E_i is additive. It will be recognized that the quotient E_i/m, of which specific volume is an example, is a member of the group of intensive variables. Such quotients are called *specific properties*. In the case of a phase which is a mixture it is also necessary, of course, to specify the composition:

$$E_i = m \times f(I_1, I_2, x_1, x_2, \ldots, x_{q-1}). \tag{1·4}$$

It may be remarked that thermodynamics provides no criterion with regard to the minimum number of variables required to fix the state of a system. There are a number of instances in which the remarks above with regard to two intensive variables fixing the remainder are inapplicable. For example, we can find pairs of states of liquid water, one member of each pair being on one side of the point of maximum density and the other member on the other side, each of which have the same density and the same pressure (chosen as greater than the vapour pressure) and yet do not have identical values of other properties, such as viscosity. This is because the density does not vary monotonically with the other variables, but passes through a maximum.

In other instances it is necessary to introduce an extra variable of state. For example, in the case of a magnetic substance it will be necessary to specify, say, the field intensity together with pressure and 'hotness'. Similarly, in the case of colloids, emulsions and fine powders the properties are greatly affected both by the total inter-facial area and by the distribution of the particles over the size range.

The minimum number of variables of state, whose values determine the magnitudes of all other macroscopic variables, is thus an em-pirical fact to be determined by experience. In any particular applica-tion, if it is found that the system does not appear to obey the laws of thermodynamics, it may be suspected that an insufficient number of variables of state have been included in the equations.

1·4. Temperature and the zeroth law†

In the last section only passing reference was made to the property of 'hotness'. It is part of thermodynamics to define what is meant by this, whereas the mechanical and geometrical concepts, such as pressure and volume, are taken as being understood.

Now it is a fact of experience that a set of bodies can be arranged in a *unique* series according to their hotness, as judged by the sense of touch. That is to say, if A is hotter than B, and B is hotter than C, then A is also hotter than C. The same property is shown also by the real numbers; thus if n_a, n_b and n_c are three numbers such that $n_a > n_b$ and $n_b > n_c$, then we have also $n_a > n_c$. This suggests that the various bodies arranged in their order of hotness, can each be assigned a number such that larger numbers correspond to greater degrees of hotness. The number assigned to a body may then be called its temperature, but there are obviously an infinite variety of ways in which this numbering can be carried out.

The notion of temperature must clearly be placed on a more exact basis than is provided by the sense of touch. Furthermore, in order to avoid any circularity in the argument, this must be done without any appeal to the notion of heat, which follows logically at a later stage.

The definition of temperature now to be obtained depends on what happens when two bodies are placed in contact under conditions where their pressures and volumes can be varied independently. For this purpose each body must be enclosed by an impermeable wall which can be moved inwards or outwards.

It is useful to think of two kinds of impermeable wall. The first, which will be called *diathermal* or *non-adiabatic*, is such that two bodies separated by a wall of this kind are nevertheless capable of exerting an influence on each other's thermodynamic state through the wall. The existence of diathermal materials is, of course, a matter of common experience and shortly they will be identified as materials

† The discussion of temperature and the first law is based on that of Born, *Phys. Z.* **22** (1921), 218, 249, 282; also his *Natural Philosophy of Cause and Chance* (Oxford, 1949).

capable of transmitting what will then be called *heat*. The second type of wall will be called *adiabatic*. A body completely surrounded by a wall of this kind cannot be influenced (apart from the possible effects of force fields) from outside, except by compressing or expanding the wall, or otherwise causing internal motion.

As remarked by Pippard† the adiabatic wall may be thought of as the end stage of a process of extrapolation. A metal wall is clearly diathermal in the above sense; on the other hand the type of double, and internally highly evacuated, wall used in a vacuum flask is almost completely adiabatic. The concept of the ideal adiabatic wall is thus a legitimate extrapolation from the conditions existing in the vacuum flask.

The definition which has been made does not depend on any previous knowledge of heat. Similarly, we shall speak of any change taking place inside an adiabatic wall as being an *adiabatic process*. Bodies will also be said to be in *thermal contact* when they are either in direct contact (e.g. two pieces of copper) or in contact through a non-adiabatic wall (e.g. two samples of gas). Their final state, when all observable change has come to an end, is called *thermal equilibrium*.

Now it is a fact of experience that if bodies A and B are each in thermal equilibrium‡ with a third body, they are also in thermal equilibrium with each other. This result is so familar that it is regarded almost as a truism. However, there is no self-apparent reason why it should be so, and it must be regarded as an empirical fact of nature and has become known as the *zeroth law of thermodynamics*. It is the basis of the scientific concept of temperature, which may now be outlined as follows.

We consider two bodies, each of them a homogeneous phase in a state of internal equilibrium, which are in contact through a non-adiabatic wall. The thermodynamic state of each body may be completely specified by means of two variables only, and these may conveniently be chosen as the volume per unit mass and the pressure. These variables will determine the property called 'hotness', together with all other properties. Let the variables be p and v for the one body and P and V for the other. When they are brought into contact in this way, at initially different degrees of hotness, there is a slow change in the values of the pressures and volumes until the state of thermal

---

† Pippard. *Elements of Classical Thermodynamics*, Cambridge, 1961.

‡ It has been pointed out by I. P. Bazarov (*Thermodynamics*, p. 4, Pergamon, 1964) that the fact that isolated systems do reach a state of equilibrium, and do not depart from it spontaneously, is essentially a basic postulate of thermodynamics. Hatsopoulos and Keenan (*Principles of General Thermodynamics*, Wiley, 1965) have further explored the meaning of the equilibrium concept, and have put forward a *law of stable equilibrium* from which the first and second laws may be derived.

equilibrium is attained. Let p', v', P' and V' be the values of the variables at the state of equilibrium. If the first body is momentarily removed it will be found that its pressure and volume can be adjusted to a second pair of values, p'' and v'', which will again give rise to a state of equilibrium with the other body, which is still at P' and V'. In fact, there are a whole sequence of states (p', v'), (p'', v''), (p''', v'''), etc., of the first body, all of which are in equilibrium with the state (P', V') of the other.

We can thus draw a curve (Fig. 1 a), with co-ordinates p and v, which is the locus of all points which represent states of the first body which are in equilibrium with the state (P', V') of the second. According to the zeroth law all states along such a curve are also in equilibrium with each other; that is to say, two replicas of the first body would be in equilibrium with each other if their pressures and volumes correspond to any two points along this curve.

A similar curve (Fig. 1 b) can be drawn for the second body in its own P, V co-ordinates—that is, there is a curve which is the locus of all states which are in equilibrium with a given state of the first body, and therefore in equilibrium with each other. There are thus two

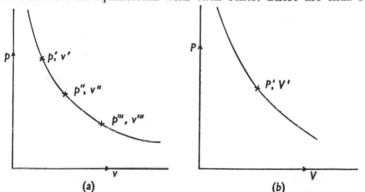

Fig. 1. The curves are not drawn to the equation $pv = $ constant because the discussion is not limited to perfect gases.

curves, one for each body, and every state on the one curve can be put into equilibrium with every other state on this curve, and also with every state on the other curve. Such states thus have a property in common and this will be called their *temperature*. It is consistent with our more intuitive ideas that such states are found also to have the same degree of hotness, as judged by the sense of touch. However, from the present point of view, this may now be regarded as being of physiological rather than of thermodynamic interest.

The existence of the common property along the two curves—which may now be called *isothermals*—can be seen more clearly as follows. The equation to a curve concerning the variables p and v can be expressed in the form†

$$f(p, v) = \theta,$$

where θ is a constant for all points along the curve of Fig. 1a. Similarly for the other body we have $F(P, V) =$ constant, along the curve of Fig. 1b. By inclusion of a suitable pure number in one or other of these functions the two constants can be made numerically equal. Therefore

$$f(p, v) = F(P, V) = \theta. \tag{1.5}$$

There is thus a certain function, θ, of the pressure and volume of a phase which has the same value for all states of the phase which are in thermal equilibrium with each other and it is equal also to a function (not necessarily of the same form) of the pressure and volume of a second phase which is in thermal equilibrium with the first. It is this function which is called the temperature.

For example, if the first phase is a perfect gas the equation to the curve of Fig. 1a is simply $pv =$ constant, and if the second phase is a gas which departs slightly from perfection, the curve of Fig. 1b can be represented by the equation $P(V - b) =$ constant, where b is also a constant. Thus (1.5) may be expressed

$$\frac{pv}{\text{constant}} = \frac{P(V - b)}{\text{constant}} = \theta.$$

and such equations are called *equations of state*.

Returning to the earlier discussion, there are, of course, states of the first phase which are not in thermal equilibrium with the second one in its state (P', V'). Let equation (1.5) be rewritten

$$f(p_1, v_1) = F(P_1, V_1) = \theta_1,$$

then there are a whole sequence of new states (p_2', v_2'), (p_2'', v_2''), etc., each of which gives rise to a state of thermal equilibrium with the second body in the states (P_2', V_2'), (P_2'', V_2''), etc. These states define two new curves, one for each body, which satisfy the relation‡

$$f(p_2, v_2) = F(P_2, V_2) = \theta_2,$$

and thus define a different value of the temperature θ_2.§ In fact, there is an infinite family of isothermals, some of which are shown in Fig. 2, such that all states on corresponding curves are in equilibrium with each other and thus define a particular temperature.

† For example, the equation to a straight line is $y - ax =$ constant; the equation to a circle whose centre is at the origin is $x^2 + y^2 =$ constant.

‡ It is tacitly assumed that the form of the function $f(p, v)$ is the same at the two temperatures θ_1 and θ_2. (The same is assumed of $F(P, V)$.) Fortunately there exist substances—perfect or near perfect gases—for which this is true and one of these, such as nitrogen may be taken as the thermometric reference substance.

§ $f(p_2, v_2)$ is not necessarily quite the same function of p_2 and v_2 as $f(p_1, v_1)$ is of p_1 and v_1. The functional relationship may change over the temperature interval, and similarly with regard to $F(P, V)$. However this does not affect the argument.

It follows that if we choose a suitable substance, e.g. a sample of gaseous nitrogen, as a reference phase we can define an empirical temperature scale by means of its appropriate function, $F(P, V)$, together with some convention for the numbering of its isothermals. The reference phase is called a *thermometer*.

The temperature is therefore any function of the pressure and volume of the reference phase which remains constant when the phase passes through a sequence of states at equilibrium with each other. For example, if we choose 1 g of nitrogen as defining our standard thermometer it would be permissible to take the value of any of the following functions as being the numerical value of the temperature, since all of them are constant along a given isothermal (provided that the thermometer is operated at a fairly low pressure):

(a) PV, (b) PV/constant, (c) $1/PV$, (d) $(PV)^{\frac{1}{2}}$, (e) $-PV$, and so on.

It is a matter of convention which of these we adopt, and also the way in which the isothermals are numbered. It may be noted that temperatures defined on scale (c) would correspond to decreasing

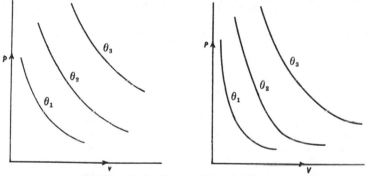

Fig. 2. Isotherms of two bodies.

values of θ with increasing values of the 'hotness', whilst the scale (e) would give values of θ which are negative. There is no objection to such scales, except as a matter of convenience. On the other hand, a function such as sine (PV) must be rejected, since it does not increase monotonically with increase in hotness.

This question need not be pursued any further because we shall shortly obtain a definition of a *thermodynamic temperature* having the important properties: (a) it is independent of any particular substance (b) the temperatures are always positive numbers and increase with increasing degrees of hotness. This scale may be shown to be identical with the scale (b) above.

One further point may be made at this stage. The relation $\theta = F(P, V)$ can be rewritten as

$$V = F'(P, \theta)$$

(where F and F' are different functions). Thus as soon as the value of the temperature has been defined by means of the reference phase, we can start to use the temperature as a variable of state. For practical purposes it is customary to regard pressure and temperature as being the independent variables, rather than pressure and volume.

1·5. Work

The notion of work is not regarded as being in need of definition in thermodynamics, since it is a concept which is already defined by the primary science of mechanics.

An infinitesimal amount of work, dw, can either be done by the system of interest on its environment, or by the environment on the system. The 1970 I.U.P.A.C. recommendation† is to take dw as being positive in the latter case and negative in the former, that is, positive work is done *on* the system. Provided the changes take place slowly and without friction, work can usually be expressed in the form

$$dw = y dx,$$

or as a sum of such terms $\quad dw = \sum_i y_i dx_i,$ \hfill (1·6)

where the y_i and x_i are generalized forces and displacements respectively.

For example, the work done *on* a body in an infinitesimal increase of its volume, dV, against an opposing pressure, p, is $-p dV$. Similarly the work done *on* a homogeneous phase when it increases its surface area by dA is $+\gamma dA$, where γ is the surface tension against the particular environment. If a system such as a galvanic cell causes dQ coulombs of electricity to flow into a condenser between whose plates the potential difference is E volts, the work done *on* the galvanic cell is $-E dQ$ joules. (Simultaneously, the atmosphere does an amount of work $-p dV$ *on* the cell, where dV is the change in volume of the cell during the chemical process in question.) Similar expressions may be obtained for the stretching of wires, the work of magnetization, etc.‡

It may be noted that all forms of work are interchangeable by the use of simple mechanical devices such as frictionless pulleys, electric motors, etc. When the term 'work' is correctly used, whatever form of work is under discussion may always be converted (because of this interchangeability) to the raising of a weight. In most kinds of chemical system, apart from the galvanic cell, the work of volume

† The adoption of this recommendation in the present edition implies that all work terms are changed in sign as compared to earlier editions.

‡ A very clear account of work terms is given by Zemansky, *Heat and Thermodynamics* (New York), McGraw-Hill, 1957, and by Pippard, *loc. cit.* It is very important to remember in what follows that all differential changes dV, dA, dU, etc., are positive quantities when they are increments, and are negative quantities when they are decrements. This is the normal convention of the calculus.

change is the only form of work which is of appreciable magnitude. However, the possibility that other forms may be significant must always be borne in mind in approaching a new problem. In such cases it may be necessary to introduce additional variables of state, e.g. the surface area of the system or the magnetic field strength.

In using expressions of the form $dw = y\,dx$ it is usually necessary to specify that the process in question is a slow one, as otherwise there may be ambiguity about the value of the force y. For example, when a gas rapidly expands or contracts, its internal pressure is not equal to the external force per unit area, and in fact the pressure varies from one region of the gas to another. This difficulty disappears when the changes are very slow and when friction is absent, since the opposing forces then approach equality. This point will be discussed more fully in a later section.

1·6. Internal energy and the first law

It will be remembered that an adiabatic wall was defined as one which prevents an enclosed body being influenced from beyond, except by the effect of motion. (We are not concerned here with force fields.) Experience shows that when there *is* motion of the wall, or parts of it, the state of the adiabatically enclosed body can be changed; for example by compressing or expanding the enclosing wall, or by shaking the body inside. The first law of thermodynamics is based on a consideration of such processes, which correspond to the performance of work.

The first law is based mainly on the series of experiments carried out by Joule between 1843 and 1848. The most familiar of these experiments is the one in which he raised the temperature of a quantity of water, almost completely surrounded by an adiabatic wall,* by means of a paddle which was operated by a falling weight. The result of this experiment was to show an almost exact proportionality between the amount of work expended on the water and the rise in its temperature. This result, considered on its own, is not very significant; the really important feature of Joule's work was that the paddle-wheel experiments gave *the same proportionality* as was obtained in several other quite different methods of transforming work into the temperature rise of a quantity of water. These were as follows:

(a) the paddle-wheel experiments;†

(b) experiments in which mechanical work was expended on an

* For each of the experimental systems now to be described, the student should ask himself, What exactly is the system which is regarded as being enclosed by adiabatic and *moving* walls? This is perhaps least obvious in experiment (b).

† *Phil. Mag.* **31** (1847), 173; **35** (1849), 533.

electrical machine, the resultant electric current being passed through a coil immersed in the water;‡

(c) experiments in which mechanical work was expended in the compression of a gas in a cylinder, the latter being immersed in the water;§

(d) experiments in which the mechanical work was expended on two pieces of iron which rubbed against each other beneath the surface of the water.‖

The scheme of Joule's experiments was thus as follows:

Mechanical work
— Turbulent motion of water → Temperature rise of water (773)

Electric current → Temperature rise of water (838)

Compression of gas → Temperature rise of water (795)

— Friction of iron blocks → Temperature rise of water (775)

The figures in brackets show the number of foot-pounds of work needed to raise the temperature of 1 lb. of water by 1 °F. The result of the most accurate series of experiments, those with the paddles, is equivalent to 4.16 joules/cal (15 °C), which is close to the present accepted value of 4.184. However, the significant conclusion is that each of the four different methods of transforming work into a temperature rise gives essentially the same result—at any rate to within the accuracy which could be attained in these early experiments.†

Now in each of the experiments the water was enclosed by an approximately adiabatic wall (as previously defined), apart from certain openings necessary for the introduction of the paddle shaft, etc. By including not only the water but also certain other items in the total thermodynamic system under discussion, each of Joule's experimental arrangements may be regarded as consisting of an adiabatic enclosure together with an external source of work; the adiabatic wall is set into motion and creates a change of state within the enclosure. Therefore, the conclusion to be drawn is that the performance of a given amount of work, on a quantity of adiabatically enclosed water, causes the same temperature rise (i.e. the same change of state), by whatever method the process is carried out.

‡ Ibid. **23** (1843), 435.   § Ibid. **26** (1845), 369.   ‖ Ibid. **35** (1849), 533.

† The student should consider carefully why it is that the work expended on the water in Joule's experiments is not equal to the integral of $p\,dV$, where p is the pressure exerted by the atmosphere on the water and dV is the change in volume of the water accompanying its rise in temperature. The answer is implicit in the last paragraph of §1·5.

This important empirical result has, of course, been confirmed by many later workers and using several substances other than water. Without discussing the great mass of evidence, we shall therefore make a preliminary formulation of the *first law of thermodynamics* as follows: the change of a body inside an adiabatic enclosure from a given initial state to a given final state involves the same amount of work by whatever means the process is carried out.†

It will be recognized that statements similar to this are met with in other branches of physics. Thus the work required to lift a weight between two points in a gravitational field, or to move a charge between two points in an electric field, is the same whatever is the path. In both of these examples it is possible to define a potential function, ϕ, such that the work done on a body in taking it from an initial state A to a final state B is equal to $\phi_B - \phi_A$, where ϕ_B and ϕ_A depend only on the states A and B—i.e. *they are independent of the path*.

In taking the work as equal to the change in potential energy it is tacitly assumed, of course, that the body does not undergo any change of its internal state; the potential energy is simply that part of the energy of a body which is due to its *position*. But the essential result of Joule's experiments is that it now allows an application of the same kind of idea to the field of thermal phenomena, involving changes of temperature and pressure. For this purpose we construct a new function U, the *internal energy*, which depends only on the internal state of a body, as determined by its temperature, pressure and composition (and in certain cases additional variables such as surface area). Thus consider the change of a body within an adiabatic enclosure from a state A to a state B, its kinetic and potential energies remaining constant.‡ Then the first law, as stated above, allows us to write
$$w = (U_B - U_A), \tag{1·7}$$

† The student should convince himself that if this law were erroneous it would be possible to construct a perpetual motion machine.

The question whether the law of the conservation of energy is soundly based on experiment or whether it is really an act of faith has been discussed in detail by Meyerson, *Identity and Reality*, transl. Loewenberg (London, Allen and Unwin, 1930) and by Bridgman, *The Nature of Thermodynamics* (Harvard, 1941).

‡ In a composite process involving changes dU, $d\phi$ and dT of the internal, potential and kinetic energy of a body respectively, the total energy change is
$$dE = dU + d\phi + dT.$$

The law of conservation of energy in its complete form refers, of course, to the constancy of E in an isolated system. In thermodynamics attention is concentrated on the changes in U, since it is this form of energy which depends on the internal state of the system. In most processes of interest it also occurs that ϕ and T are constant, due to an absence of bulk motion.

where U_A and U_B *depend only on the states A and B*—this follows immediately from the fact that w is independent of the path and may thus be written as a difference.

The *first law* may now be re-expressed as follows: *the work done on a body in an adiabatic process is equal to the increase in a quantity U, which is a function of the state of the body.* It follows that if a body is completely isolated (i.e. it does no work, as well as being adiabatically enclosed) the function U remains constant. The internal energy is thus conserved in processes taking place in an isolated system.

1·7. Heat

In the discussion of Joule's experiments we were concerned with the change in state of a body contained within an adiabatic enclosure. It would have been wrong to have spoken of the temperature rise of the water as having been due to heat (although this is sometimes done in a loose way); what we were clearly concerned with were changes of state due only to work. However it is also known from experience that the same changes of state can be produced, without the expenditure of work, by putting the body into direct contact (or through a non-adiabatic wall) with something hotter than itself. That is to say the change of internal energy, $U_B - U_A$, can be obtained without the performance of work. We are therefore led to postulate a *mode of energy transfer* between bodies different from work and it is this which may now be given the name *heat*. Our senses and instruments provide us with no direct knowledge of heat (which is quite distinct from hotness). The amount of heat transferred to a body can thus be determined, in mechanical units, only by measuring the amount of work which causes the same change of state.

For example, from experiments such as those of Joule it is found that the expenditure of 4.184 joules of work on 1 g of water causes the change of state: (1 atm, 14.5 °C) → (1 atm, 15.5 °C). The same change can be obtained by contact of the water with a hotter body, and it became customary to describe this process as the transfer of 1 cal (15 °C) of heat into the water. The work equivalent of this arbitrarily defined heat unit is thus 4.184 joules.

In brief, if we obtain a certain increase of internal energy $U_B - U_A$ when a quantity of work w is done on a body adiabatically, and if we can obtain the same increase $U_B - U_A$ (as shown by the same changes in temperature and pressure) without the performance of work, then we take $U_B - U_A$ as equal to q, the heat absorbed by the body. It follows also that the same amount of heat has been given up by some other body. This follows because the two bodies together may be regarded as forming an isolated system, whose total internal energy is therefore constant; the increase in the internal energy of the first body must thus be equal to the decrease in the internal energy of the second, and it follows from the above definition of heat that the heat gain of the first body is equal to the heat loss of the second.

Consider now a type of process $A \rightarrow B$ in which a body X both absorbs heat and has work w done on it. In this process let its change of internal energy be $U_B - U_A$. We shall suppose that the heat comes from a *heat bath*, that is, a system of constant volume which acts as a reservoir for processes of heat transfer, but performs no work (e.g. a quantity of water at its temperature of maximum density). Let its change of internal energy in the above process be $U'_B - U'_A$. Then for the body X and the heat bath together we have

$$w = [(U_B - U_A) + (U'_B - U'_A)],$$

the right-hand side being the total change of internal energy. But according to the above definition $U'_B - U'_A$ is equal to the negative of the heat lost by the bath and is equal therefore to the negative of the heat gained by X. If we denote this heat by q, we therefore have $q = -(U'_B - U'_A)$. Substituting in the previous equation we obtain

$$U_B - U_A = q + w \qquad (1·8)$$

as a statement of the first law for a body which absorbs heat q and has work w done on it, during the change $A \rightarrow B$. This law therefore states that the algebraic sum of the heat and work effects of a body is equal to the change of the function of state, U, i.e. the algebraic sum is independent of the choice of path $A \rightarrow B$.

1·8. Expression of the first law for an infinitesimal process

The differential form of equation (1·8) is

$$dU = dq + dw, \qquad (1·9)$$

which means that the infinitesimal *increment* of the internal energy of the body is equal to the algebraic sum of the infinitesimal amount of heat which it absorbs and the infinitesimal amount of work which is done on it. Thus, whereas dU is the increment of the already existing internal energy of the body, dq and dw do not have a corresponding interpretation, and for this reason some authors prefer to use a notation such as $dU = Dq + Dw$.

The essential point is that dU is the differential of a function of state and its integral is $U_B - U_A$, which depends only on the initial and final states A and B; dq and dw, on the other hand, are not the differentials of a function of state, and their integrals are q and w

respectively, whose magnitudes depend on the particular path which is chosen between the A and B states—unless, of course, one of them is zero as in Joule's experiments.

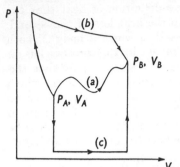

Fig. 3. (*a*) is an arbitrary path. (*b*) is a path made up of two adiabatics and an isothermal. (*c*) is also a special type of path—the gas is first cooled at constant V, then heated at constant p, and finally heated at constant V. If the initial cooling were such that p was reduced effectively to zero, this would be a zero work path.

As an example, consider the change of state of a substance: (1 kPa, 5 m³)→(2 kPa, 15 m³).† This change can be made in an infinite variety of ways, each occurring with different heat and work effects, but all having the same algebraic sum of these effects. Some of the possible paths are shown in Fig. 3, and for each of them, if carried out slowly and without friction, the work is equal to $-\int p\,dV$, the area under the curve. This work obviously varies enormously between one path and another.

As we have seen, the integral of dU between states A and B is $U_B - U_A$. If we consider the change $A{\rightarrow}B{\rightarrow}C$ we have

$$\int_A^B dU + \int_B^C dU = (U_B - U_A) + (U_C - U_B)$$
$$= U_C - U_A$$
$$= \int_A^C dU.$$

The overall change of U is thus independent of the intermediate state B, as noted previously. If we now choose the final state C as being identical with the initial state A we therefore have

$$\oint dU = 0,$$

where $\oint$ denotes integration round a cycle. The same statements are not correct for the integrals of dq and dw.

† There will, of course, be a definite change in temperature, as well as of all other intensive and extensive properties. Thus if the substance in question is a perfect gas, the ratio of the initial to the final temperatures on the absolute scale are as $1 \times 5 : 2 \times 15 = 5 : 30$. It may be noted that a change of state, (1 kPa, 5 m³, 250 K)→(2 kPa, 15 m³, 100 K), is an impossibility—too much has been specified.

The above property of U, which is the essential content of the first law, may also be expressed by the statement: dU is an *exact* differential in the variables of state. The word 'exact' merely means that the integral is independent of path.†

There is clearly nothing in the above treatment of the first law which requires us to think of energy as a 'thing'—it is the fact of conservation which tempts us to regard it as some kind of indestructible fluid. In dealing with the second law we meet a second quantity, the entropy, which is also an extensive quantity and a function of state, but is not conserved. In this case, therefore, the notion of a thing-like quality is quite inappropriate and would lead to errors. As Bridgman‡ has remarked, it would be preferable, but for the need for economy of words, to speak always of the 'energy function' and the 'entropy function' rather than of the energy and entropy. They are not material entities but are mathematical functions having certain properties.

However, it is always permissible to speak of the energy or entropy content of a body (relative to some other state), in a way in which it is not permissible to speak of its heat or work content. Heat and work are *modes of transfer* of energy between one body and another.

1·9. Adiabatically impossible processes

In this section we shall give a preliminary discussion on the basis of the second law. This basis is the impossibility of making a heat transfer and obtaining an equivalent amount of work, which would be the same as trying to carry out Joule's experiments in reverse. It is not merely that we never observe, as a spontaneous event, the water becoming cooler and the weight rising from the floor, but that this can be made to occur in no manner whatsoever, *without making a heat transfer into a second body which is at a lower temperature.*

If this were not the case it would be possible to construct a device for making a heat transfer from a region of the earth at a uniform temperature and using it as a continual source of mechanical power. This has never been achieved, and there seems good reason from molecular and statistical considerations to believe that it never will be achieved.

---

† This is not the same thing as saying that a differential, or differential expression, is 'complete'. For example,
$$dU = (\partial U/\partial T)\, dT + (\partial U/\partial p)\, dp$$
would be complete only if T and p completely determined the state of the system; in general, U depends also on the size and composition and extra terms must be added.

‡ Bridgman, *The Nature of Thermodynamics* (Harvard, 1941).

The impossibility of which we have spoken may also be expressed as follows. Let state A of Joule's system correspond to a height h_A of the weight and temperature θ_A of the water. Similarly, let state B correspond to h_B and θ_B. If $\theta_B > \theta_A$ (and $h_B < h_A$) then it is possible for state A to precede state B, but *it is not possible for state B to precede state A, within an adiabatic enclosure.*

There are, of course, a great many other examples of processes which are impossible under adiabatic conditions, and some of these are shown in Table 1. The first column shows a particular state A of the system, as defined by its temperature, volume and composition, and the second column shows another state B, characterized by different values of some or all of these variables. In each case experience shows that it is impossible to carry out the process $B \to A$ without a transfer of heat to some other body.

Now all of the processes which we have said are impossible are entirely consistent with the first law. To take the first example in Table 1, the transfer of heat between two blocks of metal takes place

TABLE 1

State A	State B
Two equal blocks of copper are connected by a wire. One block is at 20 °C and the other at 30 °C	The blocks are each at 25 °C
A dilute gas at a temperature θ occupies one half of an adiabatically enclosed vessel and the other half is a vacuum	The gas at the same temperature θ occupies the whole of the vessel
A dilute gas X occupies one half of an adiabatically enclosed vessel and a dilute gas Y occupies the other half. The temperature is θ	The gases are uniformly mixed throughout the vessel and the temperature has the same value θ
An adiabatically enclosed vessel contains hydrogen and oxygen and a catalyst. The volume is V and the temperature is θ	The vessel contains the same amount of hydrogen and oxygen, combined as water, together with the catalyst. The volume is V and the temperature exceeds θ by an amount corresponding to the heat of reaction

at constant total energy and thus is possible *in either direction*, according to the first law. It seems therefore that there must be some condition, in addition to the conservation of energy, which must be fulfilled by those processes which actually take place. In fact, it is the knowledge that certain processes are impossible in an adiabatic system which enables us to discover a new function of state, the *entropy*, which provides a criterion of whether a process is a possible one or not. Anticipating somewhat the more systematic discussion of §1·11, the reason for the existence of this function may be seen as follows: *The possibility or impossibility of the processes $A \to B$ and $B \to A$ depends entirely on the nature of the states A and B and is thus determined by the values of the variables, such as temperature and pressure, at the beginning and end of the process. We may therefore expect to find a function of these variables whose special characteristic is to show whether it is the process $A \to B$ which is the possible one, or the converse process $B \to A$.* Thus if S is the expected function it will be found that $S(p_A, \theta_A)$ differs in value from $S(p_B, \theta_B)$.

The impossibility of the processes $B \to A$ of Table 1 is by no means independent of the impossibility of reversing Joule's experiment. In fact, if the latter is made the basis of the second law, it may be shown as a deduction (using also certain equations of state in some of the examples) that the other processes are also impossible.

Finally, it may be remarked that when we speak of a possible or impossible process the notion of time's passage is implicit. It is a question of whether a state A can *precede* or *succeed* another state B, in an adiabatic enclosure. The decision which of these states is 'later than' or 'earlier than' the other is based primarily on the subjective time sense of the human observer—which is not to say however that, once the second law has been seen to be true for all isolated systems, we cannot choose one such system as *defining* the time direction for all the rest. This can be done without reducing the law to a tautology.†

1·10. Natural and reversible processes

Changes which take place in a system spontaneously and of their own accord are called *natural processes*. Examples are the equalization of temperature between two pieces of metal, the mixing of two gases and all processes which can occur spontaneously within an adiabatic enclosure. From what has been said in the last section it seems that such changes can never be reversed in their entirety, for it is known from experience that the system in question can be restored to its original condition *only* by transferring a quantity of heat elsewhere. In this respect natural processes are said to be *irreversible*. In brief, a cycle of changes $A \to B \to A$ on a particular

† Denbigh, *Brit. J. Phil. Sci.* **4** (1953), 183.

system, where $A \to B$ is a natural process, cannot be completed without leaving a change in some other part of the universe.†

It seems possible, however, that there may exist other types of change, involving a different pair of specified states, say, A and B', of the system under discussion, such that it is possible to carry out the cycle $A \to B' \to A$ without leaving more than a negligible change in any other body. Such changes of state will be said to be *reversible*.

A reversible process will therefore be defined as one which can be reversed without leaving more than a vanishingly small change in any other system. If the change $A \to B'$ involves a heat absorption q it must be possible to carry out the reverse change $B' \to A$ rejecting an equal quantity of heat into the same heat bath. (The work effects on the two paths will then be automatically equal, since $\oint dU = 0$.)

For the purpose of the detailed discussion of the second law in the next section it is necessary to postulate that reversibility can be approached, under limiting conditions, in real processes. We must now give some justification that such a state of affairs is a possible one.

Consider the isothermal expansion of a fluid in contact with a heat bath. The work done *by* the fluid is $\int p_e dV$, where p_e is the external pressure on the piston. If the fluid is now brought back to its original state by isothermal compression the work done *on* the fluid is $-\int p_f dV$, where p_f is now the pressure of the fluid itself. These two integrals will be equal only if the expansion and compression are carried out slowly and without friction, so that the pressure of the fluid and the pressure on the piston never differ more than infinitesimally. Under these conditions the work is equal along the two paths and, since $\oint dU = 0$, the quantities of heat absorbed from and rejected to the heat bath are also equal and the whole process is reversible.

Another example is a chemical reaction in a galvanic cell, the e.m.f. being balanced almost exactly against some mechanical force, through the agency of a frictionless motor. Under such ideal conditions the reaction can be carried out, first in the one direction and then in the reverse, by infinitesimal alteration in the opposing force, and without leaving more than a vanishingly small change in the environment.

In view of these examples we may tentatively expect that rever-

† For example, after hydrogen and oxygen have spontaneously combined the water can be decomposed once again into these gases—but only by leaving changes in other bodies such as electrical accumulators and heat reservoirs. These can be restored only by leaving changes somewhere else and so *ad infinitum*.

sible processes are those in which the internal forces exerted by the system in question differ only infinitesimally from the external forces acting on the system. However the matter will be discussed further in §1·13. For the present all we need is the postulate that such processes are conceivable as limiting instances of natural processes.

Much confusion in the discussion of irreversibility, whenever we are concerned with systems which are not adiabatically isolated, is due to the forgetting of the changes in the *environment*. Between the same initial and final states of such a system the change can be carried out either reversibly or with varying degrees of irreversibility; *such processes differ in regard to the resulting changes in state of the environment.*

For the complete specification of a process it is necessary to state the initial and final state of all bodies which are affected (i.e. all of those which form an adiabatic total system); when this is done it remains true, as stated in the last section, that certain changes, those in which there is appreciable displacement from equilibrium at each instant, can never be reversed.

1·11. Systematic treatment of the second law

Much of the preliminary discussion above has been lacking in precision. However, all that we require for a systematic treatment of the second law is the fact, which is securely based on experience, that it is impossible to carry out Joule's experiment in reverse (Statement A below). As a postulate, to be justified later, we shall also assume the existence of reversible paths. On this basis the development of the second law may be carried out in several different ways. In the following we shall describe the traditional method, as used by Clausius† and Poincaré,‡ which depends on the use of heat engines. The main alternatives are the methods of Planck,§ which depends on the existence of perfect gases, and of Born and Carathéodory,‖ which is based on the properties of Pfaffian differential expressions.

† Clausius, *The Mechanical Theory of Heat*, transl. Hirst (1867).

‡ Poincaré, *Thermodynamique* (Paris, Carré, 1892). See also Keenan, *Thermodynamics* (New York, Wiley, 1941).

§ Planck, *Treatise on Thermodynamics*, transl. Ogg (London, Longmans Green, 1927).

‖ Born, *Phys. Z.* **22** (1921), 218, 249, 282. An excellent account in English is given by Chandrasekhar, *Introduction to the Study of Stellar Structure* (Chicago, 1939). Shorter versions of the Carathéodory treatment are given by Buchdahl, *Amer. J. Phys.* **17** (1949), 41, 44, 212; **22** (1954), 182; **23** (1955), 65; **28** (1960), 196; by Born, *Natural Philosophy of Cause and Chance* (Oxford, 1949) and by Eisenschitz, *Science Progress*, **43** (1955), 246, by Turner, *Amer. J. Phys.* **28** (1960), 781; **29** (1961), 40; Thomson, ibid. **29** (1961), 300 and by Sears, ibid. **31** (1963), 747.

STATEMENT A. *It is impossible to make a transfer of heat from a heat bath, at a uniform temperature, and obtain an equivalent amount of work, without causing a change in the thermodynamic state of some other body.*

Some comments on the phrasing of this statement may be added.

(*a*) '...at a uniform temperature....' A heat bath which consists of two or more parts, each at different temperatures, is really equivalent to two or more heat baths. By taking heat from one of them and transferring a smaller amount of heat into a second one which is colder, it would be possible to operate a heat engine, as in the Carnot cycle shortly to be described. The net heat extracted from the compound system is equal to the work done; the reservation concerning uniform temperature is therefore essential.

(*b*) '...equivalent....' This word is redundant and has been inserted only for clarity. If no other body is affected, all the heat removed must appear as work, by the first law.

(*c*) The last clause of Statement A is essential and is equivalent to the previous statement that certain processes are impossible *under adiabatic conditions.* For example, it is entirely possible to take heat from a heat bath and convert it completely into work, provided we have an auxiliary body, namely, a perfect gas, in thermal contact with the heat bath. The gas is expanded reversibly, and the transfer of heat, referred to in the first clause of Statement A, takes place between the bath and the gas. Since at constant temperature the internal energy of a perfect gas remains constant, the whole of the heat withdrawn from the reservoir is converted into work. However, in this process the 'other body', namely, the perfect gas, has changed its thermodynamic state, as shown by its increase of volume.

Statement A will now be used to prove the following four propositions:

(1) If a body performs a reversible cycle, absorbing heat q_1 at temperature θ_1 and heat q_2 at temperature θ_2, then the ratio $q_1 : q_2$ is a function only of θ_1 and θ_2. Thus

$$q_1/q_2 = f(\theta_1,\ \theta_2).$$

(2) The last-named ratio can be expressed in the form

$$q_1/q_2 = T_1/T_2,$$

where T_1 and T_2 are defined as the thermodynamic temperatures.

(3) The entropy, S, which is defined by

$$dS \equiv (dq/T)_{\text{rev.}},$$

is a function of state.

(4) The entropy change of an adiabatically isolated system is always positive for a natural process and zero for a reversible one.

The first and second of these propositions are associated with the names of Carnot and Kelvin respectively. The third and fourth are due mainly to Clausius and are the final form of the second law.

For the proof of these propositions it will be supposed that we are concerned with closed systems in a state of internal equilibrium such that their thermodynamic states can be specified by means of two variables, such as pressure and volume or pressure and temperature. The proofs depend also on certain properties of the adiabatic and isothermal curves on a p-V diagram. These properties may be readily proved by means of Statement A and are quoted as problems below.†

(a) Proof of proposition 1. Consider a body X. In Fig. 4 let AB and CD be portions of its isothermals corresponding to the 'empirical' temperatures θ_2 and θ_1 respectively, where $\theta_2 > \theta_1$.‡ (By 'empirical' temperature is meant that which is measured by use of a reference substance as described in § 1·4.) Let AD and BC be portions of two adiabatics. If the cycle $ABCD$ is carried out under reversible conditions, it is said to be a Carnot cycle,§ and it follows from the propositions in the problem that the cycle must be of the form shown in the diagram, i.e. consisting of *four* parts.

Let q_1 and q_2 be the quantities of heat taken in by X from the heat baths needed to maintain it at constant temperature along the isothermals θ_1 and θ_2 respectively. Now q_1 and q_2 must have opposite signs. That is to say one or other of them (depending on whether the cycle is in the direction of the arrows in Fig. 4, or in the reverse direction) corresponds to heat actually taken in and the other to heat actually rejected. For if they were both of the same sign, the cycle could be carried out in a direction such that both were positive (thus yielding a positive quantity of work as done on some other body); after having done this, the hotter bath could be put into

† *Problem.* Prove the following statements concerning the adiabatics and isothermals of a particular body (which may be a gas, a liquid or a solid):

(1) an isothermal and an adiabatic cannot intersect more than once:

(2) an isothermal and an adiabatic cannot touch each other without intersection;

(3) two adiabatics cannot intersect;

(4) in each infinitesimal portion of an isothermal the heat absorbed by a body has the same sign for the same direction of movement along the curve. (*Hint.* Assume that the statements are false and show that in this case Statement A would be contravened.)

‡ The temperature scale is here chosen such that a larger numerical value of θ corresponds to greater hotness.

§ Being reversible, a cycle of this kind, if carried out in the reverse direction, will exactly nullify the effects on the heat baths of carrying it out in the forward direction.

h the cooler one for a time long enough to give the latter
h heat as it had, by supposition, given up to the body X.
ʄect would have been the removal of a quantity of heat
ᴕtter bath and its complete conversion into work, without
a ᴄʜaɴɢᴇ ɪɴ any other body and this is contrary to Statement A.

By the same argument it may be seen that q_2 is positive and q_1
is negative for the cycle occurring in the direction of the arrows in
Fig. 4.

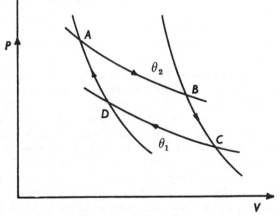

Fig. 4   Carnot cycle in p-V co-ordinates.

It must now be shown that the ratio q_1/q_2 depends only on θ_1
and θ_2. For this purpose let it be supposed that we have two different
bodies, X and Y, each of which makes Carnot cycles in an arbitrary
region of its p, V co-ordinates. The only limitation is to suppose
that the two bodies share the same pair of heat baths, and therefore
have the same upper and lower operating temperatures, θ_1 and θ_2.

If q_1 and q_2 are the heat effects for body X, let Q_1 and Q_2 be the
corresponding heat effects for body Y. Now we can always find
integers n and N such that by putting X through n complete cycles
and Y through N complete cycles *in the opposite direction* we satisfy
the equality $nq_1 = -NQ_1$. This means that, whatever be the size and
nature of the two bodies and the regions in their ᴖ p, V co-ordinates,
the number of cycles can be chosen so that the total heat taken in
by the one body from the cooler bath is equal to the total heat
rejected by the other body into the same bath, the cycles of the
bodies being in opposing directions.

Therefore, after completion of these cycles, the cooler bath is in
the same state as originally. So also are the bodies X and Y which
have completed integral numbers of cycles. It follows that we must

also have $nq_2 = -NQ_2$, i.e. there must be zero net heat effect relating to the hotter bath. For if this were not so, we could have chosen the directions of the cycles (which it will be remembered are carried out *reversibly*) so that a net positive amount of heat would have been taken from the hotter bath and completely converted into work, contrary to Statement A.

In brief, if we choose the number of cycles to satisfy

$$nq_1 = -NQ_1,$$

then we must also satisfy

$$nq_2 = -NQ_2.$$

Dividing the one equation by the other

$$q_1/q_2 = Q_1/Q_2.$$

The ratio q_1/q_2 is thus the same for all bodies making reversible cycles between the same pair of isothermals, and is independent also of the particular pressures and volumes occurring in the cycles, since we could evidently have chosen these variables arbitrarily without affecting the result. The ratio can therefore be a function only of the temperatures θ_1 and θ_2. This we shall write as

$$\frac{|q_1|}{|q_2|} = f(\theta_1, \theta_2) \qquad (1·10)$$

in order to avoid ambiguities of sign. Since the work done on some external system by the body X in each cycle is given by $|w| = |q_2| - |q_1|$, the result we have obtained may also be written in the form

$$\frac{w}{q_2} = \frac{|q_2| - |q_1|}{|q_2|} = 1 - f(\theta_1, \theta_2),$$

and the fraction of the heat taken in by the hotter bath which is converted into work is thus seen to depend only on the temperatures of the two baths.

(b) Proof of proposition 2. Fig. 5 shows three isothermals θ_1, θ_2 and θ_3 (numbered to correspond to increasing degree of hotness), and also two adiabatics, of a particular body. Using the result of equation (1·10) we have the relations

$$\text{cycle } DCEF \quad |q_1|/|q_2| = f(\theta_1, \theta_2),$$

$$\text{cycle } ABCD \quad |q_2|/|q_3| = f(\theta_2, \theta_3),$$

$$\text{cycle } ABEF \quad |q_1|/|q_3| = f(\theta_1, \theta_3)$$

Dividing the third of these relations by the second we eliminate q_3

and obtain

$$\frac{|q_1|}{|q_2|} = \frac{f(\theta_1, \theta_3)}{f(\theta_2, \theta_3)},$$

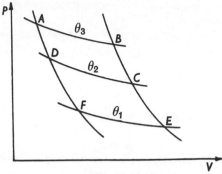

Fig. 5.

and comparing this with the first relation we have

$$f(\theta_1, \theta_2) = \frac{f(\theta_1, \theta_3)}{f(\theta_2, \theta_3)}.$$

Now this equation is true whatever is the value of θ_3. It follows therefore that θ_3 must cancel between the numerator and denominator of the right-hand side of this equation. This requires that the functions $f(\theta_1, \theta_3)$, etc., are of the forms $f(\theta_1)/f(\theta_3)$, etc.,[†] that is, they are expres-

---

† A partial differentiation of the above equation with respect to θ_3 gives

$$f(\theta_1, \theta_2) \frac{\partial f(\theta_2, \theta_3)}{\partial \theta_3} = \frac{\partial f(\theta_1, \theta_3)}{\partial \theta_3},$$

and by substituting in this relation from the original equation we obtain

$$\frac{\partial \ln f(\theta_2, \theta_3)}{\partial \theta_3} = \frac{\partial \ln f(\theta_1, \theta_3)}{\partial \theta_3}.$$

It is clear that each side of this equation must be a function of θ_3 only. Writing each side separately equal to the same function of θ_3 and integrating, we obtain

$$\ln f(\theta_2, \theta_3) = \phi(\theta_3) + \ln f(\theta_2),$$
$$\ln f(\theta_1, \theta_3) = \phi(\theta_3) + \ln f(\theta_1),$$

where the second terms on the right-hand sides are integration constants for integration at constant θ_2 and θ_1. Hence

$$f(\theta_2, \theta_3) = f(\theta_2) \exp \phi(\theta_3),$$
$$f(\theta_1, \theta_3) = f(\theta_1) \exp \phi(\theta_3),$$

and dividing the second equation by the first

$$\frac{f(\theta_1, \theta_3)}{f(\theta_2, \theta_3)} = \frac{f(\theta_1)}{f(\theta_2)}.$$

Therefore $f(\theta_1, \theta_2)$ is of the form $f(\theta_1)/f(\theta_2)$.

sible as a ratio of functions of each separate temperature. Thus we obtain

$$|q_1|/|q_2| = f(\theta_1)/f(\theta_2),$$

$$|q_2|/|q_3| = f(\theta_2)/f(\theta_3),$$

$$|q_1|/|q_3| = f(\theta_1)/f(\theta_3).$$

It is also evident that $|q_3| > |q_2| > |q_1|$,

for if we choose to go round one of the cycles in the clockwise direction, a positive amount of work is done by the body and thus the heat absorbed from the warmer bath is greater than the heat rejected into the colder one. If we go in the anti-clockwise direction, work is done *on* the body and again a greater amount of heat is rejected into the warmer bath than is removed from the colder one (refrigeration cycle). Hence from the above equations we have

$$f(\theta_3) > f(\theta_2) > f(\theta_1). \tag{1·11}$$

Now θ_1, θ_2 and θ_3 are merely 'empirical' temperatures, as defined by any chosen reference body as a thermometer, in accordance with §1·4. It is evident that $f(\theta_1)$, $f(\theta_2)$ and $f(\theta_3)$ are *just as good a measure of temperature* as are θ_1, θ_2 and θ_3 themsleves.

The necessary criteria of any measure of temperature are: (*a*) it is equal for bodies which are at thermal equilibrium and (*b*) it changes *monotonously* with increase in hotness. The first of these conditions is obviously satisfied by $f(\theta)$ because if two bodies have equal values of θ they also have equal values of $f(\theta)$. The second condition is also satisfied since we have just proved the relation (1·11), that is, $f(\theta)$ increases monotonously with increase of θ and thus with increase of hotness.

The functions $f(\theta)$ may therefore be chosen as a measure of temperature and will be denoted T. We thus define a scale of *thermodynamic temperatures* to satisfy the relation

$$T_1/T_2 = |q_1|/|q_2| \tag{1·12}$$

where T_1 and T_2 are positive numbers† and q_1 and q_2 refer to the heat effects of a body carrying out a Carnot cycle between these two temperatures. Such a scale has also been called *absolute* because it has been shown in the previous proposition that the ratio q_1/q_2 is independent of the choice of the body which makes the cycle.

To complete the setting up of the thermodynamic scale it is now necessary merely to decide on the size of the degree. Let T_s be the thermodynamic temperature of saturated steam at 1 atm pressure and let T_i be the thermodynamic temperature of melting ice at the

---

† Concerning *negative* thermodynamic temperatures see Ramsey, *Phys. Rev.* **103** (1956), 20; Bazarov, I. P., *Thermodynamics* (Pergamon, 1964).

same pressure. Th[,]n the size of the degree is chosen so that

$$T_s - T_i = 100 \text{ K}.$$

We shall not seek to discuss the experimental procedures by means of which T_i and T_s are actually determined by means of equation (1·12) †
For the present purpose it is sufficient to assert that if a body were put through a Carnot cycle absorbing heat q_s from the steam and rejecting heat q_i into the ice, we should obtain the result

$$q_s/q_i = 1.366,$$

and therefore $$T_s/T_i = 1.366.$$

Solving between the above equations we find $T_i = 273.2$ K and $T_s = 373.2$ K. (More accurately, the accepted value‡ for the ice-point is 273.15 K.)

As a corollary to the above we obtain

$$\frac{w}{q_2} = \frac{|q_2| - |q_1|}{|q_2|} = \frac{T_2 - T_1}{T_2}. \qquad (1·12b)$$

This ratio should be called the 'conversion factor' and not the 'efficiency' of the cycle. The reversible Carnot cycle is as efficient as any cycle can possibly be, when operating between the given pair of temperatures, and should thus be regarded as 100 % efficient. Any irreversible cycle has a lower conversion factor and may thus be said to be inefficient relative to the reversible one.

It can also be readily shown that the thermodynamic temperature, as defined by (1·12), is the same as the perfect gas temperature. (See Problem 8 on p. 59.) However, it will be recognized that this coincidence plays no essential role in thermodynamics—the basic theory can be developed, as in the present chapter, without any reference to the existence of perfect gases.

(c) Proof of proposition 3. In this proposition we consider the properties of the differential dS which is defined by

$$dS \equiv (dq/T)_{\text{rev.}}, \qquad (1·13)$$

---

† See, for example, Zemansky, *Heat and Thermodynamics* (New York, McGraw-Hill, 1957), Roberts and Miller, *Heat and Thermodynamics* (London, Blackie, 1963).

‡ A more correct statement of the situation is that the thermodynamic temperature scale has been fixed *by definition* so that at the *triple point* of water $T = 273.16$ K. This makes the ice point 273.15 K. The designation ' degree Celsius ' is also now used in place of ' degree Centigrade '. See for example Hatsopoulos and Keenan, *Principles of General Thermodynamics*, ch. 15 (Wiley, 1965).

where dq is the heat absorbed by a body at a thermodynamic temperature T, in a reversible change. The definition applies specifically to 'bodies' and not to open systems. The properties of S were discovered by Clausius, and it was named by him 'entropy' from the Greek word for transformation. He regarded it as a measure of the 'transformation content' of a body, meaning presumably its capacity for change.

In the present proposition we seek to prove that S is a function of state. Consider first of all the type of reversible cycle already discussed, consisting of two isothermals T_1 and T_2 and two adiabatics. The entropy changes along the isothermals are obtained from (1·13) and are

$$\int \frac{dq_1}{T_1} = \frac{q_1}{T_1}$$

and

$$\int \frac{dq_2}{T_2} = \frac{q_2}{T_2} \quad \text{respectively.}$$

Along the adiabatics it follows from (1·13) that the entropy changes are zero. Hence over the whole cycle we have

$$\oint dS = \frac{q_1}{T_1} + \frac{q_2}{T_2}.$$

Now according to (1·12) our thermodynamic temperatures are chosen to satisfy

$$\frac{q_1}{q_2} = -\frac{T_1}{T_2}.$$

Therefore

$$\oint dS = 0.$$

It must now be proved that the same result applies to any type of reversible cycle, not necessarily consisting of two isothermals and two adiabatics.

A cycle involving the generalized force y and the generalized displacement x is shown by the closed curve in Fig. 6. Let PQ be any portion of the path. Through P and Q the lines RS and TU represent two adiabatics. Between these adiabatics we can draw an isothermal —the curve VW—in such a position that the area under the zigzag path $PVWQ$ is the same as under the actual path PQ. The work is thus the same along both paths, and so also must be the heat effect, since the change of internal energy is independent of the path. Thus the heat absorbed by the body along the isothermal VW is the same as along the actual path PQ. Similarly, an isothermal YX can be drawn between the same adiabatics in such a position that the heat absorbed along this isothermal is the same as along the actual path NM.

The whole of the cycle represented by the closed curve can thus be made equivalent, as regards its heat and work effects,† to a number of Carnot cycles. This is indicated in Fig. 7. Considering

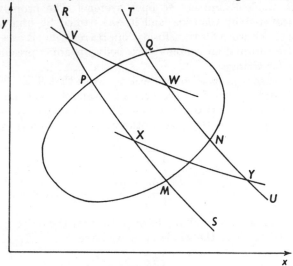

Fig. 6.

the Carnot cycle $ABCD$ farthest to the left in this diagram we have from equation (1·12)

$$\frac{\delta q_1}{T_1} + \frac{\delta q_2}{T_2} = 0,$$

where δq_2 is the heat absorbed by the body in passing from A to B and δq_1 is the heat it absorbs in passing from C to D. Similarly we have

$$\frac{\delta q_3}{T_3} + \frac{\delta q_4}{T_4} = 0,$$

$$\frac{\delta q_5}{T_5} + \frac{\delta q_6}{T_6} = 0.$$

Adding as many of these equations as the number of Carnot cycles into which we have chosen to divide up the actual cycle, we obtain

$$\Sigma\left(\frac{\delta q_i}{T_i}\right)_{\text{rev.}} = 0,$$

† However, it is only in the limiting case, where the adiabatics are only infinitesimally separated, that the heat supplied along the curved path PQ is supplied at the *same virtually constant temperature* as along the isothermal path VW. Thus the curved and zigzag paths become *completely* equivalent, as regards their thermodynamic effects, only in the limit, and it is this which justifies equation (1·14).

and this result is valid, of course, for the complete zigzag path *ABEF...A*.

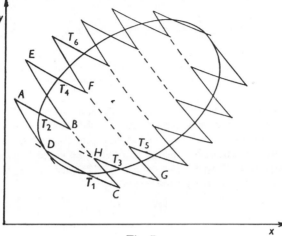

Fig. 7.

Thus in the limit, when the adiabatics are only infinitesimally far apart, we have for the whole of the closed zigzag cycle

$$\oint \left(\frac{\mathrm{d}q}{T}\right)_{\text{rev.}} = 0.$$

Returning now to Fig. 6 it is evident that when the adiabatics *RS* and *TU* are only infinitesimally distant, the temperature change along the curved path *PQ* approaches zero. Thus $\mathrm{d}q/T$ for the zigzag path *PVWQ* becomes the same as $\mathrm{d}q/T$ for the curved path *PQ*. Hence the result of the last equation applies not merely to the zigzag cycle but also to the cycle represented by the closed curve.

Using the definition (1·13), we thus obtain for the arbitrary cycle the important result

$$\oint \mathrm{d}S_s = 0. \tag{1·14}$$

The subscript s has been added in order to make it clear that $\mathrm{d}S_s$ is the entropy change of the system which absorbs the heat $\mathrm{d}q$, and it does not include the entropy change of the heat reservoir which gives up this heat.

The result (1·14) shows that the entropy is a function of state for the body. Thus if we consider the change of state, $(T_A, p_A) \rightarrow (T_B, p_B)$, by a particular path α, and the return to state A by the path β, we have for the whole cycle $A \rightarrow B \rightarrow A$

$$\int_A^B dS_{\text{path }\alpha} + \int_B^A dS_{\text{path }\beta} = 0,$$

and therefore $\qquad \displaystyle\int_A^B dS_{\text{path }\alpha} = \int_A^B dS_{\text{path }\beta}.$

The change of entropy of a body is thus the same for all paths between given initial and final states and is determined entirely by these states.

What has just been said must, of course, be true even if the chosen path does not correspond to reversible conditions, i.e. if it is a 'natural process'. The entropy change of *the body* (i.e. the system in question) is precisely the same, because it has been shown to be a function of state. However, it will be shown in §1·12, that the equality (1·13) must then be replaced by the inequality

$$\left(\frac{dq}{T}\right)_{\text{irrev.}} < dS_s.$$

In other words, for a given change of state of a body, at a uniform temperature T, the entropy change is constant, but the heat taken in is less for the irreversible path than for the reversible one. This implies, of course, that the two types of path do not give rise to the same changes of state in the *environment*.

The entropy has been shown to be a property of a body and is also clearly a member of the *extensive* class. Thus suppose that a body consists of n parts, each at the same temperature T. If these various parts absorb the quantities of heat $dq_1, dq_2, ..., dq_n$ under reversible conditions, their entropy changes are

$$\frac{dq_1}{T}, \quad \frac{dq_2}{T}, \quad ..., \quad \frac{dq_n}{T},$$

respectively. The sum of these is $\dfrac{1}{T}\Sigma dq_i$, and since Σdq_i is the total heat, dq, absorbed by the whole body, we find that the sum of the entropy changes of the various parts is equal to the overall entropy change.

From a mathematical standpoint the present proposition amounts to the demonstration that the thermodynamic temperature T is an integrating denominator for dq in a reversible change, i.e. $dS \equiv (dq/T)_{rev.}$ is an exact differential. The possibility of defining T depended, in its turn, on the use of Statement A, concerning the impossibility of achieving a conversion of heat into work by use of a heat bath and no other body.†

(d) Proof of proposition 4. In this proposition we are concerned with the magnitude of the change of entropy, $S_B - S_A$, between a state A and a state B of a system *within an adiabatic enclosure.* The change $A \rightarrow B$ may be a change of composition, due to chemical reaction, or any other change of a closed system, such as those shown in Table 1. The process $A \rightarrow B$ will be supposed to occur irreversibly— i.e. it is a spontaneous process.

Now the defining equation (1·13) for an entropy change is only applicable under reversible conditions. We shall therefore suppose that, after the original process $A \rightarrow B$ has taken place within the adiabatic enclosure, the reverse change of state $B \rightarrow A$ is carried out under conditions which are known to be reversible. In the general case, as we have noted previously, this return process cannot be carried out adiabatically. Let it be devised in such a way that any heat, $q_{B \rightarrow A}$, which must be absorbed by the system is taken in from a single‡ heat bath at a temperature T (Fig. 8).

Denoting by $w_{A \rightarrow B}$ and $w_{B \rightarrow A}$ any work which is done *by* the system (i.e. here counted as positive when done *on* the environment) during the outward and return paths respectively, we have from the first law

$$q_{B \rightarrow A} = w_{A \rightarrow B} + w_{B \rightarrow A}.$$

Now $q_{B \rightarrow A}$ must be *negative*. For if it were zero this would correspond to a complete reversal of the original process $A \rightarrow B$, contrary to the statement that we are discussing an irreversible process. And if it

† The fact that the second law is concerned with the heat and work relationships of *two* bodies is seen very clearly in Born and Carathéodory's formulation. For a single body undergoing a reversible change of temperature θ and volume V we have from the first law:

$$dq = \left(\frac{\partial U}{\partial \theta}\right)_V d\theta + \left\{\left(\frac{\partial U}{\partial V}\right)_\theta + p\right\} dV.$$

An equation of this form, in which there are only the two independent variables θ and V, always has an integrating denominator. However, for two bodies, 1 and 2, at the same temperature, the equation must be written

$$dq = \frac{\partial(U_1 + U_2)}{\partial \theta} d\theta + \left(\frac{\partial U_1}{\partial V_1} + p_1\right) dV_1 + \left(\frac{\partial U_2}{\partial V_2} + p_2\right) dV_2.$$

This equation has an integrating denominator (namely, the thermodynamic temperature) only when a certain condition is satisfied. This condition is the impossibility of certain changes of state, under adiabatic conditions, as discussed in §1·9.

‡ This condition can always be attained even if the return path cannot be devised in the simple manner of Fig. 8. For example, use may be made of a heat engine which absorbs heat from the prescribed heat bath at T and delivers heat into the system at any necessary sequence of temperatures.

were positive this would correspond to the complete conversion of heat taken from the reservoir into work, without leaving a change in any other body. (Remembering that the system itself will then have completed a cycle.)

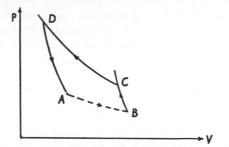

Fig. 8. Illustrating an irreversible adiabatic gas expansion, from A to B, along the dotted curve, and a return path along two reversible adiabatics BC and DA, and the reversible isothermal CD. The only heat effect in the cycle is the heat $q_{B \to A}$ absorbed along CD.

By definition the entropy change of the *system* in the reversible return path is

$$S_A - S_B = q_{B \to A}/T,$$

and must therefore be *negative*. The entropy change $S_B - S_A$ on the original outward path, being equal and opposite to that on the return path by the third proposition, must consequently be positive.

Alternatively if the original process had been carried out reversibly, $q_{B \to A}$ would necessarily have been zero,† according to the meaning of 'reversibility', since $q_{A \to B}$ is zero. Under such conditions $S_A - S_B$ would therefore be zero also—i.e. S_A and S_B are equal for a reversible adiabatic process.

Returning to the former case, the irreversibility lies in the fact that the system has been restored to its original condition only at cost of transferring heat from it into a heat bath. Thus the inevitable result of an irreversible process is the creation of entropy, and the system can be restored to its original condition only by the removal of heat. That is to say, the created entropy must be transferred to some other body, such as a heat bath, which thereby undergoes a

† If $q_{B \to A}$ is zero it means, of course, that we are able to make the return path $B \to A$ under adiabatic conditions and this may seem contrary to Statement A and the discussion of §1·9. The answer to this paradox is that if the change $A \to B$ takes place adiabatically and reversibly it does not give rise to the *same* final state B as if it took place adiabatically and irreversibly, i.e. the final state should be denoted B'. The final states B which were specified, in regard to their temperatures, etc., in Table 1 are those obtained by entirely uncontrolled processes of mixing, reaction, etc., within the adiabatic enclosure.

change of state. If the heat bath, in its turn, is restored to its original condition, it necessitates a gain in entropy of some third body, and so *ad infinitum*. As remarked already in § 1·10, *natural processes can never be reversed in their entirety*. It is only in the limiting case, where the changes are carried out under reversible conditions, that the overall entropy change of the system and its environment approaches zero.

1·12. Final statement of the second law

As a result of the last two propositions, the second law can now be expressed as follows:

(*a*) the entropy S, defined by $dS \equiv (dq/T)_{\text{rev.}}$, is a function of state;

(*b*) the entropy of a system in an adiabatic enclosure can never decrease; it increases in an irreversible process and remains constant in a reversible one.

This second result must apply to each infinitesimal stage of a process, as may be seen by applying the last proposition to an infinitesimal change of an adiabatic system, instead of to a finite change. We can therefore write† $$dS \geqslant 0,$$

where S refers to the total entropy of an adiabatic system. Alternatively, if we sum the entropy over all parts of a system which are in thermal contact, we have $\Sigma dS_i \geqslant 0$.

For many purposes it is convenient to convert this relation into an equality by defining a quantity σ:

$$\Sigma dS_i = d\sigma, \tag{1·15}$$

and $d\sigma$ is clearly the *amount of entropy which is created* in the given process. According to the second law it is positive for an irreversible process and zero for a reversible one. It can never be negative.

As a corollary to the above, consider a thermodynamic system contained within a heat bath. The two together are an adiabatically isolated system, and therefore from (1·15) we have

$$dS_s + dS_r = d\sigma,$$

where the subscripts s and r stand for the system and the heat bath respectively. Let it be supposed that any irreversible changes taking place within the system are fairly slow, so that the heat bath maintains a uniform temperature T_r. If the system absorbs heat dq, the

---

† This equation might be written more explicitly as
$$dS/dt \geqslant 0,$$
where dt is an infinitesimal increase in time. However, in thermodynamics time is always taken as increasing in the direction towards the future, as it is humanly experienced, and thus dt is always positive.

same amount of heat is lost by the bath† and its entropy change is

$$dS_r = -dq/T_r.$$

Between the last two equations we obtain

$$dS_s - dq/T_r = d\sigma, \tag{1·16}$$

or, expressed in an alternative manner,

$$dq/T_r \leqslant dS_s,$$

where the equality sign refers to the reversible path and the inequality sign to an irreversible one. Now in any change of a system between two states, A and B, the entropy change is the same irrespective of whether the path is reversible or not. It follows from (1·16) that it is the heat effect which is different, there being a smaller absorption of heat along an irreversible path than along a reversible one (or, in the case where heat is evolved, it is *greater*).

It may be remarked that the principle that entropy can only increase or remain constant applies only to a closed system which is adiabatically isolated. Whenever a system can exchange either heat or matter with its environment, an entropy decrease of that system is entirely compatible with the second law. Living organisms, for example, are 'open' to their environment and can 'build up' at the expense of the foodstuffs which they take in and degrade. Astronomical bodies, such as the sun, are continually losing thermal energy or radiation and may also be gaining matter in the form of dust. The entropy of such bodies may be either increasing or decreasing.

1·13. A criterion of equilibrium. Reversible processes

As we have seen, the only changes which can take place in an adiabatic system are those in which the entropy either increases or remains constant. The same applies, of course, if the system is completely isolated, that is, if it does no work, as well as being within an adiabatic enclosure, so that the internal energy is constant. Thus whenever a system could change from a state of lower to one of higher entropy, within an enclosure of constant volume and energy, it is possible, in a thermodynamic sense, for this change to occur.

---

† The meaning of the term *heat bath* or *heat reservoir* was discussed previously in §1·7. It signifies a body which acts merely as an acceptor or donor of heat whilst remaining of constant volume and performing no other form of work. Under these conditions the heat given up or taken in by the heat bath is equal to the change of its internal energy—i.e. it is equal to the change of a *function* of *state*. The change of state of the heat bath is thus *entirely determined* by the heat transfer dq and dS_r is therefore the same whether this transfer is reversible or not.

For example, consider the reaction

$$A + B \leftrightharpoons C$$

taking place in an isolated vessel. If we start with one mole of A and one mole of B, the formation of C can continue to occur for as

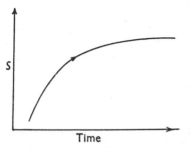

Fig. 9. The trend to equilibrium.

long as the entropy of the system can continue to increase. There will be a certain mixture of A, B and C beyond which no further increase of entropy takes place. The same mixture will be obtained, also with increase of entropy, if we were to start with pure C. The formation of an equilibrium mixture is thus attended by an increase of entropy whether it is attained from the left or right of the chemical equation. In the equilibrium state the entropy is therefore *a maximum*. In general, for systems of constant energy and volume, the condition of equilibrium is the attainment of the maximum entropy consistent with this given energy and volume. Any variation about the equilibrium state could only decrease the entropy.

The trend towards equilibrium is shown diagrammatically in Fig. 9. It does not follow, of course, that when an increase in entropy is possible, the change will necessarily take place at an observable speed. This is shown by the example of a hydrogen-oxygen mixture and many other thermodynamically metastable systems which may persist for centuries without visible change. Thermodynamics has nothing to say about the rates of processes.

In § 1·10 a path between two states A and B of a system was defined as being reversible if the cycle $A \rightarrow B \rightarrow A$ could be completed without leaving a change in any other body. The total entropy of the system and its environment must therefore remain constant throughout the cycle, any entropy change in the system being compensated by an equal and opposite change in the entropy of the environment.

The reversible path must be one for which the internal forces of the system differ only infinitesimally from the external forces, and for which all heat transfers take place over temperature differences which are only infinitesimal. That this is the case may be seen by considering two examples.†

---

† Remaining close to equilibrium is not a *sufficient* condition of reversibility, as is shown by an interesting example of Allis and Herlin (*Thermodynamics and Statistical Mechanics*, p. 85, McGraw-Hill, 1952). However it is certainly a *necessary* condition—contrary to a second example of these authors which the present author regards as fallacious.

Thermal equilibrium. If the *only* process which a body undergoes is the addition or removal of heat the effect of this addition or removal is to cause a definite change of state (since in this case $dq = dU$) and thereby a definite change of entropy. The equation

$$dS = \frac{dq}{T}$$

is therefore applicable to processes of pure heat transfer, even when they are not carried out under reversible conditions, provided that the temperature remains sensibly uniform throughout the body.† (See also the discussion on the heat bath in a footnote on p. 39.)

Consider the transfer, under the above conditions, of a quantity of heat dq from a body at a temperature T_2 to another body at the temperature T_1. The overall entropy change is

$$dS = dS_1 + dS_2$$
$$= \frac{dq}{T_1} - \frac{dq}{T_2}$$
$$= dq(T_2 - T_1)/T_1 T_2.$$

Since dS must be positive or zero, $T_2 > T_1$, and thus the heat flows from the hotter to the cooler body, in agreement with experience. The creation of entropy in the system continues for as long as T_2 exceeds T_1 and the state of equilibrium requires equality of these temperatures. This is in accordance with the meaning of temperature as discussed in § 1·4. A reversible transfer of heat thus requires that there shall be only an infinitesimal temperature difference. Thus if $T_2 - T_1$ is an infinitesimal, the increase of entropy in the above equation becomes equal to $dq dT/T^2$ and is of the second order of smallness.

The above theorem, like all those which have appeared so far, is based on the supposition that there is no exchange of matter between the bodies. If such an exchange takes place the notion of heat flow becomes ambiguous and we can speak only of the total energy flow.

Mechanical equilibrium. If a fluid, at a pressure p, expands by an amount dV against an external pressure $(p - \delta p)$, the work done *by* the fluid is $(p - \delta p)dV$. The corresponding recompression of the fluid requires a pressure, $(p + \delta p)$, which is larger than p, and the work done *on* the fluid is $-(p + \delta p)\,dV$. These two quantities of work become equal. and the processes thus satisfy the definition of reversibility, only in the limit where δp approaches zero. It will be noted that the same condition maximizes the work done by the fluid during expansion and minimizes the work done on the fluid during compression.

It will be shown in more detail in § 2·9 that the condition of mechanical equilibrium requires equality of pressure across any movable interface, provided that no changes of interfacial area are involved.

---

† This condition would be approximately satisfied by carrying out a heat transfer between two bodies through a very thin wire, whose entropy change may be neglected.

1·14. Maximum work

Consider a change of a system from state A to state B. From the first law

$$U_B - U_A = q + w.$$

As noted previously q and w vary according to the nature of the path between the two states—it is only their algebraic sum which is constant. It will now be shown that the second law places a restriction on the maximum amount of work which may be obtained.

It has been shown already that if a system absorbs heat dq from a heat reservoir at temperature T_r, the entropy change of the system is given by equation (1·16):

$$dS_s = \frac{dq}{T_r} + d\sigma,$$

where $d\sigma$ is the created entropy and is positive or zero. Subscripts s and r denote the system and the heat reservoir respectively. Now from the first law

$$dq = dU_s - dw,$$

and therefore

$$dS_s = (dU_s - dw)/T_r + d\sigma,$$

or

$$-dw = T_r dS_s - dU_s - T_r d\sigma. \tag{1·17}$$

Expressed alternatively

$$-dw \leqslant T_r dS_s - dU_s, \tag{1·18}$$

where dw is the work done *on* the system. Thus the work done *by* the system is $-dw$.

Let it be supposed that the heat bath is a large one. In this case T_r remains constant and we obtain by integration of the above relation

$$-w \leqslant T_r(S_B - S_A) - (U_B - U_A).$$

Now for passage between the assigned states A and B the right-hand side of this expression has a definite value since U and S are functions of state. It follows the the work $-w$ which is done *by* the system cannot be larger than a certain quantity, which may be denoted $-w_{\max.}$, and this is obtained in the reversible type of process, which corresponds to the equality sign in the above equation. If the process is not reversible it yields a smaller amount of work and there is a correspondingly smaller intake of heat, as follows also from equation (1·16).

Conversely, the work which must be done *on* a system, in order to obtain a given change of state, is least for the reversible path. In any irreversible change a greater amount of work must be done on the system and there is also a larger positive heat evolution—energy is 'dissipated' as heat.

Consider the case where the only form of work is a volume change dV_s, and let it be supposed that the process is carried out under reversible conditions. The pressure acting on the system is therefore equal to the pressure p_s within the system (§1·13). Similarly, the temperature T_r of the reservoir must be equal to the temperature T_s of the system in contact with it. Therefore under these conditions (1·18) may be written

$$p_s dV_s = T_s dS_s - dU_s.$$

Since all symbols now relate to the system itself, the subscript s may be deleted and the equation rearranged to give

$$\boxed{dU = T dS - p dV.} \tag{1·19}$$

As an example we shall consider the expansion of two mols of a perfect gas from the state $(0.5\ \mathrm{m^3},\ 300\ \mathrm{K})$ to the state $(5\ \mathrm{m^3},\ 300\ \mathrm{K})$.

Consider first of all the case where the expansion is carried out reversibly, the external pressure being adjusted at each moment to be only infinitesimally smaller than the pressure of the gas itself. Thus at any moment when the volume is V the pressure is nRT/V, where n is the amount of gas (mols). The temperature being held constant at 300 K, the work which is done *by* the gas is

$$w = \int p\, dV = nRT \ln V_2/V_1$$
$$= 2 \times 8.314 \times 300 \ln (5/0.5)$$
$$= 1.15 \times 10^4 \ \mathrm{J}.$$

This is equal to the heat absorbed, since the internal energy of a perfect gas remains constant at constant temperature. The entropy change of the gas is therefore $1.15 \times 10^4/300 = 38.3\ \mathrm{J\ K^{-1}}$. The corresponding entropy change of the body which supplies the heat is $-38.3\ \mathrm{J\ K^{-1}}$ and the overall entropy change is zero.

Consider the case where the gas expands against a zero pressure, e.g. by opening a tap which admits the gas into a vacuum chamber. In this case no work is done by the gas in expansion. For the same initial and final volumes and temperatures, the entropy change of the gas is the same as calculated above, $+38.3\ \mathrm{J\ K^{-1}}$, since the change of state is the same. However, in this instance, since no work has been done, no heat has been absorbed and the entropy change of the environment is zero. Therefore there has been a creation of entropy of $38.3\ \mathrm{J\ K^{-1}}$.

In any intermediate type of expansion, in which the gas expands against a pressure which is greater than zero, but less than that of the gas itself, the entropy creation will lie somewhere between zero and $38.3\ \mathrm{J\ K^{-1}}$. It would be calculable if a measurement were made of the actual amount of work done. It is only in the extreme case of *zero* work that the entropy creation can be calculated accurately by means of the type of calculation carried out above.

1·15. The fundamental equation for a closed system

Equation (1·19) is the basic equation for a closed system, i.e. one which does not exchange matter with its environment. It may be derived more directly as follows. From the first law

$$dU = dq + dw,$$

and for a reversible change we can substitute

$$dq = T\,dS,$$

$$dw = -p\,dV,$$

and thus

$$dU = T\,dS - p\,dV.$$

This equation is applicable only when volume change is the only form of work. It is also applicable only to closed systems which are in a state of internal equilibrium,† since all our considerations so far have been based upon these. However, the equation is entirely applicable to a system which changes its composition due to internal chemical reactions, provided that these take place reversibly and also that no work is performed other than that due to volume change.

As regards the application of the equation to *irreversible* paths the following may be said. The derivation was based above on a reversible change, since it is only for such a change that we can, in general, write $dq = T\,dS$. The resultant equation gives the change, dU, in the internal energy of the system, which is at a particular temperature T and a particular pressure p, in terms of the corresponding changes of entropy and volume. *All of these quantities are functions of state.* Moreover, *provided that there are no irreversible changes of composition,*‡ the choice of any two of the variables will determine the state of the system and therefore will determine the values of the other three variables. Thus, if we consider a change between a defined initial state and a defined final state, the integral of the equation must be valid even if the path is not a reversible one (but excluding irreversible changes of composition). Thus, as we go from an initial state (p_A, T_A) to a final state (p_B, T_B), the changes of U, V and S all have definite values, depending only on these states, and it is of no interest how this change takes place. This may, perhaps, be seen more clearly by rewriting the equation in terms of T and p as independent variables:

† If the system in question consists of two or more homogeneous phases whose pressures are not equal, due to the presence of rigid barriers or curved interfaces, then (1·19) must be written for each phase separately.

‡ Such as would be caused either by diffusion or chemical reaction taking place in an irreversible manner within the body, i.e. an absence of internal equilibrium.

$$dU = T\left\{\left(\frac{\partial S}{\partial T}\right)_p dT + \left(\frac{\partial S}{\partial p}\right)_T dp\right\}$$

$$-p\left\{\left(\frac{\partial V}{\partial T}\right)_p dT + \left(\frac{\partial V}{\partial p}\right)_T dp\right\}. \qquad (1\cdot20)$$

Despite what has just been said, the terms TdS and pdV can be identified as the heat absorbed and the negative of the work done *on* the system respectively, only in the case of a reversible path. Thus we can write

$$dq + dw = dU = TdS - pdV,$$

but whenever the process is irreversible dq is less than TdS and dw is less than $-pdV$. (It is to be emphasized that p in the above equation refers to the pressure of *the system itself*, and therefore, in an irreversible expansion, it is larger than p_e, the external pressure.)

Finally, it may be noted that in a cyclic process we have

$$\oint dU = 0,$$

and therefore from (1·19)

$$\oint pdV = \oint TdS.$$

The work performed in a reversible cycle may thus be obtained either as an area on the P-V diagram or as an area on the T-S diagram of the particular substance which passes through the cycle.†

1·16. Summary of the basic laws

It is useful to summarize, in outline. the argument of the present chapter. The state of a system was regarded, at the beginning, as being determined by the two variables, pressure and volume, whose nature was taken as being understood. The experimental knowledge embodied in the zeroth law was then used to show the existence of a certain function of the pressure and volume, $\theta = \theta(p, V)$,‡ which determines whether or not two bodies are in thermal equilibrium. This function was called the temperature, which thereby becomes an extra variable of state. We can invert the relationship and write $V = V(\theta, p)$ or $p = p(\theta, V)$.

It was then shown, as a result of Joule's experiments, that the total work done by a body in an adiabatic process depends only on

---

† Line integrals such as $\int pdV$ are not zero when taken round a closed path because p is not a function of V only.

‡ This symbolism means simply that θ is a function of p and V.

the initial and final states. This work can therefore be expressed as equal to the change in a quantity U, which is again a function of state:

$$U = U(p, V) \quad \text{or} \quad U = U(\theta, p).$$

Finally, we discussed the fact that certain changes are impossible under adiabatic conditions. The possibility or impossibility of the change $A \to B$ depends on the characteristics of the states A and B. There was thus shown to be a new function of state, the entropy, such that if $S_B \geqslant S_A$ the change is a possible one, within an adiabatic enclosure. The same empirical basis allowed also of the definition of a thermodynamic temperature, T, which is independent of the properties of any particular substance.

Thus, in brief, the whole of the fundamental part of thermodynamics may be regarded as the discovery of the quantities T, U and S. Their importance lies precisely in the fact that they are functions of state.† That is to say, they form exact differentials and their changes are independent of the path which is taken between assigned initial and final states. It would not be possible to develop an adequate thermodynamics on the basis of heat and work only, because their magnitudes depend on the details of the path.

From a rather different point of view, the importance of U is that it is a quantity which remains constant in an isolated system. On

TABLE 2

Law	Resulting function of state	Characteristic
0	T	Determines thermal equilibrium
1	U	Determines the algebraic sum of heat and work
2	S	Determines which of two states precedes the other

† It may be asked, what are these functions? Thermodynamics alone does not supply the answer, which can be obtained only by using additional knowledge concerning the behaviour of the molecules of which the system is composed. It is part of the programme of statistical mechanics to obtain U, V and S as explicit functions of T and p, but so far this has been achieved only in the simplest instances. For example, in the case of a perfect monatomic gas with appropriately chosen zeros for U and S, we have per mole

$$v \equiv V/n = RT/p$$

$$u \equiv U/n = 3RT/2$$

$$s \equiv S/n = R \left\{ \ln \left(\frac{M^{3/2}T^{5/2}}{p} \right) + \frac{5}{2} + \ln \left(\frac{2\pi}{h^2} \right)^{3/2} \frac{R^{5/2}}{L^4} \right\},$$

where h is Planck's constant, L is the Avogadro constant and M is the molecular weight. (See Chapter 12.)

s it is possible to establish the useful and familiar 'energy
of a process. However, the great value of thermodynamics
application to systems in which there is an additional restriction,
namely, a state of equilibrium. For such systems the entropy is also
constant, and it is from this result that the most useful relations may
be obtained.

1·17. Natural processes as mixing processes

The reason for the irreversibility of natural phenomena and the
significance of entropy can be made much clearer as soon as we have
recourse to the atomic theory of matter. A discussion on these lines
is actually foreign to pure thermodynamics, but it is so helpful
to the understanding that a preliminary account will be given in this
section, and also the following one, and will be followed up in more
detail in Part III.

As soon as it is accepted that matter consists of small particles
which are in motion it becomes evident that every large-scale natural
process is essentially a process of *mixing*, if this term is given a rather
wide meaning. In many instances the spontaneous mixing tendency
is simply the intermingling of the constituent particles, as in the
interdiffusion of gases, liquids and solids. It arises because at the
interface between two phases there are always a certain number of
atoms or molecules whose direction of movement tends to carry them
across the boundary. Similarly, the irreversible expansion of a gas
may be regarded as a process in which the molecules become more
completely mixed over the available space.

In other instances it is not so much a question of a mixing of the
particles in space as of a mixing or sharing of their total energy. For
example, if blocks of tin and copper are placed in contact there is,
on the one hand, a slow diffusional mixing of the atoms themselves
and, on the other hand, a very much more rapid exchange of energy,
taking place through the mechanism of the atomic vibrations at the
interface. The irreversible process of temperature equalization may
thus be regarded as a mixing of the available energy. Similarly,
in the irreversible process of friction, the kinetic energy of a body as
a whole is converted into the random energy of its component
molecules.

Rather less obvious is the example of chemical reaction. Here it is
a question of the mixing or 'spreading' of the total energy of the
system over the whole range of quantized energy levels of the reactants and products. The occurrence of the reaction causes a larger
number of these quantum states to become accessible, namely, those
corresponding to the products. The final equilibrium composition of

an adiabatically isolated reaction system is the composition at which the available energy is distributed over the various quantum states in the most completely random manner. By computing the state of the system for which this 'mixing' is most complete it is possible to calculate the equilibrium constant of a reaction in a perfect gas from a knowledge of the energy levels of the various molecules (Chapter 12).

There are thus two rather distinct types of mixing process; the first is the spreading of particles over positions in space, and the second is the sharing or spreading of the available energy of a system between the particles themselves. In certain instances these two factors may oppose each other. We have spoken of the tendency of two liquids, A and B, to mix, but miscibility is not always complete. If the relative magnitude of the A-A, B-B and A-B molecular forces are such that complete mixing would cause an increase in the *potential* energy of the system, this energy would have to be withdrawn from the kinetic energy of the molecules (translational, vibrational and rotational motion). There would thus be a smaller amount of energy to be distributed over the quantum levels of these states of motion. In brief, complete mixing would increase the spatial or *configurational* randomness of the system, but would decrease its *thermal* randomness. The degree of miscibility at equilibrium is determined by the composition of the two phases at which the overall randomness is a maximum.

In order to extend the mixing idea a little further it is instructive to carry out some calculations on the slow interdiffusion of two crystals. We shall suppose that we place in contact two crystals, composed of atoms A and B respectively, which are sufficiently alike in lattice structure so that the atoms may be interchanged without any change in the energy states of the crystals (in thermodynamic terms the resultant mixed crystal of A and B would be called an 'ideal' solid solution). Under such conditions we are concerned only with a mixing over positions in space, and not with any randomization of energy over energy states, since the latter are assumed to be unchanged.

For ease of calculation it will be supposed that the crystals each contain four atoms only. Initially all A atoms are to the left of the plane P, and all B atoms to the right of this plane, as shown in Fig. 10. Bearing in mind the indistinguishability of the A atoms amongst themselves and the similar indistinguishability of the B atoms, there is just *one* arrangement in the initial state of the system. Thus

$$\Omega_{4:0} = 1,$$

where $\Omega_{4:0}$ stands for the number of arrangements of the type in which there are four A atoms to the left and none to the right of the plane.

Consider the state of the system in which there is one A atom to the right of the plane and one B atom to the left. The A atom can be placed on any one of the four right-hand sites and the B atom on any

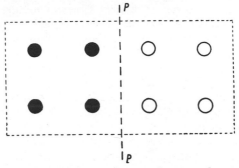

Fig. 10. Mixing of two perfect crystals.

one of the four left-hand sites. There are therefore 16 arrangements of this type, all of which are physically distinguishable. Thus

$$\Omega_{3:1} = 16.$$

The number of arrangements in which there are two A atoms to the left and two B to the right is rather less obvious. One such arrangement is shown in Fig. 11. The first A atom moved to the right can occupy any one of four sites and the second one can occupy any one

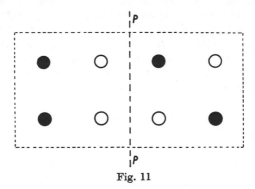

Fig. 11

of the three remaining sites. However there are not 4×3 arrangements, but only $4 \times 3/2!$, since the interchange of the two A atoms between themselves on the same sites does not give rise to a new arrangement which is physically distinguishable from the old one. The $4 \times 3/2!$ arrangements of the A atoms on the right-hand side can be combined with any of the $4 \times 3/2!$ arrangements of the B atoms on the left-hand side. Thus the total number of physically distinct arrangements or 'complexions' of this type is

$$\Omega_{2:2} = 36.$$

Proceeding in this way we can construct Table 3 as shown. The total number of complexions of *all* types is

$$\Omega = 70,$$

which is equal to $8!/4!4!$, the number of ways of distributing four A atoms and four B atoms between eight sites, allowing for the indistinguishability of the A atoms amongst themselves, and of the B atoms amongst themselves.

TABLE 3

Atoms to left of plane	Atoms to right of plane	Number of complexions
$4A$	$4B$	$1 \times 1 = 1$
$3A + 1B$	$1A + 3B$	$4 \times 4 = 16$
$2A + 2B$	$2A + 2B$	$(4 \times 3/2!)^2 = 36$
$1A + 3B$	$3A + 1B$	$(4 \times 3 \times 2/3!)^2 = 16$
$4B$	$4A$	$(4!/4!)^2 = 1$

We come now to the point of the calculation. In the absence of any reason to the contrary, it seems sensible to assume that any one of the seventy complexions is as likely to occur as any other. Therefore, after the two original crystals have been in contact for a very long time, we may expect to find the system, with equal probability, in any one of the seventy arrangements. If it is separated into two parts at the original dividing plane, the probability is $1/70$ that the left-hand crystal will contain four A atoms, it is $16/70$ that it will contain three A atoms and one B atom, and it is $36/70$ that it will contain two atoms each of A and B. The last of these is what would be called the 'most completely mixed state' (relative to the chosen reference plane), and it is seen to be the most probable end state of the mixing process simply because it is made up of a much larger number of arrangements than any of the others.

It may be noted, however, that the system having attained one of the 36 configurations belonging to the 'well-mixed' class, it can still momentarily return to any of the other configurations of the system, such as the initial one itself. All that we have shown is that we are

much more likely to find our system in a mixed condition than in an
unmixed one, after it has been left in isolation for a time long enough
to allow the atoms to wander into any of the sites. The probability
of a return to the original condition becomes progressively less as
we consider larger and larger numbers of atoms. For example with
six atoms each of A and B there are altogether 924 complexions;
400 are of the 3 : 3 type; 225 each are of the 4 : 2 and 2 : 4 types; 36 each
are of the 5 : 1 and 1 : 5 types and there is only one each of the 6 : 0
and 0 : 6 types.

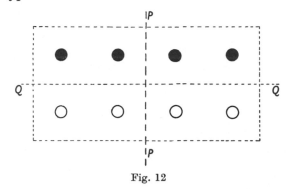

Fig. 12

Further consideration shows that the plane which was assumed to
separate the two crystals has played an important role in the above
discussion. This may be seen by considering the atomic arrangement
shown in Fig. 12. Relative to a division at the vertical plane, as used
previously, it is one of 36 equivalent arrangements of the 2 : 2 type.
Relative to the horizontal plane Q it is of the 4 : 0 type and is the only
arrangement of its kind. The calculation of the number of arrangements of a given class, e.g. $\Omega_{4:0}$, thus always presupposes a statement
on how that class is defined. On the other hand, the total number of
complexions, $\Omega = 70$, is not dependent on any choice of a reference
plane, but only on the number of atoms of each type and on the limits
of our possible knowledge concerning distinguishable arrangements.

Summarizing the above, it seems that a natural process is one in
which there is an increase in Ω, the number of complexions which are
accessible to a system. Thus, in the above example, Ω increases from
unity, corresponding to the initially known condition of the system,
to a value of 70, the number of configurations in any one of which it
might exist. As we shall see in the next section, the deeper significance
of the irreversible process is that we cannot predict which of the 70
complexions the system will actually occupy at a given moment.
Under conditions of adiabatic isolation (i.e. in the absence of a fresh

experimental determination of where the atoms actually are) there is thus a *loss of knowledge* during the irreversible change.

We come now to the relationship of the above ideas to the entropy of thermodynamics. The adiabatic mixing of the two crystals would certainly be attended by an increase in entropy and, as we have seen, it is also accompanied by an increase in Ω, which measures the decrease in our knowledge concerning the actual state of the system, as compared to the initial state. We may therefore expect a relationship between S and Ω.

The latter quantity must first be defined more closely, so that it may be applied to examples other than the one which has been discussed. Ω is the total number of distinguishable micro-states, or 'complexions', which are confined within a given macro-state of a system, this macro-state being characterized by fixed values of the energy, volume and numbers of particles of specified kinds. Thus

$$\Omega = \Omega(U, V, N_i),$$

where N_i is the number of particles of the ith kind. In other words, Ω is the number of independent quantum states which are accessible to a system. To compute changes of Ω we need to know the number of atoms of various types and the number of distinguishable positions in space which they can occupy. If it is a question of randomization over energy states, as well as over spatial positions, we shall also need to know what are the distinguishable energy states.

In the absence of any reason to the contrary, all of the Ω quantum states of a system (compatible with the given values of U, V and N_i) are assumed to be equally probable, i.e. have equal weight. This is the basic assumption of statistical mechanics.

We shall now put forward the hypothesis, which derives from Boltzmann, that the entropy of a system is related to its Ω by the equation

$$S = k \ln \Omega, \tag{1·21}$$

where k is a constant. The correctness of this hypothesis must be judged by its quantitative success (see Part III), but we shall put forward the following preliminary points in its favour. In the first place, both S and Ω are properties of the system and are functions of state, being determined by U, V and the N_i. Secondly, they both tend to increase in an irreversible process. Thirdly, the assumed logarithmic form of the relationship is the only one which will make the *multiplicative* properties of Ω compatible with the *additive* properties of entropy. Thus suppose that we consider two parts, 1 and 2, of a system, each having fixed values of its energy, volume and composition. Each of the Ω_1 complexions of the first part can be chosen in combination with any of the Ω_2 complexions of the second part. The total number of complexions for the combined system is therefore

$$\Omega = \Omega_1 \Omega_2.$$

Taking logarithms and applying equation (1·21), we obtain for the total entropy of the system

$$S = k \ln \Omega$$
$$= k \ln \Omega_1 + k \ln \Omega_2$$
$$= S_1 + S_2.$$

The functional relationship (1·21) thus makes the entropy additive, as it should be.

If (1·21) is applied to the process of mixing of two crystals as discussed above we obtain

$$\Delta S = k \ln 70 - k \ln 1 = k \ln 70.$$

In the more general case where one crystal contains N_a atoms of A and the other contains N_b atoms of B, we obtain

$$\Delta S = k \ln \frac{(N_a + N_b)!}{N_a! N_b!}. \tag{1·22}$$

It will be shown in a later chapter that this result is in exact agreement with the value obtained by pure thermodynamics for the entropy of mixing of an ideal solution. It must be noted, however, that the above calculation is based on the supposition that the interdiffusion of the A and B atoms does not affect the energy states of the system. The calculated entropy change, $k \ln 70$, is the change in the *configurational entropy* of the system, and it is a correct result for the total change in entropy only when there is no additional change in Ω arising from a redistribution of the energy of the system over changed energy levels.† Thus in the initial state of the system, as shown in Fig. 10, it would be incorrect to write $S = k \ln 1 = 0$ as an expression for its *total* entropy; the atoms are in a state of vibration about their lattice points, and the crystal has a value of Ω due to the number of ways in which the energy can be arranged over the energy levels. This number is very large at room temperature.

Another preliminary example of the application of (1·21) is the isothermal expansion of a perfect gas from a volume V_A to a volume V_B. For the ith molecule let Ω_{Ai} be the number of complexions or quantum states accessible to this molecule before the expansion and let Ω_{Bi} be the corresponding number after the expansion. It is reasonable to suppose (and provable in quantum theory) that the ratio Ω_{Ai}/Ω_{Bi} is

† For further discussion on this point see §11·12 and Chapter 14.

equal to V_A/V_B. In fact, the former ratio may be regarded as equal to the probability that the given molecule is still present in the smaller volume V_A after the larger volume V_B has become available to it. Since the molecules are independent we can obtain the overall value of Ω_A/Ω_B by multiplying the value of Ω_{Ai}/Ω_{Bi} for the ith molecule by the corresponding values for all other molecules. Thus

$$\frac{\Omega_A}{\Omega_B} = \frac{\Omega_{A1}}{\Omega_{B1}} \frac{\Omega_{A2}}{\Omega_{B2}} \cdots \frac{\Omega_{AN}}{\Omega_{BN}} = \left(\frac{V_A}{V_B}\right)^N ,$$

where N is the total number of molecules. Thus from (1·21)

$$S_B - S_A = k \ln \frac{\Omega_B}{\Omega_A} = Nk \ln \frac{V_B}{V_A}.$$

Now in the problem discussed in §1·14 it was seen that a purely thermodynamic result for the entropy increase in the isothermal expansion of one mole of a perfect gas is

$$S_B - S_A = R \ln V_B/V_A.$$

The equation (1·21) is therefore consistent with the thermodynamic result when we identify Boltzmann's constant, k, with R/L, where L is the Avogadro constant.

As a result of the above discussion what can be said about the meaning of entropy if our basic assumptions are correct? It is evidently a measure of the 'mixed-up-ness' of a system—a phrase used by Gibbs. It can also be said that high entropy states are those which have a high probability. The quantity Ω is sometimes referred to as the 'thermodynamic probability', because the ratio Ω_A/Ω_B for the distributions A and B of a system is a measure of the relative frequency with which we should expect to find these distributions, under adiabatic conditions. (But note that Ω is always greater than unity.)

Another common interpretation of entropy is in terms of order and disorder, but this is not entirely satisfactory. A counter-example to the idea that an entropy increase implies an increase of 'disorder' is due to Bridgman. This is concerned with the spontaneous crystallization of a supercooled liquid; if this takes place under adiabatic conditions the entropy of the resulting crystal will be *greater* than that of the supercooled liquid, but it would be difficult to claim that there has been an increase in 'disorder'. The answer to the paradox lies in the fact that the thermal or kinetic energy is increased during the process of crystallization, due to the decrease of potential energy on formation of the lattice. Thus, although there is a decrease of configurational entropy, consequent on the more orderly arrangement of the lattice as compared to the liquid, there is a more than compensating increase in *thermal entropy*, due to the randomization of the liberated potential energy over the vibrational motions of the atoms in the crystal

Perhaps one of the most useful verbalisms is 'spread', as used by Guggenheim;† an increase of entropy corresponds to a 'spreading' of the system over a larger number of possible quantum states. This interpretation is often more appropriate than the one in terms of mixing when the irreversible process in question is concerned less with configurational factors than with a change into a state where there is a greater density of energy levels.

On the molecular scale the internal energy, U, consists partly of the potential energy of the particles and partly of their kinetic (or 'thermal') energy. In the case of two motionless hydrogen atoms at some distance apart, the energy is entirely potential, if we exclude the energy of the nuclei themselves. If these atoms are allowed to approach each other, there is a decrease in the potential energy, and this is converted partly into translational energy and partly into the rotational, vibrational and electronic energy of the hydrogen molecule, if it is formed. Now the separation of neighbouring quantized levels is in the order:

electronic > vibrational > rotational > translational,

and the energy states belonging to the latter are so close together that translational energy may be regarded, for many purposes, as continuous or unquantized. For a given total amount of energy, the highest value of Ω is obtained when this energy is 'spread' over the molecules in the most random manner, due allowance being made for the relative separation of the levels. The 'dissipation' of mechanical energy into heat is to be interpreted in this kind of way.

Similarly, the increase in the entropy of a body when it takes in heat is essentially an increased spread over the energy states. Due to the added increment of energy, energy levels can become occupied which were previously empty. This results in an increase in randomness and therewith an increase in our ignorance of the precise molecular state of the system.

1·18. The molecular interpretation of the second law

The groundwork of this interpretation has been covered in the last section, and it remains only to express it in a more concise form. Now it is evident, in the first place, that the second law arises from the fact that matter consists of moving particles, which tend to intermingle, to collide and to share their kinetic energy. The real crux of

† Guggenheim, *Research*, **2** (1949), 450.

the matter lies, however, in the limits of our knowledge concerning the precise state of these particles.

Consider a system on which we make a set of observations at a moment t_0. Thereafter we carefully isolate the system from its environment, so that it becomes a system to which the second law is applicable. The observations at t_0 will be supposed to be the volume, composition and energy (relative to a standard state) of the various parts of the system. This information completely determines the initial *thermodynamic state* of the system and it also partially determines its *molecular state*; for example, it may be known that all of the A atoms are to the left of a certain plane, or that they are all present as a particular chemical compound. Let Ω_0 be the total number of complexions of the system each of which are compatible with the initial thermodynamic state. As time goes on it may be that the isolated system either (*a*) remains thermodynamically unchanged, or (*b*) that there occur various chemical reactions, or diffusions between the different macroscopic parts of the system, or internal transfers of energy, etc. In the latter case it evidently means that the initial state of the system was not one of true equilibrium and that Ω_0 did not include the additional complexions which are realized by the internal changes resulting from the motion of the atoms. It follows that *the final value of Ω is greater than Ω_0*—greater because (as has been illustrated by the example of the two crystals) in any natural process the system in question changes into a thermodynamic state which is made up of an increased number of complexions.

Thus with passage of time the value of Ω of an isolated system either remains constant or it increases, and the latter is equivalent to a decrease of our information concerning the molecular state of the system (since there are now more possibilities). Against this it might be argued that our original information I_0 might have been so detailed that we knew the precise position and velocity of every particle. Thereby, on the basis of the laws of mechanics, we might have hoped to make a precise prediction of the future state of the system, i.e. a prediction concerning the particular complexion in which it would exist at a given moment. If this were possible there would be no sense in speaking of an increase in randomness. Also the supposed information could be used, in conjunction with devices such as 'micro-reflectors', for returning the particles along their original paths, and thereby restoring the original state of the system.

However such completely detailed information can never be achieved. In the first place, it is an obvious practical impossibility to make an exact and simultaneous measurement of the position and velocity components of all the fundamental particles of a system. Secondly, according to Heisenberg's principle of uncertainty, it is a fundamental impossibility, not merely a practical one. Indeed, according to this principle, the very notion of a particle having a simultaneously defined position and velocity is a meaningless one.†

From this point of view, the real basis of the second law is *the impossibility of knowing the precise mechanical state of an atomic system*. As Born‡ has emphasized, it is sufficient that there should be a lack of complete information about the position and velocity of only a single particle—as a result of its subsequent collisions, the position and velocities of all the other particles becomes uncertain.

The fact that we cannot know the precise mechanical state of an atomic system implies that it is impossible to take heat from a heat bath and obtain an equivalent amount of work, without a compensating change in some other body. This is the Statement A, on which our treatment of the second law has been based, and its origin may be seen as follows. The transfer of heat from a body of constant volume implies a decrease in the kinetic energy of its component molecules, all of which are in motion in *random directions*. Because we cannot know the fine details of this random motion at any moment, it is impossible to devise a means of taking away some of the kinetic energy of the molecules and applying this energy, in its entirety, to the lifting of a weight, which is to say, causing a *co-ordinated* change in the positions of another set of molecules. In brief, what we call 'heat' is the form of energy of whose details we have the least information on the molecular scale; if heat were completely convertible into work it would be equivalent to a gain of information and this is impossible in a system which is isolated.§

† It may be noted that Bohm, and certain other physicists, do not regard Heisenberg's principle as being a final and irreducible statement about reality. However Bohm's views, far from detracting from the interpretation of the Second Law given above, tend to reinforce it. Bohm seeks to replace mechanistic principles in physics by the notion of the *qualitative infinity of nature*; according to this the motions of the particles at any given level (e.g. atoms, electrons etc.) can never be fully known because they are subject to the fluctuations existing at lower levels, and there is no reason to suppose that there is not an infinity of such lower levels. [D. Bohm, *Causality and Chance in Modern Physics*, Routledge & Kegan Paul, 1957.]

‡ Born, *Natural Philosophy of Cause and Chance* (Oxford, 1949).

§ For further discussion on the Second Law in relation to information theory see Jaynes, E. T., *Phys. Rev.* **106** (1957), 620; **108** (1957), 171, L. Brillouin, *Science and Information Theory* (Academic Press, 1962) and Tribus, *Thermostatics and Thermodynamics* (van Nostrand, 1961). As pointed out by von Neumann (*Automata Studies*, ed. Shannon and McCarthy Princeton, 1956), the essential point is that Ω corresponds to the amount of microscopic information that is missing in the merely macroscopic, i.e. thermodynamic, specification of a physical system. It follows that the system's entropy is equal (apart from a dimensional factor $k \ln 2$) to the number of 'bits' of information we should have about the system *if* we knew its actual complexion.

What has been said does not mean, of course, that we cannot increase our information about a system above the value I_0, or that we cannot reverse a natural process. Both are entirely possible, if we transfer our ' ignorance ', i.e. entropy, to some other body. The removal of heat from a system, with reduction of its entropy, involves a decrease in the randomness of its component particles and is therefore equivalent to an increase of knowledge. But this can only be achieved at the expense of an equal or larger increase in the randomness of some other system (equal for the reversible process and larger for the natural one).

Although it is never possible to reverse a natural process in its entirety, at a chosen moment, the statistical interpretation of the second law shows that very occasionally a system may be expected to reverse itself. An isolated system of finite size has only a finite number, Ω, of complexions, and it may be expected to pass through many of these during any long period. Thus, in the example of the A and B crystals there were seventy complexions (counting only those which are due to configurational factors); if the system were repeatedly separated at the plane P, on one in every seventy occasions we shall expect to find it in the original configuration. The moment at which it occurs is quite unpredictable. In the case of systems containing an appreciable number of atoms, it becomes increasingly improbable that we shall ever observe the system in a non-uniform condition. For example, it is calculated† that the probability of a relative change of density, $\Delta\rho/\rho$, of only 0.001% in 1 cm³ of air is smaller than 10^{-10^8} and would not be observed in trillions of years. Thus, according to the statistical interpretation, the discovery of an appreciable and spontaneous decrease in the entropy of an isolated system, if it is separated into two parts, is not impossible, but is merely exceedingly improbable.‡ We repeat, however, that it is an *absolute impossibility* to know when it will take place.

The matter may thus be briefly summed up as follows. Whenever a system undergoes a spontaneous change under isolated conditions, our knowledge concerning the actual atomic state of the system inevitably diminishes. This is because the particles are in motion and we can never have sufficiently detailed initial information to predict its atomic state at some later moment by use of the laws of mechanics. This decrease in our knowledge is equivalent to an increase in Ω, the number of quantum states which the system can take up, each of

† Epstein, *Commentary on the Scientific Writings of Willard Gibbs*, **2** (New Haven, Yale Univ. Press, 1936), 112.

‡ For detailed calculations, see Chapter IV of Mayer and Mayer, *Statistical Mechanics* (New York, Wiley, 1940).

them equally compatible with what knowledge we do have. The equilibrium state is that for which Ω is a maximum. †

PROBLEMS ‡

1. (*a*) Calculate the numerical value of the conversion factor between work expressed in cm³ atmospheres and work expressed in joules. Tabulate for future use the conversion factors between the following units: cm³ atm, ergs, joules, calories.

(*b*) The gas constant R has the value 8.314 J K⁻¹ mol⁻¹. Calculate its values when energy is expressed in cm³ atm, and in calories.

2. Calculate the work which is done when 1 mol of water is vaporized at 100 °C and at 1 atm pressure. In carrying out the calculation: (*a*) use the observed increase in volume of 3.019×10^{-2} m³ per mol vaporized; (*b*) use the volume increase which would occur if the water vapour were a perfect gas. Compare the results with the latent heat of vaporization $(4.061 \times 10^4$ J mol⁻¹), and consider, in terms of the molecular forces, why the heat is so much larger than the work.

3. Show by calculation that the work involved in process (*a*) below is very much greater than in process (*b*):

(*a*) an increase in volume of 1 cm³ against a steady pressure of 1 atm;

(*b*) an increase of interfacial area of 1 cm², the interfacial tension being 73×10^{-3} Nm⁻¹ (water-air at 20 °C).

4. Show in what respects the following statements require amplification so that they shall not be contradictory:

(*a*) in a reversible process there is no change in the entropy;

(*b*) in a reversible process the entropy change is $\int \dfrac{dq}{T}$.

5. 200 g of mercury at 100 °C are added to water at 20 °C in a calorimeter. The mass of the water is 80 g and the water equivalent of the calorimeter is 20 g. Find the entropy change of (*a*) the mercury, (*b*) the

---

† Some books and papers dealing with these questions are as follows: Landau and Lifschitz, *Statistical Physics*, transl. Peierls (Pergamon, 1958), Chapter II; Born, *Natural Philosophy of Cause and Chance* (Oxford, 1949); Weyl, *Philosophy of Mathematics and Natural Science* (Princeton, 1949); Margenau, *The Nature of Physical Reality* (New York, McGraw-Hill, 1950); Bridgman, *Reflections of a Physicist* (New York, Philosophical Library, 1950); Schrödinger, *Proc. Roy. Irish Acad.* **53**A (1950), 189; Demers, *Canad. J. Res.* **22** (1944), 27; **23** (1945), 47; Brillouin, *J. Appl. Phys.* **22** (1951), 334; Raymond, *Amer. J. Phys.* (1951), 109; Rothstein, *Phys. Rev.* **85** (1952), 135; Denbigh, *Brit. J. Phil. Sci.* **4** (1953), 183; Popper, K., *Nature* **177** (1956), 538; **178** (1956), 382; **179** (1957), 1297; Penrose and Percival, *Proc. Phys. Soc.*, **79** (1962), 605; Costa de Beauregard, *Le Second Principe de la Science du Temps* (Editions du Seuil, 1963).

‡ For answers to Problems, and comments, see Appendix, p. 458.

water and calorimeter, (c) the water, calorimeter and mercury together. The specific heat capacity of water and mercury may be taken as constant at 4.184 and 0.140 J K^{-1} g^{-1} respectively.

6. In a heat exchanger, air is heated from 20 to 80 °C by means of a second air stream which enters the exchanger at 150 °C. The two streams flow at equal molar rates. Calculate the entropy change per mole of the two streams and also the total entropy change.

For simplicity assume that the air has a constant heat capacity of 29 J K^{-1} mol^{-1}. Repeat the calculations for the case in which the cooler air stream is raised to a temperature of 120 °C. Why could this be achieved only in a counter-current system?

7. An electric current is passed through a resistance coil which is kept at a constant temperature by being immersed in running water. Obtain an expression for the rate of creation of entropy in the system at the steady state. Show that the electrical conductivity of a substance is always expressed by a positive number.

8. A perfect gas is a fluid which satisfies the conditions: (a) its temperature θ is proportional to pV; (b) its internal energy depends only on the temperature. By performing a Carnot cycle on such a gas, prove that the perfect gas temperature θ is proportional to the thermodynamic temperature T, as defined in §1·11.

9. A gas is initially in a state T_1, p_1 and is changed to a state T_2, p_2. Devise two or more reversible paths between these states and show that the heat absorbed by the gas is not the same along the different paths, but that its entropy change is the same. The gas may be assumed to be perfect and also to have constant heat capacity.

10. (a) Show that Statement A of §1·11, together with the knowledge that heat flows spontaneously from hot to cold, are equivalent to the following statement (which is similar to the form used by Clausius); it is impossible to transfer heat from a colder to a hotter reservoir, leaving no change in the thermodynamic state of any other body, without at the same time transforming a certain amount of work into heat.

(b) As an aspect of the first proposition in §1·11 show that no engine can have a higher conversion factor than a reversible engine working between the same temperatures.

11. Calculate the entropy change (a) of the water, (b) of all other bodies, when 1 mol of supercooled water freezes at −10 °C and 1atm. Take the heat capacity of water and ice as constant at 75 and 38 J K^{-1} mol^{-1} respectively and the latent heat of fusion at 0 °C as 6026 J mol^{-1}.

12. A cylinder of free volume 10 ft^3 is divided into two compartments by a piston. Initially the piston is at one end and the cylinder contains 2 lb of steam at 150 °C. Air from a supply main at 200 lb/in^2 and 15 °C is then admitted slowly via a throttle valve to the other side of the piston. Estimate the temperature of the air when pressure equilibrium is reached.

Neglect any heat loss from the cylinder and assume no heat transfer across the piston.

State clearly any assumptions made.   [C.U.C.E. Qualifying, 1950]
[Note. Use steam tables.]

13. In a manufacturing process the process materials change between defined initial and final states such that their overall increments of internal energy, entropy and volume are ΔU, ΔS and ΔV respectively. The only input of energy to the process materials is from condensing steam at a temperature T_s. The only input of energy is the heat transfer to the environment and the work involved in the displacement of the environment.

The quantity of heat taken from the steam is q. Starting from the basis of the first and second laws outline the steps in the proof of the statement that any irreversibility in the process gives rise to a wastage of the thermal energy of the steam by the amount

$$q - \frac{T_s}{T_s - T_0}[\Delta U + p_0 \Delta V - T_0 \Delta S],$$

where T_0 and p_0 are the temperature and pressure respectively of the environment.   [C.U.C.E. Tripos, 1954]

CHAPTER 2

AUXILIARY FUNCTIONS AND CONDITIONS OF EQUILIBRIUM

2·1. The functions H, A and G

The whole of the physical knowledge on which thermodynamics is based has already been embodied in the properties of T, U and S, and these functions alone form a sufficient basis for the development of chemical thermodynamics. It is a matter of convenience only that we introduce certain additional functions. These are defined as follows:

$$\text{the enthalpy,} \qquad H \equiv U + pV, \qquad (2\cdot1)$$

$$\text{the Helmholtz function,} \quad A \equiv U - TS, \qquad (2\cdot2)$$

$$\text{the Gibbs function,} \qquad G \equiv U + pV - TS$$
$$\equiv H - TS$$
$$\equiv A + pV. \qquad (2\cdot3)$$

These new quantities are combinations of the previous functions of state, U, p, V, T and S, and are therefore functions of state themselves. They are also extensive properties. Their value is simply that they are easier to use in certain applications, e.g. processes at constant pressure, volume or temperature, and in such circumstances they also have an easily visualized physical meaning.

2·2. Properties of the enthalpy

For the change of a system between states 1 and 2, we have from $(2\cdot1)$ above

$$H_2 - H_1 = U_2 - U_1 + p_2 V_2 - p_1 V_1. \qquad (2\cdot4)$$

Now for a closed system $\quad U_2 - U_1 = q + w, \qquad (2\cdot5)$

and therefore $\qquad H_2 - H_1 = q + w + p_2 V_2 - p_1 V_1. \qquad (2\cdot6)$

In the special case where the system is under a constant pressure p the last equation may be written

$$H_2 - H_1 = q + w + p(V_2 - V_1).$$

The last term is the work of displacing the environment of the system at pressure p. If this is the only form of work, the last two terms cancel and therefore

$$H_2 - H_1 = q. \qquad (2\cdot7)$$

The enthalpy change of a closed system is thus equal to the heat absorbed under *two* restrictive conditions, namely, that there is constancy of pressure and that the only form of work is due to the volume change of the system. In a galvanic cell, where electrical work is done, the heat absorbed is not equal to the change in enthalpy.

Unlike the internal energy, enthalpy is not a quantity which is necessarily conserved under conditions of isolation. Consider, for

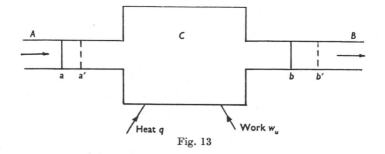

Fig. 13

example, a reaction taking place inside a thermally insulated vessel of constant volume. Since $U_2 = U_1$ and $V_2 = V_1$, equation (2·4) reduces to

$$H_2 - H_1 = V(p_2 - p_1),$$

and the enthalpy of the system changes in accordance with any change in the pressure due to the occurrence of the reaction.

The enthalpy has certain important applications to steady-flow systems. In Fig. 13 the central rectangle represents an apparatus C (e.g. turbine, reaction vessel) through which there is a steady flow of material, which enters through pipe A and leaves through pipe B. Such processes can be discussed most clearly by concentrating attention on a fixed quantity of the moving fluid, enclosed between the two imaginary pistons a and b.

Let the pressures at a and b be p_1 and p_2 respectively, and let V_1 and V_2 be the corresponding volumes per unit mass of the fluid. In order to move a mass δm through C, the left-hand piston must be displaced to a', through a volume $V_1 \delta m$, and the right-hand piston to b', through a volume $V_2 \delta m$. Provided that the pressures p_1 and p_2 are steady, the net work done *on* the sample of fluid contained between the two imaginary pistons is

$$w = -p_2 V_2 \delta m + p_1 V_1 \delta m + w_u \delta m.$$

The first term on the right-hand side is the work done *by* the sample in displacing fluid to the right of b, and the second term is the work done on the sample by fluid to the left of a. w_u is the 'useful' work done *on* unit mass of the sample as it passes through C, e.g. the 'shaft' work of a pump, or the negative of the work done *by* unit mass of the fluid if it drives a turbine or engine. Let q be the heat absorbed per unit mass of the fluid between a and b, and let E_1 and E_2 be the *total* energy† per unit mass at a and b respectively. Then from the first law

$$(E_2 - E_1)\,\delta m = q\,\delta m - (p_2 V_2 - p_1 V_1 - w_u)\,\delta m,$$

or $$E_2 - E_1 = q - p_2 V_2 + p_1 V_1 + w_u.$$

Under conditions where the changes of kinetic and potential energies of the fluid are trivial, as is usually the case, $E_2 - E_1 = U_2 - U_1$, and thus

$$U_2 - U_1 = q - p_2 V_2 + p_1 V_1 + w_u,$$

and combining this with (2·1) we finally obtain

$$H_2 - H_1 = q + w_u. \tag{2·8}$$

Comparing this equation with (2·5) it is seen that, in a steady-flow process, the enthalpy takes the place, in a certain sense, of the internal energy. However, this is only so because we choose to concentrate attention on w_u, and this is not the total work which is done *on* the fluid in passing through the system.

If the processes taking place in C are reversible

$$q = \int_1^2 T\,\mathrm{d}S,$$

and thus the maximum useful work,‡ $-w_{u,\text{max.}}$, which is done *by* the fluid is§

$$-w_{u.\,\text{max.}} = -(H_2 - H_1) + \int_1^2 T\,\mathrm{d}S. \tag{2·9}$$

In the special case where the only transfer of heat takes place reversibly at the constant temperature T', the remainder of the process being adiabatic and reversible, (2·9) reduces to

$$-w_{u.\,\text{max.}} = -(H_2 - H_1) + T'(S_2 - S_1). \tag{2·10}$$

For example, if a steam engine operates adiabatically and the waste steam passes into a condenser where it is condensed at a temperature T', the last equation gives the maximum attainable work. H_1 and S_1 refer to the inflowing steam and H_2 and S_2 to the outgoing

† See footnote on p. 17.

‡ But note that this is the work done by the *fluid* in its assigned change of state. Any heat which is given up by the fluid can also be used as a source of work, by supplying it to an external heat engine. See the discussion in §2·5*a* on 'Availability'.

§ Note that the convention of §1·5 that work counts as positive when done *on* the system of interest requires careful consideration of signs. The meaning of the L.H.S. of (2·9) may perhaps be made clearer by use of brackets: thus the useful work done *by* the fluid is $(-w_u)$ and its maximum value is $(-w_u)_{\text{max}}$.

condensate at the temperature T'. (The inclusion of any changes of kinetic and potential energy usually makes only a trivial correction, except in regions of very high velocity as in the interior of a turbine.)

2·3. Properties of the Helmholtz free energy

For the change of a system between states 1 and 2 we have from (2·2)

$$A_2 - A_1 = U_2 - U_1 - (T_2 S_2 - T_1 S_1). \tag{2·11}$$

But for a closed system $U_2 - U_1 = q + w,$

and therefore $\qquad A_2 - A_1 = q + w - (T_2 S_2 - T_1 S_1). \tag{2·12}$

Consider the special case (a) that the only heat transferred to the system is from a heat reservoir which remains at the constant temperature T; (b) that the initial and final temperatures, T_1 and T_2, of the system are equal, and are equal to the temperature T of the reservoir.

From (1·16) we have for the heat taken in by the system in terms of its entropy change

$$\frac{\mathrm{d}q}{T} \leqslant \mathrm{d}S, \tag{2·13}$$

or, since T is constant during the intake of the heat q from the reservoir, $\qquad q \leqslant T(S_2 - S_1). \tag{2·14}$

Substituting in (2·12) and putting $T_2 = T_1 = T$, we obtain after rearrangement†

$$-w \leqslant -(A_2 - A_1). \tag{2·15}$$

It may be noted that this relation remains valid even if the system passes through intermediate temperatures different from T_1 ($=T_2$), provided that the only heat reservoir which is affected is also at T_1.

In relation (2·15) the inequality sign refers to an irreversible process and the equality sign to a reversible one. Between the assigned initial and final states the value of $A_2 - A_1$ is, of course, the same whether the path is a reversible one or not, since A is a function of state. Relation (2·15) may thus be interpreted as follows. The work $-w$ done by a system during a process in which the initial and final temperatures and the temperature of the heat reservoir are all equal, is either less than or equal to the decrease in A. Thus, for such a process when carried out reversibly.

$$-w_{\mathrm{max.}} = -(A_2 - A_1), \tag{2·16}$$

and any irreversible path between the same two states will give less

† This relation could also have been obtained from (1·17). Concerning the sign of w see the last footnote to §2·2.

work. The change in A is therefore a measure of maximum attainable work under the above conditions of operation. This result applies only to a closed system.†

The reason why A has been called a free energy may be seen as follows. Equation (2·11) may be written

$$-\Delta U = -\Delta A - \Delta(TS),$$

and the decrease $-\Delta U$ in the internal energy of a system may thus be regarded as composed of two parts: the 'free-energy' decrease, $-\Delta A$, which is made available as work, and the remaining part, $-\Delta(TS)$, which may be thought of as being 'dissipated', since in a reversible isothermal change it is equal to the heat evolved. However, this point of view is rather deceptive. For example, in a reversible endothermic process $T\Delta S$ is positive and the maximum work, $-\Delta A$, which the system will perform under isothermal conditions is therefore numerically greater than the decrease of its internal energy, on account of the positive heat absorption.

Another important characteristic of A is that it provides a criterion of equilibrium. For an infinitesimal change at constant temperature we have from (2·15)

$$-dw \leqslant -dA$$

and in particular if $dw = 0,\quad dA \leqslant 0.$ \hfill (2·17)

The Helmholtz function can thus only decrease or remain constant. The criterion of equilibrium of a system enclosed in a rigid container and held at constant temperature is therefore that A has reached its minimum possible value.

2·4. Properties of the Gibbs function

For the change of a system between states 1 and 2 we have from (2·3)

$$G_2 - G_1 = (U_2 - U_1) + (p_2 V_2 - p_1 V_1) - (T_2 S_2 - T_1 S_1).$$

But for a closed system $\quad U_2 - U_1 = q + w$
and therefore

$$G_2 - G_1 = q + w + (p_2 V_2 - p_1 V_1) - (T_2 S_2 - T_1 S_1). \quad (2·18)$$

Consider the special case (a) that the only heat transferred to the system is from a reservoir which remains at the constant temperature

---

† That is, to a system containing a fixed amount of matter, even though this matter may change its chemical composition or may be transferred from one container to another within the overall system under discussion.

T; (b) that the initial and final temperatures, T_1 and T_2, of the system are equal, and are equal to the temperature T of the reservoir; (c) apart from the system, the only other body which has undergone a change of volume at the end of the process is at a constant pressure p (e.g. a surrounding fluid such as the atmosphere); (d) the initial and final pressures, p_1 and p_2, are equal, and are equal to p.

As in (2·14), we have for the heat taken in by the system

$$q \leqslant T(S_2 - S_1). \tag{2·19}$$

Substituting in (2·18) and putting $T_2 = T_1 = T$, and also $p_2 = p_1 = p$, we obtain after rearrangement

$$-w - p(V_2 - V_1) \leqslant -(G_2 - G_1). \tag{2·20}$$

The term $p(V_2 - V_1)$ is the work done *by* the system in displacing its environment at the steady pressure p. This is not necessarily the whole of the work, $-w$, done *by* the system, and therefore we define and additional work time $-w'$ by the relation

$$-w = -w' + p(V_2 - V_1). \tag{2·21}$$

For example, if the system in question were a galvanic cell, w' would stand for the electrical work.

Substituting (2·21) in (2·20) we obtain*

$$-w' \leqslant -(G_2 - G_1). \tag{2·22}$$

This relation remains valid even if the system passes through intermediate temperatures and pressures which are different from T and p respectively, provided that the only heat intake is from the reservoir at T and that the only body, other than the system itself, which has undergone a volume change at the end of the process is the environment at the pressure p.

In (2·22) the inequality sign refers to an irreversible path and the equality sign to a reversible one. In either case the value of $G_2 - G_1$ is the same. The relation may thus be interpreted as follows. In a process in which the initial and final temperatures, and the temperature of the heat bath, are all equal and in which the environment is displaced at the constant pressure p, the work done *by* the system of interest (but not including the work of displacement) is either less than or equal to the decrease of G of the system. Thus for such a process when carried out reversibly

$$-w'_{\text{max.}} = -(G_2 - G_1), \tag{2·23}$$

and any irreversible path between the same two states will give less work. The change in G is therefore a measure of the maximum attain-

* Concerning the sign of w see the last footnote to §2·2.

able work not including the work of displacement.† This result applies only to a closed system.

It may be noted from (2·21) that if $V_2 < V_1$ the work w' is numerically larger than w (and the decrease in G is larger than the decrease in A). For example, if a reaction in a galvanic cell takes place with decrease of volume, the work done *on* the cell by the atmosphere contributes (very slightly) to the amount of electrical energy which is obtainable from the reaction.

Like the Helmholtz function, G can be used to provide a criterion of equilibrium. For an infinitesimal change at constant temperature and pressure, we have from (2·22)

$$-\mathrm{d}w' \leqslant -\mathrm{d}G \tag{2·24}$$

and in particular if $\mathrm{d}w' = 0$, $\mathrm{d}G \leqslant 0$.   (2·25)

Under such conditions the Gibbs function can only decrease or remain constant. The criterion of equilibrium of a system which is held at constant temperature and pressure is therefore that G has reached its minimum value. It is, of course, precisely these conditions of constancy of temperature and pressure which are of the greatest practical interest in the laboratory, and it is for this reason that G is so important in the theory of phase equilibrium and reaction equilibrium.

Another important characteristic of G lies in its application to steady-flow processes. For such processes we have obtained equation (2·10):

$$-w_{\text{u.max.}} = -(H_2 - H_1) + T''(S_2 - S_1),$$

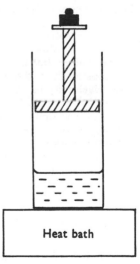

Heat bath

† It is a common misconception that
$$p(V_2 - V_1),$$
the work of volume change, is never a 'useful' form of work. Consider, for example, the device shown in the diagram where a liquid-vapour system is enclosed in a piston and cylinder. If p is the pressure exerted by the piston, then when the liquid and vapour expand from V_1 to V_2 they do work $p(V_2 - V_1)$. Let p_0 be the pressure of the atmosphere. Then the work done on the atmosphere is $p_0(V_2 - V_1)$ and the 'useful' work done in lifting the weight is $p(V_2 - V_1) - p_0(V_2 - V_1)$. (Of course the latter is negative if $p < p_0$ (and $V_2 > V_1$) corresponding to the case where a tension has to be exerted on the piston, due to the vapour pressure of the liquid being less than that of the atmosphere.)

as an expression for the maximum useful work of an engine through which there is a steady flow (e.g. of steam), when the only transfer of heat takes place reversibly at the temperature T'. If T_1 and T_2, the inlet and outflow temperatures, are both equal to T', the equation can be written
$$-w_{u.\,max.} = -(H_2 - H_1) + T_2 S_2 - T_1 S_1$$
$$= -(G_2 - G_1). \tag{2·26}$$

Thus, between initial and final states which are at the same temperature, and with a heat reservoir which is also at this temperature, the maximum useful work which may be obtained during the steady flow of a fluid is equal to the decrease of its Gibbs function.

Equation (2·26) may be compared with (2·16), just as (2·8) is to be compared with (2·5). The reason why G and H have the same significance for flow systems as A and U have for non-flow systems, lies in the fact that ΔG and ΔH each exceed ΔA and ΔU respectively, by the amount $\Delta(pV)$, which is the 'non-useful' work, $p_2 V_2 - p_1 V_1$, which is involved at the outlet and inlet of the flow system at the steady state.

2·5 *a*. Availability

In equations (2·16) and (2·23) we have obtained expressions for the maximum work which may be obtained from a process which begins and ends at the same temperature and in which the only heat reservoir which undergoes any overall gains or losses of heat is also at this temperature. We shall now consider the work which may be obtained when the heat reservoir is not necessarily at the temperature of the system.

Sources of heat, such as are obtained by the burning of fuel, are transitory in character, and the only heat reservoir (or sink) which is permanently available for man's use is the surface of the earth and the atmosphere. This permanent reservoir will be called the *medium* and its temperature T_0 (c. 290 K) has a decisive influence on the amount of work which may be obtained from any process. (If it were actually as low as $0°$ K it would be possible to operate heat engines at conversion factors close to unity, as may be seen from equation (1·12 *b*).)

It is clearly of great practical interest to calculate the maximum amount of work which may be obtained from a given process, such as a reaction, when at its conclusion *there are no changes except in the reaction system and in the medium* (and in the position of the weight which measures the performance of work). That is to say, the maximum amount of work which may be attained after all other bodies, such as intermediate sources of heat, have been restored to their original states.

From equation (1·17) we have
$$-dw = T_0 dS - dU - T_0 d\sigma, \tag{2·27}$$
where T_r has been replaced by T_0 and dS and dU refer to the changes in the system of interest. Since T_0 is constant this equation may be integrated to give
$$-w = T_0(S_2 - S_1) - (U_2 - U_1) - T_0 \sigma, \tag{2·28}$$
where σ is the total entropy created.

It will be evident that this equation remains valid even though our system may have taken in heat from some heat reservoir, at a temperature T different from T_0, provided that this heat is ultimately restored to the reservoir by taking it from the medium. This restoration of heat to the reservoir may be thought of as being carried out by means of a heat engine, and the work of this engine is included in (2·28).

Not all of the work w in equation (2·28) can necessarily be regarded as 'useful'. Since our system has been defined so that it includes all bodies in which there is any change of volume apart from the medium itself, it is evident that any change in volume of the system causes an equal displacement of the atmosphere. If the latter has the pressure p_0, the 'useful' work, as available for the lifting of a weight, is obtained from (2·28) as

$$-w_u = T_0(S_2 - S_1) - (U_2 - U_1) - p_0(V_2 - V_1) - T_0\sigma. \tag{2·29}$$

Since σ, the created entropy, cannot be negative, the useful work obviously has its maximum value when σ is zero, corresponding to reversible operation of the process. The quantity $T_0\sigma$ may be called the 'irreversibility' of the process, or alternatively the 'dissipated energy' or the 'wasted work'. In any industrial process it is obviously desirable to reduce it as far as is possible by an approach to reversible conditions.

Putting σ equal to zero we have

$$\begin{aligned} -w_{u.\,max.} = {} & T_0(S_2 - S_1) - (U_2 - U_1) - p_0(V_2 - V_1) \\ & - (U_2 + p_0 V_2 - T_0 S_2) + (U_1 + p_0 V_1 - T_0 S_1), \end{aligned} \tag{2·30}$$

and this is therefore the maximum mechanical work which may be obtained when a system changes between assigned initial and final states and when the only other body which undergoes an overall gain or loss of heat, and change of volume, is the medium. Expressed alternatively, (2·30) gives the *minimum* amount of work which must be done *on* the system, in order to obtain the given change of state, using only the medium and some source of mechanical work such as a falling weight.

The right-hand side of (2·30) has been expressed in terms of the difference of two brackets. The first of these is determined by the final state of the system, together with p_0 and T_0, and the second is similarly determined by the initial state. For fixed temperature and pressure of the medium, each of these brackets is therefore *a function of state* of the system. We can therefore rewrite (2·30) as

$$-w_{u.\,max.} = -(B_2 - B_1), \tag{2·31}$$

where B is called by Keenan† the 'availability' of the system.

† Keenan, *Thermodynamics*, Chapter XVII. Keenan's definition of the availability of a system is

$$B = (E + p_0 V - T_0 S) - (E_0 + p_0 V_0 - T_0 S_0), \tag{2·32}$$

where E denotes total energy (see footnote on p. 17) and E_0, V_0 and S_0 refer to the system when it has attained equilibrium with the medium. A somewhat similar function known as 'exergy' has been used by Rant, (*Forsch. Ing-Wes.* **22** (1956), 36) and Grassmann, (*Allg. Wärmetechn.* **9** (1959), 79), and has been applied to desalination problems by Evans and Tribus.

It will be noted that the quantities $(S_2 - S_1)$, etc., in equation (2·30) are completely determined by the temperature, pressure and composition of the initial and final states of the system of interest. These quantities may be obtained from tabulated values in the literature, and therefore it is possible to calculate the numerical value of $w_{u.max.}$. Of course in the special case where $T_1 = T_2 = T_0$ and $p_1 = p_2 = p_0$ the change in the availability is the same as the change in the Gibbs free energy at atmospheric temperature and pressure. Under these conditions equation (2·30) reduces to equation (2·23).

One other special case of (2·30) may be noted at this stage. If the system in question is simply a heat reservoir which remains at a virtually constant temperature T and does not change in volume, then (2·30) may be written

$$-w_{u.max.} = T_0(S_2 - S_1) - (U_2 - U_1)$$

$$= \frac{-T_0 q}{T} + q$$

$$= \frac{T - T_0}{T} q,$$

where q is the amount of heat given up by the reservoir in order to obtain the stated amount of work. This equation is simply the familiar Carnot expression for the maximum amount of work which may be obtained by use of two reservoirs at temperatures T and T_0 (§ 1·11).

2·5b. Digression on the useful work of chemical reaction

It follows from what has been said in § 2·4 that the maximum amount of useful work which might be obtained from, say, the combustion of carbon will be obtained when it is converted into that oxide, namely, carbon dioxide, which has the lowest value of G. It is also evident that the maximum work will be obtained if the carbon dioxide is released, at the completion of the combustion, at the same temperature, T_0, and the same pressure, p_0, as that of the atmosphere. For if it is released at some other temperature, T, then useful work has been lost which could have been obtained by heating or cooling the carbon dioxide from T to T_0, by using it as a heat reservoir to drive a heat engine with the atmosphere as the second reservoir. Similarly, if it leaves the combustion system at a pressure which is either smaller or greater than p_0, then useful work has been lost which could have been obtained by allowing the pressure to equalize with that of the atmosphere. Therefore, if we use fuel which is initially at atmospheric pressure and temperature (say 20 °C), the maximum useful work which may be obtained is the decrease of G†
in the process:

$$C \ (1 \ atm, \ 20 \ °C) + air \rightarrow CO_2 + N_2 \ (1 \ atm, \ 20 \ °C).$$

This amounts to about 395 kJ mol^{-1} of carbon.

† Equal to the decrease in availability.

In order to achieve direct combination of carbon and oxygen at an appreciable speed it is necessary, of course, to operate at an elevated temperature. However, this does not affect the issue, because it was shown in § 2·4 that the maximum useful work is independent of whether or not the system passes through intermediate temperatures different from 20 °C. The real difficulty is to achieve reversibility in the reaction process itself. At any chosen operating temperature the partial pressure of oxygen and carbon dioxide above the carbon, at equilibrium, are in a constant ratio as determined by the equilibrium constant

$$p_{CO_2}/p_{O_2} = K_p.$$

Whenever the ratio differs from its equilibrium value the reaction proceeds irreversibly and there is a decrease in the output of work.

The device which may be used—at least in principle—for attaining reversible conditions is the van't Hoff 'equilibrium box'. The carbon is to be thought of as being contained in a reaction vessel which is equipped with two windows. One of these is permeable to oxygen only and the other to carbon dioxide only. By means of a piston and cylinder oxygen is slowly fed into the system through one window, and carbon dioxide is removed through the other window by means of a second piston and cylinder. The ratio p_{CO_2}/p_{O_2} is maintained at a value only infinitesimally less than the equilibrium ratio, so that the carbon slowly burns away under reversible conditions. At the same time a net amount of work is obtained through the movement of the pistons, and, allowing for the processes of heating up and cooling down, the overall maximum work which is obtained is the value of $-\Delta G$ at atmospheric temperature.

The practical difficulties in the way of achieving such a process are sufficiently obvious. Although semi-permeable membranes for certain gases are known to exist, the rate of permeation through them is always very low—at any rate when they have the thickness which is necessary for the purpose of withstanding the pressure difference. A second difficulty arises from the largeness of the equilibrium ratio p_{CO_2}/p_{O_2}. At 900 °C this has a value of the order 10^{15}. Thus if the pressure in the carbon-dioxide cylinder were maintained at 1 atm, the pressure in the oxygen cylinder would need to be about 10^{-15} atm, in order to attain reversibility!

It is for reasons such as these that the raising of power under commercial conditions is always carried out in a manner which is much less efficient from the thermodynamic standpoint. It consists in allowing the fuel to burn freely and irreversibly, thereby obtaining a quantity of heat equal to the decrease in enthalpy in the reaction (if the combustion is at constant pressure). The change in H does not usually differ very greatly from the change in G, since the entropy change in such reactions is not very large. For example, in the reaction $C + O_2 = CO_2$ the decreases in G and H are 395 and 394 kJ mol^{-1} respectively, at 25 °C and 1 atm. Thus almost as much energy is obtained as heat as would have been obtained as work, if the process had been carried out reversibly.† In the

† Note that in the *reversible* combustion at 25 °C, there would be an *absorption* of heat equal to 1 kJ mol^{-1}.

subsequent conversion of the heat into mechanical energy, the maximum fraction of the former which can be converted into the latter is given by the Carnot conversion factor

$$w/q_2 = (T_2 - T_1)/T_2,$$

as discussed in § 1·11, and any imperfect engine will give even less work. In the case of a steam engine or turbine the value of the conversion factor, w/q_2, is determined by the highest convenient temperature T_2 of the steam and by the temperature T_1 of the cooling water in the condenser. Of the heat q_2 taken from the steam, the greater part therefore passes wastefully into the cooling water. Under power-station conditions as much as 92% of the heat of combustion of coal can be transferred to the steam, but only about 40% of the heat taken from the steam in an engine can be recovered as mechanical work. The overall efficiency is thus about 37%.[†]

For certain types of reaction it is possible to obtain a direct conversion of chemical energy into electricity and thereby into mechanical work. Processes of heat transfer are thereby avoided and the conversion factor, w/q_2, of the heat engine does not come into the picture. This can be achieved whenever the reaction has an ionic mechanism and can be set up as a galvanic cell. The electrodes act in much the same manner as the semi-permeable membranes of the equilibrium box. That is to say, the reagents can be added to the system, and the reaction products withdrawn, under reversible conditions. The two great practical advantages of this method of operation are as follows:

(*a*) The reactants and products can be brought to equilibrium by maintaining a potential difference between the electrodes of only a few volts.[‡] This is enormously easier than the maintenance of a pressure ratio of, say, 10^{15} (as in the carbon combustion) by use of pistons and semi-permeable membranes.

(*b*) The processes taking place at the electrodes usually occur with considerable speed as compared to the permeation of a gas through a membrane. For this reason it is possible to maintain the reaction fairly close to equilibrium, and thereby to obtain almost the maximum work, under practical conditions of operation. Efficiencies in the region 60 % are readily attainable under industrial conditions.

An example is the reaction of the Daniell cell,

$$Zn + Cu^{2+} \rightleftharpoons Zn^{2+} + Cu,$$

† The inefficiency of the process may be roughly interpreted in molecular terms as follows. The reaction process $C + O_2 = CO_2$ takes place with a decrease in potential energy of the particles, i.e. the ground state of the CO_2 molecule is at a lower level than those of the two reactants. In the process of steam raising this potential energy is transferred into an increased kinetic energy of the water molecules. As discussed in § 1·18 this random kinetic energy cannot be completely reconverted into potential energy, i.e. the lifting of a weight.

‡ This arises, in the first place, from the logarithmic form of the relation

$$zFE = -\Delta G = RT \ln K$$

(see below), and, secondly, from the magnitude of the conversion factor between energy expressed as volt Faradays and energy expressed as cm^3 atmospheres.

in which there is a decrease of G, from left to right, of about 51 000 cal at room temperature. This process may be carried out by placing plates of zinc and copper in solutions of zinc sulphate and copper sulphate respectively, the solutions being separated by means of a porous barrier.†
The overall reaction may be regarded as consisting of the two electrode processes:

$$Zn \rightleftharpoons Zn^{2+} + 2e,$$

$$Cu^{2+} + 2e \rightleftharpoons Cu.$$

Under conditions where the reaction takes place spontaneously, the zinc atoms tend to ionize, leaving an excess of electrons on the zinc plate. The zinc ions migrate through the porous barrier and displace copper ions. The latter unite with electrons taken from the copper plate and are deposited as copper atoms. The zinc and copper plates thus develop an excess and a deficiency of electrons respectively. An electric current can therefore be made to pass through an external circuit, with performance of work.

What is called the electromotive force (e.m.f.) of the cell is the potential difference (p.d.) between the plates when the current is zero. When the p.d. is smaller than the e.m.f. the reaction proceeds from left to right, with flow of electrons from the zinc plate to the copper. If a p.d. greater than the e.m.f. is applied, the reaction proceeds in the reverse direction. By careful adjustment the reaction can thus be carried out in either direction under conditions which approximate very closely to perfect reversibility.

Maximum useful work is obtained under the reversible conditions and is equal to the decrease in G if the system operates at constant temperature and pressure. Thus from (2·24)

$$-dG = -dw'_{max}.$$

Let F denote the Faraday equivalent, the charge carried by a gram-ion of unit positive valency. This has the value 0.6487×10^4 coulombs mol⁻¹ or 2.3060×10^4 cal V⁻¹ mol⁻¹. The positive charge carried by dn gram-ions of valency z crossing the porous plate is thus $zF\,dn$.‡ Provided that the cell is in a stationary state with regard to electric charge (as is normally always the case) the same quantity of electricity will pass through the external circuit. It can be used, for example, to drive a motor whose 'back e.m.f.' is only infinitesimally smaller than that of the cell. This work is equal to the product of the charge and the potential difference through which it moves. Therefore

$$-dG = zFE\,dn, \tag{2·33}$$

† The function of the porous barrier is to retard the mixing of the two solutions. If the copper ions existed in the proximity of the zinc plate they would be deposited on this electrode instead of on the copper one. The electrochemical process would thus take place in a localized region and it would not be possible to drive a current round an external circuit. The chemical energy would appear only as a liberation of heat, equal to the decrease in enthalpy in the reaction process.

‡ z is taken as positive for positively charged ions; e.g. it is $+2$ for Ba^{2+} and -1 for Cl^-.

where E is the e.m.f. and dG is the change in the Gibbs function of the total system, i.e. the plates, the solutions and the external circuit. However, since the circuit is unchanged, the change in G is simply that of the reaction system alone.

This change of G is, of course, the same for the same initial and final states of the reaction system, whether the process is conducted reversibly or not. It is the output of work which varies and, in the limiting case where the cell is short-circuited, the work falls to zero and all of the energy is liberated as heat. This is equivalent to carrying out the reaction irreversibly without the use of a galvanic cell, as when a piece of zinc is dropped into copper sulphate solution. Such conditions are those under which the majority of reactions are normally carried out. It is unfortunate that many of the large-scale operations of the chemical industry, such as the oxidation of ammonia, cannot be conveniently set up in the form of a cell.

The application of equation (2·33) for the purpose of measuring the free-energy change of a chemical reaction will be referred to in detail at a later stage.

2·6. The fundamental equations for a closed system in terms of *H*, *A* and *G*

The equation (1·19),
$$dU = T dS - p dV, \qquad (2·34)$$
was described in §1·15 as the fundamental equation for a closed system. This equation really contains the whole of the knowledge obtained from the basic laws—at any rate when it is taken in conjunction with the properties of U (in determining the algebraic sum of heat and work effects), of T (in determining thermal equilibrium) and of S (in relation to heat and the permissible direction of a change). The reformulation of this equation in terms of H, A and G, as is done below, therefore introduces no new knowledge.

By definition
$$H = U + pV,$$
and therefore
$$dH = dU + p dV + V dp$$
Combining this with (2·34) we obtain
$$dH = T dS + V dp, \qquad (2·35)$$
and similarly
$$dA = - S dT - p dV, \qquad (2·36)$$
$$dG = - S dT + V dp. \qquad (2·37)$$

2·7. The chemical potential

The thermodynamic theory which has been developed so far is applicable, as it stands, only to closed systems. This applies, for example, to the important relations
$$dU = dq + dw,$$
$$dS \geqslant dq/T.$$

These were based on the discussion of a 'body' which was supposed not to exchange material with its environment. In general, these relations are incorrect, or at least ambiguous,† if an exchange of matter takes place (but they may always be applied to the *total* system comprising the body and its environment).

The same remarks apply also to the 'fundamental' equations of the last section. These are not applicable to open systems, or to closed systems which undergo irreversible changes of composition. Consider, for example, the equation

$$dU = T dS - p dV.$$

In an isolated system we have $dU = 0$ and $dV = 0$, and this would therefore imply $dS = 0$. However, it is certainly possible for the entropy to increase, either on account of chemical reaction or on account of the diffusive mixing of various substances which were initially separate. Similarly from the equation (2·37) it might be concluded that if the temperature and pressure of a system are held constant its Gibbs function cannot change. This would clearly be erroneous; G is an extensive quantity, and its value can be increased merely by increasing the amount of matter in the system, at constant temperature and pressure. For example, if we increase the amount of the reagents in a galvanic cell, it is clearly able to perform a larger amount of useful work and this must be due to an increase in its Gibbs function.

These difficulties are due to the assumption, implicit in most of Chapter 1, that two variables alone may be sufficient to fix the state of a system. This can only be true for bodies of fixed composition. We must therefore introduce into the equations the variables which determine the composition and size of the system, i.e. the mole numbers.

Consider a *homogeneous phase* in which there are k different substances. Let n_1 (mols) be the amount of substance 1 in the whole of the phase, n_2 (mols) the amount of substance 2, etc. According to (2·34) if $n_1, n_2, ..., n_k$ are constant, the internal energy U of the phase depends only on S and V. However, for variable composition we must have
$$U - U(S, V, n_1, n_2, ..., n_k),$$

and thus the total differential of U is

$$dU = \left(\frac{\partial U}{\partial S}\right)_{V, n_i} dS + \left(\frac{\partial U}{\partial V}\right)_{S, n_i} dV + \sum_{i=1}^{i=k} \left(\frac{\partial U}{\partial n_i}\right)_{S, V, n_j} dn_i. \quad (2·38)$$

† If there is a transfer of matter between two systems it is no longer possible to decide unambiguously what is meant by a 'heat transfer' between these systems. This is because the matter carries with it an associated energy. See, for example, the author's *Thermodynamics of the Steady State* (London, Methuen, 1951).

In this expression the subscript n_i to the first two partials implies that the amounts of all species are constant during the variation in question. On the other hand, the last term in the equation is a sum of partials in each of which the entropy and volume are constant, together with all but one of the amounts of substances.

Now *for constant* amounts, (2·34) is entirely valid, and from this equation we can obtain

$$\left(\frac{\partial U}{\partial S}\right)_{V, n_i} = T, \quad \left(\frac{\partial U}{\partial V}\right)_{S, n_i} = -p.$$

Let μ_i be defined by

$$\mu_i \equiv \left(\frac{\partial U}{\partial n_i}\right)_{S, V, n_j} \tag{2·39}$$

Thus (2·38) may be written

$$dU = T\,dS - p\,dV + \sum_i \mu_i dn_i. \tag{2·40}$$

The quantity μ_i, called the *chemical potential*, was introduced into thermodynamics by Willard Gibbs, and it greatly facilitates the discussion of open systems, or of closed ones in which there are changes of composition. It was defined by Gibbs as follows:[†] 'If to any homogeneous mass we suppose an infinitesimal quantity of any substance to be added, the mass remaining homogeneous and its entropy and volume remaining unchanged,[‡] the increase of the energy of the mass divided by the quantity of the substance added is the *potential* for that substance in the mass considered.' This is the same as is implied by equation (2·39).

The chemical potential has an important function analogous to temperature and pressure. A temperature difference determines the tendency of heat to pass from one body to another and a pressure difference determines the tendency towards bodily movement. It will be shown shortly that a difference of chemical potential may be regarded as the cause of a chemical reaction or of the tendency of a substance to diffuse from one phase into another. The chemical potential is thus a kind of 'chemical pressure' and is an intensive property of a system, like the temperature and pressure themselves.

The chemical potential may be introduced by an alternative method

[†] Gibbs, *Collected Works* (New Haven, Yale Univ. Press, 1928), vol. i, p. 93.

[‡] The question may be asked, how may the entropy and volume be held constant when the dn_i mols are added? As regards the volume, this merely requires a readjustment of the pressure of the system. As regards the entropy of the system, the dn_i mols carry with them their own appropriate amount of entropy. After the addition the entropy of the system can therefore be considered as being brought back to its original value by the removal of an appropriate amount of heat.

which is sometimes found more congenial. This is based on equation (2·37):
$$dG = -S\,dT + V\,dp,$$

which we have seen to be inadequate as it stands. In general G must be determined by the amounts of substances, as well as by T and p. Thus
$$G = G(T, p, n_1, n_2, ..., n_k),$$
and therefore
$$dG = \left(\frac{\partial G}{\partial T}\right)_{p,\,n_i} dT + \left(\frac{\partial G}{\partial p}\right)_{T,\,n_i} dp + \sum_{i=1}^{i=k} \left(\frac{\partial G}{\partial n_i}\right)_{T,\,p,\,n_j} dn_i.$$

Now the first partial in this expression refers to the change in G at constant pressure and amounts. Since (2·37) is valid for constant amounts we obtain
$$\left(\frac{\partial G}{\partial T}\right)_{p,\,n_i} = -S,$$

and similarly
$$\left(\frac{\partial G}{\partial p}\right)_{T,\,n_i} = V.$$

Substituting in the above expression
$$dG = -S\,dT + V\,dp + \Sigma\mu_i dn_i, \tag{2·41}$$

where μ_i is now defined by
$$\mu_i \equiv \left(\frac{\partial G}{\partial n_i}\right)_{T,\,p,\,n_j}$$

The chemical potential of a component of a phase is thus the amount by which the capacity of the phase for doing work (other than work of expansion) is increased per unit amount of substance added, for an infinitesimal addition at constant temperature and pressure. For example, the chemical potential of copper sulphate in its aqueous solution in a Daniell cell is equal to the increased capacity of the cell to provide electrical energy, per unit amount of copper sulphate added (the addition actually being an infinitesimal).

It remains to be shown that μ_i, as defined above in terms of G, is identical with its definition, equation (2·39), in terms of U. For this purpose it is sufficient to add to each side of (2·40) the quantity $d(pV - TS)$. This addition transforms (2·40) into (2·41), since G is defined as $U + pV - TS$. The μ's which appear in these equations are therefore identical.

The μ's can also be expressed in terms of H and A and in fact the complete set of equations is
$$dU = T\,dS - p\,dV + \Sigma\mu_i dn_i, \tag{2·40}$$
$$dG = -S\,dT + V\,dp + \Sigma\mu_i dn_i, \tag{2·41}$$
$$dH = T\,dS + V\,dp + \Sigma\mu_i dn_i, \tag{2·42}$$
$$dA = -S\,dT - p\,dV + \Sigma\mu_i dn_i, \tag{2·43}$$

together with the identities

$$\mu_i = \left(\frac{\partial U}{\partial n_i}\right)_{S,\,V,\,n_j} = \left(\frac{\partial G}{\partial n_i}\right)_{T,\,p,\,n_j} = \left(\frac{\partial H}{\partial n_i}\right)_{S,\,p,\,n_j} = \left(\frac{\partial A}{\partial n_i}\right)_{T,\,V,\,n_j}. \quad (2\cdot44)$$

The equations (2·40)–(2·43) form the basis of chemical thermodynamics. The first of them may be regarded as the fundamental relation which contains the physical information embodied in the properties of U, T and S, whilst the other three are derived from it by virtue of the definitions of H, A and G and contain no additional information. At this stage it may also be noted that the most convenient choice of the independent variables in the solving of thermodynamic problems is determined by the structure of the above equations. Thus

$$\left.\begin{aligned} U &= U(S, V, n_1, ..., n_k),\\ G &= G(T, p, n_1, ..., n_k),\\ H &= H(S, p, n_1, ..., n_k),\\ A &= A(T, V, n_1, ..., n_k). \end{aligned}\right\} \quad (2\cdot45)$$

For example, considering the last of these equations, if A is known experimentally or theoretically as a particular function ϕ, of T, V and the n_i, then the entropy, pressure and chemical potentials can all be evaluated, as functions of the same variables, by differentiation of ϕ, in accordance with (2·43). For example, $S = -(\partial\phi/\partial T)_{V,\,n_i}$. But let it be supposed that A is known as a function ϕ' of a different set of independent variables, such as T, p and the n_i. Now from (2·43) $p = -(\partial A/\partial V)_{T,\,n_i}$. The knowledge of the function ϕ' is therefore equivalent to a knowledge of A as a function of T, $\partial A/\partial V$, and the n_i. From this function it is obviously *not* possible to determine the value of V without an unknown integration constant. The function ϕ' is thus much less useful than the function ϕ. For this reason Gibbs† refers to functional relations in which the independent variables are as in (2·45) as being *fundamental* equations.

Another important point concerns the number of terms which are to be included in the summation in (2·40)–(2·43). As discussed above these terms originate from the fact that U, etc., depend on the composition and size of the system and therefore as many dn_i's must be included as are necessary to describe all possible changes in the composition and size. Provided that we are not concerned with nuclear transformations, it will therefore be satisfactory to extend the summation over all distinct chemical species whose amounts may vary by a significant amount. In doing this it may occur that we shall be using more composition variables than are actually necessary to

† *Collected Works*, vol. I (New Haven, Yale Univ. Press, 1928), p. 88.

specify the state of the system. This occurs whenever there is a state of chemical equilibrium between some or all of the species. For example, liquid water contains the molecules H_2O, H_3O^+, OH^-, etc. For the present we shall regard the summation as being over all such species, but in a later section it will be shown that the existence of a chemical equilibrium between them implies that there is only a single independent chemical potential in the system consisting of water. In brief, there is only a single component; but this statement is based on the empirical knowledge that there is a state of chemical equilibrium between H_2O, H_3O^+ and OH^-.

If for any component i, its amount remains constant during the process in question, the corresponding term $\mu_i dn_i$ is zero. If the amount of i diminishes, dn_i is to be taken as a negative quantity.

Finally, it is perhaps useful to make some remarks concerning the expression $\Sigma\mu_i dn_i$ as a work term. Consider a *reversible* change of composition in a *closed* system. On account of the defining equation for entropy the term $T dS$ in the equation

$$dU = T dS - p dV + \Sigma\mu_i dn_i,$$

is the heat absorbed by the system. It follows that the remaining terms on the right-hand side represent the total work done. Thus $\Sigma\mu_i dn_i$ is a form of work which can be done by a system in the absence of any change of volume and due to its change of composition. This is, in fact, the 'chemical work' which has been discussed already in § 2·5b, and which can be attained by use of the van't Hoff equilibrium box or the galvanic cell. It is important to realize that it is possible for a thermodynamic system to perform work, i.e. to achieve the lifting of a weight, without there being any difference in the initial and final volume of the system.†

This interpretation of $\Sigma\mu_i dn_i$ as a work term is no longer possible if the changes dn_i are due to transfer of material into the system from outside, i.e. if the system is 'open'. For whenever there is a simultaneous transfer of matter as well as of energy the notion of heat

---

† The van't Hoff equilibrium box, discussed in § 2·5b, is a closed system if we include in the term 'system' the supplies of CO_2 and oxygen which are connected to the box through the semi-permeable membranes. If we start with 1 mol of oxygen at 1 atm pressure in the one cylinder and convert it reversibly to 1 mol of CO_2 at the same pressure in the other cylinder, there is no difference between the initial and final volumes of the total system. However, in order to carry out this process reversibly, it is necessary to pass through *intermediate* volumes which are much larger, i.e. to expand the oxygen to the very low equilibrium pressure. It is because of this intermediate expansion that the reaction system can be made to perform work by use of pistons. The student is advised to work out the process in detail, using the laws of perfect gases (Chapter 3).

becomes ambiguous. Thus the above equation remains valid, but the term $T\,\mathrm{d}S$ is no longer interpretable as 'heat' and therefore the remaining terms are not interpretable as 'work'.

2·8. Criteria of equilibrium in terms of extensive properties†

For any completely isolated system we have from the second law

$$\Sigma\,\mathrm{d}S_i \geqslant 0,$$

where the summation extends over all parts of the system. In this expression the differentials refer to the possible changes in the forward direction of time. It follows that the criterion of equilibrium of such a system is that the entropy has reached the maximum value which is consistent with the fixed value of its energy, and volume, as discussed already in §1·13.

Criteria of equilibria may also be expressed in terms of U, H, A · and G. We shall briefly repeat the derivation of the criteria relative to A and G, as given already in §§2·3 and 2·4. For a system in a thermostat, the above relation may be written

$$\mathrm{d}S + \mathrm{d}S_r \geqslant 0,$$

where the subscript r refers to the thermostat at temperature T. If $\mathrm{d}q$ is the heat absorbed by the system during an infinitesimal change of state, then $\mathrm{d}S_r = -\,\mathrm{d}q/T$, and thus

$$\mathrm{d}S - \frac{\mathrm{d}q}{T} \geqslant 0.$$

In view of the first law this may be written

$$\mathrm{d}S - \frac{(\mathrm{d}U - \mathrm{d}w)}{T} \geqslant 0, \tag{2·46}$$

and therefore, since T is constant,

$$\mathrm{d}(U - TS) \leqslant \mathrm{d}w. \tag{2·47}$$

Under conditions of constant temperature and when no work is performed *on* the system, (2·47) shows that the change in the Helmholtz free energy, $A = U - TS$, can only be negative. Therefore the criterion of equilibrium, under conditions of prescribed temperature and volume, is that A has reached its minimum possible value.

Adding $\mathrm{d}(pV)$ to the last expression

$$\mathrm{d}(U + pV - TS) \leqslant \mathrm{d}w + \mathrm{d}(pV).$$

† For a more complete discussion see Gibbs, *Collected Works*, vol. I, pp. 55 et seq. A very clear exposition, based on Gibbs, is given by Keenan, *Thermodynamics* (New York, Wiley, 1941), Chapters XXIII, XXIV and XXVI.

If the pressure on the system is constant $d(pV)=pdV$, and the right-hand side of the above expression is therefore equal to dw', the work not including that due to volume change.‡ Thus

$$dG \leqslant dw'. \tag{2·48}$$

Therefore under conditions of constant temperature and pressure, and when no work of this kind is performed *on* the system, the change in G can only be negative. The criterion of equilibrium for a system of prescribed temperature and pressure is that G has reached its minimum possible value.

Similarly, it may be shown that criteria of equilibrium for a system of prescribed S and V is a minimum of U, and for a system of prescribed S and p is a minimum of H. These are less useful than the criteria relating to A and G.

It is of interest to consider in molecular terms what is implied by a minimum value of A (or G). Now $A=U-TS$, and therefore low values of A will occur when U is small and when TS is large. As an example consider a crystal in equilibrium with its vapour in a vessel of fixed volume surrounded by a thermostat. In this system there are two opposing tendencies. On the one hand, the attractive forces between the molecules tend to make the vapour condense, with reduction in the potential energy of the system and with passage of heat outwards into the heat bath. If this occurred completely, i.e. if all of the molecules were present as crystal and none as vapour, the internal energy of the system would be low, but so also would be the entropy. (According to the statistical interpretation $S=k\ln\Omega$, and Ω is smaller for the orderly arrangement of the crystal than for the gas at the same temperature.) On the other hand, the tendency towards randomness in the system would be most completely satisfied if the crystal were to vaporize completely, with absorption of heat from the heat bath in order to overcome the attractive forces. Such a system would be one of high internal energy and high entropy.

If the values of U and TS of the total crystal-vapour system are calculated by statistical methods and plotted against the fraction of the substance which is present as a vapour, curves are obtained as shown in Fig. 14.† The minimum value of A, for a particular temperature and volume of the system, occurs when there is a certain fraction of the substance present in the vapour phase. This is the equilibrium state, at prescribed temperature and volume, and the corresponding pressure in the vapour phase is the saturation vapour pressure.

‡ Remember that pdV is a negative amount of work done *on* the system when dV is positive. The point can be seen more clearly by writing $dw+pdV=dw-(-pdV)$ which shows that the work done on the system due to its volume increase is subtracted from the total work done on the system.

† Problem 4 at the end of Chapter 13 is concerned with the calculation of these curves.

The fact that there *is* a minimum value of the free energy is because the values of U and TS do not increase proportionately to each other with increase of the fraction of substance in the vapour phase. When this fraction is small an increment of vapour causes the randomness to increase much more rapidly than the energy, as may be seen from the figure. Rather pictorially the equilibrium state may be said to be due to a balance between order and disorder in the system.

In the above discussion the heat bath, which maintains the system at constant temperature, has played an important role. For the whole

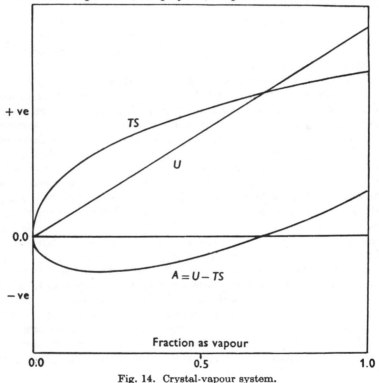

Fig. 14. Crystal-vapour system.

isolated assembly, comprising the crystal-vapour system together with the heat bath, it remains true that the entropy tends to a maximum. However, any tendency of the substance to pass more completely into the vapour phase, with increase of its entropy, implies an absorption of heat from the heat bath, which thereby diminishes in entropy. In brief, the state of the crystal-vapour system for which its A function is at a minimum is the state of the whole assembly for which the total entropy is at a maximum. Similar considerations apply to G, if the system is at constant temperature and pressure.

2·9. Criteria of equilibrium in terms of intensive properties†

(a) Thermal and mechanical equilibrium. It is the property
of temperature that whenever two systems α and β are in *thermal*
equilibrium, $T_\alpha = T_\beta$, as was discussed in §§1·4 and 1·13. In §1·13 it
was also indicated briefly that the condition of *mechanical* equilibrium requires equality of pressure across a plane interface between
the fluids. This result may now be demonstrated as a consequence of
the criteria of equilibrium of the last section.

Consider two fluids α and β which are enclosed in a rigid container
and held at the constant temperature T. Under such conditions the
total Helmholtz free energy of the two fluids must be a minimum at
equilibrium. Thus in any variation about the equilibrium state

$$\mathrm{d}A_\alpha + \mathrm{d}A_\beta = 0. \tag{2·49}$$

Consider a possible variation $\mathrm{d}V_\alpha$ in the volume of the α phase and a
variation $\mathrm{d}V_\beta$ in the volume of the β phase, the temperature and compositions remaining unchanged. Using (2·43) we have

$$\mathrm{d}A_\alpha = -p_\alpha \mathrm{d}V_\alpha,$$
$$\mathrm{d}A_\beta = -p_\beta \mathrm{d}V_\beta.$$

But also $\mathrm{d}V_\alpha = -\mathrm{d}V_\beta$, since the total volume is constant. Thus the
condition of equilibrium (2·49) reduces to

$$(p_\alpha - p_\beta)\,\mathrm{d}V_\beta = 0,$$

which can only be satisfied if

$$p_\alpha = p_\beta. \tag{2·50}$$

This conclusion is, of course, incorrect if the two fluids are separated
from each other by an immovable barrier, since in this case $\mathrm{d}V_\alpha$ and
$\mathrm{d}V_\beta$ are each individually zero. It is also in general incorrect if the
interface between the two fluids is curved. A variation $\mathrm{d}V_\alpha$ in the
volume will then imply a variation in the interfacial area between the
two fluids. That is to say, for such a system the area and surface
tension are significant additional variables of state. Under such
conditions it can be shown that the condition of mechanical equilibrium is

$$(p_\beta - p_\alpha) = \gamma\left(\frac{1}{r_1} + \frac{1}{r_2}\right), \tag{2·51}$$

where r_1 and r_2 are the principal radii of curvature and are taken as
positive when they lie in the β phase.

† For a more complete discussion see Gibbs, *Collected Works*, vol. I, and
also *Commentary on the Scientific Works of Willard Gibbs* (Yale Univ. Press,
New Haven, 1936), vol. I.

(b) Equilibrium for transfer between phases. Consider a system of several phases which are in thermal equilibrium at constant temperature. Each phase is a typical open system, since it can exchange chemical substance with its neighbouring phases. It will now be shown that the condition of equilibrium with respect to these transfers is that each substance has the same value of its chemical potential in all the phases.

We consider a variation of the system in which an amount $dn_{i\beta}$ of the substance i passes from phase β into phase α. According to equation (2·43), the temperature being constant, the overall change in A for the system as a whole is

$$dA = \Sigma dA_\delta = -\Sigma\, p_\delta\, dV_\delta + (\mu_{i\alpha} - \mu_{i\beta})\, dn_{i\beta},$$

where the summation is over all the phases and the last term arises from the fact that $dn_{i\alpha} = -dn_{i\beta}$. If the pressures p_δ, etc., are steady values, as will occur in a sufficiently slow process, the first term on the right-hand side is dw, the total work done on the system. Thus

$$dA = dw + (\mu_{i\alpha} - \mu_{i\beta})\, dn_{i\beta}.$$

Now from (2·47) we have $\quad dA \leqslant dw$,

and therefore $\qquad (\mu_{i\alpha} - \mu_{i\beta})\, dn_{i\beta} \leqslant 0.$ $\qquad\qquad$ (2·52)

It follows that the sign of $(\mu_{i\alpha} - \mu_{i\beta})$ is the opposite of the sign of $dn_{i\beta}$. Thus if $dn_{i\beta}$ is a positive transfer from β to α, the chemical potential of the substance i must be less in the phase α than in β. In general, any substance tends to pass from regions of higher to regions of lower chemical potential,† and it is to this property that the chemical potential owes its name.

On the other hand, for a reversible change $dA = dw$, and therefore the above inequality is replaced by

$$\mu_{i\alpha} = \mu_{i\beta}. \qquad\qquad (2\cdot53)$$

The condition of chemical equilibrium between phases is therefore that each substance shall have an equal value of its chemical potential in all phases between which this substance can freely pass.† On the other hand, if two phases are separated by a membrane which is permeable to some substances but not to others, the above result applies only to those substances which can pass freely through the membrane. This has important applications in the phenomenon of osmosis.

Differences of chemical potential may thus be regarded as the origin of all processes of diffusion. It is erroneous to regard diffusion as necessarily taking place in the direction of decreasing concentration. For example, suppose that a solute i is distributed between two

---

† These remarks must be qualified in view of the next subsection.

solvents α and β and let $c_{i\alpha}=2$ and $c_{i\beta}=10$ be two values of its concentration (in arbitrary units) in α and β respectively when there is a state of equilibrium. In these two solutions the chemical potentials of i are equal and there is no diffusion despite the large difference of concentration. If, at some moment during the approach to equilibrium, the concentrations were $c_{i\alpha}=4$ and $c_{i\beta}=7$, then at this moment there would have been a tendency for spontaneous diffusion from phase α to phase β, i.e. in the direction of *increasing* concentration (but of decreasing μ). However, in physical situations where there is no discontinuity of the medium, the direction of decreasing μ usually coincides with the direction of decreasing concentration.

(c) Transfer equilibrium in a potential field.

We consider first the extent to which the relation (2·53) needs to be modified if we allow for the effect of gravity. Now in the last chapter the convention was adopted that the internal energy, entropy, etc., of a body depend only on its internal state, as characterized by its temperature, pressure and composition. The functions U, S, H, A and G are thus taken as being independent of the position of the system in a gravitational field.† If the system changes its potential energy by $d\phi$ and its internal energy by dU, the first law must therefore be written

$$dU + d\phi = dq + dw.$$

Thus, if the system is held at constant temperature T we shall have in place of (2·46)

$$dS - \frac{(dU + d\phi - dw)}{T} \geqslant 0,$$

and therefore the corresponding condition of equilibrium is

$$dA + d\phi = 0, \tag{2·54}$$

where $\qquad A \equiv U - TS$ as before.

For simplicity let the system under discussion be a single homogeneous phase of prescribed temperature and volume. We consider a variation in which an amount dn_i of substance i passes from a layer β at height h_β into a layer α at height h_α. Then if M_i is the molecular weight in grams the increase in potential energy of the system is

$$d\phi = M_i g(h_\alpha - h_\beta)\, dn_i,$$

and from (2·43) the corresponding change in A of the system as a whole is

$$dA = \mu_{i\alpha} dn_i - \mu_{i\beta} dn_i.$$

Substituting these expressions in (2·54) we obtain

$$\mu_{i\alpha} + M_i g h_\alpha = \mu_{i\beta} + M_i g h_\beta. \tag{2·55}$$

† See footnote on p. 17. In some text-books, U, H, A and G are taken as including the potential energy of the system.

Thus, in so far as we need to allow for the effect of the gravitational field, the chemical potential of a substance is not equal throughout the depth of a phase, but it is the sum of the terms μ_i and $M_i gh$ which has this property of constancy.

The considerations relating to the electrostatic field are rather different.† Now in the gravitational case, as discussed above, the potential energy per unit of mass has the same value at a particular height whatever is the temperature, pressure and composition of this unit of mass. That is to say, changes of U and ϕ are independent and each has a significant meaning. This is no longer the case when we are concerned with an electrostatic field, at any rate within the *interior* of a body. It is, of course, entirely feasible to measure the change in ϕ in bringing a positive charge from infinity, through empty space, to a point *just outside* the body. However, in the process of bringing the charged particle across the phase boundary into the interior we can make no experimental distinction between that part of the total energy change which might be regarded as a change in ϕ and another part, which might be called the 'chemical work', *due to the change in the composition of the environment of the particle.*

Therefore, for movement of a charge through regions of varying composition, it is conventional to include the whole energy change in the change in U. The treatment of the previous subsection then holds without modification and μ_i, defined as $(\partial U/\partial n_i)_{S,V,\,n_j}$, the change of U with composition, is equal between any two phases α and β which are in contact and between which the species i can freely pass. Thus

$$\mu_{i\alpha} = \mu_{i\beta}, \qquad (2·56)$$

where i may refer to an electron, an ion or a neutral molecule. Unlike the case of equation (2·55), the μ now includes both the 'chemical' and the electrical effects, which we have seen to be indistinguishable.

Consider now two phases α and β which are not in contact. If they have *identical* temperature, pressure and composition their internal states are the same. In this case the *difference* in the work required to bring a unit positive charge from infinity into the interior of each of the two phases has a meaningful value, since all 'chemical' work involved when the charge enters the bodies will cancel when we take the difference. Their electrical potential difference, $\phi_\beta - \phi_\alpha$, is therefore measurable. (It may be objected that this potential difference can only exist if the two bodies contain different concentrations of electrons, or other charged species, and therefore the two bodies do *not* have precisely the same composition, as previously supposed. However, a simple calculation shows that a quite negligible difference of electron concentrations is sufficient to account for all values of $\phi_\beta - \phi_\alpha$ which are met with in practice.)

The difference of electrical potential between two identical phases α and β is therefore defined by the work done by these phases in the

† For a more complete discussion of the subtleties concerning the electrostatic field see Adam, *The Physics and Chemistry of Surfaces* (Oxford, 1941), Chapter VIII and Guggenheim, *Thermodynamics* (Amsterdam, North-Holland Publ. Co. 1949), Chapter X.

reversible transference of an amount dn_i of charged species from the α phase to the β phase at constant temperature and volume

$$dw = -z_i F(\phi_\beta - \phi_\alpha) \, dn_i, \tag{2·57}$$

where z_i is the valence of the species, being taken as positive for a positive charge, and F is the Faraday equivalent (§ 2·5b). But also by (2·43) and (2·47)

$$dw = -\,dA = -(\mu_{i\beta} - \mu_{i\alpha}) \, dn_i,$$

and therefore

$$\mu_{i\beta} - \mu_{i\alpha} = z_i F(\phi_\beta - \phi_\alpha). \tag{2·58}$$

Of course if the two phases are actually in equilibrium with each other, then (2·56) must hold and therefore $\phi_\beta = \phi_\alpha$. Thus *identical* phases in contact have the same potential. But the potential between *different* phases, whether they are in contact or not, has no meaning.

(d) Reaction equilibrium. This will be discussed in more detail at a later stage. For the present it is sufficient to note that in any *closed* system which is in a state of complete equilibrium the term $\Sigma \mu_i dn_i$ in equations (2·40)–(2·43) must be zero:

$$\Sigma \mu_i \, dn_i = 0. \tag{2·59}$$

This follows from the criteria of equilibrium of § 2·8.

2·10. Mathematical relations between the various functions of state

(a) Some identities. The basic equations (2·40)–(2·43) are

$$dU = T \, dS - p \, dV + \Sigma \mu_i \, dn_i, \tag{2·60}$$

$$dG = -S \, dT + V \, dp + \Sigma \mu_i \, dn_i, \tag{2·61}$$

$$dH = T \, dS + V \, dp + \Sigma \mu_i \, dn_i, \tag{2·62}$$

$$dA = -S \, dT - p \, dV + \Sigma \mu_i \, dn_i. \tag{2·63}$$

From these the following identities may be obtained:

$$T = \left(\frac{\partial U}{\partial S}\right)_{V, n_i} = \left(\frac{\partial H}{\partial S}\right)_{p, n_i}, \tag{2·64}$$

$$p = -\left(\frac{\partial U}{\partial V}\right)_{S, n_i} = -\left(\frac{\partial A}{\partial V}\right)_{T, n_i}, \tag{2·65}$$

$$S = -\left(\frac{\partial G}{\partial T}\right)_{p, n_i} = -\left(\frac{\partial A}{\partial T}\right)_{V, n_i}, \tag{2·66}$$

$$V = \left(\frac{\partial G}{\partial p}\right)_{T, n_i} = \left(\frac{\partial H}{\partial p}\right)_{S, n_i}, \tag{2·67}$$

and the corresponding identities relating to μ_i were given in equation (2·44).

By combining (2·64)–(2·67) with the defining equations for H, A and G, further expressions may be obtained. For example,

$$A = U - TS = U + T\left(\frac{\partial A}{\partial T}\right)_{V,\,n_i}, \qquad (2·68)$$

$$G = H - TS = H + T\left(\frac{\partial G}{\partial T}\right)_{p,\,n_i}, \qquad (2·69)$$

and it is readily confirmed that these two expressions may be re-arranged into the more compact forms

$$\left(\frac{\partial A/T}{\partial T}\right)_{V,\,n_i} = -\frac{U}{T^2}, \qquad (2·70)$$

$$\left(\frac{\partial G/T}{\partial T}\right)_{p,\,n_i} = -\frac{H}{T^2}. \qquad (2·71)$$

The expressions (2·68)–(2·71) are known as Gibbs–Helmholtz equations.

(b) Maxwell's relations. An additional crop of useful identities, known as Maxwell's relations, is obtained by applying a theorem of the calculus concerning exact differentials. Consider (2·60) in which U is expressed as a function of the variables, S, V and the n_i. From this equation we obtain

$$T = \left(\frac{\partial U}{\partial S}\right)_{V,\,n_i}, \qquad p = -\left(\frac{\partial U}{\partial V}\right)_{S,\,n_i}.$$

Now U is a function of state and thus forms an exact differential. It is also known experimentally to be a smooth function of the variables S, V and n_i, except at points of phase change. If we disregard such points, the second partial differential of U with respect to any pair of variables is independent of the order of differentiation. Thus†

$$\left[\frac{\partial}{\partial V}\left(\frac{\partial U}{\partial S}\right)_{V,\,n_i}\right]_{S,\,n_i} = \left[\frac{\partial}{\partial S}\left(\frac{\partial U}{\partial V}\right)_{S,\,n_i}\right]_{V,\,n_i},$$

and therefore from the previous relations

$$\left(\frac{\partial T}{\partial V}\right)_{S,\,n_i} = -\left(\frac{\partial p}{\partial S}\right)_{V,\,n_i}.$$

† The meaning of $\left[\dfrac{\partial}{\partial V}\left(\dfrac{\partial U}{\partial S}\right)_V\right]_S$ may be seen more clearly by imagining U, S and V as three Cartesian co-ordinates. U, as a function of S and V, is a surface in the three-dimensional space. $(\delta U/\delta S)_V$ is the gradient, g_V of this surface with respect to S, at any plane of constant V. We can now think of g_V plotted as a function of S and V, again forming a surface. $\left[\dfrac{\partial}{\partial V}\left(\dfrac{\partial U}{\partial S}\right)_V\right]_S$ is the gradient of the gradient g_V, with respect to V, at any plane of constant S.

In general, if $d\phi$ is an exact differential and is given by

$$d\phi = x_1 dy_1 + x_2 dy_2 + x_3 dy_3 + \dots,$$

then
$$\left(\frac{\partial x_i}{\partial y_j}\right)_{y_k} = \left(\frac{\partial x_j}{\partial y_i}\right)_{y_k}, \tag{2·72}$$

where the subscript y_k denotes constancy of all y's other than the one considered in the differentiation. Conversely, a necessary and sufficient condition for $d\phi$ to be an exact differential is that (2·72) shall hold for all independent combinations of the variables y_i. Equation (2·72) is known as the reciprocity relation or cross-differentiation identity.

The full set of Maxwell's relations, obtained by applying (2·72) to (2·60)–(2·63), are as follows:

$$\left.\begin{aligned}
\left(\frac{\partial T}{\partial V}\right)_{S,\,n_i} &= -\left(\frac{\partial p}{\partial S}\right)_{V,\,n_i}, \\[1mm]
-\left(\frac{\partial S}{\partial p}\right)_{T,\,n_i} &= \left(\frac{\partial V}{\partial T}\right)_{p,\,n_i}, \\[1mm]
\left(\frac{\partial T}{\partial p}\right)_{S,\,n_i} &= \left(\frac{\partial V}{\partial S}\right)_{p,\,n_i}, \\[1mm]
\left(\frac{\partial S}{\partial V}\right)_{T,\,n_i} &= \left(\frac{\partial p}{\partial T}\right)_{V,\,n_i}
\end{aligned}\right\} \tag{2·73}$$

Additional reciprocity relations which give the temperature and pressure coefficients of μ_i may also be obtained. Thus from (2·61)

$$\left.\begin{aligned}
\left(\frac{\partial \mu_i}{\partial T}\right)_{p,\,n_i,\,n_j,} &= -\left(\frac{\partial S}{\partial n_i}\right)_{T,\,p,\,n_j}, \\[1mm]
\left(\frac{\partial \mu_i}{\partial p}\right)_{T,\,n_i,\,n_j} &= \left(\frac{\partial V}{\partial n_i}\right)_{T,\,p,\,n_j}.
\end{aligned}\right\} \tag{2·74}$$

A large number of new equations may be built up by use of the Maxwell relations. Thus† from (2·60)

$$\left(\frac{\partial U}{\partial V}\right)_{T,\,n_i} = T\left(\frac{\partial S}{\partial V}\right)_{T,\,n_i} - p, \tag{2·75}$$

† In more detail, the derivation of (2·75) is carried out by expanding (2·60) as a function of T, V and the n_i:

$$dU = T\,dS - p\,dV + \Sigma \mu_i dn_i$$
$$= T\left\{\left(\frac{\partial S}{\partial T}\right)_{V,\,n_i} dT + \left(\frac{\partial S}{\partial V}\right)_{T,\,n_i} dV + \Sigma\left(\frac{\partial S}{\partial n_i}\right)_{T,\,V,\,n_j} dn_i\right\} - p\,dV + \Sigma\mu_i dn_i,$$

and therefore (2·75) is obtained by considering a change of state such that T and the n_i are constant.

and substituting in this one of the Maxwell equations we obtain

$$\left(\frac{\partial U}{\partial V}\right)_{T,\,n_i} = T\left(\frac{\partial p}{\partial T}\right)_{V,\,n_i} - p. \qquad (2\cdot76)$$

This particular equation gives a useful relationship between the variables U, V, T and p and is known as the 'thermodynamic equation of state'.

(c) Change of variable. In a number of problems it is necessary to transform formulae from one set of variables to another. Suppose, for example, that it is required to calculate $(\partial U/\partial T)_p$ from experimental data. Now U is more readily obtained from experiments in terms of T and V rather than in terms of T and p. For example, $(\partial U/\partial V)_T$ may be obtained from (2·76) and $(\partial U/\partial T)_V$ is a measurable heat capacity. We therefore commence by expressing U as a function of T and V (assuming the amounts of substances are constant):

$$U = U(T, V),$$

$$dU = \left(\frac{\partial U}{\partial T}\right)_V dT + \left(\frac{\partial U}{\partial V}\right)_T dV.$$

Therefore
$$\left(\frac{\partial U}{\partial T}\right)_p = \left(\frac{\partial U}{\partial T}\right)_V + \left(\frac{\partial U}{\partial V}\right)_T \left(\frac{\partial V}{\partial T}\right)_p, \qquad (2\cdot77)$$

which is the required expression.

Another useful type of transformation is obtained from the penultimate equation by considering a change of state at constant internal energy. Thus

$$0 = \left(\frac{\partial U}{\partial T}\right)_V \left(\frac{\partial T}{\partial V}\right)_U + \left(\frac{\partial U}{\partial V}\right)_T,$$

and therefore
$$\left(\frac{\partial U}{\partial T}\right)_V \left(\frac{\partial V}{\partial U}\right)_T \left(\frac{\partial T}{\partial V}\right)_U = -1. \qquad (2\cdot78)$$

The reader may note the cyclic order of the variables in this expression. Each of them occurs in all three of the positions, 'upstairs, downstairs and outside'.

(d) Integration of the basic equations. Consider the equation (2·60)
$$dU = T\,dS - p\,dV + \Sigma\mu_i dn_i.$$

We wish now to prove that the variables which occur in this equation must also satisfy the relation

$$U = TS - pV + \Sigma\mu_i n_i.$$

In order to show this, let it be supposed that the phase under discussion is enlarged in size, its temperature, pressure and the relative proportions of its components remaining unchanged. Under such conditions the μ_i, which are intensive variables like T and p, must also remain unchanged. Thus the direct integration of (2·60) gives

$$\Delta U = T\Delta S - p\,\Delta V + \Sigma\mu_i\Delta n_i.$$

Let the original values of the internal energy, entropy, etc., of the system be U, S, V, n_i. If the system is enlarged to k times its original size then the final values of the internal energy, etc., are kU, kS, kV and kn_i. This is because the extensive properties are proportional to the size of the system. Thus

$$\Delta U = kU - U = (k-1)\,U,$$

etc. Substituting in the previous equation we obtain

$$(k-1)\,U = T(k-1)\,S - p\,(k-1)\,V + \Sigma\mu_i(k-1)\,n_i,$$

and therefore

$$U = TS - pV + \Sigma\mu_i n_i, \tag{2·79}$$

which is the required result. By an analogous integration of equations (2·61)–(2·63) we obtain

$$G = \Sigma n_i\mu_i, \tag{2·80}$$

$$H = TS + \Sigma n_i\mu_i, \tag{2·81}$$

$$A = -pV + \Sigma n_i\mu_i, \tag{2·82}$$

and these equations are clearly consistent with the relations which define G, H and A in terms of U, pV and TS.

It may be remarked that the process by which equations (2·79)–(2·82) are obtained is not a purely mathematical one but depends on an item of *physical knowledge*; namely, that the intensive variables are not affected by the size of the system, whereas the extensive properties are directly·proportional to its size.

The same physical information implies that there is a certain relationship between the simultaneous changes in the *intensive* variables. This may be obtained by taking the complete differential of any one of the equations (2·79)–(2·82) and comparing with its 'parent' equation. Thus from (2·80) we have

$$dG = \Sigma n_i\,d\mu_i + \Sigma\mu_i\,dn_i,$$

and, comparing this with (2·61), we obtain

$$-S\,dT + V\,dp - \Sigma n_i\,d\mu_i = 0. \tag{2·83}$$

This relation, known as the Gibbs–Duhem equation, shows the necessary relation between simultaneous changes of temperature, pressure and the chemical potentials. Thus, if there are n substances in the particular phase, out of the $n+2$ intensive variables only $n+1$ can vary independently.

Equation (2·80) helps to give additional significance to the concept of the chemical potential. If each chemical potential is multiplied by the amount (mols) of the particular substance and summed over all the species, the result is the Gibbs function of the phase. (Of course the absolute value of G, like U, H, S and A, is unknown and the same is therefore true of μ_i. Thermodynamics deals only with the changes in these quantities, or their values relative to some standard state.)

For a pure substance we obtain from (2·80) $\mu_i = G/n_i$, and the chemical potential is thus simply the value of G per mole. Let it be supposed that this pure substance is in equilibrium with a mixture through a membrane which is permeable only to the substance i. On account of the condition of equilibrium, $\mu_{i\alpha} = \mu_{i\beta}$, the chemical potential of i in the mixture is equal to its value in the pure substance. We thus obtain a useful new interpretation of the meaning of μ_i: the chemical potential of a component of a mixture is equal to the value of G per mole of the same substance in its pure state, under the conditions of pressure which would put the mixture into equilibrium with the pure substance through a semi-permeable membrane.

2·11. Measurable quantities in thermodynamics

The measurable quantities are mainly as follows:
  heat capacities;
  compressibility and expansion coefficients;
  enthalpies of phase change, of mixing and of chemical reaction,
    and the corresponding changes of volume;
  equilibrium constants of reactions and e.m.f.'s of cells;
  vapour pressures, solubilities, etc.

The present section is concerned mainly with the first two groups above, and in particular it will be shown how the various partial differential coefficients of the previous section may be expressed in terms of experimental magnitudes.

(a) Heat capacities. If a body absorbs a finite quantity of heat, q, it usually rises in temperature by a finite amount, ΔT. The average *heat capacity* of the system over the range of temperature is defined by

$$C_{\text{av.}} \equiv q/\Delta T. \qquad (2·84)$$

The instantaneous heat capacity at the temperature T is the limiting value of this ratio as the quantities q and ΔT become infinitesimals:

$$C \equiv dq/dT. \qquad (2·85)$$

There are several points to be noted about this ratio. In the first place C becomes infinitely large at points of phase change, for at such points heat is absorbed without giving rise to any change of temperature. For this reason the term heat capacity is usually applied only to changes of state not involving a phase change.

Secondly, if any chemical reaction were to take place in a body during the temperature change, dT, heat would be absorbed or rejected by the body on account of this reaction, in addition to that required to change the temperature of the various substances present. If the heating were carried out so rapidly that reaction equilibrium did not have time to establish itself, a different value for C would be obtained than if the heating were carried out slowly.† In order that C shall have a definite value it must be specified that the heating is slow enough for the internal equilibrium to be maintained. Thus what is called the heat capacity of water includes the heat absorbed in the change in proportions of the various species. For similar reasons the measurement must always refer to a closed system.

Finally the value of C remains indefinite until the *path* of heating is specified. This may be seen by noting that, in the above expression for C, the heat dq may be replaced by $dU+dw$, in accordance with the first law. However, the mere statement that there is a temperature change dT is insufficient to fix the values either of dU or of dw; some other variable must be changed in a known manner, or held constant. It is therefore customary to define two particular kinds of heat capacity, C_V and C_p, which refer to constancy of volume and of pressure respectively. Thus

$$C_V \equiv \left(\frac{dq}{dT}\right)_V = \left(\frac{\partial U}{\partial T}\right)_V,$$

since in this case $dw=0$. In the case of constant pressure

$$dq = dU + p\,dV = dU + d(pV) = dH,$$

and therefore

$$C_p \equiv \left(\frac{dq}{dT}\right)_p = \left(\frac{\partial H}{\partial T}\right)_p,$$

provided that $p\,dV$ is the only form of work.

The heat capacities may also be expressed in terms of the entropy, since $dS = dq/T$ for a reversible heating process in a closed system. Thus

$$\left.\begin{aligned} C_V &= \left(\frac{\partial U}{\partial T}\right)_V = T\left(\frac{\partial S}{\partial T}\right)_V, \\ C_p &= \left(\frac{\partial H}{\partial T}\right)_p = T\left(\frac{\partial S}{\partial T}\right)_p. \end{aligned}\right\} \tag{2·86}$$

† One of the methods of measuring the rate of very rapid reactions, e.g. $2NO_2 = N_2O_4$, is based on the determination of the velocity of sound in the system, which depends on the same effect.

It may be noted that C_p and C_V refer to the heat capacity of the whole system whatever its size. The *mean molar heat capacities* are obtained by dividing by the total amounts of the various substances which are present:

$$c_V = \frac{C_V}{\Sigma n_i}, \quad c_p = \frac{C_V}{\Sigma n_i}, \tag{2·87}$$

and these are intensive properties of the system. If the latter contains only one substance we speak of its *molar heat capacity*.

(b) Expansivity and isothermal compressibility coefficients. These coefficients are defined by the relations

$$\alpha \equiv \frac{1}{V} \left(\frac{\partial V}{\partial T} \right)_p, \tag{2·88}$$

$$\kappa \equiv -\frac{1}{V} \left(\frac{\partial V}{\partial p} \right)_T. \tag{2·89}$$

Both α and κ refer to *fractional* changes in volume, dV/V, and are thus intensive properties which are independent of the actual volume of the system. The reciprocal of the isothermal compressibility, κ, is called the *bulk modulus* of the material.†

The 'pressure coefficient', $\frac{1}{p} \left(\frac{\partial p}{\partial T} \right)_V$, of a body may be calculated from values of α and κ. Using a transformation formula similar to (2·78):

$$\left(\frac{\partial p}{\partial T} \right)_V \left(\frac{\partial V}{\partial p} \right)_T \left(\frac{\partial T}{\partial V} \right)_p = -1,$$

and therefore

$$\frac{1}{p} \left(\frac{\partial p}{\partial T} \right)_V = -\frac{1}{p} \frac{(\partial V/\partial T)_p}{(\partial V/\partial p)_T} = \frac{\alpha}{p\kappa}. \tag{2·90}$$

For example, in the case of liquid mercury $\alpha = 1.81 \times 10^{-4}$ and $\kappa = 3.9 \times 10^{-6}$, in units of atmospheres and kelvin degrees. Hence $(\partial p/\partial T)_V = 46$, and the heating of a sample of mercury through $1\ ^\circ\text{C}$ at constant volume would cause the pressure to rise by 46 atm, if α and κ remain constant over this small range.

(c) Relations between C_p, C_V, α and κ. Between some or all of these quantities a number of relations can be established. As an example we shall prove the relation

$$C_p - C_V = TV\alpha^2/\kappa.$$

† It can be shown that κ, and also C_V and C_p, must always be positive quantities. See, for example, Landau and Lifshitz, *Statistical Physics*, transl. Shoenberg (Oxford, 1938), p. 100 and also Guggenheim, *Thermodynamics* (Amsterdam, North-Holland Publ. Co., 1957), Section 4·01.

Using a transformation equation of the type of (2·77),

$$\left(\frac{\partial S}{\partial T}\right)_p = \left(\frac{\partial S}{\partial V}\right)_T \left(\frac{\partial V}{\partial T}\right)_p + \left(\frac{\partial S}{\partial T}\right)_V,$$

and therefore from (2·86) and (2·73)

$$\frac{C_p}{T} = \left(\frac{\partial p}{\partial T}\right)_V \left(\frac{\partial V}{\partial T}\right)_p + \frac{C_V}{T},$$

and finally using (2·88) and (2·90)

$$C_p - C_V = T\left(\frac{\partial p}{\partial T}\right)_V \left(\frac{\partial V}{\partial T}\right)_p$$

$$= TV\frac{\alpha^2}{\kappa}. \tag{2·91}$$

It will be noticed that the proof depends on the existence of entropy as a function of state. The experimental verification of (2·91) is thus, in a certain sense, a confirmation of the second law, and the same may be said of any of the other quantitative relations based on this law. For practical purposes the value of equations such as (2·91) is the calculation of one quantity, say C_V, from known values of the others.

Another important relation which is easily proved is

$$-\frac{1}{V}\left(\frac{\partial V}{\partial p}\right)_S = \frac{C_V}{C_p}\kappa. \tag{2·92}$$

The quantity on the left of this equation is the *isentropic compressibility*, and it measures the fractional change in volume of a system in a reversible adiabatic compression.

The derivation of the relationship between the temperature coefficient of κ and the pressure coefficient of α is instructive:

$$dV = \left(\frac{\partial V}{\partial p}\right)_T dp + \left(\frac{\partial V}{\partial T}\right)_p dT$$

$$= -\kappa V\, dp + \alpha V\, dT,$$

or

$$d\ln V = -\kappa\, dp + \alpha\, dT.$$

Now $\ln V$ is a function of state and therefore the reciprocity relation (2·72) may be applied:

$$-\left(\frac{\partial \kappa}{\partial T}\right)_p = \left(\frac{\partial \alpha}{\partial p}\right)_T. \tag{2·93}$$

(*d*) **Heats of phase change and of reaction.** When these quantities are measured under conditions of constant volume they are equal to the change in internal energy in the process in question.

More usually, however, they are measured at constant pressure and are then equal to the change in enthalpy (provided that $p\,\mathrm{d}V$ is the only form of work). For example, in the vaporization of a pure liquid at constant pressure, the latent heat, as it used to be called, is

$$L = H_g - H_l, \tag{2·94}$$

where the subscripts g and l denote gas and liquid respectively. The corresponding entropy change is L/T, provided that the vaporization takes place under equilibrium conditions. (Under such conditions $\mu_g = \mu_l$ and therefore $\mathrm{d}S = \mathrm{d}H/T$, from equation (2·62).)

2·12. Calculation of changes in the thermodynamic functions over ranges of temperature and pressure

In order to compute the change of, say, the enthalpy it is desirable to use temperature and pressure as the independent variables and also to express the various partial differential coefficients in terms of experimental magnitudes. This may be illustrated as follows.

For a phase which is a closed system and remains in internal equilibrium.

$$\mathrm{d}H = T\,\mathrm{d}S + V\,\mathrm{d}p$$

$$= T\left\{ \left(\frac{\partial S}{\partial T}\right)_p \mathrm{d}T + \left(\frac{\partial S}{\partial p}\right)_T \mathrm{d}p \right\} + V\,\mathrm{d}p$$

$$= T\left(\frac{\partial S}{\partial T}\right)_p \mathrm{d}T + \left\{ V + T\left(\frac{\partial S}{\partial p}\right)_T \right\}\mathrm{d}p.$$

Using (2·73), (2·86) and (2·88) this gives

$$\mathrm{d}H = C_p\,\mathrm{d}T + \left\{ V - T\left(\frac{\partial V}{\partial T}\right)_p \right\}\mathrm{d}p$$

$$= C_p\,\mathrm{d}T + V(1 - \alpha T)\,\mathrm{d}p. \tag{2·95}$$

Therefore for a finite change $(T_1, p_1) \rightarrow (T_2, p_2)$

$$H_2 - H_1 = \int_{T_1}^{T_2} C_p\,\mathrm{d}T + \int_{p_1}^{p_2} V(1 - \alpha T)\,\mathrm{d}p. \tag{2·96}$$

Similarly
$$S_2 - S_1 = \int_{T_1}^{T_2} \frac{C_p}{T}\,\mathrm{d}T - \int_{p_1}^{p_2} V\alpha\,\mathrm{d}p. \tag{2·97}$$

By use of these expressions it is possible to calculate the enthalpy and entropy of a fluid relative to any chosen reference state. In the steam tables this reference state is usually chosen as liquid water at 0 °C. In calculating the relative value of the enthalpy or entropy of steam it is necessary, of course, to carry out two distinct integra-

tions, one for the liquid water and the other for the vapour. To these must be added the discontinuous increments of enthalpy or entropy which occur at the phase change between the liquid and vapour states (§ 2·11 *d*).

It is to be noted that the quantities C_p, V and α which occur in the above integrals are themselves functions of temperature and pressure, and these functions must be known before the integrations can be carried out. For example, in (2·97), the first integral refers to constant pressure and the second to constant temperature; it is necessary to know *either* C_p as a function of T at the lower limit p_1 of the second integral, together with $V\alpha$ as a function of pressure at the upper limit T_2 of the first integral, *or* C_p as a function of T at the upper limit p_2,

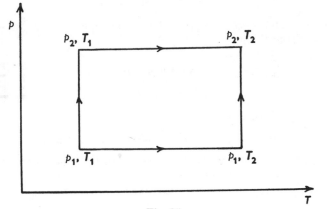

Fig. 15

together with $V\alpha$ as a function of p at the lower limit T_1 (or the corresponding quantities for any other alternative path). This may be seen from Fig. 15.

The calculation of the steam tables, or comparable tables for other fluids, thus requires a very complete knowledge of the p-V-T relations of the fluids and also of C_p as a function of T and p. Graphical integration may be adopted, or alternatively it may be possible to carry out a direct integration if the experimental data can be expressed as empirical power series. e.g.

$$C_p = a + bT + cT^2 + \ldots.$$

2·13. Molar and partial molar quantities

The symbols U, S, V, H, A and G, as used previously, all refer to the whole of the homogeneous phase which is under discussion. Since

they are all extensive properties, their values are all doubled whenever the phase is doubled in size. For this reason it is convenient to introduce an additional set of thermodynamic quantities which represent the total internal energy, entropy, etc., as equal to the sum of the *specific* contributions of each particular species.

Consider first of all a phase which consists of a single component. For such a system equation (2·60) reads

$$dU = T\,dS - p\,dV + \mu\,dn, \tag{2·98}$$

the last term being the increase in U due to the addition of dn mols of the substance at constant entropy and volume. Let the *molar* values of the internal energy, etc., be defined by

$$u = U/n, \quad s = S/n, \quad v = V/n, \tag{2·99}$$

where n (mols) is the amount of the particular substance in the system. From (2·99)

$$dU = u\,dn + n\,du,$$

$$dS = s\,dn + n\,ds,$$

$$dV = v\,dn + n\,dv,$$

Substituting these in (2·98) we obtain

$$n\,du = Tn\,ds - pn\,dv + (\mu - u + Ts - pv)\,dn.$$

The quantity in brackets is clearly zero, since it was shown in connexion with equation (2·80) that the value of μ for a pure substance is equal to G/n. Thus finally we obtain (as may be obvious)

$$du = T\,ds - p\,dv, \tag{2·100}$$

and similarly

$$dh = T\,ds + v\,dp, \tag{2·101}$$

$$da = -s\,dT - p\,dv, \tag{2·102}$$

$$dg = d\mu = -s\,dT + v\,dp, \tag{2·103}$$

where $h = H/n$, $a = A/n$ and $g = G/n$. Equation (2·103) is seen to be the same as the Gibbs–Duhem equation, (2·83), when the latter is applied to a single component phase. The equations (2·100)–(2·103) are, of course, fully applicable to open systems (of a single component) because they refer to unit quantity of the substance in question and not to the system as a whole.

In the above we have defined the molar internal energy etc. by the relations (2·99), $U = un$ etc., which express the fact that the total internal energy etc. is proportional to the size of the phase. These molar quantities, u, etc., depend only on temperature and pressure. Turning now to the case of multicomponent phases, we wish to define

similar quantities, which will be denoted U_i, etc. such that the total internal energy etc. of the phase may be expressed as the sum of the contributions of the various species. Thus $U = \Sigma n_i U_i$ etc. However, it is evident that this equation could serve to define only one of the quantities U_i in terms of the remainder. We therefore proceed rather differently as follows.

Let E stand for any of the extensive properties, such as internal energy, enthalpy, etc. Then the *partial molar* value of the property E, for the ith substance, is defined by

$$E_i \equiv \left(\frac{\partial E}{\partial n_i}\right)_{T,\,p,\,n_j} \tag{2·104}$$

For example, in the case of the partial molar volume

$$V_i \equiv \left(\frac{\partial V}{\partial n_i}\right)_{T,\,p,\,n_j},$$

which may be expressed in words as follows: the partial molar volume of component i in the system is equal to the infinitesimal increase in the volume of the system divided by the infinitesimal number of moles of this substance which are added, *the temperature, pressure and quantities of the other substances remaining constant.* Similarly with regard to the partial molar enthalpies, etc.

Let the property E be expressed as a function of T, p and the amounts of substances (mols). Then

$$dE = \left(\frac{\partial E}{\partial T}\right)_{p,\,n_i} dT + \left(\frac{\partial E}{\partial p}\right)_{T,\,n_i} dp + \Sigma \left(\frac{\partial E}{\partial n_i}\right)_{T,\,p,\,n_j} dn_i$$

$$= \left(\frac{\partial E}{\partial T}\right)_{p,\,n_i} dT + \left(\frac{\partial E}{\partial p}\right)_{T,\,n_i} dp + \Sigma E_i dn_i. \tag{2·105}$$

This equation may be integrated in the same way as in §2·10d. We obtain

$$E = \Sigma n_i E_i, \tag{2·106}$$

which expresses, in the desired manner, the value of E for the whole phase, in terms of the contributions of each substance. In detail we have

$$\left. \begin{array}{ll} U = \Sigma n_i U_i, & S = \Sigma n_i S_i, \\ H = \Sigma n_i H_i, & A = \Sigma n_i A_i, \\ V = \Sigma n_i V_i, & G = \Sigma n_i G_i. \end{array} \right\} \tag{2·107}$$

In the case of a single component system the partial molar quantities are clearly identical with the molar quantities (equation (2·99)).

The absolute values of the various extensive properties (except V) are, of course, never known, and the same applies to the partial molar quantities E_i. Therefore the latter must all be calculated with respect to the same reference state as in the case of E itself. Moreover, the E_i, although they are intensive and therefore independent of the size of the system, are still dependent on the relative proportions of the various components (e.g. on the mole fractions), and also on the temperature and pressure.

It may be noted that the partial molar value of G is identical with the chemical potential. This arises from the fact that in the equation (2·61)

$$dG = -S\,dT + V\,dp + \Sigma\mu_i\,dn_i,$$

the independent variables are T, p and the n_i, which are the same as are used in the definition of the partial molar quantities. Thus

$$\mu_i \equiv \left(\frac{\partial G}{\partial n_i}\right)_{T,\,p,\,n_j} \equiv G_i. \qquad (2·108)$$

The symbol μ_i will always be used in place of G_i.

Between the partial molar quantities there are relations entirely analogous to those between the parent extensive quantities. For example,

$$H \equiv U + pV,$$

and differentiating with respect to n_i at constant T, p and amounts of other species,

$$H_i \equiv U_i + pV_i. \qquad (2·109)$$

Similarly $$\mu_i \equiv G_i \equiv H_i - TS_i \qquad (2·110)$$
and other identities.

Two particularly important relations are obtained as follows. From the equation (2·61),

$$dG = -S\,dT + V\,dp + \Sigma\mu_i\,dn_i,$$

we obtain the two reciprocity relations

$$\left(\frac{\partial\mu_i}{\partial p}\right)_{T,\,n_i,\,n_j} = \left(\frac{\partial V}{\partial n_i}\right)_{T,\,p,\,n_j} = V_i, \qquad (2·111)$$

$$\left(\frac{\partial\mu_i}{\partial T}\right)_{p,\,n_i,\,n_j} = -\left(\frac{\partial S}{\partial n_i}\right)_{T,\,p,\,n_j} = -S_i. \qquad (2·112)$$

Combining (2·110) and (2·112)

$$\mu_i = H_i + T\left(\frac{\partial\mu_i}{\partial T}\right)_{p,\,n_i,\,n_j},$$

and this may be rearranged into the more compact form

$$\left(\frac{\partial \mu_i/T}{\partial T}\right)_{p,\,n_i,\,n_j} = -\frac{H_i}{T^2}. \tag{2·113}$$

The equations (2·111)–(2·113) will be used very extensively in the following chapters. They can also be derived from (2·67), (2·66) and (2·71) respectively, by differentiation with respect to n_i.

In the case of a pure substance the partial molar quantities are, of course, identical with the molar quantities and therefore (2·111)–(2·113) may be written

$$\left(\frac{\partial \mu_i}{\partial p}\right)_T = v_i, \tag{2·111b}$$

$$\left(\frac{\partial \mu_i}{\partial T}\right)_p = -s_i, \tag{2·112b}$$

$$\left(\frac{\partial \mu_i/T}{\partial T}\right)_p = -\frac{h_i}{T^2}. \tag{2·113b}$$

Another relation between the partial molar quantities may be obtained by taking the differential of (2·106),

$$dE = \Sigma n_i\,dE_i + \Sigma E_i\,dn_i,$$

and comparing this with (2·105), we obtain

$$\left(\frac{\partial E}{\partial T}\right)_{p,\,n_i} dT + \left(\frac{\partial E}{\partial p}\right)_{T,\,n_i} dp - \Sigma n_i\,dE_i = 0. \tag{2·114}$$

For example

$$\left(\frac{\partial H}{\partial T}\right)_{p,\,n_i} dT + \left(\frac{\partial H}{\partial p}\right)_{T,\,n_i} dp - \Sigma n_i H_i = 0.$$

As applied to G, equation (2·114) is clearly the same as the Gibbs–Duhem equation, (2·83), since $\partial G/\partial T = -S$ and $\partial G/\partial p = V$. At constant temperature and pressure, and after dividing through by Σn_i, (2·114) reduces to

$$\Sigma x_i\,dE_i = 0. \tag{2·115}$$

The various partial molar quantities, including the chemical potentials, may be regarded as functions of temperature, pressure and the mole fractions. If there are n substances in the system, there are $n-1$ independent mole fractions, and thus the total differential of μ_i is given by

$$d\mu_i = \frac{\partial \mu_i}{\partial T}\,dT + \frac{\partial \mu_i}{\partial p}\,dp + \sum_{i=1}^{i=n-1} \frac{\partial \mu_i}{\partial x_i}\,dx_i$$

$$= -S_i\,dT + V_i\,dp + \sum_{i=1}^{i=n-1}\left(\frac{\partial \mu_i}{\partial x_i}\right)_{T,\,p,\,x_j} dx_i, \tag{2·116}$$

using (2·111) and (2·112). Similar equations may be obtained for the other partial molar quantities. Thus from (2·110)

$$H_i = TS_i + \mu_i,$$

and therefore

$$dH_i = T\,dS_i + S_i\,dT + d\mu_i,$$

and combining this with (2·116)

$$dH_i = T\,dS_i + V_i\,dp + \sum_{i=1}^{i=n-1}\left(\frac{\partial\mu_i}{\partial x_i}\right)_{T,\,p,\,x_j} dx_i, \qquad (2\cdot117)$$

and similarly

$$dU_i = T\,dS_i - p\,dV_i + \sum_{i=1}^{n-1}\left(\frac{\partial\mu_i}{\partial x_i}\right)_{T,\,p,\,x_j} dx_i. \qquad (2\cdot118)$$

These equations may be compared with (2·62) and (2·60) respectively.

2·14. Calculation of partial molar quantities from experimental data

The most useful partial molar quantities are those which occur in equations (2·111) and (2·113), namely, the volumes and enthalpies. These may be computed from the volume change on mixing the pure components and from the corresponding absorption of heat at constant pressure, equal to the enthalpy increase on mixing. One of the methods will be discussed briefly for the case of two components.

Let h_1 and h_2 be the enthalpies per mole of the two pure components and let H_1 and H_2 be the partial molar enthalpies in a solution containing n_1 moles of component 1 and n_2 moles of component 2. Then the heat absorbed at constant pressure in making this mixture is

$$(n_1 H_1 + n_2 H_2) - (n_1 h_1 + n_2 h_2) = n_1(H_1 - h_1) + n_2(H_2 - h_2).$$

The heat absorbed per mole will be denoted by Δh and is obtained by dividing through by $(n_1 + n_2)$:

$$\Delta h = (1 - x_2)(H_1 - h_1) + x_2(H_2 - h_2), \qquad (2\cdot119)$$

where x_2 is the mole fraction of component 2. If experimental values of Δh are plotted against x_2 the result may be of the form shown as the curve AGB in Fig. 16.

At a particular mole fraction, x_2', the gradient CD of the curve is obtained from (2·119) as†

$$\left(\frac{\partial\Delta h}{\partial x_2}\right)_{x_2=x_2'} = -(H_1' - h_1) + (H_2' - h_2), \qquad (2\cdot120)$$

† The term $(1-x_2)\dfrac{\partial H_1}{\partial x_2} + x_2\dfrac{\partial H_2}{\partial x_2}$, in the complete differentiation of (2·119), is zero on account of (2·115). (In (2·120) the partial differential coefficient refers to constancy of temperature and pressure.)

where H_1' and H_2' are the particular values of H_1 and H_2 at the composition x_2'. Eliminating $(H_2'-h_2)$ between equations (2·119) and (2·120):

$$(H_1'-h_1)=\Delta h'-x_2'\left(\frac{\partial \Delta h}{\partial x_2}\right)_{x_2=x_2'}$$

$$=EG-AE\frac{FG}{CF}$$

$$=EG-FG$$

$$=AC.$$

Similarly $\qquad\qquad (H_2'-h_2)=BD.$

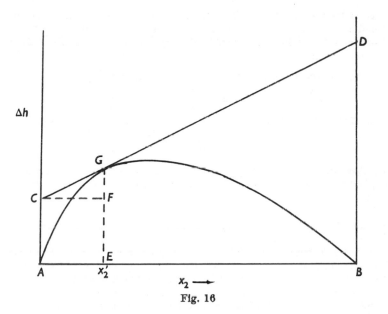

Fig. 16

The values of the partial molar enthalpies in the mixture, relative to the values h_1 and h_2 for the pure components, are thus given by the intercepts AC and BD, of the tangent CD.

A similar procedure may be used to obtain the partial molar volumes from measurements of the volume change on mixing, i.e. from the density of the mixture. In this instance the absolute values of the volumes, v_1 and v_2, of the pure components are known, and it is therefore possible to obtain the absolute values of the partial molar volumes. In some mixtures it may occur that the partial molar volume of a particular component is negative, implying that an infinitesimal addition of the substance causes a *decrease* in the volume of the mixture. Of course, in any two-component system the effect of the composition on the value of one of the partial molar quantities is not independent of its effect on the other. Thus from (2·115)

$$(1-x_2)\left(\frac{\partial E_1}{\partial x_2}\right)_{T,\,P} + x_2\left(\frac{\partial E_2}{\partial x_2}\right)_{T,\,P} = 0. \tag{2·121}$$

For more elaborate methods of calculating partial molar quantities from experimental data the reader is referred to the literature.†

PROBLEMS

1. Calculate the changes in S, U and H for the process of converting 1 mol of liquid water at 0 °C and 1 atm into steam at 200 °C and 3 atm. Assume water to have a constant density and heat capacity. Assume the steam to behave as a perfect gas and to have a heat capacity given by

$$c_p = 8.81 - 1.9 \times 10^{-3}T + 2.2 \times 10^{-6}T^2 \ \text{(cal K}^{-1}\ \text{mol}^{-1}).$$

The enthalpy of vaporization at 100 °C is 40.6 kJ mol^{-1}.

2. Obtain the following form of the Gibbs–Helmholtz equation:

$$\Delta G = \Delta H + T\frac{\partial \Delta G}{\partial T}.$$

What condition limits the type of change to which this equation may be applied? Obtain the following expression for the temperature coefficient of the e.m.f. of a galvanic cell

$$\frac{\partial E}{\partial T} = \frac{E}{T} + \frac{\Delta H}{zFT}.$$

3. A cell in which the reaction $Pb + Hg_2Cl_2 = PbCl_2 + 2Hg$ takes place has an e.m.f. at 25 °C of 0.535 7 V, and this increases by 1.45×10^{-4} V/ °C. Calculate (*a*) the maximum work available from the cell at 25 °C per mol of Pb dissolved; (*b*) the heat of reaction at 25 °C; (*c*) the entropy change of the reaction at 25 °C; (*d*) the heat absorbed by the cell at 25 °C per mol of Pb dissolved reversibly.

4. The energy of the radiation which is at equilibrium within an enclosure depends only on the volume and on the wall temperature T. It is known also that the pressure of the radiation is equal to one-third of the energy per unit volume.

† Lewis and Randall, *Thermodynamics* (New York, McGraw-Hill, 1923), Chapter IV; Sosnick, *J. Amer. Chem. Soc.* 49 (1927), 2255; Glasstone, *Thermodynamics for Chemists* (New York, Van Nostrand, 1947), Chapter XVIII; Klotz, *Chemical Thermodynamics* (New York, W. A. Benjamin Inc., 1964), Chapters XIII and XIV; Van Ness and Mrazek, *Am. Inst. Chem. Eng. J.* 5 (1959), 209.

Show that the energy u and entropy s per unit volume of the radiation are given by

$$u = \alpha T^4$$
$$s = \tfrac{4}{3}\alpha T^3,$$

where α is a constant. [C.U.C.E. Qualifying, 1954]

5. Prove equation (2·92).

If a gas obeys the equation of state $p(v-b) = RT$, where b is a constant, show that

$$\left(\frac{\partial v}{\partial p}\right)_s = -\frac{c_V}{c_V + R}\frac{v-b}{p}.$$

[Question modified from C.U.C.E. Qualifying, 1950]

6. An extended strip of rubber has a length l when subjected to a tensile force F. If the volume change on extension may be neglected, show that

$$\left(\frac{\partial U}{\partial l}\right)_T = F - T\left(\frac{\partial F}{\partial T}\right)_l.$$

Show that the small temperature rise, ΔT, which takes place in a slow adiabatic stretching is given by

$$\frac{\Delta T}{T} = \int_{l_0}^{l}\frac{1}{c_l}\left(\frac{\partial F}{\partial T}\right)_l dl,$$

where l_0 and l refer to the initial and final lengths respectively, and c_l is the heat capacity of the rubber at constant extension.

At moderate degrees of extension it is found that the tensile force at constant length is approximately proportional to the thermodynamic temperature. Show that $(\partial U/\partial l)_T = 0$ and point the analogy to the behaviour of a perfect gas. [C.U.C.E. Tripos, 1952]

7. The thermodynamic temperature T and an empirical temperature θ are both used in measuring the properties of a substance.

(a) Show that, provided $(\partial p/\partial T)_V \neq 0$,

$$\left(\frac{\partial H}{\partial V}\right)_T = -V^2\left(\frac{\partial p}{\partial T}\right)_V\frac{\partial}{\partial V}\left(\frac{T}{V}\right)_p.$$

(b) Show also that $(\partial U/\partial V)_\theta = 0$ and $(\partial H/\partial V)_\theta = 0$ are sufficient, independent and necessary conditions for the substance to obey the equation $pV = cT$. [C.U.C.E. Tripos, 1952]

8. With temperature expressed on an empirical scale θ, a gas exists for which

$$pV = nc\theta^2 \quad \text{and} \quad \left(\frac{\partial H}{\partial p}\right)_\theta = 0.$$

Establish from the first and second laws of thermodynamics that for *any* substance

$$\left(\frac{\partial U}{\partial V}\right)_\theta = \frac{\theta^3}{2}\frac{\partial}{\partial\theta}\left(\frac{p}{\theta^2}\right)_V.$$

[C.U.C.E. Qualifying, 1952]

9. The following data, taken from the National Bureau of Standards compilation, *Selected Values of Chemical Thermodynamic Properties*, gives the standard enthalpy of formation of hydrogen chloride from its elements at 25 °C in kg cal mol^{-1}. (Thus the difference between the first

and third results gives the heat of dissolving 1 mol of HCl gas in 2 mol
of water.) Show that the partial molar enthalpy of hydrogen chloride in
a 10 mol dm^{-3} solution, relative to the elements, is -35.25 kg cal mol^{-1},
and that the partial molar enthalpy of water in the same solution,
relative to pure liquid water, is -0.43 kg cal mol^{-1}.

g	-22.063	in $6H_2O$	-37.811
in $1H_2O$	-28.331	$8H_2O$	-38.371
$2H_2O$	-33.731	$10H_2O$	-38.671
$3H_2O$	-35.651	$50H_2O$	-39.577
$4H_2O$	-36.691	$100H_2O$	-39.713
$5H_2O$	-37.371	∞H_2O	-40.023

10. The diagram refers to the enthalpy relationships of a completely
miscible binary system at a constant pressure of 1 atm. The mole
fraction of component B is plotted horizontally from left to right and
the enthalpy of the mixture, relative to the pure components in chosen
reference states, is plotted vertically. Curve CD represents the enthalpy
of the liquid phase, at its boiling-point, as a function of composition.
Curve EF represents the enthalpy of the vapour above the boiling
liquid as a function of its own composition. (Thus CE and DF are the
enthalpies of vaporization of A and B respectively.) GH and IJ are
typical tie-lines, i.e. a boiling liquid of composition G is in equilibrium
with vapour of composition H.

A feed liquid at a temperature and composition corresponding to
point K is separated, by continuous flash distillation, into a residual
liquid of composition G and a vapour of composition H. Show that
the heat required, per mole of the vapour, is represented by the
length PH, where P is the intersection of the vertical through H with
the straight line GK.

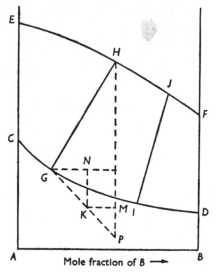

Mole fraction of B ⟶

PART II

REACTION AND PHASE EQUILIBRIA

CHAPTER 3

THERMODYNAMICS OF GASES

3·1. Models

In thermodynamics, as in other branches of science, it is convenient to set up model systems to which the behaviour of real systems approximates under limiting conditions. The value of this procedure is that simple and exact relations may be established for the model, to an extent which is impossible for the real systems themselves. These relations form a convenient standard against which actual phenomena may be compared.

The important models with which we are concerned are the perfect gas, the perfect gas mixture and the ideal solution (gaseous, liquid or solid). These may be defined in either of two ways which are *entirely equivalent*: (1) in terms of limiting experimental laws such as the gas equation and Raoult's law; (2) in terms of expressions for the chemical potentials of the various components. These expressions are as follows:

$$\text{a perfect single gas} \qquad \mu = \mu^0 + RT \ln p,$$

$$\text{a perfect gas mixture} \qquad \mu_i = \mu_i^0 + RT \ln p_i,$$

$$\text{an ideal solution} \qquad \mu_i = \mu_i^* + RT \ln x_i,$$

where μ^0 and μ_i^0 are functions of temperature only and μ_i^* is a function both of temperature and pressure. The advantage of the second procedure is that the mutual relationships of the three models may be seen in a much clearer manner.

3·2. The single perfect gas

Let μ and μ^0 be the chemical potentials of a pure gas at the pressures p and p^0, respectively, and *the same temperature* T. The gas will be said to be perfect if

$$\mu = \mu^0 + RT \ln p/p^0, \tag{3·1}$$

where R is a constant. It is convenient to choose p^0 as unity, in the same system of units as p. Then the chemical potential at any other pressure, p, and the same temperature is

$$\mu = \mu^0 + RT \ln p, \tag{3·2}$$

and this equation defines the perfect gas. Since it refers to the chosen pressure, p^0, μ^0 *is a function of temperature only*, and it may be called the *standard chemical potential* (or Gibbs free energy per mole) at the temperature T. In the logarithm of (3·2) it is to be noted that p stands for p/p^0 (where p^0 is unity) and is therefore dimensionless.

Since (3·2) is an identity it is permissible to equate the partial differential coefficients of the left-hand and right-hand sides.† Thus from the equation we obtain the pressure and temperature coefficients of the chemical potential of a perfect gas

$$\left(\frac{\partial \mu}{\partial p}\right)_T = RT \frac{\mathrm{d} \ln p}{\mathrm{d} p} = \frac{RT}{p}, \tag{3·3}$$

$$\left(\frac{\partial \mu}{\partial T}\right)_p = \frac{\mathrm{d} \mu^0}{\mathrm{d} T} + R \ln p. \tag{3·4}$$

Either of these two relations leads immediately to the normal experimental criterion of the perfect gas. Now for any single component fluid we have from (2·111 b) and (2·112 b)

$$\left(\frac{\partial \mu}{\partial p}\right)_T = v, \tag{3·5}$$

$$\left(\frac{\partial \mu}{\partial T}\right)_p = -s, \tag{3·6}$$

where v and s are the molar volume and entropy respectively. Comparing (3·3) with (3·5) we obtain

$$\left. \begin{array}{c} v = RT/p \\ pV = nRT, \end{array} \right\} \tag{3·7}$$

or

where n is the amount of substance (mols) in the total volume V of the gas. *Our definition of the perfect gas in terms of its chemical potential therefore entails the gas law* (equation (3·7)).

Similarly, comparing (3·4) and (3·6) we obtain

$$-s = \frac{\mathrm{d} \mu^0}{\mathrm{d} T} + R \ln p. \tag{3·8}$$

Now $H = G + TS$ and therefore, dividing through by n,

$$h = \mu + Ts. \tag{3·9}$$

---

† Compare, for example, the phase equilibrium relation $\mu_{i\alpha} = \mu_{i\beta}$, which is not an identity and holds only at particular pairs of temperatures and pressures, in a one-component system. In this instance we cannot write $\frac{\partial \mu_{i\alpha}}{\partial T} = \frac{\partial \mu_{i\beta}}{\partial T}$, but only

$$\frac{\partial \mu_{i\alpha}}{\partial T} \mathrm{d}T + \frac{\partial \mu_{i\alpha}}{\partial p} \mathrm{d}p = \frac{\partial \mu_{i\beta}}{\partial T} \mathrm{d}T + \frac{\partial \mu_{i\beta}}{\partial p} \mathrm{d}p \,.$$

Substituting (3·2) and (3·8) into (3·9)

$$h = \mu^0 + RT \ln p + T\left(-\frac{d\mu^0}{dT} - R \ln p\right)$$

$$= \mu^0 - T\frac{d\mu^0}{dT} = -T^2\frac{d\mu^0/T}{dT}, \tag{3·10}$$

which is a function of temperature only. Similarly

$$u = h - pv = h - RT, \tag{3·11}$$

and this also is a function of temperature only. *The enthalpy and internal energy of a perfect gas thus depend only on the temperature.* This may be expressed as

$$\left.\begin{array}{l}\left(\dfrac{\partial u}{\partial v}\right)_T = \left(\dfrac{\partial u}{\partial p}\right)_T = 0,\\[3mm] \left(\dfrac{\partial h}{\partial v}\right)_T = \left(\dfrac{\partial h}{\partial p}\right)_T = 0.\end{array}\right\} \tag{3·12}$$

The following elementary theorems concerning the perfect gas are easily verified:

(a) The molar heat capacities, c_p and c_V, are functions of temperature only.

(b) $$c_p - c_V = R. \tag{3·13}$$

(c) The heat absorbed and the work done when an amount n (mols) of a perfect gas changes reversibly and isothermally from a state 1 to a state 2, are given by

$$q = -w = nRT \ln\frac{p_1}{p_2},$$

$$= nRT \ln\frac{V_2}{V_1}. \tag{3·14}$$

It may be noted that the definition of a perfect gas, as given above, does not imply that the heat capacities are independent of temperature. This occurs only in the special case of the monatomic gas. However, over small ranges of temperature, especially in regions where the rotational degrees of freedom are fully excited but where the vibrational modes are hardly excited at all, it is often permissible to take the heat capacities as being approximately constant. Under such conditions an important equation may be obtained as follows.

Since the internal energy per mole is a function of temperature only,

$$c_V \equiv \left(\frac{\partial u}{\partial T}\right)_V = \frac{du}{dT},$$

or $du = c_V dT$ for any change of state. The basic equation

$$du = Tds - pdv$$

may thus be rewritten $c_V dT = T ds - p dv$.

Dividing through by T and using (3·7)

$$c_V \frac{dT}{T} = ds - R \frac{dv}{v}.$$

This equation may be integrated directly, because c_V is a function of temperature only. In particular, if c_V is constant over the range from T_1 to T_2 we obtain

$$c_V \ln \frac{T_2}{T_1} = s_2 - s_1 - R \ln \frac{v_2}{v_1}$$

or

$$\frac{T_2}{T_1} \left(\frac{v_2}{v_1}\right)^{R/c_V} = e^{(s_2 - s_1)/c_V}.$$

This equation may be rearranged in terms of pressures by use of (3·7) and (3·13)

$$p_1 v_1^\gamma e^{-s_1/c_V} = p_2 v_2^\gamma e^{-s_2/c_V}, \tag{3·15}$$

where $\gamma = c_p/c_V$. In the special case of an isentropic change (e.g. a reversible adiabatic), the last equation reduces to

$$p_1 v_1^\gamma = p_2 v_2^\gamma, \tag{3·16}$$

which is to say pv^γ is constant along the curve of an isentropic of a perfect gas over a range in which γ is approximately independent of temperature.

Again, if c_V is approximately constant, the work done *by* a gas in an adiabatic change of state is

$$-w = -(U_2 - U_1)$$

$$= -nc_V(T_2 - T_1)$$

$$= -\frac{nc_V}{R}(p_2 v_2 - p_1 v_1). \tag{3·17}$$

If the process is reversible (3·16) may be used to eliminate one of the four quantities p_1, p_2, v_1 and v_2 from this equation.

3·3. The perfect gas mixture

Certain gaseous mixtures approximate to simple behaviour in the following respects: (*a*) the gas mixture as a whole obeys the equation of state $pV = nRT$, where n is the total amount (mols) of all substances; (*b*) two such mixtures are at equilibrium with each other through a semi-permeable membrane when the partial pressure is the same on each side, for each component which is able to pass through the membrane; (*c*) there is no heat of mixing. The molecular conditions which must exist in order that the mixture shall have these properties are the same as in the case of the perfect single gas; namely, the gas must consist of freely moving particles of negligible volume and having negligible forces of interaction.

The above properties may be taken as defining the perfect gas mixture. Alternatively, we can proceed as in the previous section and put forward a definition in terms of the chemical potential. Proceeding in this manner a gaseous mixture will be said to be perfect if the chemical potential of each of its components is given by the following relation, in which μ_i^0 *is a function of temperature only*,

$$\mu_i = \mu_i^0 + RT \ln p + RT \ln y_i, \tag{3·18}$$

where p is the total gas pressure and y_i is the mole fraction of component i.† Now since μ_i^0 is independent of composition it retains the same value when y_i is brought up to unity. It is thus precisely the same as μ^0 in equation (3·2) and is the value of the Gibbs free energy per mole of the gas i in its pure state at unit pressure.‡

The defining equation (3·18) can be put in a more compact form by means of the *partial pressure* p_i. Thus

$$\mu_i = \mu_i^0 + RT \ln p_i, \tag{3·19}$$

where $$p_i \equiv y_i \, p. \tag{3·20}$$

This choice of definition of the partial pressure makes the sum of all the partial pressures equal to the total pressure, even if the mixture is not perfect. Thus $$\sum_i p_i = \sum y_i \, p = p \sum y_i = p. \tag{3·21}$$

The properties of the perfect mixture are as follows.

(a) The equation of state. From (3·18) we obtain the pressure coefficient of μ_i, at constant temperature and composition:

$$\left(\frac{\partial \mu_i}{\partial p}\right)_{T,\,n_i,\,n_j} = RT \frac{\mathrm{d} \ln p}{\mathrm{d}p}$$

$$= \frac{RT}{p}. \tag{3·22}$$

But for any type of substance there is the general relation (2·111)

$$\left(\frac{\partial \mu_i}{\partial p}\right)_{T,\,n_i,\,n_j} = V_i,$$

and therefore $$V_i = \frac{RT}{p}. \tag{3·23}$$

† The symbol x_i is reserved for a mole fraction in a liquid or solid mixture.
‡ As in the case of equation (3·1), the equation (3·18) is more correctly written $$\mu_i = \mu_i^0 + RT \ln p/p^0 + RT \ln y_i,$$ where p^0 is unit pressure.

Now the total volume of the mixture is given by (2·107)

$$V = \Sigma n_i V_i,$$

and therefore, provided that (3·18) holds for all components, as is essential if the mixture is perfect,

$$V = \Sigma n_i \frac{RT}{p}$$

$$= \frac{nRT}{p}, \tag{3·24}$$

where n is the total amount (mols) of all species. The defining equation thus ensures that the mixture shall obey the gas law.

It may be noted from (3·23) that V_i is the same for all components and is equal also to V/n, by (3·23) and (3·24). The pure gas i at the pressure p would also have a volume per mole, v_i, equal to RT/p. Thus

$$V_i = V/n = v_i, \tag{3·25}$$

and this implies that there is no volume change when the separate gases, each at pressure p, are put together to form a mixture at total pressure p (and the same temperature).

The 'law of partial pressures' is obtained by eliminating p between equations (3·20) and (3·24),

$$p_i = y_i p = y_i n \frac{RT}{V} = n_i \frac{RT}{V}, \tag{3·26}$$

where n_i is the amount (mols) of species i. This is, of course, the same relation as would hold if the gas i were contained in the volume V in a pure state. Thus each component behaves as if no other gases were present; Gibbs remarks, 'It is in this sense that we should understand the law of Dalton, that every gas is as a vacuum to every other gas.' It arises because the molecules all move independently in the whole volume of the container.

It may be noted that (3·20) is true by definition and (3·26) is a deduction which holds for a perfect gas mixture. On the other hand, some authors use (3·26) as the defining equation for the partial pressure and thereby deduce (3·20). The choice is of no consequence as regards the perfect mixture, but in the case of imperfect mixtures the alternative definitions of partial pressure, (3·20) and (3·26), are not equivalent and would imply different numerical values.

(b) Membrane equilibrium. Consider two perfect gas mixtures, α and β, which are at the same temperature and separated from each other by means of a membrane. For each gas i which is able to pass through the membrane, the equilibrium relation is (equation (2·53))

$$\mu_{i\alpha} = \mu_{i\beta}.$$

Using (3·19), and noting that μ_i^0 depends only on temperature, we obtain

$$RT \ln p_{i\alpha} = RT \ln p_{i\beta}$$

or

$$p_{i\alpha} = p_{i\beta}. \qquad (3·27)$$

At equilibrium the gas i must therefore have the same partial pressure on each side of the membrane. It seems that this relation has not been very adequately confirmed, although Planck[†] has reported a partial verification using a hydrogen-palladium system. It is possible that a more complete test of (3·27) could be carried out by use of the remarkable zeolites which have been shown by Barrer[‡] to act as molecular sieves.

(c) *The enthalpy of mixing.* Equation (3·18) may be written

$$\frac{\mu_i}{T} = \frac{\mu_i^0}{T} + R \ln p + R \ln y_i,$$

and therefore

$$\left(\frac{\partial \mu_i / T}{\partial T} \right)_{p,\, n_i,\, n_j} = \frac{d\mu_i^0 / T}{dT}, \qquad (3·28)$$

Therefore, using (2·113)

$$-\frac{H_i}{T^2} = \frac{d\mu_i^0 / T}{dT}. \qquad (3·29)$$

The right-hand side is a complete differential, since μ^0 is a function of temperature only. The partial molar enthalpy H_i of any component i of the mixture is thus independent of composition and pressure. It follows that H_i remains unchanged as y_i is brought up to unity and in fact the right-hand side of (3·29) is equal to $-h_i/T^2$, where h_i is the enthalpy per mole of pure i at the temperature T (cf. equation (3·10)). Hence

$$H_i = h_i, \qquad (3·30)$$

and it follows that the total enthalpy, $\Sigma n_i H_i$, of the mixture is equal to the total enthalpy of the gases before mixing, provided that they are at the same temperature. *The heat of mixing is zero.*

Similarly it can be shown that

$$U_i = u_i, \qquad (3·31)$$

and there is no change of internal energy on mixing at constant temperature.

[†] Planck, *Treatise on Thermodynamics*, transl. Ogg (3rd ed.: London, Longmans Green, 1927), p. 218. Palladium at a white heat allows the easy passage of hydrogen, but not other gases.

[‡] Barrer, *Quart. Rev. Chem. Soc.* 3 (1949), 293, and later papers.

(d) The free energy and entropy of mixing. Using (2·80) and (3·19), the total Gibbs free energy of the mixture is

$$G_m = \Sigma n_i \mu_i = \Sigma n_i \mu_i^0 + RT\Sigma n_i \ln p_i. \tag{3·32}$$

Let it be supposed that before mixing the separate gases were each at the temperature T and that their pressures were p_1', p_2', etc. Then the value of G before mixing is obtained by summing (3·2) over the various gases and is

$$G = \Sigma n_i \mu_i^0 + RT\Sigma n_i \ln p_i' \tag{3·33}$$

Subtracting (3·33) from (3·32) we obtain $\Delta_m G$, the increase in Gibbs free energy on mixing:

$$\Delta_m G = RT\Sigma n_i \ln \frac{p_i}{p_i'}. \tag{3·34}$$

Since $\Delta_m G = \Delta_m H - T\Delta_m S$, and by the result of (3·30), $\Delta_m H$ is zero, we obtain

$$\Delta_m S = -\frac{\Delta_m G}{T} = -R\Sigma n_i \ln \frac{p_i}{p_i'}. \tag{3·35}$$

In the special case where $p_1' = p_2' = \ldots = p_n' = p'$, i.e. where the original pressures of each gas are all equal to the final pressure p' of the mixture (mixing at constant volume), we have

$$\frac{p_i}{p_i'} = \frac{p_i}{p'} = y_i, \tag{3·36}$$

and thus (3·34) and (3·35) reduce to

$$\left. \begin{array}{l} \Delta_m G = RT\Sigma n_i \ln y_i, \\ \Delta_m S = -R\Sigma n_i \ln y_i. \end{array} \right\} \tag{3·37}$$

It may be noted that $\Delta_m G$ is negative and $\Delta_m S$ is positive since all the y_i are fractions. This is in accordance with the essential irreversibility of the mixing process when it occurs at constant volume.

In another special case where the partial pressures in the mixture are all equal to the original pressures of the unmixed gases, i.e. where $p_i/p_i' = 1$, equations (3·34) and (3·35) reduce to

$$\Delta_m G = \Delta_m S = 0.$$

In this instance the changes in G and S due to the irreversible mixing process may be said to be exactly cancelled by equal and opposite effects due to the reduction in volume of the system, which is necessary if $p_i/p_i' = 1$ for all components.

3·4. Imperfect gases

Gases which are not perfect do not obey the relation (3·2) and consequently do not have an equation of state of the simple form $pv=RT$. Up to an atmosphere pressure or more the behaviour of most gases and vapours can be represented with fairly good accuracy by introducing into the equation an adjustable parameter B:

$$p(v-B)=RT. \tag{3·38}$$

The quantity B, which has the dimensions of a volume, is usually negative at low temperatures, but changes sign at high temperatures. For this reason it cannot be correctly interpreted as being proportional to the volume occupied by the molecules.

Larger deviations from the gas law can be represented by means of an equation containing a greater number of adjustable constants. A typical 'two-parameter' equation is that of van der Waals:

$$\left(p+\frac{a}{v^2}\right)(v-b)=RT, \tag{3·39}$$

where the term a/v^2 has been regarded as an 'internal pressure' arising from the attractive forces between the molecules. However, this interpretation of the parameter a is intuitive rather than exact and (3·39) does not have a sound theoretical basis.

It is interesting to note that a gas obeying the simpler equation (3·38) has an internal energy (but not an enthalpy) which is a function only of temperature, in any region where B is approximately constant. Thus using (2·76)

$$\left(\frac{\partial u}{\partial v}\right)_T=T\left(\frac{\partial p}{\partial T}\right)_V-p,$$

and applying this equation to (3·38)

$$\left(\frac{\partial u}{\partial v}\right)_T=\frac{RT}{v-B}-p=0. \tag{3·40}$$

On the other hand, for the van der Waals gas, as represented by (3·39), it is easily shown that

$$\left(\frac{\partial u}{\partial v}\right)_T=\frac{a}{v^2}.$$

Large deviations from the perfect gas law may be represented by means of the power series

$$p=\frac{RT}{v}(1+B/v+C/v^2+\ldots), \tag{3·41}$$

where B and C are called the second and third virial coefficients respectively and are in general functions of the temperature. Alternatively, the volume of the gas may be expressed as a power series in the pressure, and this is often more convenient for practical purposes.

3·5. The Joule–Thomson effect

A well-known experiment, first carried out by Joule and Thomson in the period 1852–62, consists in passing a steady stream of gas through a thermally insulated tube in which there is a throttle valve or porous plug. When the conditions are steady, let p_1 and T_1 be the pressure and temperature of the gas at one side of the plug and p_2 and T_2 be the corresponding values at the other side. Let h_1 and h_2 be the enthalpies per mole of the gas under the two sets of conditions. It follows from equation (2·8) that

$$h_1 = h_2,$$

provided that the small changes in the kinetic energy of the mass motion of the gas may be neglected.

In general there is a change in temperature $(T_1 \neq T_2)$ whenever the gas is imperfect. Since the expansion takes place at constant enthalpy, this temperature change is appropriately described by the *Joule–Thomson coefficient*, defined as

$$\mu \equiv \left(\frac{\partial T}{\partial p}\right)_h. \tag{3·42}$$

(The coefficient μ is not to be confused with the chemical potential.) Using the transformation formula

$$\left(\frac{\partial T}{\partial p}\right)_h \left(\frac{\partial h}{\partial T}\right)_p \left(\frac{\partial p}{\partial h}\right)_T = -1,$$

we obtain
$$\mu = -\frac{1}{c_p}\left(\frac{\partial h}{\partial p}\right)_T. \tag{3·43}$$

The magnitude of the Joule–Thomson effect is thus determined by $(\partial h/\partial p)_T$ and not by $(\partial u/\partial v)_T$.†

On account of the second law—the existence of entropy as a function of state—the Joule–Thomson coefficient can be related to other measurable properties of the gas. Thus from (2·95)

$$\left(\frac{\partial h}{\partial p}\right)_T = v(1 - \alpha T),$$

† *Problem.* Show that a gas obeying equation (3·38) with constant B has a Joule–Thomson coefficient which is not zero, but equal to

$$-B/c_p.$$

and therefore, from (3·43) $\mu = \dfrac{v}{c_p}(\alpha T - 1)$,   (3·44)

where α is the coefficient of expansivity.†

The Joule–Thomson experiment must be carefully distinguished from the adiabatic expansion of a gas in a piston and cylinder. The latter process normally leads to a cooling, on account of the decrease in internal energy consequent on the performance of work. On the other hand, the Joule–Thomson effect gives rise to a cooling only when μ is positive, i.e. when $\alpha T > 1$. In the case of a perfect gas μ is zero. In the case of imperfect gases μ is itself a function of temperature and pressure and is only positive in certain ranges of these variables. This sets a limit to the possibility of liquefying a gas by expansion through a throttle.

Fig. 17 shows the general form of the lines of constant enthalpy of a gas.‡ Along any one of the curves, H has the same value and the gradient at any point is therefore the value of μ, in accordance with (3·42). The broken line represents the locus of the maxima of the curves, and a cooling effect is therefore only obtainable within the region to the left of this locus. It is evident that, at any fixed pressure, μ has a positive value only between two limiting temperatures, called the *upper and lower inversion temperatures*. For example, in the case of nitrogen at 100 atm. pressure, these are about $+277$ and $-156\,°C$ respectively. With rise of pressure these temperatures gradually approach each other and become coincident at the point A of the broken curve. This point thus represents the *highest pressure* at which it is possible to obtain a cooling effect by steady flow through a throttle. In nitrogen this occurs at about 376 atm and 40 °C.

Similarly, the point B is the *highest temperature* at which a cooling may be obtained. This temperature has a value of about 350 °C in nitrogen, but in hydrogen it is -78 °C. In the latter gas the Joule–Thomson effect at room temperature thus gives rise to an increase of temperature. In order to liquefy hydrogen by expansion through a throttle, the gas must first be precooled to a temperature below -78 °C. This may be done either by use of liquid air or by adiabatic expansion of the hydrogen using a cylinder and piston.

† It may be noted that (3·44) contains the thermodynamic temperature T, together with other quantities all of which are measurable. The equation may thus be used for fixing the zero of the thermodynamic scale. In this way Roebuck and Murrell obtained the value -273.17 °C, in very close agreement with values obtained from gas thermometry. For further details the reader is referred to Zemansky, *Heat and Thermodynamics* (New York, McGraw-Hill, 1957), and to Guggenheim, *Thermodynamics* (Amsterdam, North-Holland Publ. Co., 1957).

‡ Only a small number of gases have been studied experimentally but most of them seem to have similar curves. See, for example, Roebuck and Osterberg, *Phys. Rev.* **48** (1935), 450.

3·6. The fugacity of a single imperfect gas

The perfect gas, as defined by equation (3·2), is one whose chemical potential, at constant temperature, is a linear function of the logarithm of its pressure. In the case of gases which are not perfect it is convenient to define a kind of fictitious pressure, called the *fugacity*, to which the chemical potential of the gas bears the same linear relationship.

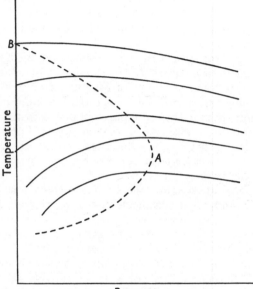

Fig. 17. Lines of constant enthalpy of a gas. N.B. The detail of the lower left-hand region has not been shown. Here the lines of constant enthalpy become very crowded and also the dotted curve intersects the vapour pressure curve of the liquid.

Let μ be the chemical potential of the pure gas at temperature T and pressure p. The fugacity f of the gas is defined by the following relations in which μ° is a function of T only:

$$\left.\begin{aligned} \mu &= \mu^0 + RT \ln f, \\ f/p &\to 1 \quad \text{as} \quad p \to 0. \end{aligned}\right\} \tag{3·45}$$

This limiting relation makes the fugacity equal to the pressure under conditions where the gas obeys the perfect gas law. Without this relation the definition would be incomplete.

The quantity μ^0 is clearly the chemical potential of the gas at unit fugacity, since in this case the logarithmic term vanishes and $\mu = \mu^0$. Now for most gases the ratio f/p does not differ appreciably from unity except at pressures greater than atmospheric. Thus, if the units of fugacity and of pressure are chosen as atmospheres, the value of μ^0 is practically the same as that of the chemical potential at unit pressure.

The fugacity is a useful function in the study of phase and reaction equilibrium, as will be shown in later chapters. It may also be used in the computation of the work of expansion or compression of a gas in a flow process. As discussed in §2·4, the maximum shaft work which may be obtained† during the steady isothermal flow of a gas is equal to the decrease of its Gibbs function:

$$w' = -(G_2 - G_1).$$

If the gas in question consists of a single component, then per mole

$$w' = -(\mu_2 - \mu_1), \tag{3·46}$$

and in the special case where the gas is perfect, the substitution of (3·2) in (3·46) gives‡

$$w' = RT \ln p_1/p_2. \tag{3·47}$$

If the gas is not perfect, the substitution of (3·45) in (3·46) gives an equation of the same form:

$$w' = RT \ln f_1/f_2. \tag{3·48}$$

The work of expansion or compression in isothermal steady-flow systems may thus be computed by use of tabulated values of f.

The fugacity is dependent on temperature and pressure, and an equation for its calculation from experimental data may be derived as follows. For any single component fluid

$$\left(\frac{\partial \mu}{\partial p}\right)_T = v,$$

where v is the molar volume. Therefore

† For reasons of clarity all of the work terms on this page refer to work done *by* the system, contrary to the I.U.P.A.C. convention of p. iv.

‡ This expression is the same as for the *total* work of isothermal expansion or compression of a perfect gas. Whether the gas is flowing or not, the total work done by a given mass of the gas is $w = \int_1^2 p \, dV$. This may also be written

$$w = \int_1^2 p \, dV = p_2 V_2 - p_1 V_1 - \int_1^2 V \, dp.$$

Consider the application of this equation to a steady flow system; the terms $p_2 V_2$ and $p_1 V_1$ represent the work at the inlet and outlet of the system and therefore the shaft work is

$$w' = -\int_1^2 V \, dp \; [= -(G_2 - G_1)].$$

In the special case of the perfect gas, $p_2 V_2 = p_1 V_1$ and therefore $w = w'$.

$$[d\mu = v dp]_T,$$

where the subscript denotes constancy of temperature. Similarly from (3·45)

$$[d\mu = RT\, d\ln f]_T,$$

and thus
$$[RT\, d\ln f = v dp]_T. \tag{3·49}$$

Subtracting $RT\, d\ln p$ from both sides of this equation and taking the constancy of temperature as being understood henceforth,

$$RT\, d\ln f/p = v dp - RT\, d\ln p$$
$$= \left(v - \frac{RT}{p}\right) dp,$$

or
$$d\ln f/p = \left(\frac{v}{RT} - \frac{1}{p}\right) dp.$$

Integrating at constant temperature from $p = 0$ to the particular pressure $p = p'$ at which it is required to calculate the fugacity, we obtain

$$\ln\left(\frac{f}{p}\right)_{p=p'} - \ln\left(\frac{f}{p}\right)_{p=0} = \int_0^{p'} \left(\frac{v}{RT} - \frac{1}{p}\right) dp,$$

and therefore, in view of the limiting relation contained in (3·45),

$$\ln\left(\frac{f}{p'}\right) = \int_0^{p'} \left(\frac{v}{RT} - \frac{1}{p}\right) d p. \tag{3·50}$$

This equation gives the fugacity at p' and T in terms of an integral which can be computed from experimental data. The equation may be expressed in a more convenient form by defining the *compressibility factor*

$$Z \equiv \frac{pv}{RT}, \tag{3·51}$$

and thus (3·50) becomes

$$\ln\frac{f}{p'} = \int_0^{p'} \frac{Z-1}{p}\, dp. \tag{3·52}$$

The ratio f/p' may therefore be evaluated either (a) by graphical integration† of $(Z-1)/p$ plotted against p, or (b) by direct integration, if v or Z is known as a power series in the pressure. For example, under conditions where (3·38) is a good approximation, the integration is particularly simple:

$$\ln\frac{f}{p'} = \frac{Bp'}{RT}. \tag{3·53}$$

† The convergency of the integral, and other important aspects of fugacity, are discussed by Tunell, *J. Phys. Chem.* **35** (1931), 2885.

In these integrations, the experimental data must, of course, refer to the same temperature T over the whole of the pressure range $0 \to p'$.

The ratio f/p may be denoted by χ,

$$\chi \equiv f/p, \tag{3.54}$$

and is sometimes called the *activity coefficient* of the gas; a better name is *fugacity coefficient*. The value of this ratio may be regarded conveniently as a function of the reduced pressure, p/p_c, and reduced temperature, T/T_c (where p_c and T_c denote the critical pressure and temperature, respectively), in accordance with the theory of corresponding states. Values of χ for some twenty-four gases have been tabulated by Newton,[†] and it was shown that the data could all be represented on a single graph, against the reduced temperature and pressure, with very fair accuracy.[‡]

The fugacity is not to be confused with another idealized pressure defined as $p_{\mathrm{id.}} \equiv RT/v$. At fairly low pressures, where (3·38) and (3·53) are valid and f/p does not differ very much from unity, it is readily shown that

$$f p_{\mathrm{id.}} \doteqdot p^2, \tag{3.55}$$

so that f and $p_{\mathrm{id.}}$ lie on opposite sides of the actual pressure p.

3·7. Fugacities in an imperfect gas mixture

For the ith component of an imperfect gas mixture, the fugacity f_i is chosen in such a way as to replace the partial pressure in equation (3·19). It is therefore defined by the following relations in which μ_i^0 is a function of T only:

$$\left. \begin{array}{l} \mu_i = \mu_i^0 + RT \ln f_i, \\ f_i/p_i \to 1 \quad \text{as} \quad p \to 0. \end{array} \right\} \tag{3.56}$$

In the imperfect mixture it is to be noted that both μ_i and f_i depend on the nature and quantities of the other species which are present, i.e. they are functions of *all* of the partial pressures, as well as of the temperature.

In equation (3·56) μ_i^0 has the same value as for the pure gas i at the same temperature (equation (3·45)). If the mole fraction

---

[†] Newton, *Industr. Engng Chem.* **27** (1935), 302. See also Hougen, Watson and Ragatz, *Chemical Process Principles*, Part II (New York, Wiley, 1959). Chapter XIV.

[‡] The principle of corresponding states has also been used extensively by Hougen, Watson and Ragatz (loc. cit. Chapter XIV). These authors give generalized charts which can be used for estimating the corrections to the enthalpy, entropy and heat capacity due to deviations from the perfect gas laws. For further discussion of the principle see Pitzer, *J. Chem. Phys.* **7** (1939), 583; Guggenheim, ibid. **13** (1945), 253; Guggenheim, *Thermodynamics*, 1957; Rowlinson, *Liquids and Liquid Mixtures* (Butterworth, 1959).

of this component is supposed to be progressively increased until it becomes equal to unity, the quantity μ_i^0 remains unchanged, because the fugacity has been defined in such a way that μ_i^0 is a function of temperature only. Therefore it is the chemical potential of pure i at unit fugacity. As noted in the last section μ_i^0 is not significantly different from the chemical potential at 1 atm pressure.

An equation for the calculation of f_i from experimental data may be derived as in the previous section. From (2·111)

$$\left(\frac{\partial \mu_i}{\partial p}\right)_{T,\,n_i,\,n_j} = V_i.$$

Therefore, for any change of pressure at *constant temperature and composition*,

$$\mathrm{d}\mu_i = V_i\,\mathrm{d}p.$$

Under the same conditions, from (3·56)

$$\mathrm{d}\mu_i = RT\,\mathrm{d}\ln f_i,$$

and therefore

$$RT\,\mathrm{d}\ln f_i = V_i\,\mathrm{d}p. \tag{3·57}$$

Subtracting $RT\,\mathrm{d}\ln p_i$ from both sides

$$RT\,\mathrm{d}\ln (f_i/p_i) = V_i\,\mathrm{d}p - RT\,\mathrm{d}\ln p_i$$

$$= V_i\,\mathrm{d}p - RT\,\mathrm{d}\ln p - RT\,\mathrm{d}\ln y_i$$

$$= \left(V_i - \frac{RT}{p}\right)\mathrm{d}p,$$

since y_i is constant.

Integrating at constant temperature and composition from $p = 0$ to the particular pressure $p = p'$ at which it is required to calculate the fugacity, and bearing in mind the limiting condition contained in (3·56) we obtain[†]

$$\ln (f_i/p_i)_{p=p'} = \int_0^{P'} \left(\frac{V_i}{RT} - \frac{1}{p}\right)\mathrm{d}p. \tag{3·58}$$

It may be noted that for a perfect gas mixture $V_i = RT/p$, according to (3·23). The integrand of (3·58) is therefore zero and the fugacity is equal to the partial pressure.

In order to calculate f_i in a mixture of chosen composition, temperature and pressure, experimental values of V_i must be available over the whole range of integration. For example, let it be supposed that the mixture in question consists of gases A and B only. At any particular temperature and pressure, the partial molar volumes V_A and V_B may be determined, as described in § 2·14, by measurements of the total volume of the gas in mixtures of several different compositions. A repetition at the same temperature and a lower

---

† Gillespie, *J. Amer. Chem. Soc.* **47** (1925), 305, 3106; *J. Amer. Chem. Soc.* **48** (1926), 28; *Chem. Rev.* **18** (1936), 359.

pressure will give a fresh set of values V_A and V_B, and so on over the pressure range. If we now pick out the values of V_A which are appropriate to a particular composition, f_A may be evaluated by graphical integration of (3·58). In this way a set of values of f_A and f_B may be obtained at any chosen temperature and total pressure. If these are consistent they must satisfy the Gibbs–Duhem equation. The temperature and pressure being the same for the particular set of fugacity values, this equation may be written

$$x_A \, \mathrm{d}\mu_A + x_B \, \mathrm{d}\mu_B = 0, \tag{3·59}$$

or using (3·56)

$$x_A \, \mathrm{d}\ln f_A + x_B \, \mathrm{d}\ln f_B = 0. \tag{3·60}$$

The procedure which has just been described requires such a large amount of experimental data on the p, V, T relations of mixed gases that it has actually been carried out in only a very small number of examples.† For this reason the fugacities of the components of a mixture are usually estimated by an approximation method to be described in § 3·9.

3·8. Temperature coefficient of the fugacity and standard chemical potential

The defining equation (3·56) may be rewritten

$$\frac{\mu_i^0}{T} = \frac{\mu_i}{T} - R \ln f_i.$$

Consider this equation as being applied to conditions of low pressure where the fugacity of the gas i becomes equal to its partial pressure. Then

$$\frac{\mu_i^0}{T} = \frac{\mu_i'}{T} - R \ln p_i',$$

where μ_i' and p_i' refer to the limiting conditions. Provided that the temperature is the same in each case, μ_i^0 has the same value in both equations, because it is a function of temperature only.

These equations are identities and can be partially differentiated with respect to temperature. Thus

$$\frac{\mathrm{d}\mu_i^0/T}{\mathrm{d}T} = \left(\frac{\partial \mu_i/T}{\partial T}\right)_{p,\,n_i,\,n_j} - R\left(\frac{\partial \ln f_i}{\partial T}\right)_{p,\,n_i,\,n_j}, \tag{3·61}$$

$$\frac{\mathrm{d}\mu_i^0/T}{\mathrm{d}T} = \left(\frac{\partial \mu_i'/T}{\partial T}\right)_{p,\,n_i,\,n_j} - R\left(\frac{\partial \ln p_i'}{\partial T}\right)_{p,\,n_i,\,n_j}. \tag{3·62}$$

† Argon and ethylene: Gibson and Sosnick, *J. Amer. Chem. Soc.* **49** (1927), 2172. Nitrogen and hydrogen: Merz and Whittaker, *J. Amer. Chem. Soc.* **50** (1928), 1522; also Bennett and Dodge, *Industr. Engng Chem.* **44** (1952), 180. Fairly complete thermodynamic data on the system N_2—H_2—NH_3 has been obtained by Michels *et al.*, *Appl. Sci. Res.* A, **3** (1951), 1.

The final term in the second of these equations is zero. Substituting from (2·113)

$$\frac{\mathrm{d}\mu_i^0/T}{\mathrm{d}T} = -\frac{H_i}{T^2} - R\left(\frac{\partial \ln f_i}{\partial T}\right)_{p,\,n_i,\,n_j}, \tag{3·63}$$

$$\frac{\mathrm{d}\mu_i^0/T}{\mathrm{d}T} = -\frac{h_i^0}{T^2}. \tag{3·64}$$

In (3·63) H_i is the partial molar enthalpy in the imperfect mixture where the fugacity is f_i. In (3·64) h_i^0 is the partial molar enthalpy in the mixture at low pressure and is equal to the enthalpy per mole of pure i at low pressure and the same temperature T (§ 3·3c). Between equations (3·63) and (3·64) we obtain

$$\left(\frac{\partial \ln f_i}{\partial T}\right)_{p,\,n_i,\,n_j} = \frac{h_i^0 - H_i}{RT^2}. \tag{3·65}$$

The pressure coefficient of f_i may be obtained in a similar manner:

$$\left(\frac{\partial \ln f_i}{\partial p}\right)_{T,\,n_i,\,n_j} = \frac{V_i}{RT}. \tag{3·66}$$

3·9. Ideal gaseous solutions and the Lewis and Randall rule

The perfect gas mixture has been defined in accordance with equation (3·18)

$$\mu_i = \mu_i^0 + RT \ln p + RT \ln y_i, \tag{3·67}$$

where μ_i^0 is a function of temperature only. The chemical potential is thus a linear function of the logarithm of the total pressure, and it is because of this that the mixture obeys the equation $pV = nRT$.

A much less restrictive model of a gas mixture is one which obeys the equation

$$\mu_i = \mu_i^* + RT \ln y_i, \tag{3·68}$$

where μ_i^* is a function both of temperature and pressure but is not necessarily equal to $\mu_i^0 + RT \ln p$ in (3·67).† In (3·68) only the dependence on composition is made explicit—in particular, the chemical potential of i depends only on its own mole fraction y_i, and not on the mole fractions of any other components, at constant total pressure and temperature.

A gas mixture for which (3·68) is valid for *every* component may be called an *ideal gaseous solution*.‡ As will be shown below, certain

---

† Standard chemical potentials which depend only on temperature are denoted μ_i^0, and those which depend both on temperature and pressure are denoted μ_i^*.

‡ As shown in Chapter 8, the equation (3·68) is also the defining equation for ideal liquid and solid solutions. This does not imply, of course, that the $p \cdot V \cdot T$ relations are the same for the three states of aggregation; it is only the

mixtures obey this relation fairly accurately at high pressures where they do not obey the gas law, $pV = nRT$. Such mixtures also have the property of having a zero volume change of mixing and a zero enthalpy of mixing at constant temperature and pressure (provided that (3·68) remains valid over the whole range of composition). Thus a partial differentiation of (3·68) gives

$$\left(\frac{\partial \mu_i}{\partial p}\right)_{T, n_i, n_j} = \left(\frac{\partial \mu_i^*}{\partial p}\right)_T.$$

The left-hand side is V_i and, since the right-hand side is independent of composition, it follows that the partial molar volume is likewise independent of composition. V_i is therefore equal to v_i, the volume per mole of pure i at the same temperature and pressure,

$$V_i = v_i. \tag{3·69}$$

By a similar argument $\qquad H_i = h_i. \tag{3·70}$

The equation (3·68) also has an important consequence in regard to fugacities in the mixture. Combining (3·56) and (3·68)

$$RT \ln \frac{f_i}{y_i} = \mu_i^* - \mu_i^0. \tag{3·71}$$

Now the right-hand side is independent of composition, and therefore the ratio f_i/y_i must remain unchanged as y_i is brought up to unity, i.e.

$$\frac{f_i}{y_i} = f'_i,$$

or $\qquad f_i = y_i f'_i, \tag{3·72}$

where f'_i is the fugacity of pure component i at the same temperature and total pressure as the mixture. (This equation may also be obtained by noting the identity of (3·50) and (3·58) when $v_i = V_i$.)

Equation (3·72) is also known as Lewis and Randall's rule: in the ideal gaseous solution 'the fugacity of each constituent is equal to its mol fraction multiplied by the fugacity which it would exhibit as a pure gas, at the same temperature and the same total pressure'.[†] Thus the problem of knowing the fugacity of a component of a mixture reduces to the much simpler problem of knowing the fugacity of the same gas in the pure state. As noted previously, extensive tabulations of single gas fugacities have been given by Newton.[‡]

composition dependence of μ_i which is the same. The molecular conditions which give rise to the ideal solution (in whatever state of aggregation) will be discussed in Chapter 8.

[†] Lewis and Randall, *Thermodynamics* (New York, McGraw-Hill, 1923), Chapter XIX.

[‡] Newton, *Industr. Engng Chem.* 27 (1935), 302.

The applicability of the ideal solution model (or of the Lewis and Randall rule which is equivalent) may be tested on those few mixtures whose true fugacities have been obtained directly from (3·58). In Gibson and Sosnick's[†] studies on argon-ethylene mixtures it was found that fugacities calculated by means of the rule were not in error by more than 20 %, up to a pressure of 50 atm. However, at a pressure of 100 atm the error was much larger and was as much as 100 % in certain mixtures.[‡] The examination of the data on the N_2—H_2 system by Merz and Whittaker[§] shows that this mixture obeys the Lewis and Randall rule fairly closely and the errors do not exceed 20 % at 1000 atm. It may be remarked that an alternative approximation for estimating the fugacities in mixtures has been put forward by Joffe.[‖]

PROBLEMS

1. Derive equation (3·37) for the entropy of mixing, as applied to two gases only, by carrying out a reversible mixing process in the apparatus shown in the diagram. Piston 1 is permeable only to gas A and piston 2 is permeable only to gas B. Initially the two pistons are in contact, so that gas A is entirely to the left and gas B is entirely to the right as shown. Finally, the two pistons are at the ends of the cylinder; the gases have thus been mixed by passage through the pistons into the space between them.

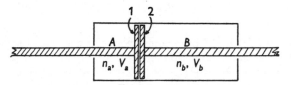

2. Show that the partial pressure p_i of a component i of the earth's atmosphere at a height h above the earth is given by

$$p_i = p_{i0}\, e^{-Mgh/RT},$$

where p_{i0} is the partial pressure at $h=0$, and M is the molecular weight. What conditions limit the applicability of the equation?

[†] Gibson and Sosnick, *J. Amer. Chem. Soc.* **49** (1927), 2172.
[‡] Ethylene seems to have a fairly general tendency to give rise to large deviations from ideality in mixtures with other gases. This may be due to the formation of association complexes arising from the π electrons. Another instance of this is discussed in §6·5.
[§] Merz and Whittaker, *J. Amer. Chem. Soc.* **50** (1928), 1522.
[‖] Joffe, *Industr. Engng Chem. Soc.* **40** (1948), 1738.

3. It is desired to make a preliminary very rough estimate for a plant producing $790 \text{ m}^3 \text{h}^{-1}$ of nitrogen and $210 \text{ m}^3 \text{h}^{-1}$ of oxygen, each at 1 atm pressure and 20 °C, from air at the same temperature. Neglecting the presence of the rare gases in the atmosphere, calculate the minimum amount of power required to operate the process and the corresponding amount of heat to be removed.

4. Obtain the critical volume, critical temperature and critical pressure in terms of the coefficients a, b and R of the van der Waals equation. Obtain a form of the van der Waals equation containing only the reduced pressure, temperature and volume together with other pure numbers.

5. Using the van der Waals equation estimate a value for the Joule–Thomson coefficient of nitrogen at 1 atm and 0 °C given the following data:

$T_c = -147$ °C, $P_c = 3450$ kPa, $c_p = 1.031 \text{ J K}^{-1}\text{g}^{-1}$ (at 1 atm and 0 °C).

6. Calculate the fugacity of nitrogen at -100 °C and pressures of 10 and 50 atm, using the following values of pV/nRT at -100 °C:

p/atm	0	14	22	30	44	58	68
$Z = pV/nRT$	1.00	0.95	0.92	0.89	0.84	0.79	0.76

7. Nitrogen is to be compressed from 10 to 50 atm at the rate of 56 kg h^{-1}. Estimate the minimum power required for isothermal compression at -100 °C (*a*) if the nitrogen behaves as a perfect gas; (*b*) allowing for deviations from the gas laws. Why is the work less in the second case and would this always be true for a non-perfect gas?

8. At 20 °C the value of pv for oxygen (in appropriate units) may be approximated at pressures up to 100 atm by the equation

$$pv = 1.074 \, 25 - 0.753 \times 10^{-3} p + 0.150 \times 10^{-5} p^2,$$

where p is in atmospheres. Calculate the fugacity of oxygen at 20 °C and 100 atm.

9. An ideal gas is expanded from state 1 to state 2. If γ, the ratio of the specific heats, may be assumed constant, show that

$$p_1 v_1^{\gamma} \, e^{-S_1/C_V} = p_2 v_2^{\gamma} \, e^{-S_2/C_V},$$

where S_1 and S_2 are the entropies in the two states.

Show that the curve $pv^{\gamma} = \text{constant}$ divides the pv diagram into two regions, one of which represents states which cannot be attained in any adiabatic expansion or compression commencing from an initial state represented by a point lying on this curve.

[C.U.C.E. Qualifying, 1951]

10. Show that

$$C_p - C_V = V\left(\frac{\partial p}{\partial T}\right)_V + \left(\frac{\partial H}{\partial V}\right)_T \left(\frac{\partial V}{\partial T}\right)_p$$

$$= \left[V - \left(\frac{\partial H}{\partial p}\right)_T\right]\left(\frac{\partial p}{\partial T}\right)_V = T\left(\frac{\partial V}{\partial T}\right)_p \left(\frac{\partial p}{\partial T}\right)_V,$$

and distinguish which of the equalities is conditional on the temperature being measured on an absolute scale.

For a certain gas $c_p - c_V = R$, while at constant volume p is proportional to T. Obtain the general form for the equation of state of the gas.

[C.U.C.E. Qualifying, 1951]

11. (a) Show that

$$\left(\frac{\partial T}{\partial p}\right)_s = \frac{T}{C_p}\left(\frac{\partial V}{\partial T}\right)_p.$$

(b) Isentropic changes of state of a certain gas are represented by equations of the form $p^\alpha = \beta T$, where α is a constant and β is a function of the initial conditions only. If C_p is assumed constant show that the general form of the equation of state of the gas is

$$p[V - f(p)] = \alpha C_p T.$$

[C.U.C.E. Qualifying, 1953]

CHAPTER 4

EQUILIBRIA OF REACTIONS INVOLVING GASES

4·1. Introduction

The subject-matter which is included under the above title is actually more comprehensive than it might seem, because a reaction which involves liquid or solid substances may be regarded, from a thermodynamic standpoint, as a reaction between the *vapours* of these substances. The condition of reaction equilibrium in the vapour phase, together with the conditions of physical equilibrium between the vapour and the condensed phases, will clearly imply reaction equilibrium between the latter phases themselves. The advantage of this point of view is that whenever the vapour phase is approximately perfect, the equilibrium constant for this phase can be expressed in a particularly simple form.

4·2. The stoichiometry of chemical reaction

In the earlier part of this chapter it will be supposed that there is only a single independent stoichiometric process in the system of interest; the extension of the theory to the case where there is more than one will be given in §4·17. But first we must consider what is meant by the *number* of such processes.

Consider a system containing hydrogen, chlorine and hydrogen chloride molecules, together with hydrogen and chlorine atoms in low concentration. Between these species it is seen on inspection that there are *three* independent stoichiometric equations. These may be chosen as

$$H_2 = 2H, \tag{A}$$

$$Cl_2 = 2Cl, \tag{B}$$

$$H_2 + Cl_2 = 2HCl, \tag{C}$$

and the form of these equations is determined by the fact that *there is conservation of atoms in the chemical processes.*†

† This is no longer true of nuclear reactions. The way in which the stoichiometric relations are written is always based on a knowledge of what types of particles are conserved in the particular process.

Of course other equations can also be written but all of them are simply linear combinations of (A), (B) and (C). For example, the equation

$$Cl + H_2 = HCl + H \qquad \text{(D)}$$

may be obtained by adding (C) to (A), followed by subtraction of (B), rearrangement and division by two. Therefore there are only three stoichiometric processes between the species of interest. These may be chosen quite arbitrarily; thus (A), (B) and (D) is an equally valid choice. The important point is that three is the number of such equations which are *independent*.†

The same point can perhaps be made more clearly without the use of chemical equations; in some respects they tend to obscure the issue. Let Δn_{H_2}, Δn_H, etc., be the changes in the amounts (e.g. as mols) of H_2, H etc., during a certain period of time. The conservation of the two types of atom, H and Cl, implies the following relations between the five Δ's:

$$\Delta n_H + \Delta n_{HCl} = -2\Delta n_{H_2},$$

$$\Delta n_{Cl} + \Delta n_{HCl} = -2\Delta n_{Cl_2}.$$

Inspection of these equations shows that the fixing of three of the Δ's, say Δn_{H_2}, Δn_{Cl_2} and Δ_{HCl}, will determine the other two. In brief the amounts of only three substances can be varied arbitrarily, and it is this, and only this, which is meant by the statement that there are three independent stoichiometric processes.

On the other hand, if the concentrations of hydrogen and chlorine atoms were regarded as being too small to be of interest in a particular problem, then Δn_H and Δn_{Cl} are effectively zero and the above relations reduce to

$$\Delta n_{HCl} = -2\Delta n_{H_2} = -2\Delta n_{Cl_2},$$

and therefore the change in any one of these variables determines the other two. This is equivalent to the statement that there is now only one significant stoichiometric process, namely, (C) above.

A useful notation is as follows. The chemical equation

$$2A + 3B = 4C \qquad \text{(E)}$$

may also be represented

$$0 = 4C - 2A - 3B,$$

or, more generally, as

$$0 = \nu_a M_a + \nu_b M_b + \nu_c M_c,$$

where M_a, etc., are the chemical symbols of the various species and ν_a, etc., are *stoichiometric coefficients*. These are negative for those substances which are on the left of the equation as usually written and are positive for those substances which are on the right. For

† On the other hand, the number of elementary reactions which is needed in order to represent the *kinetic mechanism* of the process may be larger than three. For example, it might occur that all four of the reactions (A) to (D) have significant rates.

example, in the equation (E) the coefficients are $\nu_a = -2$, $\nu_b = -3$ and $\nu_c = 4$, and the stoichiometry is expressed by

$$\frac{\Delta n_a}{\nu_a} = \frac{\Delta n_b}{\nu_b} = \frac{\Delta n_c}{\nu_c}.$$

Quite generally a single stoichiometric process may be represented equally well by the *chemical* equation

$$0 = \sum \nu_i M_i, \tag{4·1}$$

or by the *algebraic* equations

$$\frac{\Delta n_a}{\nu_a} = \frac{\Delta n_b}{\nu_b} = \ldots = \frac{\Delta n_m}{\nu_m}.$$

The summation in (4·1), or this set of equations, may be taken as extending over *all* species present in the system; however, for any which are inert, the corresponding ν is zero.

It may be remarked that the set of equations only represents the changes in the amounts of substances if the system in question is a closed one; if there is an auxiliary inflow or outflow of substances the equations apply merely to the conservation of atoms in the chemical changes, and auxiliary equations must be written for the conservation in the transfer processes.

A useful variable is the *extent of reaction*, ξ. Since the quotients $\Delta n_i/\nu_i$ are equal for all species, the set of equations may be written:

$$\frac{\Delta n_a}{\nu_a} = \frac{\Delta n_b}{\nu_b} = \ldots \frac{\Delta n_m}{\nu_m} = \Delta\xi, \tag{4·2}$$

where $\Delta\xi$ is called the change in extent of reaction. Alternatively ξ may be defined as $\xi = (n_i - n_{i0})/\nu_i$, where n_{i0} is some reference amount of species i, e.g. the amount at the commencement of reaction.

4·3. Preliminary discussion on reaction equilibrium

Consider the reaction $A + B = 2C$, and let it be supposed that at a given moment during the process n_a, n_b and n_c are the amounts (mols) of A, B and C respectively. The Gibbs function of the system is

$$G = n_a\mu_a + n_b\mu_b + n_c\mu_c.$$

Our present purpose is to give an illustration of the way in which G varies during the course of the reaction and in particular to show that it has a minimum value at a certain composition.

For simplicity it will be supposed that the substances are present as a perfect gas mixture. Using (3·18) the equation above may therefore be rewritten

$$G = [n_a\mu_a^0 + n_b\mu_b^0 + n_c\mu_c^0 + RT(n_a + n_b + n_c)\ln p]$$
$$+ RT(n_a \ln y_a + n_b \ln y_b + n_c \ln y_c),$$

where p is the total pressure and the y's are the mole fractions. Now the square bracket is the value of the total Gibbs function of the three gases if each existed in a separate vessel at the pressure p. The remaining term on the right-hand side is therefore a *free energy of mixing* (cf. equation 3·37).

If the system before reacting consisted of, say, 1 mol each of A and B, then at the given moment

$$n_b = n_a,$$

$$n_c = 2(1 - n_a).$$

Therefore n_b and n_c can be eliminated from the last equation to give

$$G = [n_a(\mu_a^0 + \mu_b^0) + 2(1 - n_a)\,\mu_c^0 + 2RT \ln p]$$
$$+ 2RT[n_a \ln \tfrac{1}{2}n_a + (1 - n_a)\ln(1 - n_a)].$$

In this expression the μ^0's are properties of the pure components, and, if the temperature is constant, they do not change during the course of the reaction. Thus, at constant temperature and pressure, G is a function of the single variable n_a, which is capable of variation between 1 and 0.

If the pressure is chosen as being, say, unity, the equation can be rearranged to give

$$G - 2\mu_c^0 = n_a(\mu_a^0 + \mu_b^0 - 2\mu_c^0) + 2RT[n_a \ln \tfrac{1}{2}n_a + (1 - n_a)\ln(1 - n_a)]. \quad (4·3)$$

The left-hand side is the amount by which the Gibbs free energy of the reaction system exceeds that of two mols of C, when there are still n_a mols of A left in the system. Of course this falls to zero when $n_a = 0$, corresponding to complete conversion to C.†

It follows that $(G - 2\mu_c^0)$ can be plotted as a function of n_a for a given value of the quantity $(\mu_a^0 + \mu_b^0 - 2\mu_c^0)$, which can be determined experimentally. A typical plot is the curve PTS of Fig. 18, and it will be seen that there is a considerable range of compositions for which the value of $(G - 2\mu_c^0)$ is negative, i.e. *compositions for which the free energy of the mixture is even smaller than if there were complete conversion into the reaction product.*

The characteristic shape of this curve is due to the second term in equation (4·3) which, it has been mentioned already, represents a free energy of mixing. (In fact, if this term were not present the value of $(G - 2\mu_c^0)$ would be given by the straight line RS and would thus fall continuously with increasing degree of reaction.)

The important point is that the curve of the Gibbs function of a

† It may be noted that when $n_a = 1$, the right-hand side of (4·3) has the value $\mu_a^0 + \mu_b^0 - 2\mu_c^0 + 2RT \ln 0·5$. The last term, which has a negative value, is due to the free energy of mixing of A with B before any reaction has occurred. The student is advised to work out Problem 1 on p. 177.

homogeneous phase in which there is a reaction always shows a minimum; its existence is due to the fact that the creation of a *mixture* is an irreversible process and gives rise to a free-energy decrease of mixing. For let it be supposed that A and B were *completely* converted into C. The end-state of the system would therefore correspond to point S on the diagram. However, as soon as this complete conversion had taken place, it would be possible for the value of G to fall to an even lower value T, if some of the C were to dissociate again, on account of the free energy of mixing of C with the A and B which would result from this dissociation.

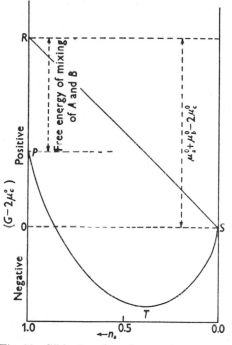

Fig. 18. Gibbs function of a reacting system.

As discussed in §§2·4 and 2·8 the most stable state of a system, if it is held at constant temperature and pressure, is the one for which G has its least value. In the example under discussion the equilibrium composition therefore corresponds to the point T. The position of this minimum is determined by the value of the term $(\mu_a^0 + \mu_b^0 - 2\mu_c^0)$ in equation (4·3) which represents the difference of free energies of reactants and products in their pure states. On the other hand, what might be called the 'driving force' of reaction, tending to make the system approach this minimum, is determined partly by this difference but also by the additional free energy of mixing, as discussed above.

In all cases the lowest value of G never occurs either at absolutely zero or at absolutely complete conversion. The term 'irreversible' as applied to a reaction is merely descriptive and implies that a reaction proceeds to such a degree that the residual amounts of unchanged reactants are almost immeasurably small. In such a case the minimum ‚value of G could not be represented at all conveniently on a diagram.

The position of the minimum may be obtained by differentiation of (4·3) and by putting

$$\left(\frac{\partial G}{\partial n_a}\right)_{T,\,p} = 0. \tag{4·4}$$

A more general but entirely equivalent procedure for finding this minimum is as follows. For the reaction system as a whole

$$dG = -S\,dT + V\,dp + \mu_a\,dn_a + \mu_b\,dn_b + \mu_c\,dn_c.$$

On account of the stoichiometry of the reaction dn_b and dn_c may be expressed in terms of dn_a:

$$-dn_a = -dn_b = \tfrac{1}{2}dn_c,$$

and therefore

$$dG = -S\,dT + V\,dp + (\mu_a + \mu_b - 2\mu_c)\,dn_a. \tag{4·5}$$

At constant temperature and pressure, G has its minimum value when $\partial G/\partial n_a = 0$ and therefore when

$$\mu_a + \mu_b = 2\mu_c. \tag{4·6}$$

This is therefore the condition of reaction equilibrium and is seen to be of the same form as the chemical equation

$$A + B = 2C,$$

with the chemical symbols of the compounds replaced by their chemical potentials.

It will be noted that the derivation of (4·6) does not imply any assumption with regard to the system being a perfect gas mixture; it is a condition of equilibrium which is applicable to any state of aggregation. However, if it is now assumed that each of the substances A, B and C has a chemical potential of the form appropriate to a perfect gas mixture, then it may be readily verified that (4·6) gives rise to precisely the same equilibrium composition as is obtained by minimization of (4·3).

Prior to the attainment of equilibrium if

$$\mu_a + \mu_b > 2\mu_c,$$

then it is evident from (4·5) that $\partial G/\partial n_a$ is positive or $\partial G/\partial n_c$ is negative. The formation of C therefore causes the Gibbs function of the system to diminish, and the reaction thus proceeds from left to right of the chemical equation. Conversely, if the direction of the inequality is changed, the reaction proceeds from right to left.

In conclusion, it may be useful to give fresh emphasis to remarks already made in §2·8 to the effect that a minimum value of G (or F) implies a compromise between the effect of the attractive forces on the one hand and of 'randomness' on the other. Since $G = H - TS$ a low value of G will occur if the value of H is *small* and if the value of TS is *large*. Consider in this connexion the simple reaction

$$Cl_2 = 2Cl.$$

Now small values of H will be attained when there is little or no dissociation. This is because the combination of chlorine atoms to form molecules is an intrinsically exothermic process, on account of the attractive forces. On the other hand, the largest value of the entropy will occur when there is almost complete dissociation; this gives rise to two particles having random translational motion in place of the single original one.†

The equilibrium degree of dissociation, which minimizes $H - TS$, thus occurs as a compromise between the effect of the molecular forces, which favour the formation of Cl_2 with decrease of enthalpy, and the tendency of the system to attain maximum 'spread', which favours the existence of free atoms.

Moreover, the effect of an increase of temperature will clearly be to increase the degree of dissociation. The temperature rise is equivalent to an increase in the energy content of the system, and this increment of energy will be most completely 'randomized' within the system if, in addition to a mere heating of the two species, it causes reaction to be displaced in the direction corresponding to an uptake of energy, i.e. towards chlorine atoms. Thus the increase of temperature causes the composition to change in the endothermic direction (cf. Le Chatelier's principle).

4·4. Concise discussion on reaction equilibrium

In the previous section the special case of the reaction $A + B = 2C$ was discussed. We shall continue to suppose that only a single reaction takes place in the system, but it will now be written in the generalized form

$$0 = \Sigma \nu_i M_i, \tag{4·7}$$

† This effect is counterbalanced to some extent by the disappearance of the randomness of the vibrations and rotations of the chlorine molecules. However, the energy levels for the latter types of motion are much more widely spaced than for translation and therefore a given amount of energy can be distributed over them in a much smaller number of ways.

as discussed in §4·2. This equation implies the algebraic equalities:

$$\frac{\mathrm{d}n_a}{\nu_a} = \frac{\mathrm{d}n_b}{\nu_b} = \ldots = \frac{\mathrm{d}n_m}{\nu_m} = \mathrm{d}\xi \tag{4·8}$$

It follows that the change of G in the system can be written

$$\mathrm{d}G = -S\mathrm{d}T + V\mathrm{d}p + \mu_a\mathrm{d}n_a + \ldots \mu_m\mathrm{d}n_m$$

$$= -S\mathrm{d}T + V\mathrm{d}p + (\nu_a\mu_a + \ldots \nu_m\mu_m)\mathrm{d}\xi$$

and therefore

$$(\partial G/\partial\xi)_{T,p} = \Sigma\nu_i\mu_i \tag{4·9}$$

The negative of the quantity $\Sigma\nu_i\mu_i$ has been called by de Donder the *affinity* of the reaction. As is evident from the discussion near the end of §4·3, the affinity must be positive if the reaction is to proceed spontaneously from left to right of the chemical equation as usually written, and it must be negative for reaction in the reverse direction.

The condition of chemical equilibrium is obtained by minimizing G with respect to ξ, at constant temperature and pressure, i.e. by putting

$$\Sigma\nu_i\mu_i = 0 \tag{4·10}$$

This is the most general condition of the equilibrium of a single reaction and is applicable whether the reactants and products are solids, liquids or gases. It is to be noted that it is obtained from the chemical equation (4·7) simply by replacing each chemical formula by the corresponding chemical potential.

4·5. The equilibrium constant for a gas reaction

We consider first of all a reaction in a perfect mixture.† For each component of such a mixture, as in equation (3·19),

$$\mu_i = \mu_i^0 + RT\ln p_i,$$

where μ_i^0 is a function of temperature only. Substituting in (4·10) we obtain

$$-RT\Sigma\ln p_i^{\nu_i} = \Sigma\nu_i\mu_i^0,$$

or‡

$$-RT\ln \prod_i p_i^{\nu_i} = \Sigma\nu_i\mu_i^0. \tag{4·11}$$

† As shown in §3·3c, in a perfect mixture there is zero enthalpy of mixing and this may seem incompatible with the existence of a heat of reaction. However, the zero enthalpy of mixing applies only when there is conservation of the mole numbers; thus in the reacting system the result (3·30), $H_i = h_i$, remains true, but the total enthalpy of the system, $H = \Sigma n_i H_i$ changes during the reaction on account of changes in the n_i. Only if the reactants and products were mixed in their equilibrium proportions would there be no heat effect.

‡ The symbol $\prod_i$ stands for a continued product. Thus $\prod_{i=1}^{i=4} x_i = x_1 x_2 x_3 x_4$.

If we define
$$K_p \equiv \prod_i p_i^{\nu_i},$$
(4·12)

then (4·11) may be written
$$-RT \ln K_p = \Sigma \nu_i \mu_i^0.$$
(4·13)

The right-hand side of this equation is a function only of temperature. Therefore the quantity K_p, called the *equilibrium constant* of the reaction, is also a function only of temperature. In brief the above theorem shows that there exists a certain function of the equilibrium partial pressures which has a constant value, at a given temperature, whatever are the individual values of these partial pressures.

For example, in the reaction $A + B = 2C$ we have $\nu_a = -1$, $\nu_b = -1$ and $\nu_c = +2$, and therefore from (4·11)†

$$-RT \ln K_p \equiv -RT \ln \frac{p_c^2}{p_a p_b}$$
$$= 2\mu_c^0 - \mu_a^0 - \mu_b^0.$$
(4·14)

It may be noted that the equilibrium constant, as conventionally defined, contains in the numerator the partial pressures of those substances which are on the right-hand side of the chemical equation, as usually written, and in the denominator the partial pressures of those substances which are on the left-hand side of this equation. Notice also that *equilibrium constants are always dimensionless* and this is so even in the case of reactions such as $A + B = C$ where the number of molecules changes. This is because of (3·1); the standard pressure p^0 has subsequently been taken as unity and has therefore not been included explicitly in the equilibrium constant; in fact each p_i in K_p is really a ratio p_i/p^0.

The significance of K_p may perhaps be seen more clearly when it is brought to mind that there is an infinite number of sets of chemical potentials and partial pressures which satisfy the condition of reaction equilibrium at a given temperature Let μ_i and p_i be members of one such set and let μ_i' and p_i' be members of another set. For the first set
$$\Sigma \nu_i \mu_i = 0,$$

and for the second
$$\Sigma \nu_i \mu_i' = 0.$$

Thus
$$\Sigma \nu_i \mu_i = \Sigma \nu_i \mu_i',$$

and therefore, using (3·19),
$$RT\Sigma \nu_i \ln p_i = RT\Sigma \nu_i \ln p_i',$$

or
$$\prod_i p_i^{\nu_i} = \prod_i p_i'^{\nu_i} = \prod_i p_i''^{\nu_i} = \ldots = K_p.$$

† The student should verify that (4·14) is equivalent to the minimizing of (4·3), when the partial pressures are expressed in terms of mole numbers. See Problem 1 on p. 177.

This shows, perhaps more clearly than in (4·11), that the product of any set of equilibrium partial pressures, each to the power of the appropriate stoichiometric coefficient, is a constant at any particular temperature.

Turning now to the case of the imperfect mixture the chemical potentials must be expressed in terms of fugacities in place of partial pressures. Using (3·56) $\mu_i = \mu_i^0 + RT \ln f_i,$

and substituting in the general condition of equilibrium (4·10), we obtain

$$-RT \ln K_f = \Sigma \nu_i \mu_i^0, \qquad (4·15)$$

where

$$K_f \equiv \prod_i f_i^{\nu_i}. \qquad (4·16)$$

For reactions in imperfect mixtures it is therefore K_f, rather than K_p, which is the true equilibrium constant, in the sense of being a function of temperature only. It is, of course, entirely permissible to continue to define

$$K_p \equiv \prod_i p_i^{\nu_i},$$

but it is a quantity which is no longer independent of pressure or composition. The numerical difference between K_p and K_f is often quite large at elevated pressures, as in ammonia synthesis.

Consider now the quantity $\Sigma \nu_i \mu_i^0$ which occurs in (4·13) and (4·15). As noted in §§3·3 and 3·7, the μ^0's are the free energies per mole of the pure components at unit pressure (or at unit fugacity in the case of the imperfect mixture). The quantity $\Sigma \nu_i \mu_i^0$ may thus be called the *standard change of free energy* in the reaction at the temperature T. It is often denoted ΔG_T^0, $\Sigma \nu_i \mu_i^0 \equiv \Delta G_T^0.$ (4·17)

For example, if the reaction is

$$A + B = 2C,$$

ΔG^0 has the significance

$$\Delta G_T^0 \equiv 2\mu_c^0 - \mu_a^0 - \mu_b^0.$$

where $2\mu_c^0$ is the Gibbs free energy of two moles of C at unit pressure, etc. ΔG_T^0 is therefore the increase in G when 1 mole of A and 1 mole of B, each at unit pressure, are *completely converted* into two moles of C, at unit pressure, at the particular temperature under discussion.

Collecting together equations (4·13), (4·15) and (4·17),

$$-RT \ln K_p = \Sigma \nu_i \mu_i^0 \equiv \Delta G_T^0, \qquad (4·18)$$

$$-RT \ln K_f = \Sigma \nu_i \mu_i^0 \equiv \Delta G_T^0. \qquad (4·19)$$

The equilibrium constants of all reactions between a set of compounds can therefore be calculated if the relative values of the μ^0's of these compounds are known. In any of these reactions a high degree of conversion from left to right corresponds to a value of K_p (or K_f)

which is large compared to unity and, in such cases, ΔG_T^0 must be negative. But even when ΔG_T^0 is positive there is always *some* conversion of the reactants, as discussed previously.

The relative values of μ^0 are usually tabulated in the literature at a standard pressure of 1 atm. With this convention it follows that the partial pressures which appear in K_p should *also be expressed in atmospheres*.† In the case of an imperfect gas mixture the quantity μ^0, as discussed in §3·7, actually refers to unit fugacity. However, for most gases and vapours this differs very little from the value at 1 atm because deviations from the perfect gas laws are small in the range 0–1 atm. In using (4·19) to evaluate K_f it is usually a sufficiently good approximation to use the tabulated μ^0's at atmospheric pressure. For the sake of comparability with other thermodynamics texts it may be added that if the equation $\mu_i = \mu_i^0 + RT \ln f_i$ had been substituted in (4·9) [rather than in (4·10), as has been done] we should have obtained

$$(\partial G/\partial \xi)_{T,\, p} = \Sigma \nu_i \mu_i = \Sigma \nu_i \mu_i^0 + RT \ln \prod_i f_i^{\nu_i}$$

and therefore at any stage of reaction, including nonequilibrium conditions,

$$\Delta G_T = \Delta G_T^0 + RT \ln \prod_i f_i^{\nu_i}$$

where $\Delta G_T \equiv \Sigma \nu_i \mu_i$. This is the negative of de Donder's affinity and of course is zero at equilibrium. It is is to be noted however that (4·16) applies only when the f_i refer to a set of fugacities at chemical equilibrium – i.e. when ΔG_T is zero.

4·6. The temperature dependence of the equilibrium constant

In equations (4·18) and (4·19) K_p and K_f necessarily refer to the temperature T and so also do the μ^0's. The extent to which the equilibrium constant varies with temperature is readily obtained in terms of the heat of reaction.

Equation (4·18) may be written

$$\ln K_p = -\frac{1}{R} \Sigma \nu_i \left(\frac{\mu_i^0}{T} \right),$$

and its differentiation gives

$$\frac{\mathrm{d} \ln K_p}{\mathrm{d}T} = -\frac{1}{R} \Sigma \nu_i \frac{\mathrm{d}(\mu_i^0/T)}{\mathrm{d}T}.$$

Substituting from (3·29) and (3·30),

$$\frac{\mathrm{d} \ln K_p}{\mathrm{d}T} = \frac{1}{RT^2} \Sigma \nu_i h_i, \tag{4·20}$$

† Of course the partial pressures could readily be expressed in SI units provided that a numerical factor were included in (4·18) and (4·19) to allow for the fact that a standard pressure of 1 atm = 101.325 kPa.

where h_i is the enthalpy per mole of pure i at the temperature T, and is a function only of temperature. The term $\Sigma \nu_i h_i$ is clearly the increase in enthalpy, ΔH, when the reaction takes place from left to right. It is the negative of the heat evolved when reaction takes place at constant temperature and pressure, provided that volume change is the only form of work. Thus (4·20) may be written

$$\frac{\mathrm{d}\ln K_p}{\mathrm{d}T} = \frac{\Delta H}{RT^2}, \tag{4·21}$$

and this is known as *van't Hoff's equation*.†

Similarly in the case of an imperfect gas mixture, a differentiation of (4·19) and substitution of (3·64) gives

$$\frac{\mathrm{d}\ln K_f}{\mathrm{d}T} = \frac{1}{RT^2}\Sigma \nu_i h_i^0$$

$$\equiv \frac{\Delta H^0}{RT^2}, \tag{4·22}$$

where h_i^0 is the enthalpy per mole of i at a pressure low enough for the gas to approach perfection and ΔH^0 is the corresponding enthalpy change in the reaction. In the application of (4·22) it is usually a satisfactory approximation to use the values of the molar enthalpies of gases at atmospheric pressure, as tabulated in the literature.

It may be noted that the differentials in (4·21) and (4·22) are complete differentials because K_p and K_f are functions only of temperature. These equations may be used either to calculate ΔH from the measured temperature coefficient of the equilibrium constant, or alternatively, if ΔH is known, to calculate the equilibrium constant at one temperature from a knowledge of its value at some other temperature.

Over small ranges it may be assumed, as an approximation, that ΔH is independent of temperature and therefore

$$\ln\frac{K_{p_2}}{K_{p_1}} = \frac{\Delta H}{R}\left(\frac{1}{T_1} - \frac{1}{T_2}\right).$$

For the purpose of integration over an extended range it is not satisfactory to assume that ΔH, or even Δc_p, is constant. The procedure for integration is as follows.

The molar heat capacity at constant pressure is

$$\left(\frac{\partial h_i}{\partial T}\right)_p = c_{p_i}.$$

† The physical significance of the equation, which implies that K_p increases with temperature in endothermic reactions, may be considered in relation to the remarks at the end of §4·3.

For the present purposes this may be written as a complete differential because the h_i and h_i^0 which appear in equations (4·20) and (4·22) refer to the gas i in a state of perfection where the enthalpy is independent of pressure (§3·2). Therefore

$$\frac{d\Delta H}{dT} \equiv \frac{d}{dT}(\Sigma\nu_i h_i)$$

$$= \Sigma\nu_i c_{p_i}, \tag{4·23}$$

which is *Kirchhoff's equation.*

The heat capacities of each of the pure gases i are usually expressed in the literature as empirical power series in the temperature:[†]

$$c_{p_i} = a_i + b_i T + c_i T^2 + \dots. \tag{4·24}$$

The use of the first three terms of this series is usually sufficient. Therefore from (4·23) and (4·24)

$$\frac{d\Delta H}{dT} = \Sigma\nu_i(a_i + b_i T + c_i T^2), \tag{4·25}$$

and the integral of this is

$$\Delta H - \Delta H_0 = \Sigma\nu_i\left(a_i T + \frac{b_i T^2}{2} + \frac{c_i T^3}{3}\right). \tag{4·26}$$

In this equation, ΔH_0 is simply an integration constant; it might be interpreted as the heat of reaction at the absolute zero if (4·24) continued to remain valid down to this temperature, but this is not usually the case.

Substituting (4·26) in (4·21)

$$\frac{R d \ln K_p}{dT} = \frac{\Delta H_0}{T^2} + \Sigma\nu_i\left(\frac{a_i}{T} + \frac{b_i}{2} + \frac{c_i T}{3}\right), \tag{4·27}$$

and the integration of this equation gives

$$R \ln K_p = C - \frac{\Delta H_0}{T} + \Sigma\nu_i\left(a_i \ln T + \frac{b_i T}{2} + \frac{c_i T^2}{6}\right), \tag{4·28}$$

where C is a second integration constant. A similar relation can be written for K_f.

Combining equations (4·18) and (4·28), we obtain an equation which gives ΔG_T^0 as a function of temperature:

$$\Delta G_T^0 \equiv \Sigma\nu_i\mu_i^0 = -CT + \Delta H_0 - \Sigma\nu_i\left(a_i T \ln T + \frac{b_i T^2}{2} + \frac{c_i T^3}{6}\right). \tag{4·29}$$

[†] For data on several gases see Spencer et al., *J. Amer. Chem. Soc.* **56** (1934), 2311; *J. Amer. Chem. Soc.* **64** (1942), 2511; *J. Amer. Chem. Soc.* **67** (1945), 1859; J. R. Partington, *An Advanced Treatise on Physical Chemistry*, vol. I, p. 807 (Longmans, 1949); and K. A. Kobe *et al.*, articles in *Petroleum Refiner*, 1949 onwards.

This equation could also have been obtained by integration of the relation

$$\frac{d(\Delta G_T^0/T)}{dT} = -\frac{\Delta H}{T^2},$$
(4·30)

which follows from (3·29).

The equations (4·26)–(4·29) may be used in a variety of ways. For example, if measured values of K_p are available at fairly widely spaced temperatures it is possible to evaluate both of the integration constants ΔH_0 and C. For this purpose (4·28) may be rearranged as follows:

$$R \ln K_p - \Sigma \nu_i \left(a_i \ln T + \frac{b_i T}{2} + \frac{c_i T^2}{6} \right) = C - \frac{\Delta H_0}{T}.$$
(4·31)

A plot against $1/T$ of the experimental values of the left-hand side of this equation should therefore be a straight line, if the data are consistent. The value of ΔH_0 may be obtained from the gradient and C from the intercept. ΔG_T^0 at any desired temperature, e.g. 25 °C, may then be obtained from (4·29).

Alternatively, if there is available a value of ΔH, for example at 25 °C, the value of ΔH_0 may be evaluated from (4·26). The constant C may then be obtained from (4·28) by use of only a single measurement of K_p, as a minimum requirement.

Of course these equations can only be applied over the range of temperatures for which the power series (4·24) is valid, with constant and known values of a_i, b_i and c_i.

4·7. Other forms of equilibrium constant for perfect gas mixtures

For the reaction $\qquad \Sigma \nu_i M_i = 0$

in a perfect mixture, the equilibrium constant has been defined in (4·12):

$$K_p \equiv \prod_i p_i^{\nu_i}.$$

For the ith component we have, by (3·26),

$$p_i = n_i RT/V$$
$$= c_i RT,$$
(4·32)

where c_i is the molar concentration. Therefore

$$K_p = (RT)^{\Sigma \nu_i} \prod_i c_i^{\nu_i}.$$
(4·33)

If we define a new equilibrium constant, K_c, in terms of the concentrations,

$$K_c \equiv \prod_i c_i^{\nu_i},$$
(4·34)

the relation between K_c and K_p is evidently†

$$K_c = K_p(RT)^{-\Sigma\nu_i}. \qquad (4\cdot35)$$

This equation shows that K_c, like K_p, depends only on temperature, and the magnitude of its dependence may be obtained from (4·35) and (4·20):

$$\frac{d\ln K_c}{dT} = \frac{d\ln K_p}{dT} - \frac{\Sigma\nu_i}{T}$$

$$= \frac{1}{RT^2}\Sigma\nu_i h_i - \frac{\Sigma\nu_i}{T}$$

$$= \frac{1}{RT^2}\Sigma\nu_i(h_i - RT). \qquad (4\cdot36)$$

Now for a perfect gas, the internal energy, enthalpy and volume per mole are related by

$$u_i = h_i - pv_i$$

$$= h_i - RT.$$

Therefore (4·36) may be written

$$\frac{d\ln K_c}{dT} = \frac{1}{RT^2}\Sigma\nu_i u_i$$

$$= \frac{\Delta U}{RT^2}, \qquad (4\cdot37)$$

where ΔU is the change of internal energy in the reaction and is the heat absorbed when reaction occurs at constant temperature and volume. It may be noted that K_p and K_c become equal when $\Sigma\nu_i = 0$, as in reactions of the type $H_2 + Cl_2 = 2HCl$.

Another equilibrium constant may be defined in terms of mole fractions:

$$K_y \equiv \prod_i y_i^{\nu_i}. \qquad (4\cdot38)$$

Since $y_i \equiv p_i/p$, where p is the total pressure, the relationship between K_y and K_p is given by

$$K_y = \prod (p_i/p)^{\nu_i}$$

$$= p^{-\Sigma\nu_i}\prod_i p_i^{\nu_i}$$

$$= p^{-\Sigma\nu_i}K_p. \qquad (4\cdot39)$$

The equilibrium constant K_y therefore depends on the pressure as well as on the temperature, and for this reason is less useful than K_p

† For example, for the reaction $2A + 3B = 4C$, $\Sigma\nu_i = -2 - 3 + 4 = -1$, and thus

$$K_c \equiv \frac{c_c^4}{c_a^2 c_b^3} = RTK_p.$$

for application to gas reactions. On the other hand, it is one of the most convenient forms of equilibrium constant for application to liquid and solid mixtures. Because of the small molar volumes of condensed phases their equilibria are very insensitive to changes of pressure, in accordance with the relation $\partial\mu_i/\partial p = V_i$.

4·8. Free energies and enthalpies of formation from the elements

Consider the following reactions, in each of which one mole of a compound is formed from its component elements:

$$\tfrac{1}{2}N_2 + \tfrac{3}{2}H_2 = NH_3,$$

$$\tfrac{1}{2}N_2 + \tfrac{1}{2}O_2 = NO,$$

$$H_2 + \tfrac{1}{2}O_2 = H_2O.$$

From measured values of the equilibrium constants the values of ΔG^0 may be calculated by use of equation (4·18). If the temperature to which this calculation refers is, say, 1000 K, the result is denoted $\Delta_f G^0_{1000}$ and is called the *standard free energy of formation* of the particular compound at 1000 K. The word 'standard' refers not to any particular temperature, but to unit pressure† of 1 atm for each of the pure reactants and products. However, by use of (4·28) or (4·29), the value of $\Delta_f G^0_T$ may be reduced to a reference temperature and this is usually chosen as 25 °C. The result is then called the standard free energy of formation at 25 °C and is denoted $\Delta_f G^0_{298}$, which is an abbreviation for $\Delta_f G^0_{298·15}$. These figures always refer to 1 g formula weight (i.e. 1 mol) of the compound in question. Thus if we write

$$NH_3(g), \quad \Delta_f G^0_{298} = -3976 \text{ cal mol}^{-1}$$

it is meant that the formation of 17.033 g of gaseous NH_3 at 1 atm pressure, from gaseous oxygen and nitrogen each at 1 atm pressure, would be accompanied by a decrease of G of 3976 cal. The extent to which these conventions may need modification if any of the elements or compounds are not gaseous at 25 °C and 1 atm (as in the third of the above reactions) will be discussed in § 4·10.

Consider now the reaction

$$4NH_3 + 5O_2 = 4NO + 6H_2O.$$

The value of ΔG^0 for this reaction will have the same value whether it occurs directly as written, or as a process involving several steps. This is because G is a function of state, and its change in any process

† More correctly unit fugacity, but see §4·5. Although in the SI system the unit of pressure is the Pascal (or N m^{-2}), most existing thermodynamic data continues to refer to a pressure of 1 atm = 101.325 kPa.

depends only on the initial and final states. Now the reaction can be written as compounded of the three reactions already discussed:

$$4NH_3 \qquad = 2N_2 \ + 6H_2$$
$$2N_2 \ + 2O_2 = 4NO$$
$$\underline{6H_2 \ + 3O_2 = 6H_2O}$$
$$4NH_3 + 5O_2 = 4NO \ + 6H_2O$$

The value of ΔG_T^0 for the overall process is therefore

$$\Delta G_T^0 = -4\Delta_f G_T^0(NH_3) + 4\Delta_f G_T^0(NO) + 6\Delta_f G_T^0(H_2O),$$

where $\Delta_f G_T^0(NH_3)$, etc., denote the standard free energies of formation of the compounds.

From this example it will now be clear that ΔG_T^0, and therefore also the equilibrium constant, may be calculated for any conceivable reaction provided that the free energies of formation of the various compounds are already known. This property of additivity applies also, of course, to internal energies, entropies, etc., and in the case of the enthalpies it is known as *Hess's law*.

It is to be noted that we do not assert that the Gibbs function or enthalpy of the elements is zero. This may be adopted as a convention, if it is desired, but it is quite unnecessary and is often misleading. The reason for choosing the elements against which to refer the values of G and H is because the atoms remain unchanged in chemical reactions (excluding nuclear reactions). However, it is clearly necessary to specify the physical state of the elements in question; for example, the above figure for the free energy of formation of NH_3 refers to its formation, not from atomic nitrogen and hydrogen, but from the stable molecules N_2 and H_2. *In general, the reference state is the normal condition of the element—gaseous, liquid or solid—as it occurs at 25 °C and 1 atm.*

In place of a list of standard free energies and enthalpies of formation, it is equally satisfactory to tabulate the standard entropies and enthalpies of formation because at any particular temperature

$$\Delta_f G = \Delta_f H - T\Delta_f S.$$

4·9. Some examples

The following examples have been chosen, not so much for the purpose of exemplifying the most recent or most accurate data, but in order to illustrate some of the approximations which must often be used in the application of the equations of §§ 4·5 and 4·6.

(a) *The free energy of formation of nitric oxide.* In the work of Briner, Boner and Rothen† air was heated in a vessel to a temperature of 1873 K for a time sufficient to approach equilibrium in the reaction

$$\tfrac{1}{2}N_2 + \tfrac{1}{2}O_2 = NO.$$

Samples of the gas were rapidly cooled and were allowed to remain in contact with concentrated potassium hydroxide solution for a day. The nitrite and nitrate which were formed were subsequently estimated. The volume percentage of nitric acid which is in equilibrium with air at 1873 K was thus found to be 0.73–0.85 %, the mean value being 0.79 %.

Air contains 78.03 % N_2 and 20.99 % O_2. The mean composition of the equilibrium gas must therefore have been 77.63 % N_2, 20.59 % O_2, 0.79 % NO. Hence

$$K_p = 1.98 \times 10^{-2}.$$

The enthalpy of reaction is known to be

$$\Delta_f H = 21\ 600 \text{ cal. mol}^{-1}$$

With regard to the heat capacities there seems to be no accurate data on nitric oxide at high temperatures. However, N_2, O_2 and NO are all diatomic molecules and may be expected to have almost equal heat capacities at any particular temperature (see § 12·12). Since there is no change in the number of molecules when reaction takes place, it is reasonable to suppose that Δc_p is small or zero. With this assumption it follows from equation (4·23) that ΔH is approximately independent of temperature.

Using (4·28) we obtain

$$R \ln K_p = C - \frac{21\ 600}{T},$$

and substituting $K_p = 1.98 \times 10^{-2}$ at $T = 1873$ K, it is found that $C = 3.7$. Finally, from (4·29)

$$\Delta_f G^0_{298} = -3.7 \times 298.16 + 21\ 600$$

$$= 20\ 500 \text{ cal mol}^{-1}$$

the standard free energy of formation of nitric oxide at 25 °C. By an independent spectroscopic method, to be described in Chapter 12, Giauque and Clayton‡ obtained a value of $\Delta_f G^0_{298} = 20\ 650$ cal mol⁻¹

It may be remarked that in some earlier experimental work Nernst had found the equilibrium percentage of NO in air at 1877 K to be

† *J. Chim. Phys.* **23** (1926), 788.
‡ Giauque and Clayton, *J. Amer. Chem. Soc.* **55** (1953), 4875.

only 0.42%, about half the value used in the above calculation. If this figure had been used the value of $\Delta_f G^0_{298}$ would have been calculated as 20 850 cal/mol^{-1}, which is very little different from the value based on Briner's experiments. This is a good example of what Lorentz (in a different context) called 'l'insensibilité des fonctions thermodynamiques', and it arises from the logarithmic relation between K_p and ΔG^0. Conversely, in order to make a close estimate of K_p it is necessary to have available very accurate values of ΔG^0.

(b) The free energy of formation of NO_2 *and* N_2O_4. Having established the free energy of formation of NO, it is now possible to calculate the value for NO_2 by utilizing the experimental results of Bodenstein and Katayama on the equilibrium of the reaction

$$NO + \tfrac{1}{2}O_2 = NO_2.$$

The calculations were carried out by Lewis and Randall. The heat capacities of NO and NO_2 were not available, but were assumed to be equal to those of O_2 and CO_2 respectively, on account of similarities in molecular structure. Equation (4·28) was used and the constants C and ΔH_0 were evaluated as described in §4·6. They obtained

$$NO + \tfrac{1}{2}O_2 = NO_2, \quad \Delta G^0_{298} = -8930 \text{ cal mol}^{-1}$$

Combining this with the value, 20 500 cal, for the free energy of formation of NO, we obtain

$$\tfrac{1}{2}N_2 + O_2 = NO_2, \quad \Delta_f G^0_{298} = 11\ 570 \text{ cal mol}^{-1}.$$

This figure is therefore the standard free energy of formation of NO_2 at 25 °C.

By utilizing the data on the equilibrium of

$$2NO_2 = N_2O_4,$$

it now becomes possible to calculate the free energy of formation of N_2O_4. The process by which a table of standard free energies is built up needs no further elaboration.

(c) Ammonia synthesis equilibrium. This example is chosen in order to illustrate the necessity of using fugacities in the case of high-pressure gas reactions and also to show the use of the Lewis and Randall rule. The following discussion is based on Dodge.†

At 450 °C the experiments of Larson and Dodge gave the values of $K_p \equiv p_{NH_3}/p_{N_2}^{\frac{1}{2}} p_{H_2}^{\frac{3}{2}}$ shown in Table 4. It is evident that the deviations from constancy of K_p increase rapidly as the pressure is raised above 100 atm.

† Dodge, *Chemical Engineering Thermodynamics* (New York, McGraw-Hill, 1944).

TABLE 4

Total pressure/atm	10	30	50	100	300	600	1000
$K_p \times 10^3$	6.59	6.76	6.90	7.25	8.84	12.94	23.28

The true equilibrium constant for imperfect gas mixtures is not K_p but K_f. By using the fugacity coefficient, $\chi_i \equiv f_i/p_i$, K_f may be expressed in the form

$$K_f = \frac{f_{NH_3}}{f_{N_2}^{\frac{1}{2}} f_{H_2}^{\frac{3}{2}}}$$

$$= \frac{p_{NH_3}}{p_{N_2}^{\frac{1}{2}} p_{H_2}^{\frac{3}{2}}} \frac{\chi_{NH_3}}{\chi_{N_2}^{\frac{1}{2}} \chi_{H_2}^{\frac{3}{2}}}$$

$$= K_p \frac{\chi_{NH_3}}{\chi_{N_2}^{\frac{1}{2}} \chi_{H_2}^{\frac{3}{2}}}.$$

Let it be assumed, for purposes of trial, that the mixture N_2—H_2—NH_3 behaves as an *ideal gaseous solution*, equivalent to assuming that it obeys the Lewis and Randall rule, equation (3·72). This rule may be expressed

$$\frac{f_i}{p_i} = \frac{f_i'}{P}$$

or $$\chi_i = \chi_i',$$

where the left-hand sides of each equation refer to the mixture and the right-hand sides to substance i in its pure state at the same temperature and pressure as the mixture.

As discussed in § 3·6 the values of χ_i' for a considerable number of pure gases can all be represented by approximately the same function of the reduced temperature and pressure. The critical data of hydrogen, nitrogen and ammonia are given in Table 5, and these will be used, together with Newton's† graph, to estimate the fugacity coefficients of these gases.

TABLE 5

	H_2	N_2	NH_3
T_c/K	33.2	126.0	406
P_c/atm	12.8	33.6	111.6

Consider the ammonia synthesis equilibrium at 450 °C (723 K) and 300 atm. The reduced temperatures‡ and pressures are as given in Table 6, together with the interpolated values of χ' from the graph.

† Newton, *Industr. Engng Chem.* **27** (1935), 302.
‡ In the case of H_2 and He Newton uses as reduced temperature $T_r = T/(T_c+8)$ and as reduced pressure $P_r = P/(P_c+8)$.

TABLE 6

	H₂	N₂	NH₃
T_r	17.53	5.73	1.78
P_r	14.4	8.94	2.69
χ'	1.09	1.14	0.91

At 450 °C and 300 atm the value of K_p from Table 4 is 8.84×10^{-3}. The value of K_f, as estimated by use of the Lewis Randall rule, is therefore

$$K_f = 8.84 \times 10^{-3} \times \frac{0.91}{1.14^{\frac{1}{2}} \times 1.09^{\frac{3}{2}}}$$

$$= 6.6 \times 10^{-3}$$

Proceeding in the same way with the other figures from Table 4, we obtain the values of K_f given in Table 7. These are seen to be nearly constant up to 300 atm. The assumption of the ideal solution model therefore represents a considerable advance on the supposition that the mixture is perfect.

TABLE 7

Total pressure/atm	10	30	50	100	300	600	1000
$K_f \times 10^3$	6.5	6.6	6.6	6.6	6.6	7.4	10.3

4·10. Free energies of formation of non-gaseous substances or from non-gaseous elements

As defined in §4·8, the standard Gibbs free energy of formation of a compound at 25 °C is the increase in G when 1 mol (1 g formula weight) of the substance at 1 atm pressure is formed from its elements, each at 1 atm pressure. The question arises whether any modification needs to be adopted when the substance in question, or the elements from which it is formed, are not in a stable gaseous state at 25 °C and 1 atm. This can best be discussed by use of an example, and we shall take the case of the free energy of formation of water.

At sufficiently high temperatures water vapour is appreciably dissociated (e.g. about 1 % at 2000 K) and measurements on the equilibrium of the reaction

$$H_2 + \tfrac{1}{2}O_2 = H_2O \, (g)$$

were made by Nernst, Langmuir and others. Using experimental results of this kind, together with the available data on the heat capacities, Lewis and Randall used equation (4·29) to obtain $\Delta G'$ as a function of T:

$$\Delta_f G_T^0 = 3.92T - 57\,410 + 0.94T \ln T + 1.65 \times 10^{-3}T^2 - 3.7 \times 10^{-7}T^3.$$

In this equation if we put $T=298.16$ K we obtain

$$\Delta_f G^0_{298} = -54\ 507\ \text{cal mol}^{-1}.$$

This is therefore the value† at 25 °C of

$$\mu^0_{H_2O(g)} - \mu^0_{H_2} - \tfrac{1}{2}\mu^0_{O_2},$$

and it refers specifically to *gaseous* water at 1 atm pressure. This is because the result has been obtained by extrapolation from the high-temperature data and because K_p for the reaction has been expressed in atmospheres.

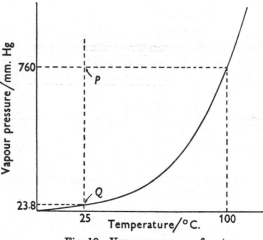

Fig. 19. Vapour pressure of water.

Thus whenever the above equation for $\Delta_f G^0$ is applied to a temperature below 100 °C it refers automatically to a metastable state where the water vapour is supersaturated. However, the free energy of formation of water in its stable state as a liquid at 25 °C and 1 atm may be readily calculated by considering a change of state in three successive steps.

A diagrammatic representation of the vapour pressure of water is shown in Fig. 19; the stable state to the left of the curve is liquid and to the right is vapour. The metastable vapour state to which the figure $\Delta_f G^0_{298} = -54\ 507$ refers is marked P. Consider the isothermal expansion of one mole of this vapour from the pressure 760 mm to the pressure 23.8 mm (point Q) at which it would be in stable equilibrium with liquid water at 25 °C. Using the relation (2·111b),

† A more recent value is $-54\ 636$ cal mol^{-1}.

$(\partial\mu/\partial p)_T = v$, and assuming the water vapour to be a perfect gas, we obtain for the change of G in this process

$$\Delta G = \int_{760}^{23\cdot 8} v\,dp = RT\ln\frac{23.8}{760}$$

$$= -2053\text{ cal mol}^{-1}$$

The second step is the condensation of the vapour at 23.8 mm and 25 °C to give liquid water at the same temperature and pressure. Since this is a phase change at equilibrium there is no change of G ($\mu^g = \mu^l$, as shown in §2·9b).

The final step is the compression of the liquid water from 23.8 mm to the standard pressure of 1 atm. In this process the change of G is negligible. Thus

$$\Delta G = \int_{23\cdot 8}^{760} v\,dp,$$

where v is now the molar volume of liquid water. Its value may be taken as 18 cm³ and independent of pressure. The value of the integral is thus

$$\Delta G = 18\,(760 - 23.8)/760 = 17\text{ cm}^3\text{ atm mol}^{-1}$$

which is equivalent to a mere 0.43 cal. It may be noted as a general result that *a change in pressure of a few atmospheres has a negligible effect on the free energy of a condensed phase*, on account of its small molar volume.

As a result of these calculations we can now obtain the free energy of formation of liquid water at 25 °C and 1 atm:

$$
\begin{array}{lll}
\text{H}_2 + \tfrac{1}{2}\text{O}_2 & = \text{H}_2\text{O}\ (g,\ 1\,\text{atm}), & \Delta_f G = -54\,507, \\
\text{H}_2\text{O}\ (g,\ 1\,\text{atm}) & = \text{H}_2\text{O}\ (g,\ 23.8\,\text{mm}), & \Delta G = -2\,053, \\
\text{H}_2\text{O}\ (g,\ 23.8\,\text{mm}) & = \text{H}_2\text{O}\ (l,\ 23.8\,\text{mm}), & \Delta G = 0, \\
\text{H}_2\text{O}\ (l,\ 23.8\,\text{mm}) & = \text{H}_2\text{O}\ (l,\ 1\,\text{atm}), & \Delta G = 0.43
\end{array}
$$

Thus adding the equations

$$\text{H}_2 + \tfrac{1}{2}\text{O}_2 = \text{H}_2\text{O}\ (l,\ 1\,\text{atm}),$$

$$\Delta_f G_{298}^0 = -56\,560\text{ cal mol}^{-1}$$

Exactly similar considerations apply if one or more of the elements from which the compound is formed is solid or liquid at 25 °C and 1 atm. In general, the values of the standard free energies of formation which are quoted in the literature are equal to the increase in G when 1 g formula weight of the substance in question is formed from its elements at 25 °C, the substance and the elements each being in their normal stable states at 1 atm pressure. Similar considerations apply to the enthalpy change of reaction.

4·11. Preliminary discussion on reaction equilibria involving gases together with immiscible liquids and solids

We now consider chemical reactions in which one or more of the products or reactants is a solid or liquid. In the present chapter the discussion will be limited to cases where each of these solids or liquids is present in the system as a pure phase, i.e. when they do not take into solution appreciable amounts of the other components. Under these conditions the free energy of mixing, which has been shown to be an important part of the 'driving force' of reaction, is limited to the gaseous phase. (The discussion of the case where there is an additional free energy of mixing in the condensed phases depends on a knowledge of the chemical potentials in solutions and will be deferred to Chapter 10.)

Examples of the type of reaction under discussion are

$$ZnS + \tfrac{3}{2}O_2 = ZnO + SO_2,$$
$$CaCO_3 = CaO + CO_2.$$

Now every substance has a finite vapour pressure and therefore the reaction equilibrium may be thought of as being established in the gaseous phase, according to the principles already discussed. If the various solids or liquids which take part in the reaction are in phase equilibrium with their vapours, there will be reaction equilibrium throughout the entire system.

For example, in the second of the above reactions, the condition of reaction equilibrium is

$$\mu_{CaCO_3} = \mu_{CaO} + \mu_{CO_2}, \tag{4·40}$$

The conditions of phase equilibrium for the oxide and carbonate between their solid phases and their vapours are

$$\mu_{CaCO_3(g)} = \mu_{CaCO_3(s)},$$
$$\mu_{CaO(g)} = \mu_{CaO(s)}.$$

In equation (4·40) it is therefore immaterial whether μ_{CaCO_3} and μ_{CaO} are taken to refer to the solids, or to the vapours in equilibrium with these solids.

Consider the second of these alternative ways of regarding the problem. If the gaseous phase is assumed to be a perfect gaseous mixture, then we can write, in accordance with (3·19),

$$\mu_{CaCO_3(g)} = \mu^0_{CaCO_3(g)} + RT \ln p_{CaCO_3},$$

together with similar equations for the CaO and CO_2. Substituting these in (4·40) we obtain the equilibrium constant in its normal form:

$$-RT \ln \frac{p_{CaO} p_{CO_2}}{p_{CaCO_3}} = \mu^0_{CaO(g)} + \mu^0_{CO_2} - \mu^0_{CaCO_3(g)}. \tag{4·41}$$

Therefore
$$K_p \equiv \frac{p_{CaO} \, p_{CO_2}}{p_{CaCO_3}} \qquad (4\cdot42)$$

is a function only of temperature.

Now the quantities $\mu^0_{CaO(g)}$ and $\mu^0_{CaCO_3(g)}$ which appear in equation (4·41) refer to highly metastable states, that is, to *gaseous* calcium oxide and carbonate each at 1 atm pressure. To be sure these quantities are calculable (by the methods of §4·10), but only if the vapour pressures of the oxide and carbonate are known, and in many instances of this type of reaction the vapour pressures may be immeasurably small.

This difficulty may be avoided by noting that the pressures p_{CaO} and p_{CaCO_3} which occur in (4·41) are the *saturation vapour pressures*, because of the supposition of phase equilibrium. These terms therefore have *fixed values at any particular temperature*.† Therefore it is possible to remove these particular terms from K_p and to define a *partial equilibrium constant, K'_p*, which does not contain them. Thus

$$K'_p \equiv p_{CO_2} \equiv K_p \frac{p_{CaCO_3}}{p_{CaO}}, \qquad (4\cdot43)$$

and this has a definite value at any particular temperature.†

The value of K'_p may be expressed in terms of standard free energies by returning to (4·40). As mentioned previously it is immaterial whether μ_{CaCO_3} and μ_{CaO} are taken as referring to the solids or to the vapours of the carbonate and oxide. We now adopt the first of these alternatives and in equation (4·40)

$$\mu_{CaCO_3} = \mu_{CaO} + \mu_{CO_2}$$

we substitute
$$\mu_{CO_2} = \mu^0_{CO_2} + RT \ln p_{CO_2},$$

but no longer make the corresponding substitution for the oxide and carbonate. Hence

$$\mu_{CaCO_3} = \mu_{CaO} + \mu^0_{CO_2} + RT \ln p_{CO_2}$$

or
$$-RT \ln K'_p = \mu_{CaO} + \mu^0_{CO_2} - \mu_{CaCO_3}. \qquad (4\cdot44)$$

In this equation only the term $\mu^0_{CO_2}$ refers to the standard state of 1 atm; the terms μ_{CaO} and μ_{CaCO_3} refer to the chemical potentials of solid oxide and carbonate at the particular pressure p of the reaction

---

† These statements are correct only because CaO and CaCO₃ do not form with each other a solid solution, i.e. there is no free energy of mixing in the condensed phases. Also it may be noted that vapour pressures are affected *very slightly* by the total pressure on the system, as will be discussed in Chapter 6. This dependence of p_{CaO} and p_{CaCO_3} on the total pressure is equivalent to the pressure coefficient of the chemical potential of a condensed phase, as discussed in the present section, and is quite trivial.

system. However, it will now be shown that these quantities may usually be taken as equal to the chemical potentials of solid oxide and carbonate respectively at *unit* pressure. Let these latter quantities be denoted $\mu^0_{CaO(s)}$ and $\mu^0_{CaCO_3(s)}$ (and these are to be distinguished from the quantities appearing in (4·41)). Now since the calcium oxide is present in the system as a pure phase, and not as a solution, we have from (2·111 b)

$$\left(\frac{\partial \mu_{CaO}}{\partial p}\right)_T = v_{CaO}, \qquad (4·45)$$

where v_{CaO} is the volume per mole. Integrating this equation at constant temperature and disregarding any change of v_{CaO},

$$\mu_{CaO} - \mu^0_{CaO(s)} = v_{CaO}(p-1), \qquad (4·46)$$

where p is the pressure of the reaction system.

As noted in the calculations on water in § 4·10, this kind of integral is usually trivial compared to the free-energy change of the reaction. For example, $v_{CaO} = 16.5$ cm³ mol⁻¹, and therefore at 100 atm $\mu_{CaO} - \mu^0_{CaO(s)}$ has a value of only 40 cal mol⁻¹. Similar considerations apply to the calcium carbonate. Thus, provided the pressure p is not excessive, (4·44) may be written as follows with negligible error:

$$-RT \ln K'_p = \mu^0_{CaO(s)} + \mu^0_{CO_2} - \mu^0_{CaCO_3(s)}. \qquad (4·47)$$

As compared to (4·44), the useful result has been to obtain quantities on the right-hand side of this equation which refer to the stable states at unit pressure. These standard chemical potentials may be replaced by the standard free energies of formation at the appropriate temperature, in accordance with the discussion of § 4·8.

It follows from the above that $K'_p = p_{CO_2}$ is a function only of temperature to a normally sufficient degree of approximation. In particular, it is independent of the quantities of calcium oxide and carbonate which are present in the system (provided that at least some of each of these solids is present, as otherwise there would not be an equilibrium).

Let it be supposed that p_{CO_2} is below its equilibrium value K'_p, for example, by means of a current of hot air as in limestone burning. This would result in *complete* reaction from left to right of the equation

$$CaCO_3 = CaO + CO_2.$$

Conversely, if p_{CO_2} were maintained above its equilibrium value, the oxide would be completely converted into the carbonate.

4·12. Concise discussion on reaction equilibria involving gases together with immiscible liquids and solids

The discussion of the preceding section will now be presented in more general terms. For a reaction

$$\sum_{i=1}^{N} \nu_i M_i = 0,$$

involving N species, the condition of equilibrium is

$$\sum_{i=1}^{N} \nu_i \mu_i = 0,$$

and this is applicable to phases in any state of aggregation. In this equation there is evidently no need to substitute any more relations of the form†

$$\mu_i = \mu_i^0(T) + RT \ln p_i,$$

than is desired or is found convenient. Let it be supposed that the substances 1 to n are present in the gaseous phase only, whilst the substances $n+1$ to N are also present as solids or liquids. The condition of equilibrium may then be written

$$\sum_{i=1}^{n} \nu_i \mu_i + \sum_{i=n+1}^{N} \nu_i \mu_i = 0. \tag{4.48}$$

In this expression we make the substitution in terms of partial pressures (or fugacities if the vapour is not perfect) only for the substances 1 to n. This gives

$$RT \sum_{i=1}^{n} \ln p_i^{\nu_i} + \sum_{i=1}^{n} \nu_i \mu_i^0 + \sum_{i=n+1}^{N} \nu_i \mu_i = 0. \tag{4.49}$$

Defining the *partial equilibrium constant*

$$K_p' \equiv \prod_{i=1}^{n} p_i^{\nu_i}, \tag{4.50}$$

the equation (4·49) may be written

$$-RT \ln K_p' = \sum_{i=1}^{i=n} \nu_i \mu_i^0 + \sum_{i=n+1}^{N} \nu_i \mu_i. \tag{4.51}$$

The first term on the right-hand side refers to the standard chemical potentials, at the temperature T, of those substances which are present only in the gaseous phase and whose partial pressures are contained in K_p'. The second term refers to the chemical potentials of all the remaining substances, and these chemical potentials are the values as they actually occur in the reaction system, that is, at a particular pressure, composition and temperature.

† The notation $\mu_i^0(T)$ means that μ_i^0 is a function only of T.

So far no assumption has been made with regard to the various solids and liquids, substances $n+1$ to N, being present in a pure state. If this assumption is now made, each of the μ_i which occur in the second summation of (4·51) is a function of temperature and pressure only and is independent of the composition of the system. Thus for each of the substances $n+1$ to N we have

$$\left(\frac{\partial \mu_i}{\partial p}\right)_T = v_i, \qquad (4·52)$$

where v_i is the volume per mole. As shown by a numerical example in §4·11, the integral of (4·52) between unit pressure and the pressure p of the reaction system is usually quite trivial. It follows that each of the μ_i in (4·51) may be replaced by μ_i^0, the chemical potential of the pure liquid or solid at unit pressure. Equation (4·51) may therefore be written

$$-RT \ln K'_p = \sum_{i=1}^{N} \nu_i \mu_i^0 \equiv \Delta G_T^0, \qquad (4·53)$$

the term on the right-hand side being a summation of the standard chemical potentials of all the substances in their normal states of aggregation at unit pressure and the temperature T, and is equal to the corresponding summation of the standard free energies of formation at this temperature.

The above theorem shows that, to a satisfactory degree of approximation, K'_p is a function only of temperature and also that its value may be calculated from the free energies of formation of the various substances at the temperature T. The temperature coefficient of K'_p is the same as in equation (4·21) and the process of extrapolating from one temperature to another, or of utilizing the tabulated free energies of formation at 25 °C, is the same as was discussed in §4·6.

It is perhaps worth discussing in more detail the significance of the assumption that the various solids or liquids are each present as pure phases. If this were not the case it would not be possible to neglect the difference between μ_i and μ_i^0 for these substances, as has been done above. For example, in the reaction

$$CaCO_3 = CaO + CO_2,$$

let it be supposed that the calcium oxide and carbonate are miscible with each other and form a solid solution. Then, in addition to their dependence on pressure and temperature, μ_{CaO} and μ_{CaCO_3} will also depend on the mole fractions of the two components of the solution, and this may be quite a large effect. It gives rise to a free energy of mixing in the condensed phase, and in such instances it is quite inappropriate to use a partial equilibrium constant. Instead it is necessary to adopt a different procedure, involving the thermodynamics of solutions, as will be described in Chapter 10.

In general, it may be remarked that whenever a reaction takes place in a *mixed* phase, the chemical potentials of the various components of this phase modify themselves during the course of the reaction on account of the change in composition. For example, in the perfect gas mixture the chemical potential of a component is related to its partial pressure by the relation

$$\mu_i = \mu_i^0 + RT \ln p_i,$$

and is infinitely negative for zero p_i. There would therefore be an infinitely large 'driving force' in any gas reaction which tends to raise this partial pressure above zero.

A similar effect occurs in solutions. In general, in any mixed phase the magnitude of the 'driving force', $-\Sigma \nu_i \mu_i$, alters during the course of the reaction because the μ's are themselves dependent on the composition. As discussed in § 4·3 there is always some composition, short of complete conversion, at which the driving force falls to zero and gives rise to a state of equilibrium.

On the other hand, in a process such as the dissociation of calcium carbonate, the two solid phases are practically immiscible and their chemical potentials depend only on the temperature (and to a very slight extent, the pressure). These potentials are therefore unable to modify themselves as reaction proceeds. The condition of equilibrium is

$$\mu_{CaCO_3} = \mu_{CaO} + \mu_{CO_2},$$

and if the pressure of the CO_2 is maintained at such a value that

$$\mu_{CaCO_3} > \mu_{CaO} + \mu_{CO_2}.$$

the carbonate will disappear entirely. Conversely, the oxide will be completely lost if the CO_2 pressure is maintained above equilibrium. Such reactions have an 'all-or-nothing' character.

In brief, whenever a *pure* solid or liquid phase participates in a reaction it may disappear entirely, and this is because its chemical potential is not affected by the composition of the system. On the other hand, the concentration of a component of a *mixture* cannot fall to zero, since this would give rise to an infinite driving force tending to form this substance.

4·13. Example on the roasting of galena

For the purpose of illustrating the principles discussed above it is of interest to calculate the equilibrium constant of the reaction

$$PbS + \tfrac{3}{2}O_2 = PbO + SO_2,$$

at a temperature of, say, 800 K.

The free energies of formation at 25 °C, and the enthalpies of formation at 18 °C are as follows:

	$\Delta_f G_{298}^0$	$\Delta_f H_{291}$
$SO_2(g)$	$-69\ 660$	$-70\ 920$
$PbO(s)$	$-44\ 900$	$-52\ 060$
$PbS(s)$	$-18\ 000$	$-22\ 300$

Assuming the gas phase to be perfect, the partial equilibrium constant is

$$K'_p = p_{SO_2}/p_{O_2}^{\frac{3}{2}},$$

and its value is given by (4·53):

$$-RT \ln K'_p = \mu^0_{PbO} + \mu^0_{SO_2} - \mu^0_{PbS} - \tfrac{3}{2}\mu^0_{O_2}.$$

At 25 °C, using the data on the free energies of formation, we obtain

$$-R \ln K'_p = \frac{-96\,560}{298} = -324.$$

Equations for the molar heat capacities of a number of gases have been collected by Spencer.[†] The equations for oxygen and sulphur dioxide are

$$O_2, \; c_p = \;\;\, 6.148 + 3.102 \times 10^{-3}T - 9.23 \;\; \times 10^{-7}T^2 \; \text{cal} \, K^{-1} \, \text{mol}^{-1},$$

$$SO_2, \; c_p = 11.895 + 1.089 \times 10^{-3}T - 2.642 \times 10^{5}/T^2 \; \text{cal} \, K^{-1} \, \text{mol}^{-1}$$

In order to illustrate a useful approximation we shall assume that the heat capacities of the lead compounds are not available and instead apply Kopp's rule.[‡] This states that the molar heat capacity of a solid compound is the sum of the atomic heats of its component elements ($6.4 \, \text{cal} \, K^{-1} \, \text{mol}^{-1}$ for most elements). With this approximation the heat capacities of the lead oxide and the lead sulphide clearly cancel.

Noting that the last term in the heat capacity equation for SO_2 is in T^{-2}, the equation analogous to (4·26) is

$$\Delta H - \Delta H_0 = T(11.895 - \tfrac{3}{2} \times 6.148) + \tfrac{1}{2}T^2 \times 10^{-3}(1.089 - \tfrac{3}{2} \times 3.102)$$

$$+ \tfrac{3}{2} \times \tfrac{1}{3}T^3 \times 9.23 \times 10^{-7} + 2.642 \times 10^5 T^{-1}$$

$$= 2.673T - 1.782 \times 10^{-3}T^2 + 4.61 \times 10^{-7}T^3 + 2.642 \times 10^5 T^{-1}.$$

From the data on the enthalpies of formation we have $\Delta H = -100\,680$ cal at $T = 291.15$ K. Substituting in the last equation we obtain $\Delta H_0 = -102\,230 \, \text{cal} \, \text{mol}^{-1}$. Van't Hoff's equation,

$$\frac{R \mathrm{d} \ln K'_p}{\mathrm{d}T} = \frac{\Delta H}{T^2},$$

can therefore be put in the form

$$\frac{R \mathrm{d} \ln K'_p}{\mathrm{d}T} = -\frac{102\,230}{T^2} + \frac{2.673}{T} - 1.782 \times 10^{-3} + 4.61 \times 10^{-7}T$$

$$+ 2.642 \times 10^5 T^{-3}$$

† Spencer and Justice, *J. Amer. Chem. Soc.* **56** (1934), 2311; Spencer and Flannagan, *J. Amer. Chem. Soc.* **64** (1942), 2511; Spencer, *J. Amer. Chem. Soc.* **67** (1945), 1859.

‡ This rule is also known by the names of Woestyn and Neumann. For a discussion of its basis see §13·9.

This may be worked out as a definite integral between the limits 298.16 and 800 K. Using the value $R \ln K_p' = 324$ at $T = 298.16$ K we finally obtain
$$K_p' = 10^{24} \text{ at } 800 \text{ K.}$$

It may be concluded that reaction from left to right of the equation
$$PbS + \tfrac{3}{2}O_2 = PbO + SO_2$$
is essentially complete at 800 K. For example, if oxygen were heated with an excess of lead sulphide, the partial pressure of oxygen would fall to a value given by $(p_{SO_2}/10^{24})^{\frac{2}{3}}$. Alternatively, if the oxygen were in excess, the lead sulphide would disappear entirely†, even though the concentration of sulphur dioxide in the gaseous phase enormously exceeded the concentration of oxygen.

It is instructive to consider the reverse reaction. In principle, if a steady stream of pure SO_2 were passed over lead oxide, it would be possible in time to achieve a complete conversion to lead sulphide. However, the concentration of oxygen as an impurity in the inflowing SO_2 would need to be less than one part in $(10^{24})^{\frac{2}{3}}$, if the SO_2 were at 1 atm !

As far as the roasting of galena is concerned it may be remarked that the above is not a complete discussion, because alternative products such as lead sulphate and metallic lead need to be considered. By utilizing the known free energies of these substances it may be established by similar reasoning that the main products of reaction at 800 K are lead oxide and lead sulphate.

4·14. Measurement of the free energy of reaction by use of galvanic cells

Four important methods of measuring the free energy of reaction are as follows:

(a) the direct measurement of equilibrium constants as discussed above;

(b) the determination of the e.m.f. of galvanic cells;

(c) the measurement of the spacing of molecular energy levels by spectroscopy;

(d) the use of the 'third law'.

The third and fourth of these methods will be discussed in Part III. The present section is concerned with a short outline of the method based on the galvanic cell.

The essential aspect of the cell is the fact that the equilibrium composition of the reaction system is not the same as it would be in the absence of a potential difference between the two electrodes; if the potential difference is E the reactants and products come to a

† Always assuming that the lead oxide and lead sulphide are not mutually soluble.

certain state of equilibrium, whereas if the cell is short-circuited they come to a quite different state of equilibrium. This state of affairs is in no way contrary to the correctness of the relation $\Sigma \nu_i \mu_i = 0$, but arises from the fact that electrons participate in the reaction process as it occurs in the cell. Whenever the cell is on open circuit the electrons in the one electrode do not have the same chemical potential as those in the other electrode, and this difference of chemical potential can exist only because the liquid phase of the cell is a very poor *electronic* conductor. The existence of the e.m.f. thus depends on the cell being in a state of metastable equilibrium as regards the internal passage of electrons.

These points will be made clearer in §4·15. For the moment it may be remarked that the important feature of the cell is simply this: it is possible to discover what value of the potential difference across the electrodes will cause any *arbitrary composition* of the chemical system to be a state of equilibrium.

As an example we consider the reaction

$$\tfrac{1}{2}H_2 \ (g,\ 0.9 \ \text{atm}) + HgCl \ (s) = HCl \ (g,\ 0.01 \ \text{atm}) + Hg \ (l).$$

This process could be carried out electrochemically in a cell in which one of the electrodes consists of mercury covered with a layer of solid mercurous chloride. The latter is in contact with a solution of hydrochloric acid of a concentration such that its partial pressure of hydrogen chloride is 0.01 atm† Dipping into the acid is a piece of platinum (or other inert metal) and hydrogen, at a partial pressure of 0.9 atm, is released as gas bubbles in close proximity to the platinum to form the hydrogen electrode.

The cell is represented symbolically in the sequence of the electrical contacts:

Pt, H_2 (0.9 atm) | HCl (0.01 atm) HgCl | Hg; $E_{298} = 0.011 \ 0$ V,

where each vertical line represents an interface betweeen the phases. E_{298} is the measured potential difference across the electrodes at 25.00 °C when the cell operates reversibly (the current flowing through an external circuit being infinitesimally small). This reversible potential difference is called the electromotive force (e.m.f.).

Convention. The cell is written in the above sequence, and not in the reverse, on account of a generally accepted convention; the electrode at the right-hand side is the one which would tend to become positively charged if the chemical reaction were to take place spontaneously, i.e. without opposing potential difference.

---

† The cell reaction is here regarded as involving *gaseous* hydrogen chloride in equilibrium with its aqueous solution. This is for reasons of simplicity since the standard states of substances in solution have not yet been discussed.

Now, in the above reaction, the direction of spontaneous change is from left to right; that is, the partial pressure of 0.9 atm of hydrogen is greater than would be in equilibrium with a partial pressure of 0.01 atm of hydrogen chloride, together with mercury and mercurous chloride, if there were no electrical constraints. Thus, if the cell were short-circuited, the reactions

$$\tfrac{1}{2}H_2 \rightarrow H^+ + e,$$

$$H^+ + HgCl \rightarrow HCl + Hg^+,$$

$$Hg^+ + e \rightarrow Hg,$$

would occur at the hydrogen electrode, in the body of the solution and at the mercury electrode respectively. The first and third of these would give rise to an excess of electrons on the platinum and a deficit of electrons on the mercury. The latter electrode would thus become positively charged. It may be noted that as soon as the above convention has been adopted there is no further need to state the direction of the potential difference, and the e.m.f. is always taken as being positive.

Let it be supposed that the cell performs work by driving a current through an external circuit under reversible conditions, the rate of flow of electricity being vanishingly small. Over a certain period of time let dn mols of hydrogen pass into solution as ions. In order to preserve electroneutrality† in the system, dn mols of Hg^+ ions are simultaneously discharged at the mercury electrode, and a corresponding amount of electricity flows through the external circuit. As discussed in § 2·5 b

$$-dG = -dw' = zFE\,dn, \tag{4·54}$$

and in this equation z may be taken as the valence of any ion of which dn mols *pass from left to right* through a cross-section of the cell.‡ dG is the change in the Gibbs function of the entire system including

---

† Even a very small disparity in the number of positive and negative ions would give rise to an enormous coulombic force tending to restore electroneutrality. According to Schriever and Reed (*Nature, Lond.* **165** (1950), 108) the disparity between the number of positive and negative ions in an electrolyte solution does not exceed one in 10^6, even when an appreciable current passes through the solution.

‡ Suppose that the cell reaction is such that the two electrode processes are $\tfrac{1}{2}H_2 \rightarrow H^+ + e$ and $Zn^{2+} + 2e = Zn$, the latter involving a divalent ion. Then one zinc ion is discharged for every *two* hydrogen ions passing into solution. Thus $dn_H = 2dn_{Zn}$. Also $z_{H^+} = \tfrac{1}{2}z_{Zn^{2+}}$. Thus the expression $zFE\,dn$ is the same for both ions. If the electrode reactions are $\tfrac{1}{2}H_2 \rightarrow H^+ + e$ and $\tfrac{1}{2}Cl_2 + e \rightarrow Cl^-$, then for dn hydrogen ions passing from left to right, dn chloride ions pass *from right to left*, i.e. $-dn$ chloride ions pass from left to right. Also $z_{Cl^-} = -z_{H^+}$, since z must be taken as negative for negative ions. Thus the expression $zFE\,dn$ is again the same for both ions.

the external circuit, but, since the latter is unchanged, dG is simply the change in G of the reactants and products in the chemical process.

In the reaction in question dn mols each of mercury and hydrogen chloride are produced, and dn and $\frac{1}{2}dn$ mols of mercurous chloride and gaseous hydrogen respectively are used up. Therefore, at constant temperature and pressure, we have from (2·61)

$$dG = (\mu_{Hg} + \mu_{HCl} - \mu_{HgCl} - \tfrac{1}{2}\mu_{H_2})\, dn. \qquad (4·55)$$

Between the last two equations we obtain

$$\mu_{Hg} + \mu_{HCl(0.01\,atm)} - \mu_{HgCl} - \tfrac{1}{2}\mu_{H_2(0.9\,atm)} = -zFE, \qquad (4·56)$$

and z is $+1$ in the reaction in question. Inserting the numerical values $E = 0.0110$ V and $F = 23\,052$ cal V^{-1}, we obtain

$$\Delta G_{298} = -254 \text{ cal mol}^{-1} \qquad (4·57)$$

where ΔG_{298} is merely a symbol for the left-hand side of (4·56) at the particular temperature (and is *not* ΔG_{298}^0). It is the increase in the Gibbs function of the system when half a mol of gaseous hydrogen at 0.9 atm completely reacts with 1 mol of solid mercurous chloride to give 1 mol of liquid mercury and 1 mol of gaseous† hydrogen chloride at 0.01 atm. The negative value of ΔG corresponds to the fact that this reaction takes place spontaneously, in the absence of any potential difference. This is consistent with the convention.

In (4·56), μ_{Hg} and μ_{HgCl} may be taken as equal to μ_{Hg}^0 and μ_{HgCl}^0, even if the quoted e.m.f. refers to the cell operating at a pressure somewhat different from 1 atm; this is because the free energy of condensed phases is insensitive to pressure, as discussed already in §§4·10 and 4·11. The chemical potentials of the hydrogen and hydrogen chloride may be corrected to atmospheric pressure by assuming that they are perfect gases. Thus from equation (3·2)

$$\mu_{H_2}^0 = \mu_{H_2(0.9\,atm)} \quad - RT \ln 0.9$$

$$= \mu_{H_2(0.9\,atm)} \quad + 62 \text{ cal mol}^{-1}$$

$$\mu_{HCl}^0 = \mu_{HCl(0.01\,atm)} - RT \ln 0.01$$

$$= \mu_{HCl(0.01\,atm)}. \quad + 2730 \text{ cal. mol}^{-1}$$

Combining these equations with (4·56) and (4·57) we finally obtain the *standard change in free energy* in the reaction

$$\tfrac{1}{2}H_2 (g, 1 \text{ atm}) + HgCl (s) = HCl (g, 1 \text{ atm}) + Hg (l),$$

$$\Delta G_{298}^0 = 2445 \text{ cal mol}^{-1}$$

---

† Alternatively, one mol of hydrogen chloride in solution at the concentration which is in equilibrium with gaseous hydrogen chloride at 0.01 atm. The chemical potentials are equal.

It is of interest to use the above figure for ΔG^0 to calculate the partial pressure of hydrogen which would be necessary to achieve chemical equilibrium in the absence of the electrical constraint. We have at 25 °C

$$- RT \ln K'_p = \Delta G^0_{298} = 2445 \text{ cal mol}^{-1}$$

and thus

$$K'_p \equiv p_{\text{HCl}}/p_{\text{H}_2}^{\frac{1}{2}} = 0.016.$$

For the same hydrochloric acid solution, equilibrium would therefore have been attained with a hydrogen pressure of

$$(0.01/0.016)^2 = 0.39 \text{ atm}.$$

In brief, the effect of the potential difference of 0·0110 V is to raise the equilibrium partial pressure of hydrogen from 0·39 to 0·9 atm for a constant concentration of acid.

In general (4·56) may be written in the form

$$\mu^0_{\text{Hg}} + \mu^0_{\text{HCl}} + RT \ln p_{\text{HCl}} - \mu^0_{\text{HgCl}} - \tfrac{1}{2}\mu^0_{\text{H}_2} - \tfrac{1}{2}RT \ln p_{\text{H}_2} = -zFE_T,$$

or

$$- RT \ln \left(\frac{p_{\text{HCl}}}{p_{\text{H}_2}^{\frac{1}{2}}} \right) = \Delta G^0_T + zFE_T, \qquad (4 \cdot 58)$$

where ΔG^0_T has the value 2445 cal mol^{-1} at 25 °C. Thus for the *particular value* of the electrical constraint $E_{298} = 0.0110$ V, the equilibrium condition is satisfied by an infinite set of values of p_{H_2} and p_{HCl}, of which $p_{\text{H}_2} = 0.9$ atm and $p_{\text{HCl}} = 0.01$ atm is merely one possible choice. For $E_T = 0$, (4·58) reduces to the normal condition of reaction equilibrium in the absence of a potential difference within the system.

4·15. Alternative discussion of the galvanic cell

In the last section the cell was discussed under conditions where it is in process of performing external work reversibly and the equations were based on the relation $-\mathrm{d}G = -\mathrm{d}w_{\text{max}}$. of §2·4. An alternative method is to consider the cell on *open circuit*, not performing work. This method has the advantage of showing that the cell obeys the condition of reaction equilibrium

$$\Sigma \nu_i \mu_i = 0,$$

provided that proper allowance is made for the electrons.

Consider the same cell as in the last section and let $\mathrm{d}n$ be any variation in the number of moles of hydrogen chloride. Because of the stoichiometry of the reaction

$$\tfrac{1}{2}\text{H}_2 + \text{HgCl} = \text{HCl} + \text{Hg},$$

the corresponding variations are $\mathrm{d}n$ mols of mercury, $-\mathrm{d}n$ mols of mercurous chloride and $-\tfrac{1}{2}\mathrm{d}n$ mols of gaseous hydrogen. In addition, the two electrode processes involve the gain of $\mathrm{d}n$ mols of electrons on the platinum and the loss of $\mathrm{d}n$ mols of electrons from the mercury. The reaction should thus strictly be written

$$e_{\text{Hg}} + \tfrac{1}{2}\text{H}_2 + \text{HgCl} = \text{HCl} + \text{Hg} + e_{\text{Pt}}, \qquad (4 \cdot 59)$$

where e_{Hg} and e_{Pt} are symbols for 1 g mol of electrons† in the mercury and the platinum respectively. These two terms do not cancel in the chemical equation because the electrons are not in an equivalent state in the two metals. In particular, they have different chemical potentials and these may be denoted μ_e^{Hg} and μ_e^{Pt}.

When the cell is held at constant temperature and pressure, but performs *no work*, the condition of equilibrium is a minimum of G. Using equation (2·61) we have for the change in G in the above process

$$dG = -S\,dT + V\,dp + (\mu_{\text{Hg}} + \mu_{\text{HCl}} + \mu_e^{\text{Pt}} - \mu_{\text{HgCl}} - \tfrac{1}{2}\mu_{\text{H}_2} - \mu_e^{\text{Hg}})\,dn.$$

Putting
$$\left(\frac{\partial G}{\partial n}\right)_{T,\,p} = 0,$$

we obtain
$$\mu_{\text{Hg}} + \mu_{\text{HCl}} + \mu_e^{\text{Pt}} - \mu_{\text{HgCl}} - \tfrac{1}{2}\mu_{\text{H}_2} - \mu_e^{\text{Hg}} = 0, \tag{4·60}$$

which is seen to be the normal condition of reaction equilibrium

$$\Sigma \nu_i \mu_i = 0,$$

when proper allowance is made for the electrons as in (4·59). The reason why the electrons did not appear explicitly in the discussion of the last section was that the cell was there supposed to be in process of performing work; the passage into solution of hydrogen gas did not cause any actual variation in the number of electrons on the two electrodes, because these electrons were able to pass round the external circuit with the performance of work.

Equation (4·60) may be written

$$\mu_{\text{Hg}} + \mu_{\text{HCl}} - \mu_{\text{HgCl}} - \tfrac{1}{2}\mu_{\text{H}_2} = \mu_e^{\text{Hg}} - \mu_e^{\text{Pt}}.$$

Let it be supposed that the mercury is connected to a copper terminal I. Since electrons are able to pass freely between the mercury and the copper they have the same chemical potential in each. Similarly, let it be supposed that the platinum is connected to a second copper terminal II. Then the last equation may be written

$$\mu_{\text{Hg}} + \mu_{\text{HCl}} - \mu_{\text{HgCl}} - \tfrac{1}{2}\mu_{\text{H}_2} = \mu_{\text{I}} - \mu_{\text{II}}, \tag{4·61}$$

where μ_{I} and μ_{II} denote the chemical potentials of electrons in the two terminals. Since these have identical composition their electrical potential difference has a definite meaning, as discussed in § 2·9c. Using (2·58) we obtain

$$\mu_{\text{I}} - \mu_{\text{II}} = -F(\phi_{\text{I}} - \phi_{\text{II}}), \tag{4·62}$$

the valence of the electrons being -1. The question arises whether we should take $\phi_{\text{I}} - \phi_{\text{II}} = E$ or $\phi_{\text{II}} - \phi_{\text{I}} = E$. The first of these choices is consistent with the convention of the last section. For in this case we obtain from the above equations

$$\mu_{\text{Hg}} + \mu_{\text{HCl}} - \mu_{\text{HgCl}} - \tfrac{1}{2}\mu_{\text{H}_2} = -FE, \tag{4·63}$$

and it is evident that, since E is always to be taken as positive, this corresponds to spontaneous reaction from left to right of the chemical

† I.e. the Avogadro number of electrons.

reaction, whereby the terminal I becomes positive and the termir becomes negative.

Equation (4·63) is the same as (4·56) of the last section. This impd equation has therefore been obtained by a method which is in harmony with the theory of reaction equilibrium of the present chapter.

4·16. Number of independent reactions

The discussion so far in this chapter has been based on the supposition that there is only a single stoichiometric process in the system of interest. As noted in §4·2, this means simply that the changes in the amounts of *all* substances in the system can be expressed as small multiples or submultiples of the change in the amount of any *one* of them. For example, if the only stoichiometric process is

$$H_2 + Cl_2 = 2HCl,$$

the decrease in the amounts of hydrogen and chlorine are each equal to one-half the increase in the amount of hydrogen chloride. This would no longer be true if bromine were also present in the system.

When only a small number of substances are present it is usually quite easy to write down by inspection the minimum number of chemical equations which will represent the complete stoichiometry. In more complex systems it is desirable to use a more systematic procedure. For example, it would be time-consuming to determine, by trial and error, what are the minimum number of independent chemical equations involving, say, a dozen hydrocarbons, as in a cracking process.

A simple rule for determining this minimum number is as follows. Chemical equations are first written down for the formation from their component atoms of all compounds which are regarded as being present in the system.† These equations are then combined in such a way as to eliminate from them any free atoms which are not actually present. The result is the minimum number, R, of chemical reactions which are sufficient to represent the stoichiometry (although not necessarily the kinetics)‡ of the system.

By way of a simple example let it be supposed that the system of

† By 'present in the system' is meant 'present in significant amount'. Within the context of any particular problem it is always necessary to decide which of the various possible species may be neglected.

‡ Free radicals or other intermediates may be important in regard to the kinetic mechanism, but these may be insignificant in regard to the overall stoichiometry.

interest comprises the species H_2, CH_4, C_2H_6 and C_3H_8. We write down the equations

$$2H = H_2,$$

$$C + 4H = CH_4,$$

$$2C + 6H = C_2H_6,$$

$$3C + 8H = C_3H_8.$$

Since, by hypothesis, the system is not regarded as containing free hydrogen atoms in significant amount, the first equation may be used to eliminate H from the other three. This gives

$$C + 2H_2 = CH_4,$$

$$2C + 3H_2 = C_2H_6,$$

$$3C + 4H_2 = C_3H_8.$$

Similarly, eliminating the carbon atom, which is also not present in the above sense, we finally obtain

$$2CH_4 - H_2 = C_2H_6,$$

$$3CH_4 - 2H_2 = C_3H_8.$$

There are thus only two independent equations and $R = 2$. Any third equation made by combination of the last two, for example,

$$5CH_4 - 3H_2 = C_2H_6 + C_3H_8,$$

does not represent an independent stoichiometric process. It is, of course, quite immaterial which two out of all the possible linear combinations are chosen as the independent ones.

The following method for determining the number of independent reactions is very instructive, although it is less simple in practice than the method already described. Any *chemical* equation is simply a concise statement of the conservation of the atoms. This conservation may be expressed equally well by means of *algebraic* equations and the number, R, of independent reactions can be determined by finding the number of independent variables in these equations. Consider, for example, a closed system consisting of the five species CO, H_2, CH_3OH, H_2O and C_2H_6, and let Δ_{CO}, Δ_{H_2}, etc., be the changes in the mole numbers due to reaction over a given interval of time. The three equations

$$\left.\begin{array}{l} \Delta_{CO} + 0\Delta_{H_2} + 0\Delta_{H_2O} + \Delta_{CH_3OH} + 2\Delta_{C_2H_6} = 0, \\ 0\Delta_{CO} + 2\Delta_{H_2} + 2\Delta_{H_2O} + 4\Delta_{CH_3OH} + 6\Delta_{C_2H_6} = 0, \\ \Delta_{CO} + 0\Delta_{H_2} + \Delta_{H_2O} + \Delta_{CH_3OH} + 0\Delta_{C_2H_6} = 0, \end{array}\right\} \qquad (4\cdot64)$$

are based on the conservation of C, H and O atoms respectively. An examination of these equations shows that any *two* of the Δ's will determine the other three.† Thus if we choose Δ_{CH_3OH} and $\Delta_{C_2H_6}$ as the two independent variables, the solving of the equations (most conveniently by determinants) gives the changes in the dependent variables Δ_{CO}, Δ_{H_2} and Δ_{H_2O}:

$$\left.\begin{aligned}\Delta_{CO} &= -\ \Delta_{CH_3OH} - 2\Delta_{C_2H_6},\\ \Delta_{H_2} &= -2\Delta_{CH_3OH} - 5\Delta_{C_2H_6},\\ \Delta_{H_2O} &=\ \ 2\Delta_{C_2H_6}.\end{aligned}\right\} \quad (4\text{·}65)$$

Because *two* of the Δ's determine the other three, there are just *two* independent reactions. On account of the above choice of the independent variables, these two reactions may be conveniently chosen as those which represent the formation of CH_3OH and C_2H_6 from the other three substances:

$$\left.\begin{aligned}CO + 2H_2 &= CH_3OH,\\ 2CO + 5H_2 - 2H_2O &= C_2H_6.\end{aligned}\right\} \quad (4\text{·}66)$$

The question may now be asked, what are the *number of components* of the above system? By the number of components is meant the minimum number of substances which must be available in the laboratory in order to make up any chosen equilibrium mixture of the system in question. Thus, suppose it is required to prepare a mixture containing n_{CO} moles of CO, n_{H_2} moles of H_2, etc. It would be possible to do this by starting with the *three* substances CO, H_2 and H_2O, whose mole numbers we have chosen to regard as the dependent variables in the above equations. For if we make up a mixture of these substances only, and allow it to react, the mole numbers n_{CH_3OH} and $n_{C_2H_6}$ of CH_3OH and C_2H_6, when reaction is complete, are equal to the *increases* Δ_{CH_3OH} and $\Delta_{C_2H_6}$ respectively, because neither of these substances were present originally. Equations (4·65) show that these increases completely determine the amounts of CO, H_2 and H_2O which must react.

Expressed alternatively, the equations (4·65) may be written

$$\left.\begin{aligned}n^0_{CO} - n_{CO} &=\ \ \ n_{CH_3OH} + 2n_{C_2H_6},\\ n^0_{H_2} - n_{H_2} &=\ \ 2n_{CH_3OH} + 5n_{C_2H_6},\\ n^0_{H_2O} - n_{H_2O} &= -2n_{C_2H_6},\end{aligned}\right\} \quad (4\text{·}67)$$

where the n^0's are the initial mole numbers of CO, H_2 and H_2O. It is evident from these equations that the choice of the particular mixture‡ which it is desired to prepare (i.e. the choice of the five n's), requires definite values of the three n^0's. The mixture can therefore be prepared from *three* components.

† In this instance the number of independent Δ's, namely, two, is equal to the total number of Δ's, less the number of equations between them. The fact that *this is by no means generally true* may be seen by setting up the corresponding equations for the system composed of NH_3, HCl and NH_4Cl in which there is one independent Δ (and therefore one reaction) despite the fact that there are three Δ's and three equations between them, one for each type of atom.

‡ However, this mixture must also be 'accessible' within the conditions of reaction rate or equilibrium.

In general, it may be shown that if there are N species present in the system at reaction equilibrium, and if there are R independent reactions between them, the number of components† is $C = N - R$. For the present purposes it is not necessary to give a general proof of this relation. The matter has been discussed by Jouguet, Brinkley and others,‡ and it has been shown that the number C of components in a system of N reacting species is equal to the rank of the matrix of the subscripts to the symbols of the elements in the formulae of the N substances. The number, R, of independent reactions is then given by $R = N - C$. This procedure is mathematically the same as is involved in obtaining the solutions (4·65) of the equations (4·64), in the example already discussed, the determination of the rank of the matrix involving the same steps as in the sorting out of the dependent variables in the equations (4·64). For practical purposes it is easier to determine the number R of independent reactions by use of the rule which was formulated at the beginning of the section.

4·17. Conditions of equilibrium for several independent reactions

In § 4·4 we obtained the important condition of equilibrium

$$\Sigma \nu_i \mu_i = 0$$

for the case where there is only a single reaction in the system of interest. Its derivation was based on minimizing the total Gibbs function of *all* species which are present. Therefore it cannot be taken for granted that when there is more than one reaction there will be a relation of the above type for each of them, but it will now be shown that this is the case.

Consider the system comprising the species CO, H_2, H_2O, CH_3OH and C_2H_6, as discussed in the last section. For this system the equation for a change in G is

$$dG = - S\,dT + V\,dp + \mu_{CO}\,dn_{CO} + \mu_{H_2}\,dn_{H_2} + \mu_{H_2O}\,dn_{H_2O}$$
$$+ \mu_{CH_3OH}\,dn_{CH_3OH} + \mu_{C_2H_6}\,dn_{C_2H_6}.$$

In this equation only two of the dn's are independent; using (4·65) the changes in the mole numbers of CO, H_2 and H_2O can be expressed in terms of the changes in the mole numbers of CH_3OH and C_2H_6. The above equation may therefore be written

$$dG = - S\,dT + V\,dp + (\mu_{CH_3OH} - \mu_{CO} - 2\mu_{H_2})\,dn_{CH_3OH}$$
$$+ (\mu_{C_2H_6} + 2\mu_{H_2O} - 2\mu_{CO} - 5\mu_{H_2})\,dn_{C_2H_6}, \quad (4·68)$$

---

† But see § 5·4 concerning electroneutrality.

‡ Jouguet, *J. Éc. Polyt., Paris*, (2), **21** (1921), 62; Defay, *Bull. Acad. Roy. Belg.* **17** (1931), 940; Brinkley, *J. Chem. Phys.* **14** (1946), 563; Prigogine and Defay, *J. Chem. Phys.* **15** (1947), 614; Peneloux, *C.R. Acad. Sci., Paris*, **228** (1949), 1727. See also R. Aris, *Introduction to the Analysis of Chemical Reactors*, Prentice-Hall (1965), Chapter 1.

in which the variations dn_{CH_3OH} and $dn_{C_2H_6}$ are quite independent. This being the case, there are *two* conditions of equilibrium

$$\left(\frac{\partial G}{\partial n_{CH_3OH}}\right)_{T,p,\,n_{C_2H_6}} = 0,$$

$$\left(\frac{\partial G}{\partial n_{C_2H_6}}\right)_{T,p,\,n_{CH_3OH}} = 0.$$

These give the equations

$$\left.\begin{aligned}\mu_{CH_3OH} - \mu_{CO} - 2\mu_{H_2} &= 0,\\ \mu_{C_2H_6} + 2\mu_{H_2O} - 2\mu_{CO} - 5\mu_{H_2} &= 0,\end{aligned}\right\} \quad (4·69)$$

which correspond to the chemical reactions (4·66).

By generalizing the above it is readily seen that, for a system in which there are R independent reactions, there are R independent conditions of chemical equilibrium of the form

$$\Sigma \nu_i \mu_i = 0,$$

and there will also be R independent equilibrium constants.

It may be noted that the equilibrium constants of reactions which are not independent may be expressed as products or quotients of those which are. For example, the K_p of the reaction

$$3CO + 7H_2 = 2H_2O + CH_3OH + C_2H_6$$

is the product of the K_p's of reactions (4·66). Similarly, the value of ΔG_T^0 for this reaction is the *sum* of the values for the reactions (4·66).

4·18. General remarks on simultaneous reactions

Section 4·17 was concerned with the problem of calculating the equilibrium composition of a system when it is already known what are the substances which are present in significant amount. On the other hand, the following rather different problem will often present itself: given an initial set of reactants what are the various products which might be obtained? This question really precedes the one already discussed, for we are here concerned with finding which are the N species to be considered.

As an example we shall consider a project for the production of methanol from a mixture of carbon monoxide and hydrogen—as is actually done on the industrial scale. From these reagents it would obviously be possible to form many thousands of organic chemicals, some of which might be formed in much higher yield than the methanol itself and might be quite valueless, due to difficulties of separation. As a preliminary to such a project it is therefore necessary to carry out a thermodynamic analysis, based on available free-energy data, to determine:

(a) at what temperature and pressure methanol would be formed in appreciable yield;

(b) what other compounds might be formed under the same conditions.

An investigation of this sort is not always such a large undertaking as it might seem. In each homologous series there is usually a fairly constant change in the free energy of formation from one member to the next. Therefore it might be possible to eliminate from consideration, as possible products, all members of a particular homologous series by examining the possibility of the formation of only two widely separated members. Conversely it is evident that almost all water forming reactions, e.g.

$$CO + 3H_2 = CH_4 + H_2O,$$

will be thermodynamically favoured because the free energy of formation of water has a large negative value, corresponding to its great stability.

The result of the preliminary survey would establish the optimum conditions of temperature and pressure for the required product and also what other compounds might be formed under the same conditions.† In general let it be supposed that a total of N substances are likely to be present at approximate equilibrium. It would then be a question of determining the number, R, of independent reactions between these substances, as described in § 4·16, and finally of setting up R equations of equilibrium. For example, if the system in question is a perfect gas mixture, there will be R simultaneous equations of the type

$$-RT \ln K_p = \Delta G_T^0,$$

and their solution, together with the conservation equations such as (4·67), will give the equilibrium composition of the system.

However, the solving of this family of equations is often a matter of considerable algebraic complexity. One method is to obtain a first approximation by treating each reaction as if it were the only one to occur in the system. This rough solution then becomes the basis for a second approximation and so on by successive steps. Useful discussions on the solving of complex equilibria are given in the literature.‡

† In the methanol example a large number of alternative products are possible; successful manufacture therefore depends on the discovery of a selective catalyst which speeds up the required reaction to the exclusion of others, i.e. a kinetic rather than a thermodynamic effect.

‡ Dodge, *Chemical Engineering Thermodynamics* (New York, McGraw-Hill, 1944), p. 526; idem, *Amer. Inst. Chem. Engng*, **34** (1938), 529; Taylor and Turkevitch, *Trans. Faraday Soc.* **35** (1939), 921; Brinkley and Kandiner, *Industr. Engng Chem.* **42** (1950), 850; Hutchison, Chem. Eng. Science, **17** (1962), 703.

4·19. General remarks on maximum attainable yield

A large positive value of ΔG^0 at 25 °C implies a small value of K_p at this temperature, in accordance with

$$-RT \ln K_p = \Delta G_T^0.$$

This does not necessarily mean that the reaction in question is unsuitable for preparative or manufacturing purposes. In the first place, it may occur that the equilibrium becomes more favourable at some other temperature; in endothermic reactions, for example, the equilibrium is favoured by a rise of temperature, as follows from the equation

$$\frac{d \ln K_p}{dT} = \frac{\Delta H}{RT^2},$$

and conversely in the case of exothermic reactions.†

Secondly, although the total pressure does not have any influence on K_p (in perfect gas mixtures), it may have a large effect on the *yield*. Whenever a gaseous reaction takes place with decrease in the number of molecules, a rise in pressure increases the fraction of the reactants which are converted.

Consider, for example, the reaction

$$A + 2B = C,$$

which will be supposed to have a very unfavourable value of ΔG^0, namely, 10 000 cal mol^{-1} at 500 K. Then, at this temperature,

$$K_p = \frac{p_c}{p_a p_b^2} = 4 \times 10^{-5}.$$

Let x be the number of moles of C which are obtained at equilibrium from a reactant mixture containing originally one mole of A and two moles of B. The equilibrium gas therefore consists of $(1-x)$ moles of A, $(2-2x)$ moles of B and x moles of C and the equilibrium relation can be expressed as

$$\frac{x(3-2x)^2}{4P^2(1-x)^3} = 4 \times 10^{-5},$$

where P is the total pressure. Solving this equation, we obtain the values of P which result in a given value of the yield x. The results are given in Table 8 and they show that a yield of as much as 50 % may be obtained at 316 atm, despite the apparently unfavourable value of ΔG^0.

It may be remarked that incomplete reaction does not necessarily imply wastage of the unused reactants. Ammonia synthesis is usually

---

† It follows that *both* exothermic and endothermic reactions give rise to a reduced equilibrium yield when conducted adiabatically, as compared to isothermally. This point is discussed quantitatively by Aris, *loc. cit.*

operated at a conversion of only 20–30%, but the unused nitrogen and hydrogen are recirculated to the inlet of the reaction vessel, after condensation of the ammonia which has been formed.

TABLE 8

x	0.01	0.1	0.5
p/atm	24	82	316

The converse situation to that which has been discussed above is where there is an increase in the number of molecules. Gas reactions of this type are favoured by a decrease of pressure, or alternatively by dilution with an inert gas. Under practical conditions the advantages of such a procedure might be seriously offset by decreased *rate* of reaction.

Some other factors concerning the yield of a reaction are as follows:

(*a*) The effect of an excess of one of the reactants is to decrease the yield relative to this reactant and to *increase* the yield relative to the other. For example, in the reaction discussed above let it be supposed that recirculation of unconverted A and B is for some reason impracticable. Then, if A is a more valuable raw material than B, it might be found advantageous to operate with an excess of B, in order to raise the degree of conversion relative to A.

(*b*) Whenever it is desired to achieve *maximum partial pressure* of the reaction products in the equilibrium gas, the reagents should be present in their stoichiometric proportions. Consider, for example, the reaction $A + 2B = C$; by setting up the equilibrium equation for an arbitrary mixture, followed by differentiation and maximization, it can be readily proved that the highest partial pressure of C is obtained when $p_b = 2p_a$.†

(*c*) In the case of exothermic reactions the equilibrium becomes less favourable with rise of temperature, but it may occur that a high temperature is necessary in order to obtain a sufficient rate. There is therefore a conflict between the thermodynamic requirements for maximum yield at equilibrium and the kinetic requirements for the approaching of that equilibrium with a sufficient speed. In such instances there are considerable advantages in reducing the temperature progressively, from an initially high value, along the path of reaction. If kinetic data is available it is possible to make a quantitative evaluation of the optimum temperature at any stage of the reaction process.‡

† This result is strictly true only if the gas mixture is perfect. See Pings, *Chem. Eng. Science*, **16** (1961), 181; Sortland and Prausnitz, *ibid.* **20** (1965), 847.

‡ Denbigh, *Trans. Faraday Soc.* **40** (1944), 352; *Chemical Reactor Theory* (Cambridge, 1965).

(d) Finally it may be remarked that thermodynamic predictions of the effect of temperature and pressure on reaction yield apply accurately only under conditions of reaction equilibrium, such as are seldom actually attained in industrial practice. The extent to which such predictions are in error, as a function of the fractional degree of attainment of equilibrium, has been examined by Rastogi and Denbigh (*Chem. Eng. Science* **7** (1958), 261) and are often rather surprisingly large under typical conditions.

PROBLEMS

1. The gaseous reaction $\frac{1}{2}A_2 + \frac{1}{2}B_2 = AB$ at 500 K has a standard free-energy change of -1000 cal mol^{-1}. A reaction system consists initially of $\frac{1}{2}$ mol of A_2 and $\frac{1}{2}$ mol of B_2 at 500 K and a total pressure of 1 atm. Calculate the free energy of the system, relative to the elements A_2 and B_2 in their standard states, at 10, 20, ..., 100 % conversion to AB, and plot the values. Show that the lowest point on the curve agrees with that calculated from the equation

$$RT \ln K_p = -\Delta G_T^0.$$

2. A catalyst has been found which gives adequate velocity at 500 °C in the reaction $CO + 2H_2 = CH_3OH$. Estimate the order of magnitude of the pressure which would be required to make this reaction feasible as an industrial process. The free energy and heats of formation are as follows (cal mol^{-1}

	$\Delta_f G^0{}_{298}$	$\Delta_f H_{298}$
CO	$-32\,810$	$-26\,420$
$CH_3OH(g)$	$-38\,690$	$-48\,080$

3. Prove that the maximum concentration of NH_3 is obtained in an ammonia synthesis process when the ratio of nitrogen to hydrogen is 1:3, provided that the gas mixture may be assumed to be perfect. In what respect will deviations from perfection cause this conclusion to be modified?

4. At 450 °C the equilibrium constant for ammonia synthesis is

$$\frac{f_{NH_3}}{f_{N_2}^{\frac{1}{2}} f_{H_2}^{\frac{3}{2}}} = 6.56 \times 10^{-3}.$$

Using the Lewis and Randall rule and the fugacity data of Newton (*Industr. Engng Chem.* **27** (1935), 302) calculate the maximum percentage yield of NH_3 in a 1:3 nitrogen-hydrogen mixture at 450 °C and 200 atm pressure.

5. What are the thermodynamic conditions under which the equilibrium constant of a chemical reaction passes through a maximum or minimum value?

6. Gaseous nitrogen peroxide consists of a mixture of NO_2 and N_2O_4, and the chemical equilibrium between these substances is established very rapidly. It has been suggested that this gas should be used as a

heat-transfer medium. Show in outline that the effective heat capacity per unit mass of the mixture may be expected to be much larger than for either of the pure components and that it passes through a maximum at a certain temperature.   [Modified from C.U.C.E. Qualifying, 1954]

7. A producer gas contains 7 % CO_2, 22 % CO and 14 % H_2. In order to increase its hydrogen content the gas is mixed with steam, and the water gas reaction, $H_2O + CO = CO_2 + H_2$, is carried out over a catalyst at a temperature at which K_p equals 8.0. It may be assumed that equilibrium is attained. The hydrogen which is formed has a monetary value which is n times the value of the added steam, mole for mole. On the assumption that the material costs are predominant, obtain an equation or equations for determining the most economic ratio of steam to producer gas, as a function of n.   [C.U.C.E. Qualifying, 1949]

8. A nitrous gas obtained by oxidation of ammonia is analysed (a) for total nitrous gas content, by absorption in alkali; (b) for degree of oxidation to equivalent NO_2, by use of an oxidizing agent.

It is thus found that the total nitrogen oxides, expressed as equivalent NO, amount to 10 % by volume and that the gas is 80 % oxidized to the higher valence state. Assuming that the nitrogen oxides are entirely present as NO, NO_2, N_2O_3 and N_2O_4, calculate the percentage of each of these constituents in the gas. The temperature and pressure are 25 °C and 1 atm respectively, and the standard free energies of formation are as follows:

	NO	NO_2	N_2O_3	N_2O_4
$\Delta_f G^0_{298}$(cal/mol)	20 650	12 275	33 130	23 350

[C.U.C.E. Qualifying, 1949]

9. At 1200 K the equilibrium constants of the reactions

$$C\,(graph) + CO_2 = 2CO,$$
$$CO_2 + H_2 = CO + H_2O,$$

are 63 and 1.4 respectively (atmosphere units). The standard free energy and enthalpy of formation of water vapour at 25 °C are $-54\,640$ and $-57\,800$ cal/mol respectively. Heat-capacity data are as follows (in cal K^{-1} mol^{-1}):

$$H_2, \qquad c_p = 6.947 - 0.2 \times 10^{-3}T + 4.8 \times 10^{-7}T^2;$$
$$O_2, \qquad c_p = 6.148 + 3.1 \times 10^{-3}T - 9.2 \times 10^{-7}T^2;$$
$$H_2O\,(g), \qquad c_p = 7.256 + 2.3 \times 10^{-3}T + 2.8 \times 10^{-7}T^2.$$

Use the data to evaluate the equilibrium constant, at 1200 K, of the reaction

$$C\,(graph) + \tfrac{1}{2}O_2 = CO.$$

[C.U.C.E. Qualifying, 1952]

10. It is required to estimate the maximum percentage conversion which might be expected in the gas reaction $A + B = C + D$ when carried out in mild steel equipment.

It is known that the changes of standard Gibbs free energy and of enthalpy in the reaction are $\Delta G^0_{298} = 4000$ cal mol^{-1} and $\Delta H_{298} = 10\,000$ cal mol^{-1}. No data are available on heat capacities, but A and C are diatomic molecules, B is a linear triatomic molecule and D is a non-linear triatomic molecule.

Estimate probable values for the upper and lower limits of the difference of heat capacity of products and reactants. Calculate the corresponding upper and lower limits for the degree of conversion which might be expected under practicable conditions in a mild steel vessel. Assume that the initial gas is an equimolal mixture of A and B.

[C.U.C.E. Tripos, 1952]

11. Outline the steps necessary to establish from the laws of thermodynamics, and the minimum other necessary postulates and definitions, the conditions under which the equilibrium constant of a gaseous reaction, expressed in terms of partial pressures, is independent of total pressure.

[C.U.C.E. Tripos, 1951]

12. The equilibrium $C + 2H_2 = CH_4$ was studied experimentally by Pring and Fairlie and the following rather discordant values were obtained for $K'_p = p_{CH_4}/p^2_{H_2}$:

Temp/K	1473	1573	1573	1648	1648	1673	1723
$K'_p \times 10^3$	2.44	1.46	1.58	1.00	1.17	0.89	0.75

Carbon $\quad c_p = 1.1 + 4.8 \times 10^{-3}T - 1.2 \times 10^{-6}T^2$, (cal K^{-1} mol^{-1})
H$_2$ $\qquad\quad c_p = 6.88 + 0.07 \times 10^{-3}T + 0.28 \times 10^{-6}T^2$, (cal K^{-1} mol^{-1})
CH$_4$ $\qquad c_p = 3.38 + 17.91 \times 10^{-3}T - 4.19 \times 10^{-6}T^2$, (cal K^{-1} mol^{-1})

Estimate upper and lower limits for the standard free energy and enthalpy of formation of CH$_4$ at 25 °C.

13. A gas consisting of CO and H$_2$ in the proportion $1:2$ is heated to 500 °C at 250 atm in the presence of a catalyst. Assume that only CH$_3$OH, C$_2$H$_5$OH and H$_2$O are formed. Estimate the ratio of these products at equilibrium using the following data:

	$\Delta_f G^0_{298}$	$\Delta_f H_{298}$
CO	$-32\,810$	$-26\,420$
CH$_3$OH (g)	$-38\,900$	$-48\,100$
C$_2$H$_5$OH (g)	$-38\,700$	$-56\,300$
H$_2$O (g)	$-54\,640$	$-57\,800$

14. The standard free energy of formation of methanol vapour at 25 °C is -38900 cal. Calculate the standard free energy of formation of liquid methanol, assuming the vapour to behave as a perfect gas.

The vapour pressure is 122 mmHg.

15. A system consists initially of NO$_2$ and water at 25 °C. Consult tables of free energies and determine what substances might be formed

in significant amounts. Which of these can be excluded on account of the known slowness of reaction? What are the minimum number of relations required to determine all the equilibria which are likely to be established in a short period of time?

16. A gas obtained from ammonia oxidation enters a counter-current absorption system with the composition 10 % NO_2 and 1 % NO, the remainder being nitrogen, oxygen and water vapour. The predominant reaction is

$$3NO_2 + H_2O = 2HNO_3 + NO.$$

Estimate the maximum concentration of nitric acid which can be obtained by absorption in water at 20 °C and 1 atm total pressure. Use the following data for the free energy and heat of formation and the partial pressures of nitric acid:

	$\Delta_f G^0{}_{298}$	$\Delta_f H_{298}$
$NO_2(g)$	12 275	8 030
$NO(g)$	20 650	21 600
$HNO_3(g)$	$-17 900$	$-32 000$
$H_2O(g)$	$-54 640$	$-57 800$

% HNO_3 by weight	49.94	53.83	60.12	69.62	76.5
p_{HNO_3}/mmHg	0·183	0.345	0.93	2.86	6.89
p_{H_2O}/mmHg	7.9	6.77	4.80	2.65	1.33

[Modified from C.U.C.E. Tripos, 1951]

17. A cell in which the following reaction takes place at atmospheric pressure

$$Zn(s) + 2AgCl(s) = ZnCl_2(1\,\text{M}) + 2Ag(s),$$

has an e.m.f. of 1.005 V at 25 °C and of 1.015 V at 0 °C. Assuming the temperature coefficient to be constant, estimate the following at 25 °C and atmospheric pressure:

(*a*) the change in enthalpy in the reaction,

(*b*) the amount of heat absorbed, per mole of zinc reacting, during the reversible operation of the cell.

Explain clearly the relation between these quantities. How would they compare if the cell were not operating reversibly?

18. In the manufacture of formaldehyde a mixture of air with methanol vapour is brought into reaction on a silver catalyst. In this process the silver slowly loses its metallic lustre and partially disintegrates. Use the following data to examine whether this might be due to the formation of silver oxide:

Gas pressure: 1 atm.

Operating temperature: 550 °C

Standard molar free energy of formation of silver oxide: -2590 cal at 25 °C.

Standard molar enthalpy of formation of silver oxide: -7310 cal at 25 °C.

The following mean heat capacities may be used: silver 6.4, silver oxide 15.7, oxygen 7.5 cal K^{-1} mol^{-1}.   [C.U.C.E. Qualifying, 1953]

19. It is proposed to prepare barium oxide by heating witherite ($BaCO_3$) in a furnace open to the atmosphere. Use the data below to show that a first estimate to the lowest temperature at which the process could be carried out is about 1400 °C. It may be assumed that BaO and $BaCO_3$ do not form a solid solution. Indicate in what respects the discussion of the system would need to be modified if this were not the case.

The standard free energies and heats of formation are as follows in kg cal/g mole:

	$\Delta_f G^0_{298}$	$\Delta_f H_{298}$
CO_2	-94	-94
BaO	-126	-133
$BaCO_3$	-272	-291

The heat capacity of barium carbonate is about 21.7 cal K^{-1} mol^{-1} and that of barium oxide may be estimated as about 10.4. To the same accuracy, the heat capacity of carbon dioxide may be taken as 9.0 cal K^{-1} mol $^{-1}$.   [C.U.C.E. Tripos, 1951]

20. Zinc sulphide is roasted in a current of dry air at atmospheric pressure and at a temperature of 1700 K. Determine whether zinc oxide or zinc sulphate is the more stable solid reaction product under the above conditions. In such a process the outgoing gas is found to contain 7 % by volume of sulphur dioxide. What percentage of sulphur trioxide would this gas contain if the trioxide were in equilibrium with the dioxide at the above temperature? What statements could be made with regard to the mechanism of the reaction, if the actual content of sulphur trioxide in the gas was found to exceed the calculated equilibrium figure?

The free energies of formation in cal/g mole at 1700 K and 1 atm pressure are as in the following table:

ZnO	$ZnSO_4$	SO_2	SO_3
$-43\ 300$	$-94\ 300$	$-69\ 700$	$-55\ 900$

[C.U.C.E. Qualifying, 1950]

21. It is desired to produce the substance B by the gas reaction

$$A = B + C.$$

For technical reasons it is required that the process shall be carried out in a single-stage flow system, without recycling, at a total pressure of 1 atm and a temperature of 500 °C. The catalyst which is used in the reaction chamber is sufficiently active to bring the gas to equilibrium and no appreciable reaction takes place in the absence of the catalyst. The substance B may be readily condensed to a liquid, without dissolving appreciable amounts of A or C, by cooling the gas after it has left the reaction vessel.

By what device would it be possible to attain, from a single-stage reactor, a maximum yield of liquid B, relative to the quantity of A which enters the system? Estimate its value.

The increase in the standard Gibbs free energy in the reaction process at 500 °C is 3530 cal mol^{-1}. The vapour pressure of B at the temperature of condensation is 0.01 atm.   [C.U.C.E. Tripos, 1952]

CHAPTER 5

PHASE RULE

5·1. Introduction

Two phases which are in equilibrium must always have the same temperature (§§ 1·4 and 1·13). In addition, they must have the same pressure, provided that they are not separated by a rigid barrier or by an interface having appreciable curvature (§ 2·9a). Finally, any substance which is able to pass freely between the two phases must have the same chemical potential in each of them (§ 2·9b). These important criteria of equilibrium, expressed in terms of the *intensive* properties T, p and μ, lead directly to the phase rule of Willard Gibbs.

The origin of the phase rule may be understood most clearly if we consider, in the first place, a pure substance. Now the state of each of its phases is completely determined by temperature and pressure (at any rate when external fields have a negligible effect). The fixing of these variables completely determines all other intensive properties of a pure phase. In particular, therefore, it completely determines the value of the chemical potential. It follows that if μ is plotted as a function of T and p in three dimensions, the continuum of the μ values for a particular phase will all lie on a certain surface. Different phases of the same substance will be represented by different surfaces; for example, for the phase α the equation to the surface will be a function, $\mu_\alpha(T, p)$, of temperature and pressure, and for some other phase β it will be a quite different function, $\mu_\beta(T, p)$. In general, such surfaces will intersect, as shown in Fig. 20.

Now, as has been said, the state of phase α is determined by its temperature T_α and pressure p_α. Similarly, the state of phase β is determined by T_β and p_β. Thus the state of the combined system of two phases is completely determined by the four variables $T_\alpha, ..., p_\beta$. But, if there is equilibrium, there are three equations between these four variables. These are†

$$T_\alpha = T_\beta,$$
$$p_\alpha = p_\beta,$$
$$\mu_\alpha(T_\alpha, p_\alpha) = \mu_\beta(T_\beta, p_\beta).$$

It follows that only one of the four variables can be chosen arbitrarily; *there is one degree of freedom*. For example, if we choose the tem-

† The fact that the third equation is also a relation between the temperatures and pressures, like the first two, may perhaps be seen more clearly by imagining that $\mu_\alpha(T_\alpha, p_\alpha)$ and $\mu_\beta(T_\beta, p_\beta)$ are known explicitly, e.g. $\mu_\alpha = RT(\text{constant} - \ln T^{\frac{5}{2}}/p)$. Cf. footnote to § 1·16.

perature of one phase as being 20 °C, the other phase will have the same temperature: the pressures of the two phases will also be equal and will have a definite value determined by the nature of the particular substance.

The three relations above are obviously satisfied only along the line of intersection of the two surfaces of Fig. 20. This line therefore determines the simultaneous values of temperature and pressure which allow of the co-existence of the two phases. If one of these is a vapour phase, the projection of the line on to the p-T plane would give the ordinary vapour-pressure curve.

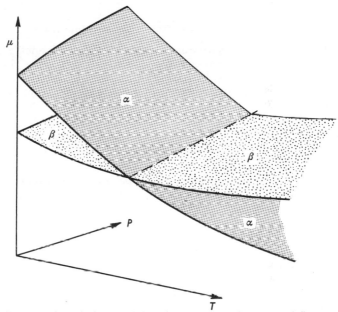

Fig. 20. Chemical potential surfaces for two phases α and β.

Similar considerations apply to the equilibrium of three phases, α, β and γ. There are six variables of state, T_α, P_α, T_β, ..., p_γ, and between these there are six independent equations of equilibrium:

$$T_\alpha = T_\beta = T_\gamma,$$

$$p_\alpha = p_\beta = p_\gamma,$$

$$\mu_\alpha = \mu_\beta = \mu_\gamma.$$

The equations entirely determine the values of the variables and therefore none of these can be chosen arbitrarily—the freedom of the system is zero. Considered geometrically, the three phases can only be at equilibrium at the point of intersection of three surfaces. This occurs at a unique temperature and pressure, known as the *triple point*.

Considered in a possibly simpler manner, it might be said that there are only two variables, the single temperature T and the single pressure p, which are assumed in the first place to be equal throughout the system. Between these two variables there are two equations of phase equilibrium:

$$\mu_\alpha(T,\,p) = \mu_\beta(T,\,p) = \mu_\gamma(T,\,p),$$

which therefore determine T and p completely.

5·2. The phase rule for non-reactive components

Consider a system containing a number, C, of distinct chemical species, none of which enter into reaction with each other. Let it be assumed, for the moment, that all of the substances are present in all of the phases, although some of the concentrations may be extremely low.

The state of each phase of a mixed system of this type requires a specification not only of its temperature and pressure but also of its composition. Since, by hypothesis, each phase contains C substances, the composition of each phase is specified by $C-1$ variables, e.g. mole fractions or weight percentages. Thus, including the temperature and pressure, each phase is completely specified by $C+1$ variables. Let the number of phases be P. Then for the entire system we have

$$\textit{number of variables} = P(C+1). \tag{5·1}$$

It may be noted that these variables specify the *state* of the phases, but not their *size*.

If the system is in complete equilibrium there are the following equalities between the variables:

$$T_\alpha = T_\beta = T_\gamma = \ldots \quad (P-1 \text{ equalities of temperature}),$$

$$p_\alpha = p_\beta = p_\gamma = \ldots \quad (P-1 \text{ equalities of pressure}\dagger),$$

$$\mu_{1\alpha} = \mu_{1\beta} = \mu_{1\gamma} = \ldots \quad (P-1 \text{ equalities of } \mu \text{ for species 1}),$$

$$\mu_{2\alpha} = \mu_{2\beta} = \mu_{2\gamma} = \ldots \quad (P-1 \text{ equalities of } \mu \text{ for species 2}),$$

etc., for the remaining components.

† The advantage of not assuming in the first place that the temperatures and pressures are necessarily equal is that it is then easier to see how the phase rule must be modified if the phases are not at the same pressure, as in osmotic equilibrium. The form taken by the phase rule for osmotic equilibrium has been discussed by Guggenheim, *Modern Thermodynamics* (London, Methuen, 1933), p. 28.

Adding these up we obtain

$$\text{number of equations between the variables} = (P-1)(C+2). \quad (5\cdot2)$$

With regard to the relations $\mu_{1\alpha} = \mu_{1\beta}$, etc., it should be noted that the chemical potentials in a mixed phase are functions of the $C-1$ composition variables, as well as of temperature and pressure. For example, if we denote the mole fractions of species i in the α and β phases by $x_{i\alpha}$ and $x_{i\beta}$ respectively, the condition of equilibrium of i between these phases, when the μ's are written out in full as functions of the independent variables, is

$$\mu_{i\alpha}(T_\alpha, p_\alpha, x_{1\alpha}, x_{2\alpha}, ..., x_{c-1,\,\alpha}) = \mu_{i\beta}(T_\beta, p_\beta, x_{1\beta}, x_{2\beta}, ..., x_{c-1,\,\beta}).$$

This is therefore a relation between the T's, the p's and the x's.

Now, in general, the number of independent equations connecting a set of variables must not be greater than the number of these variables themselves. Otherwise some of these equations will be incompatible. Therefore, from (5·1) and (5·2) we have

$$P(C+1) \geqslant (P-1)(C+2)$$

or
$$C+2-P \geqslant 0. \quad (5\cdot3)$$

We thus obtain the conclusion that the number of phases cannot exceed the number of components by more than two.† For example, in a single component system the maximum number of phases in equilibrium together is three, as occurs at the triple point.

The result (5·3) may also be expressed as an equality as follows. Let F be the amount by which the total number of variables exceeds the number of equations between them. Then subtracting (5·2) from (5·1):

$$F = C+2-P. \quad (5\cdot4)$$

This is the phase rule and F is called the *variance* or *degrees of freedom* of the system. It is the number of variables of the system whose values may be freely chosen by the experimentalist and *must* be so chosen before the system is in a determinate state.

For example, in a single component system of two phases, we have $F = 1$ and the system is said to be univariant. We can freely choose

† It is interesting that Gibbs, in his original derivation of the phase rule, expressed this conclusion rather cautiously, as being probable rather than certain. It may be that he had in mind the possibility of 'hidden' parameters of state (e.g. strain in a solid), or, alternatively, the possibility that the chemical potential of a component might be the same in two phases over finite ranges of temperature and pressure. This implies contact of the μ surfaces and seems very unlikely to occur.

An interesting photograph of seven liquid phases and a vapour phase in equilibrium in a seven-component system is given as the frontispiece to Hildebrand and Scott, *Solubility of Non-Electrolytes* (New York, Reinhold, 1950).

the temperature of the pair of phases, but not also the pressure. The latter has a definite value, at the chosen temperature, and is determined by the nature of the particular substance. In a two-component system of two phases we have $F=2$; therefore it is necessary for the experimentalist to decide on the values of two of the variables, e.g. temperature together with the mole fraction of one of the components in one of the phases, in order that the state of the system shall be completely determined.

In the above derivation it has been supposed that every component is present in every phase. Suppose, however, that a particular component j is unconditionally absent from a certain phase γ. In this case the number of composition variables which are required to specify the state of that phase (and thus of the whole system) is reduced by unity. On the other hand, the number of conditions of equilibrium is also reduced by the same amount, since the fact that j is not present in the phase implies that

$$\mu_{j\gamma} > \mu_{j\alpha} = \mu_{j\beta} = \mu_{j\delta} = \ldots.$$

The result (5·4) is therefore unchanged.

The proof of the phase rule originally given by Gibbs in 1875 is simpler but more sophisticated than the one above. In place of the mole fractions, he takes the chemical potentials as independent variables. He regards the state of each phase as being determined by its temperature and pressure and by the chemical potentials of each of its C components, i.e. $C+2$ variables in all. At equilibrium the value of each of these variables is constant through all of the phases. The state of the *whole* system is therefore determined by the $C+2$ variables, T, p and the μ_i. However, in each phase the possible changes in these variables are related to each other through a Gibbs–Duhem equation (2·83). Thus for the α phase we have

$$S_\alpha \, \mathrm{d}T - V_\alpha \, \mathrm{d}p + \Sigma n_{i\alpha} \, \mathrm{d}\mu_i = 0.$$

Since there are P such equations, the number of independent variables is only $C+2-P$. This is the same result as obtained previously.

It may be remarked that the number 2 which appears in (5·4) is a consequence of the following assumptions:

(*a*) the pressure is constant across each phase interface;

(*b*) in addition to the variables of composition, *two* other variables, namely, temperature and pressure, are sufficient to determine the state of the system.

Now in osmotic equilibrium the two liquid phases do not have the same pressure. In certain problems it may occur also that additional variables are significant, such as the magnetic field intensity. The statement $F = C + 2 - P$ is therefore by no means universally true.

5·3. The phase rule for reactive components

Consider a system in which there are altogether N chemical species, some of which may be inert and do not react at all, whilst the remainder are at reaction equilibrium with each other. Let R be the number of the independent reactions, as determined by the method described in §4·16. Let P be the number of phases.

Proceeding as in 5·2, the state of the system is specified by P values of the temperature, P values of the pressure and by $P(N-1)$ composition variables, such as the mole fractions. Thus

$$number\ of\ variables = P(N+1).$$

The conditions of equilibrium include $P-1$ equalities of temperature, $P-1$ equalities of pressure, $N(P-1)$ equalities of chemical potential of the N species between the P phases, and finally, R conditions of chemical reaction equilibrium, each of the form $\Sigma \nu_i \mu_i = 0$. Thus

$$number\ of\ equations\ between\ the\ variables = (N+2)(P-1)+R.$$

Therefore
$$F = P(N+1) - (N+2)(P-1) - R$$

or
$$F = N - R + 2 - P. \tag{5·5}$$

Comparing with (5·4), the quantity $N-R$ is seen to take the place of C, which was the total number of species in the case of a non-reactive system. The two forms of the phase rule, (5·4) and (5·5), may be made formally identical if we define the number of components of the reactive system by the relation

$$C \equiv N - R. \tag{5·6}$$

The number of components, in the sense of the phase rule, is therefore to be taken as the total number of chemical species less the number of independent reactions between them. In §4·16 it has already been shown (but not in complete generality) that $N-R$ is in fact the minimum number of substances which must be available in the laboratory in order to prepare any arbitrary equilibrium mixture of the system in question. The 'number of components' is therefore the same in both senses in which the term is used.

In the proof as given above it may be noted that there are only R conditions of reaction equilibrium and not PR such conditions. If reaction is at equilibrium in any one phase it is necessarily at equilibrium in every other phase; at any rate whenever the $N(P-1)$ equalities of chemical potential between phases are also satisfied.

5·4. Additional restrictions

In certain types of problem there are additional restrictions on the system, and the effect of each of these is equivalent to an extra equation between the variables and thus reduces the number of degrees of freedom by unity.

As a preliminary example consider a liquid A which is in contact with its vapour and also with the atmosphere at its standard pressure. Such a system may be discussed in several different ways:

(a) By disregarding the presence of the air. Then we have $C = 1$, $P = 2$ and therefore $F = 1$. Thus, as we know from experience, the liquid has a definite vapour pressure at any chosen temperature.

(b) By regarding the system as consisting of *two* components, A and air, and allowing for the restriction that the total pressure on the system is 1 atm. Thus $F = 1$, as before.

(c) By regarding the system as consisting of *three* components, A and oxygen and nitrogen, and allowing for *two* restrictions, namely the constancy of the total pressure and of the oxygen : nitrogen ratio. Thus again we have $F = 1$.

The several alternative ways of considering the problem are thus entirely equivalent.

In general, what we have called an 'additional restriction' implies that the system under discussion is of a special type, for example, that the pressure has a certain value or that the concentrations of two species in a particular phase are in a fixed ratio. An important instance occurs in the case of ionic solutions; on account of the powerful Coulombic forces, the concentrations of the ions must be such that the solution as a whole is electrically neutral.

It is evident that whenever a restriction is operative, it is equivalent to taking the number of components as being one less than is given by the relation $C = N - R$. For example, in the case of ionic solutions the number of effective components is $N - R - 1$, on account of the electroneutrality restriction.†

5·5. Example of the application of the phase rule

If the quantities of two substances are present in a fixed ratio in any phase, this constitutes an additional restriction. If however the quantities are in a fixed ratio only when they are taken as *totals* over two or more phases, this does not constitute an additional restriction.

This may be illustrated by considering the system comprising species A, B and C between which there is the equilibrium $A = B + C$.

---

† For example, a solution of acetic acid in water contains five species: H_2O, H^+, OH^-, HAc and Ac^-. There are two independent dissociation reactions and there are two components.

For this system (as follows from what was said previously) there are *two* components.

(a) Vapour phase only. If the system consists only of a single phase, we obtain $F = 3$ (i.e. temperature, total pressure and one mole fraction may all be varied independently). On the other hand, let it be supposed that *the system has been prepared from A only*, e.g. by heating pure A to the appropriate temperature. There is therefore the restriction that the partial pressures of B and C are equal and thus $F = 2$. We should evidently have obtained the same result if we had been unaware that the substance A suffered the dissociation into B and C; for we should then have regarded the system as consisting of a single component only and have obtained

$$F = 1 + 2 - 1 = 2.$$

(b) Vapour plus a liquid phase consisting of pure A. Let it be supposed that only the species A is present in the liquid phase, the vapour phase consisting of A, B and C in equilibrium. We obtain $F = 2$ in the general case and $F = 1$ in the special case where the system has been prepared from pure A only.

(c) Vapour and liquid phases each comprising all species. We now suppose that all three substances are present in both the liquid and the vapour phases. In the general case we have $F = 2$, as in the last paragraph. However, in the special case where the system has been prepared from A only *there is no longer the restriction that the partial pressures of B and C are necessarily equal*; this is because these substances will in general have an unequal solubility in the liquid phase. Although it remains true that the number of moles of B and C are equal in the total system, there is no longer a restriction on the composition of either of the single phases. Therefore F remains equal to two and does not fall to unity as in case (*b*) above.

It is this point which the present example, otherwise rather elementary, has been intended to illustrate, and it is perhaps worth considering the matter in rather more detail. We shall suppose that the reaction system, prepared from A only, is contained in a vessel equipped with a movable piston and held at constant temperature. If only the substance A is able to be present in the liquid phase, then $F = 1$ as shown above. It is therefore possible to change the relative amounts of the two phases, by causing a change in the volume of the system by means of the piston, without change in the total pressure or of the partial pressures; at a fixed temperature each of these variables has a definite value for as long as the two phases co-exist.

On the other hand, let it be supposed that either or both of the gases B and C can dissolve in the liquid A. If we draw out the piston there will be a change in the relative proportions of B and C in the two phases, due to their unequal solubilities. The composition of the phases is thus not uniquely determined by the temperature and, in fact, $F = 2$ as shown above. The situation is, perhaps, clearest if we suppose that only B is able to dissolve appreciably in the liquid A. If the piston is touching the liquid surface, so that no vapour phase is present, the reaction equilibrium, $A = B + C$, is entirely to the left, since by hypothesis no C can be present in the liquid. As the piston is drawn outwards the vapour phase enlarges and some dissociation of A is able to take place. Thus the composition of both liquid and vapour phases change progressively as vaporization is continued. In the limit, when all liquid has disappeared, the stoichiometric restriction $p_b = p_c$ becomes operative. However, since a phase has now disappeared the variance remains unaltered at $F = 2$.

In brief, the system prepared from A only is univariant if B and C are confined to the vapour phase and is divariant if one or both of them can dissolve in the liquid A.†

The phase rule is not concerned with the *quantities* of the phases which are in equilibrium, only with their intensive variables. On the other hand, whenever a stoichiometric restriction applies to a collection of two or more phases, rather than to each phase separately, it is often useful to introduce the *ratios* of the quantities of these phases as additional variables. In the above example let there be n_g moles of vapour phase, containing mole fractions y_b and y_c of components B and C, and let n_l, x_b and x_c be the corresponding quantities for the liquid phase. Then if the system has been prepared from pure A we have

$$n_l x_b + n_g y_b = n_l x_c + n_g y_c$$

or
$$(x_b - x_c) = \frac{n_g}{n_l}(y_c - y_b). \qquad (5·7)$$

The stoichiometric restriction on the system as a whole can thus be expressed in terms of an additional variable, namely, the ratio n_g/n_l. An application of this type of equation will be discussed shortly in connexion with zinc smelting.

It may be remarked that if we were unaware of the dissociation of A to form B and C the system would be regarded as consisting of a single component only. We should thus have deduced that $F = 1 + 2 - 2 = 1$; as we have seen this is incorrect whenever the products of the dissociation are not confined to a single phase. Two systems, each prepared from A only and held at the same temperature but of different ratios n_g/n_l, would not have the same physical properties.

† It may seem a paradox that a *discrete* change from $F = 1$ to $F = 2$ is determined by a quantity such as the solubility which is continuous. However, there is nothing absolute about a value $F = 1$ or $F = 2$; if the solubilities of B and C in the liquid are both small, the system is effectively univariant, although not precisely so. In the application of thermodynamics it is always necessary to idealize the system under discussion, by choosing the variables which are significant and neglecting all others.

5·6. Alternative approach

Greater physical insight into a particular problem may often be obtained by making direct application, not of the phase rule itself, but of the conditions of equilibrium on which it is based. As an example consider the system discussed under case (c) above. Let T and p be the temperature and total pressure respectively and let p_a, etc., be the partial pressures in the vapour and let x_a, etc., be the mole fractions in the liquid phase. If the gas phase may be assumed for simplicity to be a perfect mixture and if the liquid phase is an ideal solution, then the eight variables are related by the following six equations:

$$p_a + p_b + p_c = p,$$
$$x_a + x_b + x_c = 1,$$
$$p_a = f(T, p, x_a),$$
$$p_b = f(T, p, x_b),$$
$$p_c = f(T, p, x_c),$$
$$\frac{p_b p_c}{p_a} = K_p = f(T).$$

For example, the third equation states that the partial pressure of A is determined by the temperature, by the mole fraction of A in the liquid phase and also, although only to a small extent, by the total pressure (see § 6·5). The final equation states the condition of reaction equilibrium.

An examination of the equations shows that two of the variables are independent. The system is therefore divariant, in agreement with the result of applying the phase rule. If the system had not been assumed to have the properties of a perfect gas and of an ideal solution, the equations would have taken a rather more complicated form, but the same conclusion would have been reached.

5·7. Two examples from the zinc smelting industry

(a) The reduction of zinc oxide by carbon. This reaction gives rise to metallic zinc and also to the two oxides of carbon. We shall discuss an equilibrium state of the system in which the following species are all present: ZnO, C, Zn, CO, CO_2. Between these there are two independent reactions† and therefore there are three components.

† These may be written formally as
$$ZnO + C = Zn + CO,$$
$$2CO = C + CO_2,$$
but the following alternative choice probably corresponds more closely to the

Let it be supposed, in the first place, that there are three phases present, namely, zinc oxide and carbon as immiscible solids, and a vapour phase consisting of CO, CO_2 and zinc vapour. (The possibility of forming liquid zinc as a fourth phase will be discussed later.) The system would therefore be divariant if it were an entirely arbitrary mixture of the five species in question. However, the fact that it is prepared from zinc oxide and carbon implies a stoichiometric restriction on the composition of the vapour phase; for every atom of zinc vapour there must be one atom of combined oxygen as CO or CO_2. Hence

$$p_{Zn} = p_{CO} + 2p_{CO_2}. \tag{5·8}$$

The system is therefore univariant, so that at any chosen temperature it has a definite total pressure and vapour phase composition. It may be noted that if the process is operated in a furnace open to the atmosphere it is impossible for any steady evolution of zinc oxide to occur until the condition

$$p_{Zn} + p_{CO} + p_{CO_2} = 1 \text{ atm}$$

is satisfied. This occurs at a particular temperature, T'_0, and in this respect the system resembles the dissociation of calcium carbonate, which was discussed in §4·11.

The matter may also be discussed by the method of §5·6. The state of the system may be specified by means of the five variables, p, T, p_{Zn}, p_{CO}, p_{CO_2}. Between these there are four equations:

$$p_{Zn} = p_{CO} + 2p_{CO_2}, \tag{5·9}$$

$$p_{Zn} + p_{CO} + p_{CO_2} = p, \tag{5·10}$$

$$p_{Zn} p_{CO} = K'_{p1} = f_1(T), \tag{5·11}$$

$$p_{CO_2}/p_{CO}^2 = K'_{p2} = f_2(T), \tag{5·12}$$

and it is evident that only one of the variables can be chosen arbitrarily. The last two equations refer, of course, to the partial equilibrium constants of the reactions as written in the footnote.

We consider now the possibility of the presence of an additional phase, namely, liquid zinc. If the temperature of the reaction system is raised above T_0 the pressure will rise above 1 atm. It is found that the displacement to the right in the chemical reactions causes the partial pressure of the zinc vapour to rise more rapidly, with increasing temperature, than the vapour pressure of liquid zinc. The vapour

actual mechanism of the process, in which it is the gaseous CO, rather than the solid carbon, which reduces the zinc oxide:

$$ZnO + CO = Zn + CO_2,$$
$$CO_2 + C = 2CO.$$

phase therefore becomes saturated and liquid zinc makes its appearance in the system above a certain temperature and total pressure. Under such conditions there is an extra phase but the stoichiometric restriction (5·8) no longer holds. The system therefore continues to be univariant.

Under such conditions, of course, the total number of atoms of metallic zinc in the system is still equal to the number of atoms of oxygen which are present as CO and CO_2. The following equation is analogous to (5·7):

$$p_{Zn} = (p_{CO} + 2p_{CO_2})\,\phi, \qquad (5·13).$$

where ϕ is the fraction of metallic zinc which is present as vapour. This equation now replaces (5·8) but introduces the additional variable ϕ. However, there is also the condition that the zinc vapour is saturated

$$p_{Zn} = f(T, p), \qquad (5·14)$$

so that there are altogether five equations between the six variables. Therefore the system is univariant and ϕ has a definite value at any chosen temperature.

(b) The oxidation of zinc sulphide by air. In zinc manufacture the process discussed above is preceded by the roasting in air of the mineral zinc sulphide. The substances which may be formed in significant amounts are ZnO, $ZnSO_4$, SO_2 and SO_3. It is readily seen that there are three independent reactions and these may be written

$$ZnS + \tfrac{3}{2}O_2 = ZnO + SO_2,$$

$$ZnS + 2O_2 = ZnSO_4,$$

$$SO_2 + \tfrac{1}{2}O_2 = SO_3.$$

An examination of the equilibrium constants of these reactions shows that zinc sulphate is not formed above a certain temperature whose value depends on the SO_2 partial pressure. Under industrial conditions, where a steady stream of air passes through a bed of the red-hot sulphide, the formation of sulphate may therefore be avoided. However, for the present illustrative purposes it will be supposed that the conditions are such that all the species occurring in the above equations are present in the equilibrium state of the system.

Let it be assumed, in the first place, that the system is prepared from zinc sulphide and from pure oxygen, in place of air. There are thus six species in all, and, since there are three independent reactions, there are three components. Assuming that the oxide, sulphide and sulphate are present as immiscible solids, there are four phases and the system is therefore univariant.

In this example the fact that the system is prepared in a certain way does not give rise to any stoichiometric restriction on the composition of the vapour phase. There is no relationship analogous to (5·8) between p_{O_2}, p_{SO_2} and p_{SO_3}. This is because the sulphur and oxygen atoms do not originate from the same substance and because they are present in more than one phase at the equilibrium state of the system.

If it were now supposed that the system is prepared from zinc sulphide together with an oxygen-nitrogen mixture, the number of components would be increased by unity and the system would be divariant. Its state could be specified by a statement of, say, the temperature together with the total gas pressure, or the partial pressure of nitrogen. The fact that we might have used a particular nitrogen-oxygen mixture, such as air, would not affect this result; some of the oxygen, originally present in the gaseous phase, is transferred into the solids, i.e. the zinc sulphate and oxide. Therefore there is no necessary relationship between the partial pressure of nitrogen and the partial pressures of O_2, SO_2 and SO_3.

On the other hand, the variance of the system is *not* further affected by the presence in the air of argon, krypton, etc., since the partial pressures of these gases stand in a fixed ratio to the partial pressure of the nitrogen.

PROBLEMS

1. How many independent components are there in the following systems?

(a) any mixture of N_2, H_2 and NH_3 at room temperature;

(b) any mixture of N_2, H_2 and NH_3 at a temperature at which the chemical equilibrium is rapidly established;

(c) a system obtained by heating ammonia gas to a temperature at which it partially dissociates into nitrogen and hydrogen.

2. An aqueous solution contains n solutes whose mole fractions are $x_1, x_2, ..., x_n$. The solution is in equilibrium with pure water through a membrane permeable only to the solvent. The temperature of the system is T and the pressures on the water and the solution are p_w and p_s respectively. How many of these variables are independent?

3. State the degrees of freedom in the sense of the phase rule in each of the following systems. In each example state briefly the considerations on which your result is based:

(a) A pure substance at its critical point.

(b) An azeotrope in a binary system.

(c) The system obtained by heating a metallic nitrite MNO_2 to a temperature at which the substances present at equilibrium in significant amounts are as follows: MNO_2, MNO_3, M_2O, N_2, O_2, NO and NO_2. Write down sufficient independent chemical reactions to represent all the equilibria. Discuss the degrees of freedom of the system on the suppositions (i) that the first three substances above are present as immiscible phases, (ii) they are completely miscible. The solubility of the gases in the condensed phases may be neglected. [C.U.C.E. Qualifying, 1951]

4. Derive from first principles the form of the phase rule applicable to a system containing N species between which there are R independent chemical reactions.

A liquid-vapour system contains the species H_2O, NO_2, N_2O_4, NO, N_2O_3, HNO_3 and HNO_2. What are the number of independent chemical reactions and the number of degrees of freedom in the sense of the phase rule? Show that any property of this system at equilibrium can be represented by means of a surface through a triangular prism.

[C.U.C.E. Qualifying, 1952]

5. Establish the number of degrees of freedom, in the sense of the phase rule, of the system composed of solid iron, the solid oxides FeO and Fe_3O_4 and gaseous CO and CO_2.

Show that the following data on enthalpies and free energies of formation and heat capacities, all at 25 °C, suggest that the temperature at which all the above substances are at equilibrium together at 1 atm is about 1120 °K. Mention briefly the possible sources of error in the calculation. Will the equilibrium temperature be dependent on the total pressure? Discuss this in relation to the phase rule.

	$\Delta_f H_{298}$ (cal mol^{-1})	$\Delta_f G^0_{298}$ (cal mol^{-1})	c_p (cal K^{-1} mol^{-1})
Fe (s)	0	0	6·0
FeO (s)	− 63 700	− 58 400	10.4 (estimated)
Fe$_3$O$_4$ (s)	− 267 000	− 242 400	36.4
CO	− 26 400	− 32 800	7.0
CO$_2$	− 94 100	− 94 300	8.9

[C.U.C.E. Tripos, 1954]

6. A mixture of three miscible liquids X, Y and Z is subjected to a simple distillation. Component Z is almost involatile. The progressive change in the weight percentage composition of the residue is determined by sampling and is represented by a curve on a triangular diagram XYZ.

On this curve point A represents the initial composition of the liquid and point B represents the composition of the residue at a later stage in the distillation. Through A a line is drawn parallel to the XY axis and intersects BX and BY at C and D. Show that the ratio CD/XY is the fraction of the liquid which has evaporated.

Show also that the composition of the vapour which is in equilibrium with the liquid phase at any point P on the curve is determined by the intersection of the tangent at P with the XY axis.

[C.U.C.E. Qualifying, 1952]

CHAPTER 6

PHASE EQUILIBRIA IN SINGLE COMPONENT SYSTEMS

6·1. Introduction

It is a familiar fact that the phase changes

$$\text{solid} \to \text{liquid} \to \text{gas},$$

in the direction of the arrows, each require an input of heat to the system whenever these changes take place under conditions of constant temperature and pressure. Only a small part of the heat absorption is due to the work which is associated with the change of volume and by far the greater part is due to an increase in the internal energy. This is because the phase changes, in the directions considered, involve a disorientation or a separation of the component molecules.

The relative smallness of the work may be illustrated by the vaporization of water. At 100 °C and 1 atm the latent heat which must be provided is 41.1 kJ mol^{-1}. At the same time there is an increase of volume of $3.02 \times 10^4 \text{ cm}^3 \text{ mol}^{-1}$ and the work done on the atmosphere is thus $3.02 \times 10^4 \text{ cm}^3 \text{ atm mol}^{-1}$ or 3.1 kJ mol^{-1}. This is only 8% of the latent heat. The remaining 92% is accounted for by the increase in internal energy in the change from water into steam:

$$q = \Delta U + w,$$

or
$$\Delta H = \Delta U + \Delta(pV).$$

This increase in internal energy consists in a change, probably rather small, in the translational, rotational and vibrational energy† of the molecules, together with a much more substantial increase in their potential energy, consequent on an increase in their separation. Similar remarks apply to the solid → liquid transition, although in this case the change in potential energy is much smaller.

If h_s, h_l and h_g denote the enthalpies per mole of a substance in the solid, liquid and gaseous states of aggregation respectively, then at any temperature and pressure we may expect

$$h_s < h_l < h_g,$$

and this sequence is due, at least in part, to the relative magnitude of the attractive forces in the three states of aggregation. On the other hand, the processes solid → liquid and liquid → vapour are both

---

† All these are forms of kinetic ('thermal') energy at the molecular level.

characterized by an increase in the 'randomness' of the system, or degree of spread over the quantum states. Thus, if Ω is the total number of distinguishable micromolecular states or 'complexions', we have

$$\Omega_s < \Omega_l < \Omega_g,$$

or

$$s_s < s_l < s_g,$$

where s denotes the entropy per mole.

At any particular temperature and pressure the stable phase is that which has the smallest value of its chemical potential, or Gibbs free energy per mole. Thus, considering liquid and vapour phases, if

$$\mu_l < \mu_g,$$

the liquid is the more stable of the two. This condition can also be expressed in the form

$$(h_l - T s_l) < (h_g - T s_g),$$

or

$$T(s_g - s_l) < (h_g - h_l). \tag{6·1}$$

Conversely, if

$$T(s_g - s_l) > (h_g - h_l), \tag{6·2}$$

the vapour is the more stable form. Equilibrium between the phases is determined by the temperature and pressure at which these relations become equalities.

Therefore, as emphasized already in §2·8, the equilibrium between two phases is determined by a compromise between the energy and entropy factors, or, in molecular terms, between order and disorder. The transition from a stable liquid phase to a stable vapour phase, which takes place with rising temperature, is due to a transition from the condition (6·1), where the molecular forces, as reflected in the value of $h_g - h_l$, are the most significant factor, to the condition (6·2), where the increase in randomness, as measured by $s_g - s_l$, becomes the dominant effect. The transition is made possible by the increased energy available to the system, as a consequence of its being in an environment at a higher temperature.†

6·2. The Clausius–Clapeyron equation

Consider any two phases α and β of the same substance. Now in a single component system μ_α and μ_β are each functions of the temperature and pressure only. On the other hand, they are not the same functions and the two phases can co-exist only at such values of the temperature and pressure that the chemical potentials are equal.

The possible states of equilibrium correspond to the line of intersection of the two μ surfaces, as discussed in §5·1. Let μ'_α and μ'_β be

† See also the remarks at the end of §4·3.

the particular values of the potentials which occur at a certain point along this line. Then

$$\mu'_\alpha = \mu'_\beta, \tag{6·3}$$

and at a neighbouring point along the line we have

$$\mu'_\alpha + d\mu_\alpha = \mu'_\beta + d\mu_\beta.$$

Therefore†

$$d\mu_\alpha = d\mu_\beta. \tag{6·4}$$

This equation may also be written‡

$$\left(\frac{\partial \mu_\alpha}{\partial T}\right)_p dT + \left(\frac{\partial \mu_\alpha}{\partial p}\right)_T dp = \left(\frac{\partial \mu_\beta}{\partial T}\right)_p dT + \left(\frac{\partial \mu_\beta}{\partial p}\right)_T dp, \tag{6·5}$$

which expresses the relationship between a change of temperature dT and a simultaneous change of pressure dp, such that the phase equilibrium is maintained.

Substituting from (2·111 b) and (2·112 b) we obtain

$$-s_\alpha dT + v_\alpha dp = -s_\beta dT + v_\beta dp,$$

where s_α and s_β are the molar entropies of the substance in the two phases and v_α and v_β are the corresponding molar volumes. Hence

$$\frac{dp}{dT} = \frac{s_\alpha - s_\beta}{v_\alpha - v_\beta}. \tag{6·6}$$

At the point of equilibrium, where $\mu_\alpha = \mu_\beta$, the transition between the two phases is reversible and therefore

$$s_\alpha - s_\beta = \frac{L}{T} = \frac{h_\alpha - h_\beta}{T}, \tag{6·7}$$

where L is the enthalpy of vaporization. Hence

$$\frac{dp}{dT} = \frac{L}{T \Delta v}, \tag{6·8}$$

where Δv is the increase in volume in the phase change.

This equation, known by the names of Clausius and Clapeyron, determines the pressure increase dp which is necessary in order to

---

† It may be remarked that this equation is not a sufficient condition of equilibrium, because it would be satisfied by $\mu_\alpha = \mu_\beta + \text{constant}$. The equation merely asserts that between two neighbouring points along the equilibrium curve there is an equal increment of potential for each phase.

‡ dp and dT are necessarily the same for the two phases, if they are to remain at hydrostatic and thermal equilibrium. Thus when $\mu_\alpha = \mu_\beta$, we also have $T_\alpha = T_\beta$ and $p_\alpha = p_\beta$, and thus $dT_\alpha = dT_\beta$ and $dp_\alpha = dp_\beta$. The latter relation will apply even when there is a significant pressure difference across a curved interface, as in equation (2·51), provided that the radii of curvature are held constant.

maintain phase equilibrium when there is a temperature increase dT. Its derivation does not depend on any assumptions concerning the nature of the two phases, but the equation is applicable as it stands only to single-component systems, because μ_α and μ_β have been assumed to be functions of T and p only. It may be noted that a substance such as water is of this type, even though it may contain several chemical species, such as H_2O, OH^-, H_3O^+, $(H_2O)_2$, etc., *which are at reaction equilibrium.* If their total number is N, it is readily seen that between them there are altogether $N-1$ chemical reactions and stoichiometric restrictions. There is thus only one independent component† and only one independent chemical potential. As shown already, in connexion with the phase rule, such a system has one degree of freedom when there are two phases α and β. Thus an arbitrarily chosen temperature change, ΔT, will give rise to a definite pressure change Δp, as given by the integral of (6·8). For the same reason the left-hand side of this equation is a complete, and not merely a partial, differential.

Of course the equation is not applicable to a mixture of alcohol and water, between which there is no reaction equilibrium. Such a system contains two components and there are two independent chemical potentials. Each of these is a function of composition, as well as of temperature and pressure; equation (6·5), on which (6·8) is based, is therefore incomplete and incorrect. From the point of view of the phase rule a two-phase, two-component system is divariant; a change in pressure, Δp, is thus not uniquely determined by a temperature change, ΔT, but depends also on the change in composition of one of the phases. The analogue of equation (6·8) for two component systems will be discussed in § 7·2.

For application to solid-solid or solid-liquid phase changes, the Clausius–Clapeyron equation may be expressed in the inverted form

$$\frac{dT}{dp} = \frac{T\,\Delta v}{L}, \tag{6·9}$$

which shows the effect of the applied pressure on the temperature of phase transition. This is usually quite small; for example, the freezing-point of water is depressed by 0.007 K atm^{-1}.

For application to solid-vapour or liquid-vapour phase changes, the equation may be put in a more convenient approximate form by neglecting the volume of the condensed phase by comparison with the vapour and also by assuming that the latter behaves as a perfect gas. Thus

$$\Delta v \doteq v_g \doteq RT/p. \tag{6·10}$$

† The number of components here being understood in the sense of the last paragraph of § 5·4.

With this approximation, (6·8) may be written

$$\frac{\mathrm{d}\ln p}{\mathrm{d}T} \doteqdot \frac{L}{RT^2}.$$ (6·11)

The approximation is a good one only under conditions where the vapour pressure p is not too large.

6·3 The enthalpy of vaporization and its temperature coefficient

The integration of (6·8) or (6·11) over a range of temperature and pressure requires a knowledge of the temperature and pressure-dependence of the latent heat. Putting $L = h_\alpha - h_\beta$ and expressing as functions of temperature and pressure:

$$\begin{aligned}
\mathrm{d}L &= \left(\frac{\partial L}{\partial T}\right)_p \mathrm{d}T + \left(\frac{\partial L}{\partial p}\right)_T \mathrm{d}p \\
&= \frac{\partial(h_\alpha - h_\beta)}{\partial T}\,\mathrm{d}T + \frac{\partial(h_\alpha - h_\beta)}{\partial p}\,\mathrm{d}p \\
&= (c_{p\alpha} - c_{p\beta})\,\mathrm{d}T + \left(\frac{\partial h_\alpha}{\partial p} - \frac{\partial h_\beta}{\partial p}\right)\mathrm{d}p.
\end{aligned}$$

Now
$$\mathrm{d}h = T\,\mathrm{d}s + v\,\mathrm{d}p$$

and therefore
$$\begin{aligned}
\left(\frac{\partial h}{\partial p}\right)_T &= T\left(\frac{\partial s}{\partial p}\right)_T + v \\
&= -T\left(\frac{\partial v}{\partial T}\right)_p + v,
\end{aligned}$$

by one of Maxwell's relations. Substituting in the previous equation

$$\begin{aligned}
\mathrm{d}L &= (c_{p\alpha} - c_{p\beta})\,\mathrm{d}T + \left(v_\alpha - T\frac{\partial v_\alpha}{\partial T} - v_\beta + T\frac{\partial v_\beta}{\partial T}\right)\mathrm{d}p \\
&= \Delta c_p\,\mathrm{d}T + \left(\Delta v - T\frac{\partial \Delta v}{\partial T}\right)\mathrm{d}p.
\end{aligned}$$ (6·12)

Now the variations $\mathrm{d}T$ and $\mathrm{d}p$ of temperature and pressure which maintain phase equilibrium are not independent but are related by means of (6·8). This equation may therefore be substituted in (6·12) to eliminate $\mathrm{d}p$. We thus obtain the following equation for the change of L along the equilibrium curve (Fig. 20 on p. 183):

$$\begin{aligned}
\frac{\mathrm{d}L}{\mathrm{d}T} &= \Delta c_p + \left(\Delta v - T\frac{\partial \Delta v}{\partial T}\right)\frac{L}{T\Delta v} \\
&= \Delta c_p + \frac{L}{T} - L\left(-\frac{\partial \ln \Delta v}{\partial T}\right)_p.
\end{aligned}$$ (6·13)

This formula was obtained by Planck† and is usually known by his name.

In the case of solid-vapour and liquid-vapour transitions equation (6·10) may be used as an approximation. Substituting (6·10) in the last term of (6·13)

$$L\left(\frac{\partial \ln \Delta v}{\partial T}\right)_p \doteq L\left(\frac{\partial \ln RT/p}{\partial T}\right)_p$$

$$= \frac{L}{T}. \qquad (6\cdot14)$$

Therefore, for these types of phase change, the last two terms of (6·13) approximately cancel and we obtain

$$\frac{dL}{dT} \approx \Delta c_p. \qquad (6\cdot15)$$

In the case of solid-solid and solid-liquid transformations this equation is quite inapplicable and (6·13) must be used as it stands.

It is appropriate to conclude this section with a few remarks on the magnitude of the latent heats. For most liquids the enthalpy of vaporization amounts to 10–100 kJ mol^{-1}; its value at the boiling-point is related to the temperature of boiling on the absolute scale by the following approximate relation due to Trouton:

$$\Delta S_B = \frac{L_B}{T_B} = 21 \text{ cal K}^{-1} \text{ mol}^{-1} = 88 \text{ J K}^{-1} \text{ mol}^{-1} \qquad (6\cdot16)$$

The entropy of boiling is thus approximately the same for all liquids. There are, of course, appreciable deviations from this rule, particularly in the case of hydroxylic substances. Nevertheless, it is useful for the purpose of obtaining an estimate of the enthalpy of vaporization in cases where it has not been measured. A more accurate rule of a similar type is due to Hildebrand.‡

If L_M is the enthalpy of melting and T_M is the melting-point, the entropy of melting is

$$\Delta S_M = L_M/T_M. \qquad (6\cdot17)$$

This quantity is much less constant than the entropy of boiling, and its magnitude seems to depend to a considerable degree on molecular shape. In the case of substances consisting of fairly compact molecules the value of ΔS_M is usually about 8–16 J K^{-1} mol^{-1}, but in exceptional cases the value may be as high as 40 J K^{-1} mol^{-1}. On the other hand, substances such as the paraffins and their alcohols consisting of elongated molecules have entropies of melting which may reach values

† Planck, *Treatise on Thermodynamics*, transl. Ogg (3rd ed.; London, Longmans, Green, 1927), p. 154.

‡ Hildebrand and Scott, *The Solubility of non-Electrolytes* (New York, Reinhold, 1950); see also Staveley and Tupman, *J. Chem. Soc.* (1951), p. 3597.

as high as 120 J K^{-1} mol^{-1} or more. These large values may be explained as due to the great increase in the number of complexions when a long molecule, previously extended and orientated in the crystalline solid, is able to coil up, in a great many alternative ways, on passing into the liquid state.†

In the case of covalent substances, the value of the enthalpy of melting is usually about 10–30% of the corresponding enthalpy of vaporization.

6·4. Integration of the Clausius–Clapeyron Equation

As shown previously, a form of the Clausius–Clapeyron equation which is approximately valid for solid-vapour or liquid-vapour systems at not too high values of the vapour pressure is

$$\frac{\mathrm{d}\ln p}{\mathrm{d}T} = \frac{L}{RT^2}.$$

With the same approximation, and for the same types of phase change, we also have equation (6·15)

$$\frac{\mathrm{d}L}{\mathrm{d}T} = \Delta c_p.$$

Hence

$$L = L_0 + \int_0^T \Delta c_p \, \mathrm{d}T, \tag{6·18}$$

where L_0 is an integration constant and denotes the value of the enthalpy of phase change at a temperature approaching the absolute zero.

Between the two equations we have

$$\frac{\mathrm{d}\ln p}{\mathrm{d}T} = \frac{L_0}{RT^2} + \frac{1}{RT^2} \int_0^T \Delta c_p \, \mathrm{d}T,$$

and therefore

$$\ln p = C - \frac{L_0}{RT} + \int_0^T \frac{\mathrm{d}T}{RT^2} \int_0^T \Delta c_p \, \mathrm{d}T, \tag{6·19}$$

where C is a second integration constant. It will be recognized that the theory on which this integration is based is closely analogous to that used in obtaining the integrated form of van't Hoff's equation in Chapter 4.‡

† For a discussion on the entropy of fusion see Ubbelohde, *Quart. Rev. Chem. Soc.* **4** (1950), 356; *Melting and Crystal Structure* (Oxford, 1965).

‡ The condition of equilibrium between the condensed phase β and the vapour phase α may be written as follows, if α is a perfect gas:

$$\mu_\beta = \mu_\alpha = \mu_\alpha^0 + RT \ln p_\alpha.$$

Thus

$$RT \ln p_\alpha = \mu_\beta' - \mu_\alpha^0.$$

This is the analogue of equation (4·51) as applied to reaction equilibria involving condensed phases.

Over short ranges of temperature Δc_p may be taken as being approximately constant. With this assumption the integration gives

$$\ln p = -\frac{L_0}{RT} + \frac{\Delta c_p}{R} \ln T + \text{constant}, \qquad (6\cdot20)$$

but L_0 in this equation is no longer to be interpreted as the true value of the enthalpy of phase change at a temperature approaching the absolute zero.

Equation (6·20) is a useful approximate form for the purpose of expressing the vapour pressures of liquids and solids. For example, the vapour pressure of mustard gas is given by†

$$\log_{10} p = -\frac{4500}{T} - 9.86 \log_{10} T + 38.525.$$

For this substance the value of the latent heat, as determined by plotting $\ln p$ against $1/T$ and taking tangents at various points, varies from $14\,420 \text{ cal mol}^{-1}$ at $14.4\,^{\circ}\text{C}$ to $13\,260 \text{ cal mol}^{-1}$ at $104.0\,^{\circ}\text{C}$.

For many substances it has been possible to compare the values of L, as obtained by use of the Clausius–Clapeyron equation, with values obtained by direct calorimetric measurement. With sufficient care the agreement seems always to be exact, and this may be regarded as one of the confirmations of the second law.

6·5. The effect of a second gas on the vapour pressure of a liquid or solid

Consider a solid or liquid phase β in equilibrium with its vapour α. Let it be supposed that in the gas phase there is present some other gas which is not appreciably soluble in β. The total gas pressure will therefore be greater than the vapour pressure of the solid or liquid. The former will be denoted p and the latter p_α.

As in equation (6·4) we have for the equilibrium of the liquid or solid with its vapour

$$d\mu_\alpha = d\mu_\beta.$$

However, in the gaseous phase the chemical potential is now a function of the mole fraction, y, as well as of the temperature T and the total pressure p. ($1-y$ is the mole fraction of the second gas, with which we are concerned only in so far as it makes $p > p_\alpha$.) The expansion of the above expression in terms of the independent variables T, p and y gives

$$\frac{\partial \mu_\alpha}{\partial T} dT + \frac{\partial \mu_\alpha}{\partial p} dp + \frac{\partial \mu_\alpha}{\partial y} dy = \frac{\partial \mu_\beta}{\partial T} dT + \frac{\partial \mu_\beta}{\partial p} dp.$$

† Balson, Denbigh and Adam, *Trans. Faraday Soc.* **43** (1947), 42.

Substituting from (2·111), (2·112), (2·111b) and (2·112b)

$$-S_\alpha \, dT + V_\alpha \, dp + \frac{\partial \mu_\alpha}{\partial y} \, dy = -s_\beta \, dT + v_\beta \, dp, \qquad (6·21)$$

where S_α and V_α refer to the partial molar entropy and volume respectively of the substance in the gaseous phase, and s_β and v_β refer to the entropy and volume per mole respectively in the condensed phase.

Let it be supposed that the gaseous phase behaves as a perfect mixture. Then according to equation (3·18) the chemical potential of the saturated vapour is given by

$$\mu_\alpha = \mu_\alpha^0(T) + RT \ln p + RT \ln y,$$

and thus

$$\left(\frac{\partial \mu_\alpha}{\partial y} \right)_{T,\,p} = \frac{RT}{y},$$

and also, from (3·23),

$$V_\alpha = \frac{RT}{p}.$$

Substituting these relations in (6·21) and noting that $p_\alpha = yp$, we obtain

$$RT \, d \ln p_\alpha = (S_\alpha - s_\beta) \, dT + v_\beta \, dp. \qquad (6·22)$$

Thus at constant total pressure

$$\left(\frac{\partial \ln p_\alpha}{\partial T} \right)_p = \frac{S_\alpha - s_\beta}{RT}, \qquad (6·23)$$

and at constant temperature

$$\left(\frac{\partial \ln p_\alpha}{\partial p} \right)_T = \frac{v_\beta}{RT}. \qquad (6·24)$$

It will be seen that these expressions are entirely consistent with the phase rule. A two-component two-phase system is divariant, and therefore two variables such as temperature and pressure may be varied independently; in fact (6·22) shows that the changes dT and dp entirely determine the change in the third variable, $d \ln p_\alpha$, which is not independent.

In (6·23), $S_\alpha - s_\beta$ may be replaced by L/T as in (6·7). We thus obtain

$$\left(\frac{\partial \ln p_\alpha}{\partial T} \right)_p = \frac{L}{RT^2}. \qquad (6·25)$$

This equation is obviously very similar to (6·11). However, the latter is accurate only under *two* conditions, namely, that the vapour is perfect and that the molar volume of the condensed phase is negligible compared to that of the vapour. The accuracy of (6·25) depends only on the first of these conditions. Expressed in an alternative way, the approximate form of the Clausius–Clapeyron equation

$$\frac{d\ln p_\alpha}{dT} = \frac{L}{RT^2}$$

becomes more accurate if a second gas is present, insoluble in the condensed phase, in such a way that the total pressure is constant. This applies, for example, to liquids open to the atmosphere, provided the air does not dissolve in them appreciably.

The second equation (6·24) shows the effect of the total pressure p on the vapour pressure p_α at constant temperature. The magnitude of the effect, which is usually small, is seen to depend on v_β, the molar volume of the condensed phase. For example, in the case of water v_β is about 18 cm^3. Putting $R = 82$ cm^3 atm and $T = 273$ K we find that the logarithm of the vapour pressure increases by about 0.1 % per atm. If v_β may be taken as constant the integration of the equation gives

$$\ln\frac{p_\alpha'}{p_\alpha} = \frac{v_\beta}{RT}(p - p_\alpha),$$

where p_α' denotes the vapour pressure at the total pressure p and p_α is the vapour pressure of the substance under its own pressure only.

The interpretation of the increase in vapour pressure due to the presence of the second gas is roughly as follows. The surface of the liquid or solid is in process of 'bombardment' not only by the molecules of its own vapour but also by those of the second component. The effect of the higher total pressure may therefore be regarded as causing some of the molecules of the condensed phase to be 'squeezed out' into the vapour, resulting in an increase in its vapour-phase concentration until a new state of equilibrium is attained. In a perfect gas there is no compensating effect in the reverse direction; the presence of the second gas does not give rise to an increased tendency of the vapour to condense, because a perfect gas is to be interpreted kinetically as consisting of point particles.

The equations for an imperfect vapour analogous to (6·23) and (6·24) may be obtained by a similar procedure. The following method of derivation is rather simpler, and could also have been adopted for the derivation of (6·23) and (6·24). Let f be the fugacity of the volatile substance in the vapour phase, where its chemical potential is μ_α. Then from the fugacity definition (3·56)

$$\mu_\alpha = \mu_\alpha^0 + RT \ln f.$$

As before, the condensed phase will be denoted β and the condition of equilibrium $\mu_\alpha = \mu_\beta$ may be written

$$\mu_\alpha^0 + RT \ln f = \mu_\beta.$$

Hence
$$R \ln f = \frac{\mu_\beta}{T} - \frac{\mu_\alpha^0}{T}$$

and
$$R d \ln f = d\left(\frac{\mu_\beta}{T}\right) - d\left(\frac{\mu_\alpha^0}{T}\right). \tag{6·26}$$

Now μ_β is a function of temperature and pressure, but μ_α^0 is a function only of temperature. The last equation may therefore be expanded

$$R d \ln f = \frac{\partial(\mu_\beta/T)}{\partial p} \, dp + \frac{\partial(\mu_\beta/T)}{\partial T} \, dT - d\left(\frac{\mu_\alpha^0}{T}\right).$$

Using (2·111 b), (2·113 b) and (3·64) we obtain

$$R d \ln f = \frac{v_\beta}{T} \, dp - \frac{h_\beta}{T^2} \, dT + \frac{h_\alpha^0}{T^2} \, dT, \tag{6·27}$$

where h_α^0 is the enthalpy per mole of the substance in the vapour phase at a pressure low enough for it to behave as a perfect gas. Hence finally

$$\left(\frac{\partial \ln f}{\partial p}\right)_T = \frac{v_\beta}{RT}, \tag{6·28}$$

and
$$\left(\frac{\partial \ln f}{\partial T}\right)_p = \frac{L}{RT^2}, \tag{6·29}$$

where
$$L = h_\alpha^0 - h_\beta.$$

Some results of Diepen and Scheffer[†] are an example of the need for (6·28) in place of (6·24). These workers found that the concentration of naphthalene in the vapour phase is very greatly increased if ethylene is also present in the gas space. At 12 °C and 100 atm total pressure the concentration of naphthalene in the vapour phase was 25 600 times larger than would be expected from the vapour pressure of naphthalene alone. Equation (6·24) accounts for only a very small part of this increase, and the remainder must be due to large deviations from the perfect gas laws, with specific interaction between the naphthalene and ethylene molecules in the vapour phase.

This kind of effect may be called the *solubility of a solid in a gas* and it was discovered by Hannay and Hogarth in 1880.[‡] In one of their experiments they showed that a solution of cobalt chloride could be raised above the critical temperature, and the 'dissolved' salt was then present in the vapour phase. Its absorption spectrum was the same as in the normal liquid solution. More recently it has been stated that the deposition of silica on turbine blades is due, at least in part, to the solubility of silica in high-pressure steam.[§]

† Diepen and Scheffer, *J. Amer. Chem. Soc.* **70** (1948), 4085.

‡ Hannay and Hogarth, *Proc. Roy. Soc.* **30** (1879–80), 178.

§ A review of such effects is given by Booth and Bidwell, *Chem. Rev.* **44** (1949), 477.

These phenomena are only significant under conditions of high pressure where the density of the vapour phase is appreciable. Under such conditions the separation between gas molecules is quite small and many of them may be present as clusters. When there is also a strong specific interaction between the clusters and any solid substance which is present, it may be expected that the latter will show a greatly enhanced volatility. The solvent power of a gas may thus be accounted for along the same lines as for a liquid solvent. †

6·6. Lambda transitions

Simple kinds of phase change, such as melting and vaporization, are characterized by considerable changes of volume, and also of entropy and enthalpy, at the point of transition. Thus, whereas the chemical potentials of the two phases are equal when they are at equilibrium together, their volumes, entropies and enthalpies are far from equal.

For a pure substance we have from equations (2·111 b), (2·112 b), (2·89) and (2·86)

$$\left.\begin{aligned}
\left(\frac{\partial \mu}{\partial p}\right)_T &= v, \\[2mm]
\left(\frac{\partial \mu}{\partial T}\right)_p &= -s, \\[2mm]
\left(\frac{\partial^2 \mu}{\partial p^2}\right)_T &= \left(\frac{\partial v}{\partial p}\right)_T = -v\kappa, \\[2mm]
\left(\frac{\partial^2 \mu}{\partial T^2}\right)_p &= -\left(\frac{\partial s}{\partial T}\right)_p = -\frac{c_p}{T},
\end{aligned}\right\} \tag{6·30}$$

and these equations express v, s, κ and c_p as first and second derivatives of μ.

In the ordinary type of phase change, if μ, s, v, etc., are plotted as functions of temperature and pressure, we obtain curves as shown diagrammatically in Fig. 21. The chemical potential itself shows a change of gradient at the point of phase change, but *no discontinuity*. The latter is manifested in the entropy, volume and all other higher derivatives of the chemical potential. ‡

Apart from the discontinuous changes of entropy and volume, an important feature of these normal types of phase change is that the values of c_p and κ do not usually change at all rapidly as the transition

---

† For a full review of the whole field see Rowlinson and Richardson, *Advances in Chemical Physics* **2** (1959), 85–118.

‡ With regard to the heat capacity, it may be noted that its value may either increase (as in the ice-water transition) or decrease (as in the water-steam transition) in passing from the lower to the higher temperature phase.

point is approached; the observable physical properties of the material give no intimation that a change of a rather drastic nature is about to take place.

Other types of phase change have been discovered which show quite a different character. In these there seems to be no difference of volume between the two forms of the substance and also little or no difference in entropy or enthalpy, i.e. zero or almost zero latent heat. The transition is manifested simply by a sharp change in the heat capacity and compressibility. These properties also vary rather rapidly as the transition point is approached.

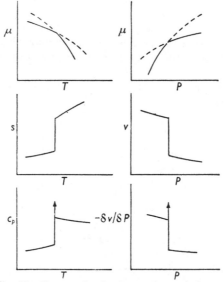

Fig. 21. Changes in the thermodynamic functions in a normal type of phase transition.

A carefully studied example is the transition which takes place in liquid helium at about 2.2 K. The heat capacity and density of liquid helium as functions of the temperature are shown in Fig. 22. In this instance at least it is known that the two 'phases' are able to co-exist, not merely at a particular temperature and pressure, but along a p-T equilibrium curve, as in the ordinary phase transitions already discussed. Similar effects are known to occur in many solids, notably in alloys, in the crystalline ammonium salts, in polymers and in solidified methane and hydrogen halides. For example, in ammonium chloride there is a sharp break in the heat capacity curve at -30.4 °C.

Now in any given example, if it were known with certainty that the latent heat and volume change were vanishingly small, the entropy and volume, as functions of temperature and pressure, would presumably be of the form shown in Fig. 23. This has led to the conception of phase changes which are of 'higher order' than the ordinary ones such as melting. According to this terminology an nth order phase change is one in which it is the nth derivative of μ, with respect to temperature and pressure, which first becomes discontinuous at the point of phase change. For example, in the ordinary first-order phase

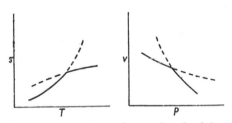

Fig. 22. Lambda transition in liquid helium.

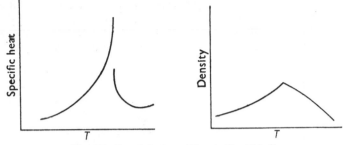

Fig. 23. Entropy and volume changes in a lambda transition.

change, such as melting, it is the first derivatives, namely, entropy and volume, which are discontinuous, and in the second-order phase change it is the second derivatives. The theory of such changes, if they exist, has been discussed by Ehrenfest and others, but it is not yet in an established position.† The discussion is based on the supposition that the μ surfaces of the two 'phases' do not intersect at a finite angle, as in ordinary phase change (see Fig. 20), but actually *make contact* over a range of temperature and pressure. On this basis Ehrenfest obtained an equation analogous to the Clausius–Clapeyron

† For discussion of the theory and literature references see Epstein, *Thermodynamics* (New York, Wiley, 1937); Roberts and Miller, *Heat and Thermodynamics* (London, Blackie, 1951); Temperley, *Sci. Progr.* **39** (1951) 27; Mayer and Streeter, *J. Chem. Phys.* **7** (1939), 1019; Pippard, *Classical Thermodynamics* (Cambridge, 1961); Guggenheim, *Thermodynamics*, 3rd edition, § 7·20–7·26.

equation,† but based on the second derivatives of μ instead of on the first. This equation is in approximate agreement with the experimental data on the transition in liquid helium, but has not yet been adequately tested in other examples.

One of the difficulties on the experimental side is to show whether any particular anomalous phase change really satisfies the theoretical criterion of being 'second order'. For this purpose it is necessary to demonstrate experimentally that the heat capacity does not rise to infinity, for if it did it would be equivalent to a latent heat. This requires heat-capacity measurements of very high accuracy taken at very small intervals of temperature close to the temperature of discontinuity. In

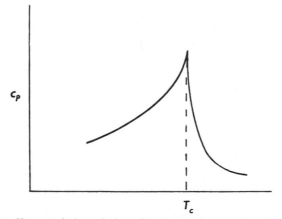

Fig. 24. Characteristic variation of heat capacity at a lambda transition.

many cases the measured heat capacity rises to extremely high values, and it is therefore difficult to eliminate the possibility that it rises to infinity.

However, the question whether or not the latent heat is zero is perhaps not the most significant aspect of many of the observed anomalous phase changes. What can be clearly demonstrated experimentally is that they differ from ordinary phase changes in regard to the way in which c_p varies as the point of apparent discontinuity is approached. The curves are of the type shown in Fig. 24 and are clearly distinct from those characteristic of normal phase change, as shown in Fig. 21, where c_p remains almost constant up to the transition point. Considering the lower temperature portion of the curve in Fig. 24, it is seen that the increase in c_p rapidly accelerates, as if a change is already taking place in the substance below the temperature T_c.

† In the Clausius–Clapeyron equation itself, if L and ΔV are both zero dp/dT is indeterminate.

The shape of the curve, similar to the Greek letter lambda, has led to the term *lambda transition*. This is preferable to the term 'higher-order phase change', which presupposes the correctness of the theory as outlined above. Indeed, the occurrence of such transitions seems really to imply a breakdown, or at least a limitation, in the usual concept of 'phase'. This concept is based on the idea that there exist entirely distinct forms of a substance, and for each of them there is a μ surface which may be thought of as being continued into an unstable region, as in Fig. 20. The most fruitful approach to the theory of the lambda transitions seems to lie in quite a different direction—namely, that there is a *gradual* transition, over a range of temperature and pressure, from one kind of arrangement of the component atoms and molecules to some other arrangement, characterized by a different degree of orderliness, the point of apparent discontinuity on the c_p curve being that at which the gradual change has completed itself.

This kind of interpretation may be made clearer by considering the lambda transition in β brass, which consists of copper and zinc atoms in almost equal numbers. The crystal structure has been established by X-ray methods, and both above and below the λ point (about 470 °C) the system has a body-centred cubic lattice, each atom having eight nearest neighbours. The structure thus consists of two interpenetrating simple cubic lattices, which may be called A and B. It seems that at temperatures well below the λ point all of the copper atoms are on one of these lattices and all of the zinc atoms are on the other, so that each copper atom is surrounded by eight zinc atoms as nearest neighbours, and vice versa. However, with rise of temperature the copper and zinc atoms tend to change places, and this process occurs to a gradually increasing extent over a fairly wide temperature range, until finally the system becomes a completely random arrangement of copper and zinc atoms over the A and B lattices. The temperature at which randomization is complete seems to be that at which c_p is observed to fall very rapidly, i.e. the λ point.

It seems that the interaction energy between a zinc and a copper atom is larger than the mean interaction energy between a copper-copper pair and a zinc-zinc pair. For if the interaction energies are as stated, the lowest potential energy of the system will occur when each atom is surrounded by eight atoms of the other kind, and this is the observed structure at low temperature. As noted in § 6·1, the energy factor dominates over the entropy factor at low temperatures and the system tends to take up the configuration in which the attractive forces are most completely satisfied. On the other hand, a completely ordered structure of this type is an intrinsically improbable one. Therefore, with rising temperature (i.e. with increasing energy available from the heat bath), there is a tendency for the completely ordered structure to pass over into one which is more random and has higher entropy. For this to occur interchange of zinc and copper atoms must take place, against the operation of the interatomic forces, and thus considerable heat must be absorbed per unit rise of temperature, i.e. c_p has a large value. However, when the structure is at last completely random, any further interchange does not increase the potential energy, and the heat capacity falls to a lower value characteristic only of the kinetic energy of vibration of the atoms in the lattice. There is therefore a fairly sharp discontinuity of c_p, as is observed. But there is no discontinuity in the internal structure of the material itself, and there is nothing which could be called a phase change in the accepted sense. The λ point seems to be simply that temperature at which the process of randomization has approximately completed itself.

According to the theory[†] the accelerating rise in c_p below the λ point is due to the fact that the greater the number of zinc and copper atoms which have already changed positions between the A and B lattices, the easier it is for the next interchange to occur. When the randomization is already partially complete any particular copper atom is surrounded on the average by fewer than eight zinc atoms; the total force on the copper atom, tending to prevent its interchange, is therefore less than at a lower temperature where the number of zinc atoms in its vicinity was larger. It is thus as if the change catalyses itself, and such effects are said to be 'co-operative'.

In brief, it seems that the discontinuity in the heat capacity is not due to an abrupt change from one well-defined phase to another, but arises from a gradual change in the internal degree of order of the material.

Similar ideas may be applied to transitions in other materials. The discontinuities in heat capacity which have been observed in solid methane and the ammonium salts have been attributed by Pauling to the gradual onset of free rotation of the CH_4 molecules or NH_4^+ ions inside the lattice. At low temperatures these particles are thought to carry out small torsional oscillations about an axis, but with increasing temperature, the amplitude of the oscillations gets larger, with absorption of heat and increase in c_p, until the molecules or ions eventually rotate freely about their lattice poin[t].An alternative explanation is that it is not so much a case of transition from vibration to rotation but rather a transition from vibration about fixed axes to vibration about disordered axes. The evidence has been discussed by Staveley.[‡]

Discontinuities in heat capacity and other properties are quite common in a number of salts which contain symmetrical ions, e.g. NH_4^+, NO_3^-, SO_4^{2-}, PO_4^{3-}, etc., and also in polymeric materials such as rubber. In the polymers it is probably a question of the side groups, or segments of the molecular chains, developing increased disorder of vibration or rotation as the temperature rises, the process becoming essentially complete at a fairly closely defined temperature, the λ point. The latter often coincides with important changes in the plastic properties.

† Bragg and Williams, *Proc.Roy. Soc.* A,.**145** (1934), 699; *Proc. Roy. Soc.* A, **151** (1935), 540.

‡ Staveley, *Quart. Rev. Chem. Soc.* **3** (1949), 65.

PROBLEMS

1. The increase, ΔT, in boiling-point of a liquid due to a *small* increase in pressure may be expressed in the following approximate form:

$$\frac{\Delta T}{T_b} = \frac{\Delta p}{10},$$

where T_b is the boiling-point at 1 atm and Δp is the excess pressure as a fraction of an atmosphere. Derive this expression, using the Trouton relation.

2. Calculate the melting point of water at a pressure of 1.80×10^3 kPa. Take the densities of ice and water as 0.917 and 1.000 g cm^{-3} respectively and the enthalpy of melting as 6008 J mol^{-1}.
[Kelvin's direct measurement gave -0.129 °C.]

3. Calculate the temperature coefficient of the enthalpy of melting of water, under the equilibrium pressure. The coefficients of thermal expansivity of water and ice at 0 °C are -6.0×10^{-5} and 11.0×10^{-5} respectively, and the specific heat capacities are 1.0 and 0.5 respectively. The specific volume of ice is 1.093 cm^3 g^{-1}.

4. For water at 100 °C, dp/dT has the value 27.12 mmHg/°C. The volume of 1 g of saturated water vapour at 100 °C is 1674 cm^3. Calculate the enthalpy of vaporization and compare with the calorimetric value of 2254 J g^{-1}.

5. Show that the change of volume with temperature of a fixed quantity of vapour which is in equilibrium with a liquid phase is given by the approximate relation

$$\frac{1}{V}\frac{dV}{dT} = \frac{1}{T}\left(1 - \frac{L}{RT}\right).$$

6. Estimate the maximum weight of oil which can be vaporized into a stream of air at 20 °C and 1 atm. The only available data on the oil are: mol. wt. $= 120$ and boiling-point $= 200$ °C.

7. A quantity of steam contains suspended droplets of water. The 'dryness' q is defined as the fraction of the total mass which is actually present as vapour. Show that if the system is heated at constant volume

$$\left(\frac{\partial q}{\partial T}\right)_V = -\frac{1}{(v'' - v')}\left(\frac{\partial V}{\partial T}\right)_q,$$

where V is the specific volume of the mixture and v'' and v' are the specific volumes of the vapour and liquid phases respectively when they are at equilibrium (the 'saturated' phases).

8. Consider a vapour-liquid system at equilibrium. The heat capacity of either phase at 'saturation' is defined as the value of its heat capacity when the temperature is raised and at the same time the pressure is

adjusted so as to maintain the two-phase equilibrium. If c_s'' and c_s' are the saturated heat capacities of vapour and liquid respectively, show that

$$c_s'' = c_p'' - \frac{L}{(v''-v')}\left(\frac{\partial v''}{\partial T}\right)_p, \quad c_s' = c_p' - \frac{L}{(v''-v')}\left(\frac{\partial v'}{\partial T}\right)_p, \quad c_s'' - c_s' = \frac{dL}{dT} - \frac{L}{T}.$$

c_p'' and c_p' are the normal heat capacities of the phases at constant pressure and v'' and v' are the specific volumes. Estimate the value of c_s'' for steam at 100 °C.

9. Show that, under certain conditions, the effect of the total pressure p on the vapour pressure p_l of a liquid is given by the equation

$$\left(\frac{\partial \ln p_l}{\partial p}\right)_T = \frac{v_l}{RT},$$

where v_l is the volume per mole of the liquid.

A gas leaving an ammonia synthesis converter contains 12 mol % of NH_3. It circulates through a cooler at 250 atm and passes out of the cooler at 30 °C. Use the data below to estimate the fraction of the entering ammonia which is condensed.

[Properties of ammonia at 30 °C:

Liquid density	0.595 g cm^{-3}.
Vapour pressure	11.5 atm.]

[C.U.C.E. Qualifying, 1951]

10. Ammonia is to be made by bringing a stoichiometric mixture of nitrogen and hydrogen to chemical equilibrium over a catalyst at a temperature T_1 and a high pressure p_1. The resulting gas mixture will then pass to a condenser where it will be cooled to a temperature T_2 at the same pressure p_1.

Explain how you would estimate the minimum value of p_1 required to cause liquid ammonia to condense at T_2, if the following information were provided:

Standard free energy $\Delta_f G^0_{298}$ and enthalpy $\Delta_f H_{298}$ of formation of NH_3 at 25 °C.

c_p for N_2, H_2 and NH_3 between 25 °C and T_1 at 1 atm.

p-V relations for pure N_2, H_2 and NH_3 at T_2 and T_1 between 1 atm and p_1, including the data for liquid NH_3 at T_2.

State any assumptions and approximations involved in making the estimate. What further thermodynamic information would you seek to obtain in order to make your estimate more reliable?   [C.U.C.E. Tripos, 1953]

11. Below the triple point (-56.2 °C) the vapour pressure of solid carbon dioxide may be expressed by the relation

$$\log_{10} p = -\frac{1353}{T} + 9.832,$$

where p is in mmHg and T in K. The enthalpy of melting is 8328 J mol^{-1}.

Make an estimate of the vapour pressure of liquid carbon dioxide at 0 °C and indicate sources of inaccuracy in the method used.

[C.U.C.E. Qualifying, 1955]

CHAPTER 7

GENERAL PROPERTIES OF SOLUTIONS AND THE GIBBS–DUHEM EQUATION

7·1. The Gibbs–Duhem equation

In the previous chapter we obtained a number of relations, such as the Clausius–Clapeyron equation, which are correct whatever the nature of the phases which are in equilibrium, provided that there is only a single component. The question arises whether any equations of a comparable generality may be obtained for the case of a multi-component system.

The present chapter is concerned with the *general* relationships concerning a binary solution, together with some discussion of its empirical behaviour.† The following chapter is concerned with the *special* relationships which arise when the solution may be assumed to be ideal.

At the outset it may be remarked that the chief difficulty in the study of solutions is that thermodynamics provides no detailed information concerning the dependence of the chemical potential (or other thermodynamic functions) on the composition. The Clausius–Clapeyron, and other equations of the last chapter, were based on the temperature and pressure coefficients of μ:

$$\left.\begin{aligned} \frac{\partial \mu_i}{\partial T} &= -S_i, \\[2mm] \frac{\partial \mu_i}{\partial p} &= V_i, \end{aligned}\right\} \tag{7·1}$$

which thereby relate the temperature and pressure dependence of the phase equilibrium to measurable latent heats and volume changes. There are no comparable equations for the dependence of μ on the composition, except in the special case of the ideal solution.

The only guidance comes from the Gibbs–Duhem equation (2·83). For a homogeneous phase of two components A and B this reads

$$S\,\mathrm{d}T - V\,\mathrm{d}p + n_a\,\mathrm{d}\mu_a + n_b\,\mathrm{d}\mu_b = 0, \tag{7·2}$$

where S and V are the total entropy and volume of the phase respectively and n_a and n_b are the amounts of the substances. The equation shows that of the four variables T, p, μ_a and μ_b, only three can be varied independently.

† For a more complete discussion see Rowlinson, *Liquids and Liquid Mixtures* (Butterworth, 1959).

If we choose the mole fractions in place of the chemical potentials as the independent variables, it is evident that the state of a binary phase can be completely specified by means of temperature and pressure together with a *single mole fraction*.

For the present purposes equation (7·2) may be expressed in an alternative form as follows. Expanding the equation in terms of the variables T, p and the mole fraction x (which will be taken to refer to component A) we obtain

$$S\,dT - V\,dp + n_a\left(\frac{\partial \mu_a}{\partial T}\,dT + \frac{\partial \mu_a}{\partial p}\,dp + \frac{\partial \mu_a}{\partial x}\,dx\right)$$
$$+ n_b\left(\frac{\partial \mu_b}{\partial T}\,dT + \frac{\partial \mu_b}{\partial p}\,dp + \frac{\partial \mu_b}{\partial x}\,dx\right) = 0.$$

Substituting from equation (7·1),

$$S\,dT - V\,dp - (n_a S_a + n_b S_b)\,dT + (n_a V_a + n_b V_b)\,dp$$
$$+ \left(n_a\frac{\partial \mu_a}{\partial x} + n_b\frac{\partial \mu_b}{\partial x}\right)dx = 0.$$

The first term cancels the third and the second term cancels the fourth. Hence

$$n_a\frac{\partial \mu_a}{\partial x} + n_b\frac{\partial \mu_b}{\partial x} = 0. \tag{7·3}$$

or, dividing through by $(n_a + n_b)$ and noting that $x = n_a/(n_a + n_b)$ and $1 - x = n_b/(n_a + n_b)$, we obtain

$$x\left(\frac{\partial \mu_a}{\partial x}\right)_{T,\,p} + (1 - x)\left(\frac{\partial \mu_b}{\partial x}\right)_{T,\,p} = 0. \tag{7·4}$$

7·2. Pressure-temperature relations†

(a) Solution in contact with two pure phases. We consider a binary solution in contact either with the vapour of component A and the solid phase of component B, or, alternatively, with the pure solids of both components. An example of the first type is an aqueous salt solution in contact with water vapour and also with solid salt. An example of the second type is an aqueous salt solution in equilibrium, at the eutectic, with solid ice and solid salt (but under a pressure greater than the vapour pressure). Such a system is univariant and we may therefore expect a unique relation between pressure and temperature analogous to the Clausius–Clapeyron equation.

† For other examples of the type discussed in this section see the article by Morey in the *Commentary on the Scientific Writings of J. Willard Gibbs*, vol. I (New Haven, Yale Univ. Press, 1936).

Let the pure phases of components A and B be denoted by single and double primes respectively. The solution phase will be denoted by symbols without a prime. The condition of equilibrium for component A is

$$\mu_a = \mu_a',$$

and therefore for any variations of temperature, pressure and composition†

$$\frac{\partial \mu_a}{\partial T} dT + \frac{\partial \mu_a}{\partial p} dp + \frac{\partial \mu_a}{\partial x} dx = \frac{\partial \mu_a'}{\partial T} dT + \frac{\partial \mu_a'}{\partial p} dp.$$

Substituting from equations (7·1), (2·111 b) and (2·112 b)

$$-S_a dT + V_a dp + \frac{\partial \mu_a}{\partial x} dx = -s_a' dT + v_a' dp, \qquad (7·5)$$

or

$$\frac{\partial \mu_a}{\partial x} dx = -(s_a' - S_a) dT + (v_a' - V_a) dp. \qquad (7·6)$$

Similarly for component B, whose mole fraction is $(1-x)$,

$$\frac{\partial \mu_b}{\partial x} dx = -(s_b'' - S_b) dT + (v_b'' - V_b) dp. \qquad (7·7)$$

Multiplying the first equation by x and the second by $(1-x)$ and adding, we obtain by use of equation (7·4)

$$-x(s_a' - S_a) dT - (1-x)(s_b'' - S_b) dT$$
$$+ x(v_a' - V_a) dp + (1-x)(v_b'' - V_b) dp = 0.$$

After rearrangement

$$\frac{dp}{dT} = \frac{(s_a' - S_a) + r(s_b'' - S_b)}{(v_a' - V_a) + r(v_b'' - V_b)}, \qquad (7·8)$$

where r is the ratio of the mole fraction of component B to the mole fraction of component A in the solution.

Since $\mu_a = \mu_a'$ and $\mu_b = \mu_b''$, the entropy differences in (7·8) may be replaced by enthalpy differences divided by the absolute temperature. Thus (7·8) becomes

$$\frac{dp}{dT} = \frac{(h_a - H_a) + r(h_b - H_b)}{T\{(v_a - V_a) + r(v_b - V_b)\}}, \qquad (7·9)$$

where the primes have been deleted as unnecessary. This equation thus gives the change of pressure with temperature along the three-phase equilibrium curve in terms of measurable heat and volume effects. Consider, for example, the application of the equation to the system: solid salt + solution + water vapour. From a quantity of the

---

† Note that μ_a is a function of T, p and x, but μ_a' is a function of T and p only, since it refers to a pure phase.

solution let 1 mole of water (component A) be evaporated, at constant temperature and pressure, and let $r = n_b/n_a$ moles of salt (component B) be simultaneously precipitated. In these proportions the composition of the solution remains unchanged. The heat absorbed in this process is $\{(h_a - H_a) + r(h_b - H_b)\}$ and may be denoted ΔH. The simultaneous volume change is $\{(v_a - V_a) + r(v_b - V_b)\}$ and may be denoted ΔV. The last equation may therefore be written

$$\frac{\mathrm{d}p}{\mathrm{d}T} = \frac{\Delta H}{T \Delta V}. \tag{7·10}$$

As in the case of the Clausius–Clapeyron equation, it is usually a good approximation to put $\Delta V = RT/p$, equivalent to neglecting all terms except v_a in (7·9) and also assuming the vapour to be perfect. Hence

$$\frac{\mathrm{d}\ln p}{\mathrm{d}T} = \frac{\Delta H}{RT^2}. \tag{7·11}$$

It is to be noted that ΔH includes both the enthalpy of vaporization of water from the given solution and the enthalpy of precipitation of r moles of the solute.

(b) Two binary phases in contact. As examples of the type of system now under discussion there are

(1) a mixture of alcohol and water in equilibrium with their vapours;

(2) a solution of ether in water in equilibrium with the conjugate solution of water in ether;

(3) a liquid solution in each other of bromobenzene and iodobenzene in equilibrium with a solid solution of the same two components.

These systems are all divariant and there is no unique relation between changes of pressure and changes of temperature. It is only when, say, the temperature and the composition of one of the phases are both held constant that the pressure and also the composition of the other phase have definite values. Therefore we cannot obtain an equation for $\mathrm{d}p/\mathrm{d}T$, but only a relation for $(\partial p/\partial T)_x$, where the subscript x denotes constant composition of one of the phases.

Let the chemical potential, partial molar volume and partial molar entropy of one phase be denoted by quantities without a prime and let x be the mole fraction of component A in this phase. Let the corresponding quantities for the other phase be denoted by primes and let y be the mole fraction of component A in this second phase. In place of (7·5), we now have for component A

$$-S_a\,\mathrm{d}T + V_a\,\mathrm{d}p + \frac{\partial \mu_a}{\partial x}\,\mathrm{d}x = -S_a'\,\mathrm{d}T + V_a'\,\mathrm{d}p + \frac{\partial \mu_a'}{\partial y}\,\mathrm{d}y,$$

since μ'_a now depends on y, as well as on T and p. After rearranging we obtain

$$\frac{\partial \mu_a}{\partial x}\,\mathrm{d}x - \frac{\partial \mu'_a}{\partial y}\,\mathrm{d}y = -(S'_a - S_a)\,\mathrm{d}T + (V'_a - V_a)\,\mathrm{d}p. \qquad (7\cdot12)$$

Similarly for component B

$$\frac{\partial \mu_b}{\partial x}\,\mathrm{d}x - \frac{\partial \mu'_b}{\partial y}\,\mathrm{d}y = -(S'_b - S_b)\,\mathrm{d}T + (V'_b - V_b)\,\mathrm{d}p. \qquad (7\cdot13)$$

In these equations not all of the variations are independent, because for the one phase we have the relation (7·4)

$$x\frac{\partial \mu_a}{\partial x} + (1-x)\frac{\partial \mu_b}{\partial x} = 0, \qquad (7\cdot14)$$

and for the other phase the analogous relation

$$y\frac{\partial \mu'_a}{\partial y} + (1-y)\frac{\partial \mu'_b}{\partial y} = 0. \qquad (7\cdot15)$$

Between the four equations above we obtain†

$$\left(\frac{y-x}{1-x}\right)\frac{\partial \mu_a}{\partial x}\,\mathrm{d}x = -\{y(S'_a - S_a) + (1-y)\,(S'_b - S_b)\}\,\mathrm{d}T$$
$$+\{y(V'_a - V_a) + (1-y)\,(V'_b - V_b)\}\,\mathrm{d}p. \qquad (7\cdot16)$$

Therefore of the three variables, x, T and p, only two are independent, as is known already from the phase rule. The thermodynamic information used in deriving (7·16) is, of course, the same as is used in obtaining the phase rule and (7·16) extends our knowledge only in showing the quantitative relationship between the variables.

The mole fraction x refers to one of the two phases. Let it be supposed that this phase is maintained at constant composition. Then from (7·16)

$$\left(\frac{\partial p}{\partial T}\right)_x = \frac{y(S'_a - S_a) + (1-y)\,(S'_b - S_b)}{y(V'_a - V_a) + (1-y)\,(V'_b - V_b)}. \qquad (7\cdot17)$$

During the temperature rise $\mathrm{d}T$ there will, of course, be a tendency for both phases to change in composition, but constancy in composition of *one* of the phases can always be maintained artificially—for example, in the first example quoted above, by addition to the liquid of an increment of alcohol, to replenish what is lost to the vapour.

† Multiply (7·12) by y and (7·13) by $(1-y)$ and add. Equation (7·15) then allows the elimination of μ'_a and μ'_b. Finally, apply (7·14) and simplify. Equations similar to (7·16) but applying to multicomponent systems are given by Srivastava and Rastogi, *Proc. Nat. Inst. Sci. India* **19** (1953), 613, 653. A number of other important papers are by Strickland-Constable, *Proc. Roy. Soc.* **A209** (1951), 14; **A214** (1952), 36; Redlich and Kister, *Ind. Eng. Chem.* **40** (1948), 341, 345; Ibl and Dodge, *Chem. Eng. Science* **2** (1953), 120.

Since $\mu_a = \mu_a'$ and $\mu_b = \mu_b'$, the entropy differences in (7·17) may be replaced by enthalpy differences divided by the absolute temperature. Thus

$$\left(\frac{\partial p}{\partial T}\right)_x = \frac{y(H_a' - H_a) + (1 - y)(H_b' - H_b)}{T\{y(V_a' - V_a) + (1 - y)(V_b' - V_b)\}}. \tag{7·18}$$

The numerator, which may be denoted ΔH, is clearly the heat absorbed, at constant temperature and pressure, when y moles of component A and $(1 - y)$ moles of component B are transferred to the primed phase from a quantity of the unprimed phase which is so great that its composition remains virtually unchanged.

In the case of vapour-liquid equilibrium this is called the *differential enthalpy of vaporization*. If only the component A is appreciably volatile, $y = 1$ and $(1 - y) = 0$, and thus ΔH is the heat of vaporization of component A from the solution in which its mole fraction is x. In general this is not the same as the latent heat of pure A.

The bracket in the denominator has a similar interpretation and may be denoted ΔV. Then (7·18) takes the form

$$\left(\frac{\partial p}{\partial T}\right)_x = \frac{\Delta H}{T \Delta V}. \tag{7·19}$$

In the case of vapour-liquid equilibria it is a good approximation to put $\Delta V = RT/p$, and therefore

$$\left(\frac{\partial \ln p}{\partial T}\right)_x = \frac{\Delta H}{RT^2}. \tag{7·20}$$

In this equation p refers to the total vapour pressure of the two components. On the other hand, in the application of (7·19) to the equilibrium with each other of a pair of liquid or solid solutions (in the absence of vapour), p must be taken as some pressure greater than the vapour pressure. $(\partial p/\partial T)_x$ is then to be interpreted analogously to equation (6·9).

Another important consequence of (7·16) relates to maxima and minima. From this equation we can obviously obtain expressions for

$$\left(\frac{\partial T}{\partial x}\right)_p \quad \text{and} \quad \left(\frac{\partial p}{\partial x}\right)_T.$$

These equations need not be written down explicitly, but both of the above differential coefficients are seen to be proportional to $(y - x)/(1 - x)$. Now in certain mixtures it may occur that there is a particular composition of one phase which is in equilibrium with a second phase *which has an equal mole fraction*, that is, $y - x = 0$ at a particular value of x. When this occurs it follows that

$$\left(\frac{\partial T}{\partial x}\right)_p = 0 \quad \text{and} \quad \left(\frac{\partial p}{\partial x}\right)_T = 0. \tag{7·21}$$

For example, a mixture of benzene and alcohol containing a mole fraction of alcohol of 0.46 has a vapour phase of equal composition, when the pressure is 1 atm. The first relation above, states that the boiling-point of the mixture will pass through either a maximum or a minimum at this composition (actually a minimum). Similarly, the second relation states that the total vapour pressure, at constant temperature, will pass through a maximum or a minimum at the same composition (actually a maximum).

Mixtures which have the above properties are known as *azeotropes*. The azeotropic composition is itself a function of temperature or pressure; for example, the proportions of alcohol and benzene which give rise to a maximum boiling mixture vary somewhat as the pressure of ebullition is changed.

Similar conclusions apply to a liquid solution in equilibrium with a solid solution of the same two components. For example,† bromobenzene and iodobenzene are completely miscible over the whole range of composition both in the liquid and solid phases; the liquid and solid solutions have the same composition when the percentage of iodobenzene is about 35 %. At this point the freezing-point passes through a minimum value, at constant pressure.

It can be readily proved that a similar conclusion applies when the solid phase is of fixed composition. For example, the solubility curve of ferric chloride shows a number of eutectics, due to the various hydrates, and between each of these eutectics the curve passes through a maximum. At each of these maxima the solution has the same composition as the solid hydrate with which it is in equilibrium at constant pressure.

7·3. Partial pressure-composition relations

Consider a liquid or solid solution of several components in equilibrium with its vapour. For the ith component we have

$$\mu_i^{\text{soln.}} = \mu_i^{\text{vap.}}.$$

At temperatures at which the vapour pressure is sufficiently small the vapour phase may be expected to approximate fairly closely to a perfect gas mixture. Thus

$$\mu_i^{\text{vap.}} = \mu_i^0 + RT \ln p_i,$$

and therefore $\qquad \mu_i^{\text{soln.}} = \mu_i^0 + RT \ln p_i, \qquad (7·22)$

† Quoted by Butler in the *Commentary on the Scientific Writings of J. Willard Gibbs*, vol. I (New Haven, Yale Univ. Press, 1936), p. 114.

where μ_i^0 depends only on the temperature. This equation gives the chemical potential of any component of a solution in terms of its partial pressure.†

It follows that the equilibria of solutions can always be discussed thermodynamically in terms of the partial pressures in the saturated vapour (or the fugacities if the vapour phase is imperfect). Consider, for example, the reaction of NO_2 with water as in nitric acid manufacture,

$$3NO_2 + H_2O = 2HNO_3 + \overset{.}{N}O.$$

The equilibrium constant for the vapour phase is

$$\frac{p_{HNO_3}^2 \, p_{NO}}{p_{NO_2}^3 \, p_{H_2O}} = K_p,$$

and this may be calculated by use of standard-free energies. In a gas containing known partial pressures of NO, NO_2 and H_2O it would therefore be possible to calculate the maximum attainable partial pressure of HNO_3. However, this would not tell us the maximum attainable concentration of this substance in the liquid phase.

Therefore the difficulty in the study of solutions arises when we wish to relate the composition of the gas phase with the composition of the solution with which it is in contact.‡ In general, the partial pressure of any component of the vapour may be expected to be a function of temperature, together with its mole fraction in solution and *also the mole fractions of all other species* (or rather a function of $N - 1$ mole fractions if there are N substances in solution). Thus

$$p_i = f(T, x_1, x_2, \ldots, x_i, \ldots, x_{N-1}). \tag{7·23}$$

The simplest possible relationship of this kind is where p_i is directly proportional to x_i and is independent of all other mole fractions:

$$p_i \propto x_i. \tag{7·24}$$

In the next section we shall examine the extent to which such behaviour is actually observed.

7·4. The empirical partial pressure curves of binary solutions

(a) Non-electrolytes. A very thorough experimental study of the partial vapour pressures of binary organic mixtures was made by Zawidski.§ Out of a large number of such mixtures only two were

† Note also that p_i must be the same for any two condensed phases which are in equilibrium with respect to the particular component.

‡ This difficulty is the same as the one already mentioned in §7·1, namely, knowing the chemical potential as a function of the composition of the solution.

§ Zawidski, *Z. Phys. Chem.* **35** (1900), 129.

found whose behaviour approximated at all closely to the relation (7·24). These were ethylene bromide with propylene bromide and benzene with ethylene chloride; in each of these mixtures the partial pressure of a component is almost exactly proportional to its mole fraction in the solution. Behaviour of this kind is shown diagrammatically in Fig. 25, where the lines AJ and BK represent the partial pressures of components B and A respectively, when plotted against the mole fraction at a particular temperature. The line JK shows the total pressure of both components.

Thus

$$p_a = p_a^* x_a, \\ p_b = p_b^* x_b, \Bigg\} \tag{7·25}$$

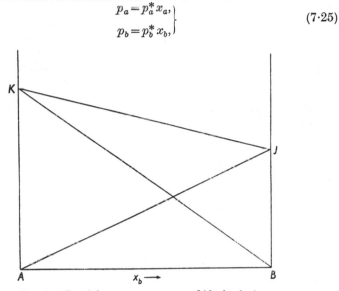

Fig. 25. Partial vapour pressures of ideal solution.

where p_a^* and p_b^* are the vapour pressures of pure A and pure B respectively, at the temperature T. These are the equations to the lines BK and AJ respectively, and the equation to the line JK is

$$p = p_a + p_b = p_a^* x_a + p_b^* x_b \\ = p_a^* x_a + p_b^* (1 - x_a). \tag{7·26}$$

The functional relationships (7·25) are known as *Raoult's law*. Mixtures which obey the law over the whole range of composition are the exception rather than the rule, but an approximation to the ideal behaviour is usually found whenever the components are closely similar in molecular structure. Much more frequently the observed partial pressures are of the types shown in Figs. 26 and 27, which illustrate *positive* and *negative deviations* from Raoult's law respectively. The former is the more common in organic mixtures.

In the case of systems which do not obey the relation $p_i = p_i^* x_i$ it is found nevertheless that the observed partial pressures are tangential to the straight line† represented by this equation as x_i approaches unity. This is shown in Figs. 26 and 27, where the lines AJ and BK are seen to be the tangents to the actual partial pressure curves at $x_b = 1$ and $x_a = 1$ respectively.

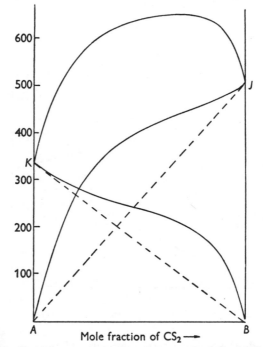

Fig. 26. Partial pressures in the mixture carbon disulphide-acetone at 35.2°C. (Note the azeotrope.)

Another important feature of these curves, closely connected thermodynamically with that which has just been mentioned, is

† Even where the law $p_i = p_i^* x_i$ is obeyed exactly, the lines AJ and BK of Fig. 25 are not *precisely* straight. This is because the total pressure on the system is not constant but varies along the line JK. As shown in equation (6·24) the vapour pressures p_a^* and p_b^* of the pure liquids are very slightly dependent on the total pressure; in the expression $p_i = p_i^* x_i$ the quantity p_i^* must therefore be taken as the vapour pressure of pure i at the pressure of the solution under discussion, and it is not quite constant as x_i changes (unless the total pressure is held constant by means of an insoluble gas). For most practical purposes the effect is trivial. See also a footnote to §8·3.

that p_i has a finite slope as x_i approaches zero. This is the essential content of *Henry's law* and it may be expressed as

$$p_i \to K_i x_i \quad \text{as} \quad x_i \to 0, \tag{7·27}$$

where K_i is independent of composition. Thus from (7·27) we have

$$\frac{\partial p_i}{\partial x_i} \to K_i \quad \text{as} \quad x_i \to 0, \tag{7·28}$$

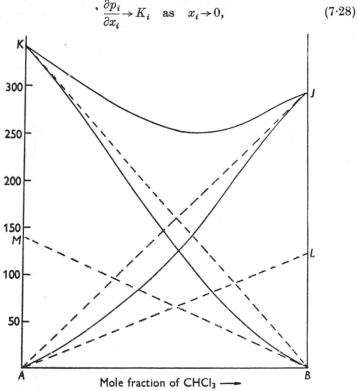

Fig. 27. Partial pressures in the mixture acetone-chloroform at 35.2 °C.
(Note the azeotrope.)

which is equivalent to a finite slope at infinite dilution of component i. An alternative law, such as $p_i \to K_i x_i^n$, would imply either an infinite slope (if $n < 1$) or a zero slope (if $n > 1$). The case $n = 1$ is known as Henry's law.

In brief the empirical study of the vapour-pressure curves of binary liquid mixtures shows that at the one side of the diagram we have

$$p_a \to p_a^* x_a, \quad p_b \to K_b x_b, \quad \text{as} \quad x_a \to 1 \quad \text{and} \quad x_b \to 0 \Big\}$$

and at the other side of the diagram
$$\left. \begin{array}{l} \\ p_a \to K_a x_a, \quad p_b \to p_b^* x_b, \quad \text{as} \quad x_a \to 0 \quad \text{and} \quad x_b \to 1. \end{array} \right\} \tag{7·29}$$

K_a and K_b have the significance of the lengths AM and BL in Fig. 27. Of course, if the mixture obeys Raoult's law over the whole range of composition, the lines AL and AJ coincide and also the lines BK and BM. Thus K_a and K_b become the same as p_a^* and p_b^* respectively.

(b) Electrolytes. An apparent exception to Raoult's law occurs in the case of electrolyte solutions unless allowance is made for the dissociation. Consider a solution made up from n_{H_2O} mols of water

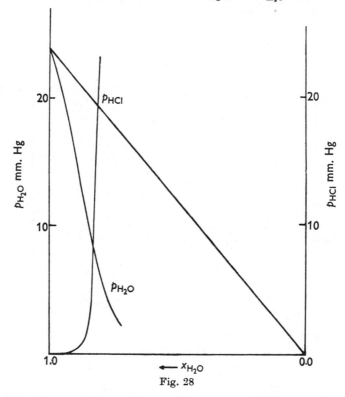

Fig. 28

and n_{HCl} mols of hydrogen chloride. The mole fractions might be chosen in either of the following ways:

$$(1) \quad x_{H_2O} = \frac{n_{H_2O}}{n_{H_2O} + n_{HCl}}, \quad x_{HCl} = \frac{n_{HCl}}{n_{H_2O} + n_{HCl}}, \quad (7\cdot30)$$

$$(2) \quad x'_{H_2O} = \frac{n_{H_2O}}{n_{H_2O} + 2n_{HCl}}, \quad x'_{H^+} = x'_{Cl^-} = \frac{n_{HCl}}{n_{H_2O} + 2n_{HCl}}. \quad (7\cdot31)$$

Figs. 28 and 29 show the observed partial pressure of water, above the solution at 25 °C, plotted against x_{H_2O} and x'_{H_2O} respectively.

There is clearly much more indication of a tangential approach to Raoult's law,

$$p_{H_2O} \to p^*_{H_2O} x'_{H_2O} \quad \text{as} \quad x'_{H_2O} \to 1, \qquad (7·32)$$

when the mole fraction is chosen as x'_{H_2O} than when it is chosen as x_{H_2O}. Therefore it seems that the most appropriate choice of the mole fraction of the solvent is that which would be computed by allowing for all independently mobile particles in the solution—in this instance H_2O molecules and Cl^- and H^+ ions.† The same considerations applied

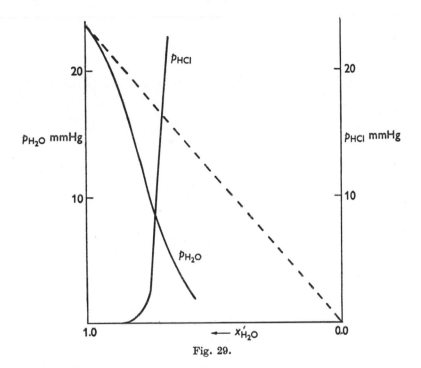

Fig. 29.

to data on the depression of the freezing-point etc., were those which originally led van't Hoff and Arrhenius to the idea of the dissociation of electrolytes.

Figs. 28 and 29 also show the partial pressure of hydrogen chloride over the solution. The curves do not appear to approach a finite slope at zero mole fraction of hydrogen chloride and therefore Henry's

† As will be shown in §8·1 it is only in this way that we obtain a correct statistical computation of the entropy of mixing. If it were assumed erroneously that the H and Cl particles are always together as a pair it would imply less randomness in the solution than corresponds to the actual situation.

law is not obeyed. On the other hand, if the same data are plotted against the product $x'_{H^+} x'_{Cl^-}$ (or, alternatively, against x^2_{HCl} which does not differ appreciably in very dilute solution), there is an indication of a finite slope at the origin. This is shown in Fig. 30 on an enlarged scale. The discussion will be taken up again in §§ 7·7 and 10·17.

(c) Solids. It is not always possible to trace smooth partial pressure curves across the whole range of mole fractions from zero to unity. For example, in the case of aqueous salt solutions, at a certain

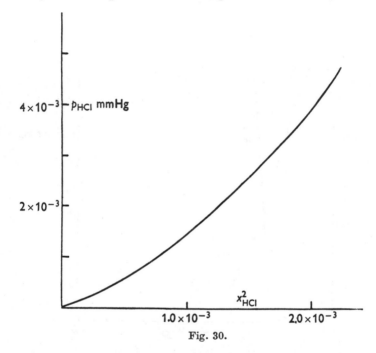

Fig. 30.

mole fraction of salt the solution becomes saturated. Any further addition of crystalline salt to the system does not change the mole fraction in the liquid phase, and the partial pressure of water thereafter remains constant, in accordance with the phase rule. At mole fractions of salt less the saturation value the partial pressure curve of the water is similar to that discussed above in connexion with hydrogen chloride and is asymptotic to the line $p_{H_2O} = p^*_{H_2O} x_{H_2O}$, provided that the ionization of the salt is allowed for in calculating x_{H_2O}. The partial pressure of the salt itself is, of course, extremely small and may be thought of as following a course almost coincident with the horizontal axis in the preceding diagrams.

(d) Gases. In Figs. 28 and 29 the partial pressure of hydrogen chloride is seen to rise very steeply when the mole fraction of the dissolved gas is appreciable. It reaches a value of 46 atm when the mole fraction of water is zero, this being the vapour pressure of pure hydrogen chloride at 25 °C

In the case of less condensible gases such as oxygen it is not possible to follow the course of the partial pressures across the whole of the diagram, due to the enormous pressures needed to obtain a large mole fraction of dissolved gas.† Fig. 31 shows the partial pressure of oxygen over aqueous solutions at 23 °C, and it is seen that Henry's law

$$p_{O_2} = K x_{O_2}$$

is obeyed fairly accurately, at least up to 3000 mmHg pressure.

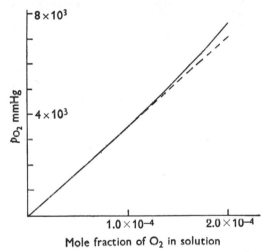

Fig. 31. Solubility of oxygen in water at 23 °C.

(e) Incompletely miscible liquids. Consider a mixture of substances A and B. Molecular considerations indicate that positive deviations from Raoult's law may be expected to occur when the interaction energy between an A and a B molecule is smaller than the mean of the interaction energy of A with itself and of B with itself. For example, in the case of a dilute solution of A in B, if the B-B interaction is larger than the A-B interaction, the 'escaping

† The solubility of oxygen in water has been measured by Krichevski and Kasarnovsky up to a pressure of 8000 atm (*J. Amer. Chem. Soc.* **57** (1935), 2168).

tendency' of the substance A will be enhanced and its partial pressure over the solution will be greater than corresponds to Raoult's law.

When positive deviations from Raoult's law reach a certain point (at which the A-A and B-B interactions are large compared to the A-B interactions) the system may break down into two liquid layers, one containing an excess of A and the other containing an excess of B. When these layers are in equilibrium they are said to be *conjugate phases*. In both of such layers the chemical potential of a particular component is, of course, the same and also its partial pressure. For example, at 25 °C a mixture of water and benzene consists of two

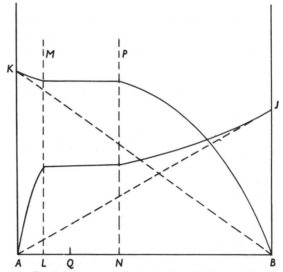

Fig. 32. Partial pressures of a pair of partially miscible liquids.

liquid phases, one containing about 0.07% of water and the other containing 99.9% of water. The partial pressure of water over both of these solutions is the same. Similarly with regard to the benzene.

Typical partial pressure curves of a system of this type are shown in Fig. 32. The dotted straight lines AJ and BK represent Raoult's law, if it were applicable. Any mixture of overall composition lying between LM and NP consists of the two liquid layers of compositions L and N respectively (and it can be proved that a mixture of overall composition corresponding to the point Q consists of the phases of compositions L and N in the molar ratio $QN : QL$ respectively).

In the region where there are two liquid phases, together with a vapour phase, the system is univariant. Therefore, at any chosen

total pressure, the partial pressures of each component have fixed
values. The isobaric distillation of such a mixture, therefore, yields a
distillate of constant composition for as long as both of the conjugate
phases continue to co-exist in the distillation vessel.

With increasing temperature the mutual solubility of the two
liquids usually increases and at a certain temperature complete
miscibility is attained. This is called the *consolute or critical mixing*

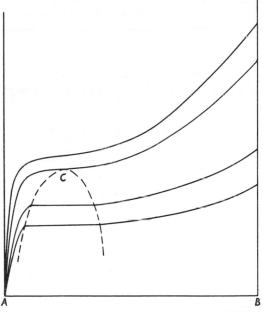

Fig. 33. Partial pressures of component B at temperatures above and below the
critical mixing temperature. (The two-liquid region is within the dotted locus.)

temperature. A family of curves showing the partial pressure of component B at a number of temperatures is shown in Fig 33. Intuitively,
it is fairly clear that the consolute point C is determined by the two
conditions

$$\left(\frac{\partial p_b}{\partial x_b}\right)_T = 0, \quad \left(\frac{\partial^2 p_b}{\partial x_b^2}\right)_T = 0, \qquad (7\cdot33)$$

but for a more complete discussion the reader is referred to the
literature.† It follows from the Gibbs–Duhem equation that, when
this condition is satisfied for the one component, it is automatically
satisfied for the other.

† Gibbs, *Collected Works*, vol. I, p. 129; *Commentary on the Scientific Writings
of J. Willard Gibbs*, vol. I, p. 163; Guggenheim, *Thermodynamics*, § 5·41 and
5·42.

7·5. Application of the Gibbs–Duhem equation to the partial pressure curves

The only information available from thermodynamics concerning the partial pressures is the Gibbs–Duhem equation, and this is essentially a limitation on the number of variables which can be varied independently. Consider a solution containing any number of components and held at constant temperature whilst its composition is varied. For such a system equation (7·2) reduces to

$$\Sigma n_i \, d\mu_i = V \, dp. \qquad (7 \cdot 34)$$

This equation may, of course, be applied to any phase, but we are here applying it specifically to the solution; V is therefore the volume of this solution and the n_i are the amounts of substances. Now

$$\mu_i^{\text{soln.}} = \mu_i^{\text{vap.}}$$

$$= \mu_i^0 + RT \ln p_i,$$

if the vapour phase is a perfect mixture. Thus, under the latter conditions and at constant temperature,

$$d\mu_i^{\text{soln.}} = RT \, d\ln p_i, \qquad (7 \cdot 35)$$

Substituting (7·35) in (7·34)

$$RT \Sigma n_i \, d\ln p_i = V \, dp. \qquad (7 \cdot 36)$$

This equation therefore gives a relationship between changes in the total pressure p and the partial pressures p_i.

It is, of course, entirely permissible to regard the temperature as being held constant during the variations considered in the above equation. In a two-phase system of C components there are C degrees of freedom, and an arbitrary choice can be made of the temperature, together with the $C - 1$ mole fractions required to fix the composition of the condensed phase. When this has been done the total pressure and the partial pressures have definite values, and the relationship between the changes in these quantities, *as the composition is varied at constant temperature*, is given by the above equation.

Dividing through the equation by Σn_i, the total amounts of all substances in the condensed phase, we obtain, after rearrangement,

$$\Sigma x_i \, d\ln p_i = \frac{V \, dp}{RT \Sigma n_i}, \qquad (7 \cdot 37)$$

or

$$\Sigma x_i \, d\ln p_i = \frac{v_l \, dp}{RT}, \qquad (7 \cdot 38)$$

where v_l is the mean molar volume of the condensed phase.† If the vapour phase is not perfect the partial pressures in this equation must be replaced by fugacities.

In (7·38) the term occurring on the right-hand side is usually quite trivial compared to the various terms which occur on the left-hand side. Thus, if we put $RT = pv_g$, where v_g is the mean molar volume of the vapour phase, the equation may be rewritten

$$\Sigma x_i \, \mathrm{d} \ln p_i = \frac{v_l}{v_g} \, \mathrm{d} \ln p. \tag{7·39}$$

The term v_l/v_g is always small compared to unity in the pressure range which is of most general interest, i.e. well below the critical pressure. The right-hand side of (7·39) is therefore usually much smaller than any of the individual terms on the left-hand side. As an approximation for practical purposes we thus obtain

$$\Sigma x_i \, \mathrm{d} \ln p_i = 0. \tag{7·40}$$

This relation becomes exact if, on the one hand, we substitute fugacities to allow for deviations from gas perfection and, on the other hand, we hold p really constant by addition to the gas phase of an extra component which is insoluble in the condensed phase.

In the case of a binary system of components A and B equation (7·40) becomes

$$x_a \, \mathrm{d} \ln p_a + x_b \, \mathrm{d} \ln p_b = 0. \tag{7·41}$$

If we give explicit recognition to the fact that the changes $\mathrm{d} \ln p_a$ and $\mathrm{d} \ln p_b$ are due to the composition change $\mathrm{d} x_a$ at constant temperature, the last equation may be written

$$x_a \left(\frac{\partial \ln p_a}{\partial x_a} \right)_T + x_b \left(\frac{\partial \ln p_b}{\partial x_a} \right)_T = 0. \tag{7·42}$$

Since $x_a + x_b = 1$ and $\mathrm{d} x_a + \mathrm{d} x_b = 0$, other forms of (7·42) are

$$x_a \left(\frac{\partial \ln p_a}{\partial x_a} \right)_T = x_b \left(\frac{\partial \ln p_b}{\partial x_b} \right)_T, \tag{7·43}$$

and

$$\frac{x_a}{p_a} \left(\frac{\partial p_a}{\partial x_a} \right)_T = \frac{x_b}{p_b} \left(\frac{\partial p_b}{\partial x_b} \right)_T. \tag{7·44}$$

These are known as *Duhem–Margules equations* and they imply a relationship between the gradients of the vapour-pressure curves. For example, equation (7·44) means that if an increment $\mathrm{d} x_a$ in mole

† In the special case where the condensed phase consists of a pure substance, (7·38) is evidently equivalent to equation (6·24) concerning the effect of total pressure on vapour pressure.

fraction causes an increase dp_a in the partial pressure of component A, it will also cause a perfectly definite change dp_b in the partial pressure of the other component; this will be a *decrease*, since x_a/p_a and x_b/p_b are always positive and $dx_a = -dx_b$.

The equations may be used for checking the consistency of experimental data. For this purpose, either the partial pressures or their

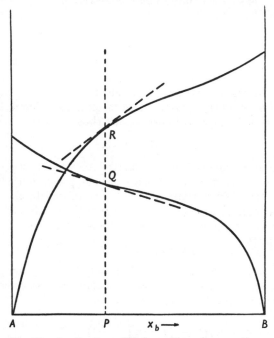

Fig. 34. Application of Duhem–Margules equation.

logarithms are plotted against the mole fraction of one of the components and the tangents are measured at various points. For example, in Fig. 34 consider the composition P where the mole fraction of B is x_b. Then

$$\left.\begin{aligned} \frac{x_b}{p_b} &= \frac{AP}{PR}, \quad \frac{\partial p_b}{\partial x_b} = \text{gradient at } R, \\[2mm] \frac{x_a}{p_a} &= \frac{BP}{PQ}, \quad \frac{\partial p_a}{\partial x_a} = \text{gradient at } Q. \end{aligned}\right\} \tag{7·45}$$

These quantities must satisfy (7·44) (and at all other compositions) if the two experimental curves are mutually consistent. The question whether it is better to plot the pressures themselves and use (7·44), or to plot their logarithms and to use (7·43), depends on the shape of the curves in the particular instance; the essential requirement is the accurate measurement of tangents.†

Departures from the above equations due to deviations from the law of the perfect gas mixture do not usually exceed a small percentage except at pressures above atmospheric or if there is association in the vapour phase, such as occurs in the case of formic and acetic acids. A procedure for applying the Duhem–Margules equation allowing for deviations from the gas law has been described by Scatchard and Raymond.‡

The other approximation, the neglecting of the term containing $d \ln p$ in (7·39), may be avoided if the partial pressures are measured in the presence of an inert gas which is insoluble in the condensed phase. If the latter contains C components, the whole system, including the inert gas, contains $C + 1$ components and has $C + 1$ degrees of freedom. Thus, having chosen the temperature and the $C - 1$ variables required to fix the composition of the condensed phase, it is still possible to vary the pressure or to hold it constant at any desired value, by use of the inert gas. This point has been discussed in more detail by Bury§ and by Krichevsky and Kasarnovsky.‖

7·6. Application of the Gibbs–Duhem equation to the total pressure curve

In the case of a binary system the Gibbs-Duhem (or Duhem–Margules) equation can be put into a useful form in terms of the total pressure.¶ Let x be the mole fraction of component A in the condensed phase and let y be its mole fraction in the vapour phase. It will be assumed that the system contains components A and B only, i.e. there is no inert gas. Then

$$p_a = yp,$$

$$p_b = (1 - y)\,p.$$

† Applications of the method are given by Adam and Guggenheim, *Proc. Roy. Soc.* A, **139** (1933), 231; Beatty and Calingaert, *Industr. Engng Chem.* **26** (1934), 504, 905; Carlson and Colburn, *ibid.* **34** (1942), 581. Other forms of consistency test have been described by Herington (*Nature*, **160** (1947), 610); Herington and Coulson, *Trans. Faraday Soc.* **44** (1948), 629; Redlich and Kister, *Ind. Eng. Chem.* **40** (1948), 345; van Ness, *Chem. Eng. Sci.* **11** (1959), 118. For a critique see Rowlinson, *Liquids and Liquid Mixtures*, § 4·5.

‡ Scatchard and Raymond, *J. Amer. Chem. Soc.* **60** (1938), 1278. See also Rowlinson, loc. cit.

§ Bury, *Trans. Faraday Soc.* **36** (1940), 795.

‖ Krichevsky and Kasarnovsky, *J. Amer. Chem. Soc.* **57** (1935), 2168.

¶ Lewis and Murphree, *J. Amer. Chem. Soc.* **46** (1924), 1.

Substituting these relations in (7·39), which applies at constant temperature,

$$x \, d \ln yp + (1-x) \, d \ln (1-y) \, p = \frac{v_l}{v_g} \, d \ln p. \tag{7·46}$$

On rearranging this equation we obtain

$$\frac{x}{y} \, dy - \frac{(1-x)}{(1-y)} \, dy = \left(\frac{v_l}{v_g} - 1 \right) \, d \ln p,$$

or

$$\left(\frac{\partial \ln p}{\partial y} \right)_T = \frac{(y-x)}{y(1-y) \left(1 - \dfrac{v_l}{v_g} \right)}. \tag{7·47}$$

This equation thus gives the change in the logarithm of the total pressure with change in the composition. Now v_l/v_g is less than unity and the denominator of (7·47) is therefore a positive quantity. It follows that $\partial \ln p/\partial y$ has the same sign as $(y-x)$. Now an increase dy in the mole fraction of A in the vapour phase may have the effect of either increasing or decreasing the total pressure. In the former case $\partial \ln p/\partial y$ is positive and therefore $y > x$, i.e. the component A is richer in the vapour phase than in the condensed phase. Conversely, a component has a lower mole fraction in the vapour than in the solution if an increase in its mole fraction in the vapour tends to reduce the total pressure. It follows also that whenever the left-hand side of (7·47) is zero, i.e. at compositions of maximum or minimum vapour pressure, we have $y = x$ and also $1 - y = 1 - x$. The same result was obtained previously in equation (7·21).

Finally, it may be noted that v_l/v_g in the denominator of (7·47) is usually very small compared to unity and therefore as a good approximation

$$\left(\frac{\partial \ln p}{\partial y} \right)_T = \frac{(y-x)}{y(1-y)}. \tag{7·48}$$

7·7. The Gibbs–Duhem equation in relation to Raoult's and Henry's laws

Any formula purporting to give the partial pressures as a function of the composition of the condensed phase must comply with the differential expression (7·39). Such formulae are, in fact, solutions of this differential expression. It will be shown in the present section that one such solution is where *all* of the components obey the relation

$$p_i = K_i x_i, \tag{7·49}$$

where K_i is a function of temperature and pressure only and may or may not be equal to p_i^*. (As discussed in connexion with equation (7·29), Raoult's law may be regarded as a special case of Henry's law with K_i equal to p_i^*, the vapour pressure of pure i at the temperature and total pressure of the solution under discussion.)

For simplicity it will be supposed that the total pressure, as well as the temperature, is held constant by the presence in the gaseous phase of a component which is insoluble in the solution. Under these conditions K_i is constant and thus from (7·49)

$$\mathrm{d}\ln p_i = \mathrm{d}\ln K_i + \mathrm{d}\ln x_i$$

$$= \mathrm{d}\ln x_i$$

$$= \frac{1}{x_i}\,\mathrm{d}x_i. \tag{7·50}$$

Thus, provided that (7·49) is satisfied for all components of the solution, we can multiply (7·50) through by x_i and sum over all these components to obtain
$$\Sigma x_i \,\mathrm{d}\ln p_i = \Sigma \mathrm{d}x_i.$$

This is clearly zero since $\Sigma x_i = 1$. Thus (7·49) satisfies the condition

$$\Sigma x_i \,\mathrm{d}\ln p_i = 0,$$

and therefore satisfies the Gibbs–Duhem equation in the form of equation (7·40), which is applicable to constant temperature and total pressure. In brief, (7·49) is a permissible relation between p_i and x_i if it applies to all components of a particular solution.

Let substance 1 be present in the solution in large mole fraction and let all the remaining components, 2 to n, be present in very small mole fractions. Substance 1 is then conventionally called the *solvent* and the remainder are called *solutes*. Under such conditions, as discussed in § 7·4, there is good empirical evidence that the partial pressures of all the solutes approach Henry's law, provided that they are in the same molecular form in solution as they are in the vapour phase. (In § 8·1, it will be shown that there are also theoretical reasons for expecting this limiting behaviour.)

Let it be supposed then that the various solutes all obey the relation

$$p_i \to K_i x_i \quad \text{as} \quad x_i \to 0 \quad (i = 2, 3, \ldots, n).$$

It follows from the above theorem that the remaining component, namely, the solvent, must obey the relation

$$p_1 \to p_1^* x_1 \quad \text{as} \quad x_1 \to 1,$$

since the value of K_i for the solvent must be the same as p_1^*. *Thus under conditions where all of the solutes obey Henry's law, the solvent obeys Raoult's law.*

It may be noted, however, that there is no thermodynamic necessity for the solution to appear to obey the ideal laws at infinite dilution.

Deviations do in fact occur when there is a change in molecular association between the gaseous and liquid phases and when the mole fractions have not been chosen appropriately to allow for this. Consider a solution of n_{HCl} mols of hydrogen chloride in n_{H_2O} mols of water and let it be supposed that the idea of dissociation is unknown. The mole fractions would therefore be chosen as in equation (7·30):

$$x_{H_2O} = \frac{n_{H_2O}}{n_{H_2O} + n_{HCl}}, \quad x_{HCl} = \frac{n_{HCl}}{n_{H_2O} + n_{HCl}}. \tag{7·51}$$

As shown in §7·4b, the evidence indicates that the partial pressure of the hydrogen chloride probably becomes proportional to the square of x_{HCl} in very dilute solutions.† Using this empirical result it will now be shown that the other component, the water, does *not* obey Raoult's law when this law is expressed in terms of the above choice of mole fractions.

From the experimental relation

$$p_{HCl} \to K x_{HCl}^2 \quad \text{as} \quad x_{HCl} \to 0, \tag{7·52}$$

we obtain

$$\frac{\partial p_{HCl}}{\partial x_{HCl}} \to 2 K x_{HCl}. \tag{7·53}$$

According to the Duhem–Margules equation (7·44)

$$\frac{x_{H_2O}}{p_{H_2O}} \frac{\partial p_{H_2O}}{\partial x_{H_2O}} = \frac{x_{HCl}}{p_{HCl}} \frac{\partial p_{HCl}}{\partial x_{HCl}}. \tag{7·54}$$

Substituting on the right-hand side of this relation from (7·52) and (7·53) we obtain

$$\frac{x_{H_2O}}{p_{H_2O}} \frac{\partial p_{H_2O}}{\partial x_{H_2O}} \to \frac{x_{HCl}}{K x_{HCl}^2} 2 K x_{HCl} \quad \text{as} \quad x_{HCl} \to 0$$

or

$$\frac{x_{H_2O}}{p_{H_2O}} \frac{\partial p_{H_2O}}{\partial x_{H_2O}} \to 2 \quad \text{as} \quad x_{H_2O} \to 1.$$

Since $p_{H_2O}^*$ is the limiting value of p_{H_2O} as x_{H_2O} tends to unity, this relation may be rearranged to give

$$\frac{\partial p_{H_2O}}{\partial x_{H_2O}} \to 2 p_{H_2O}^* \quad \text{as} \quad x_{H_2O} \to 1. \tag{7·55}$$

This is just *twice* the limiting gradient which would be expected from Raoult's law and is in accordance with the experimental results shown in Fig. 28.

Consider now the use of mole fractions assuming virtually complete ionization. These are

$$x_{H_2O}' = \frac{n_{H_2O}}{n_{H_2O} + 2 n_{HCl}}, \quad x_{H^+}' = x_{Cl^-}' = \frac{n_{HCl}}{n_{H_2O} + 2 n_{HCl}}, \tag{7·56}$$

where n_{HCl} continues to refer to the total number of mols of hydrogen

---

† But the *limiting* behaviour of a solution at infinite dilution cannot be investigated experimentally and can only be predicted on the basis of a molecular theory.

chloride in solution. From (7·51) and (7·56) we can obtain the following relation:

$$x'_{H_2O} = \frac{x_{H_2O}}{2 - x_{H_2O}}. \tag{7·57}$$

Therefore
$$\frac{dx'_{H_2O}}{dx_{H_2O}} = \frac{2}{(2 - x_{H_2O})^2}. \tag{7·58}$$

Now (7·55) may be written

$$\frac{\partial p_{H_2O}}{\partial x'_{H_2O}} \frac{dx'_{H_2O}}{dx_{H_2O}} \to 2p^*_{H_2O} \quad \text{as} \quad x_{H_2O} \to 1.$$

Substituting from (7·58) we obtain

$$\frac{\partial p_{H_2O}}{\partial x'_{H_2O}} \to p^*_{H_2O} \quad \text{as} \quad x'_{H_2O} \to 1, \tag{7·59}$$

and therefore the x' scale, which allows for the ionization, is one on which the solvent obeys Raoult's law.

In brief, there is no purely thermodynamic criterion which leads us to prefer one definition of the mole fraction to any other. The ideal solution laws may be obeyed if this choice is made in one way but not in another, and the most appropriate choice for this purpose is determined by molecular considerations.

It remains to mention very briefly why it is that the solute partial pressure varies as the *square* of its mole fraction in the solution at high dilution (equation 7·52). The reasons for this type of behaviour will be discussed in more detail in Chapter 10, but it is evidently due to the fact that the hydrogen chloride is almost entirely ionized in solution with the result that the significant equilibrium is

$$HCl\,(gas) = H^+ + Cl^-.$$

Because there are *two* ions, the partial pressure varies as the square of the concentration, or mole fraction, of the hydrochloric acid in solution.

Throughout the above discussion it has been supposed that the total pressure is held constant by use of an additional component of the vapour phase which is not soluble in the liquid. If this is not the case, the correct form of the Gibbs–Duhem equation is equation (7·39)

$$\Sigma x_i\, d \ln p_i = \frac{v_l}{v_g} d \ln p.$$

The question may be asked whether Henry's law and Raoult's law continue to be compatible with this equation when the small changes in total pressure are allowed for. This involves the pressure dependence of K_i in equation (7·49). This point will be discussed from a rather different standpoint in the next chapter where it will be shown that the thermodynamic consistency of the ideal solution laws implies constancy of the partial molar volumes.

7·8. The Gibbs–Duhem equation in relation to the Margules and van Laar equations†

As shown in the last section, Raoult's and Henry's laws are permissible solutions of the differential expression (7·40) which governs vapour-liquid equilibria at constant temperature and constant (or effectively constant) total pressure. However, these laws are not the only functional relationships connecting p_i and x_i which are compatible with (7·40), and in the present section we shall discuss some alternative relationships for the special case of a *binary* system.

In this instance the differential expression reduces to the Duhem–Margules equation (7·42)

$$x\frac{\partial \ln p_a}{\partial x} + (1-x)\frac{\partial \ln p_b}{\partial x} = 0, \qquad (7·60)$$

where x is the mole fraction in the solution of component A. A solution to this differential equation may obviously be expressed in the form of a power series as proposed originally by Margules:

$$\left.\begin{array}{l}\ln\left(p_a/p_a^* x\right) = \alpha_a(1-x) + \tfrac{1}{2}\beta_a(1-x)^2 + \tfrac{1}{3}\lambda_a(1-x)^3 + \dots, \\[4pt] \ln\left\{p_b/p_b^*(1-x)\right\} = \alpha_b x + \tfrac{1}{2}\beta_b x^2 + \tfrac{1}{3}\lambda_b x^3 + \dots, \end{array}\right\} \qquad (7·61)$$

where p_a^* and p_b^* are the vapour pressures of the two pure liquids at the particular temperature and total pressure. ‡

Even in cases where the behaviour of the solution deviates extensively from Raoult's law, the observed partial pressures may be expressed quite accurately by means of (7·61), provided that a sufficient number of terms are included. However, in order to avoid undue complication it is usually attempted to include terms up to x^3 only. When the series is limited in this way there is a relationship between the coefficients α_a, β_a, etc. By differentiation of (7·61) with respect to x and substitution in (7·60) it is found that

$$\left.\begin{array}{l}\alpha_a = \alpha_b = 0, \\[4pt] \beta_b = \beta_a + \lambda_a, \\[4pt] \lambda_b = -\lambda_a, \end{array}\right\} \qquad (7·62)$$

and the first terms in the series (7·61) are therefore zero. In view of (7·62) the series with terms up to x^3 may be expressed more conveniently in the following form, as proposed by Carlson and Colburn: §

---

† The author has decided (against advice) to retain this section in the present edition. Rather than the Margules or van Laar equations, a power series expression for the excess free energy is now widely preferred as a means of expressing non-ideality and this is described in §9·9 below.

‡ As elsewhere in this chapter, partial pressures and vapour pressures must be understood as being replaced by fugacities, in cases where the vapour phase departs appreciably from being a perfect mixture.

§ Carlson and Colburn, *Industr. Engng Chem.* **34** (1942), 581.

$$\log(p_a/p_a^* x) = (2B - A)(1-x)^2 + 2(A-B)(1-x)^3,$$
$$\log\{p_b/p_b^*(1-x)\} = (2A - B)x^2 + 2(B-A)x^3, \qquad (7.63)$$

where A and B are coefficients which depend on the temperature.

Zero values of the coefficients in (7·61) or (7·63) are equivalent to Raoult's law. Non-zero values imply that Raoult's and Henry's laws are approached asymptotically. Thus as $x \to 1$ we obtain for component A

$$\ln(p_a/p_a^* x) \to 0$$

or

$$p_a \to p_a^* x, \qquad (7.64)$$

and for component B,

$$\ln\{p_b/p_b^*(1-x)\} \to \tfrac{1}{2}\beta_b + \tfrac{1}{3}\lambda_b,$$

or

$$p_b \to p_b^*(1-x)\, e^{\frac{1}{2}\beta_b + \frac{1}{3}\lambda_b}. \qquad (7.65)$$

Since $x_b = 1 - x$ this equation may be expressed as

$$p_b \to \text{constant} \times x_b, \qquad (7.66)$$

which is Henry's law. In view of these results, it is evident that the equations (7·61) or (7·63) are suitable for the expression of partial pressure data only when the mole fractions are chosen in such a way that the ideal laws are approached asymptotically at either end of the composition scale (cf. the discussion on hydrochloric acid solution in § 7·7).

Another 'two-constant' equation for the representation of partial pressure data in a binary system is known by the name of van Laar. The equation was originally put forward as the result of a theory based on the van der Waals equation of state. This theory is probably erroneous, but the van Laar equation is nevertheless very useful for the empirical representation of partial pressure data. For the component whose mole fraction is x the equation is

$$\log\frac{p_a}{p_a^* x} = \frac{A}{\left\{1 + \dfrac{Ax}{B(1-x)}\right\}^2}, \qquad (7.67)$$

and for the other component

$$\log\frac{p_b}{p_b^*(1-x)} = \frac{B}{\left\{1 + \dfrac{B(1-x)}{Ax}\right\}^2}. \qquad (7.68)$$

The relative merits of (7·63) and the van Laar equation have been discussed by Carlson and Colburn. The determining factor seems to be the ratio of the molar volumes of the two components; when this is near to unity the equation (7·63) may give the better results and vice versa. These authors also discuss the most convenient methods for evaluating the coefficients A and B. For example, A may be determined by plotting $\log p_a/p_a^* x$ against x and extrapolating to $x = 0$. Similarly, B may be determined by plotting $\log p_b/p_b^*(1-x)$ against x and extrapolating to $x = 1$.

Whenever the mixture in question forms an azeotrope, approximate values of A and B may be calculated even in the absence of published data on partial pressures. All that is required is a knowledge of the total vapour pressure and the composition of the azeotropic mixture, at the temperature under discussion, together with the vapour pressures of the pure components. At the azeotropic point we have

$$x_a = y_a = p_a/p,$$

where p is the total pressure. Rearranging this equation

$$\frac{p_a}{x_a} = p$$

and dividing through by p_a^*, the vapour pressure of pure A, we obtain

$$\frac{p_a}{p_a^* x_a} = \frac{p}{p_a^*}. \tag{7·69}$$

Similarly for component B $\qquad \dfrac{p_b}{p_b^* x_b} = \dfrac{p}{p_b^*}. \tag{7·70}$

The substitution of these relations in (7·63), or in the van Laar equations, makes possible an evaluation of the coefficients A and B. These may then be used for the approximate prediction of the partial pressures at compositions other than that of the azeotrope.†

PROBLEMS

1. Confirm the relations (7·62). Show that equations (7·67) and (7·68) are consistent with the Gibbs–Duhem equation.

2. The table below gives published data on the partial pressures over aqueous nitric acid solution at 20 °C. Examine whether the data are self-consistent:

HNO₃ % by wt.	50	60	70	80	90	100
p_{H_2O}/mmHg	7.53	4.93	2.83	1.35	0.46	0.00
p_{HNO_3}/mmHg	0.49	0.89	3.08	10.49	26.03	48.0

3. The substances A and B form liquid and solid solutions with each other, and in both types of solution there is complete miscibility over the whole range of composition. Make sketches of the various forms of the phase diagram on the temperature-composition diagram.

In a particular instance the melting-point of the solid solution passes through a minimum near 0 °C at a certain composition C. Show that when a liquid of any composition other than C is pumped through a system of pipes and tanks, the outlet liquid will differ in composition from

---

† An extensive compilation of azeotropic data has been given by Lecat (*L'Azeotropisme*, Brussels, 1918) and by Horsley (*Industr. Engng Chem.*, Anal. ed., **19** (1947), 508; **21** (1949), 831).

the inlet liquid whenever there is partial freezing in the system at the onset of cold weather.

4. Give a clear account of the process which takes place when salt is added to ice, causing the temperature to fall below 0 °C. Illustrate your answer with a diagram of the chemical potentials as functions of temperature.

5. A continuous flow of salt solution passes into a salting evaporator and crystallization takes place as the water evaporates.

The pressure p of water vapour above the solution is maintained by a vacuum system. Show that the effect of a change in pressure on the temperature of the boiling liquor is given approximately by

$$\frac{\mathrm{d}T}{\mathrm{d}\ln p} = \frac{RT^2}{Q},$$

where Q is the total heat absorbed in the evaporation of one mole of water and the crystallization of the salt formerly dissolved in it.

[C.U.C.E. Tripos, 1953]

6. A salt S dissolves in water and the solid phases which can be in equilibrium with this solution are ice, S and the hydrate $S.2H_2O$. No solid solutions are formed.

On the freezing-point diagram, at constant pressure, there are two eutectics and between them the curve passes through a maximum. Show that at this maximum the composition of the solution is identical with that of the solid phase which is in equilibrium with it, namely $S.2H_2O$.

[C.U.C.E. Qualifying, 1954]

7. A solution consists of components 1, 2 and 3 and is in equilibrium with the pure solid phase of component 3 and with the vapours of components 1 and 2. This vapour phase may be assumed to be a perfect gaseous mixture.

Show that the variation of the total pressure with the vapour phase composition, at constant temperature, may be expressed in the form

$$\left(\frac{\partial \ln p}{\partial y_1}\right)_T = \frac{(y_1 - y_1 x_3 - x_1)}{y_1(1 - y_1)} \frac{v_g}{\{v_g - v_l - x_3(v_g - v_s)\}},$$

where v_g, v_l and v_s are the molar volumes of the gas, liquid and solid phases respectively.

[C.U.C.E. Tripos, 1955]

8. For the case of a binary solution, components A and B, prove the relation

$$\mu_a = g - x_b \left(\frac{\partial g}{\partial x_b}\right)_{T,p}$$

$$= -x_b^2 \left(\frac{\partial g/x_b}{\partial x_b}\right)_{T,p},$$

where $g \equiv G/(n_a + n_b)$ is the free energy per mole of the mixture.

CHAPTER 8

IDEAL SOLUTIONS

8·1. Molecular aspects of solutions

In this section we shall outline certain points of view, actually quite extraneous to pure thermodynamics, concerning the molecular interpretation of solutions. The reader will observe that the arguments are far from precise but they will be referred to again in Chapter 14.

(a) Causes of solubility. It is often said that solubility is due to the molecular forces. That this is incorrect may be seen from the fact that two gases mix in all proportions and have infinite mutual solubility; the mixing is due, not to any interaction, but to the motion of the molecules and to the fact that a mixed state is immensely more probable than an unmixed one. The mutual solubility of gases is therefore an aspect of the statistical origin of the second law.

However, as soon as the molecules are brought into close proximity, as in condensed phases, the molecular forces have a decisive influence and may either increase or decrease the tendency towards mixing. Let it be supposed that two liquids A and B are contained in an isolated vessel so that the condition of equilibrium is a maximum of the entropy or of the number Ω of complexions. The intermingling of the molecules, if it were the only effect to be considered, would certainly cause an increase in the entropy because there are a vastly larger number of complexions for the mixed state than for the un-mixed one. However, it may occur that the A-B attractive force is less than the attractive force of A molecules with themselves or of B molecules with themselves. In this case complete mixing would imply an increase in the potential energy of the whole assembly and —since the system is isolated—this could only be achieved by a withdrawal of energy from the thermal motion of the molecules, which is to say from the kinetic energy of their translation, vibration and rotation. This would cause a decrease in the randomness of the thermal motion with consequent decrease of the entropy. Thus, when the two factors are considered together, it may occur that the overall maximum of the entropy is attained when miscibility is not complete.

The same ideas may be expressed in terms of the Gibbs free energy which is the appropriate function if we wish to discuss equilibrium at constant temperature and pressure, in place of conditions of isolation. Since $G = H - TS$ a decrease of G is favoured by a decrease of H and by an increase of S.

From the discussion above, it seems probable that the highest value of the entropy may now occur at complete, or almost complete, mixing, since the effect of the heat bath, which maintains constant temperature, is to prevent the decrease in thermal energy. The effect of the molecular forces, in determining the mutual solubility of the two liquids, is now largely a question of their influence on the enthalpy.

Let w_{aa} be the increase in potential energy when a pair of A molecules are brought together from infinite separation—and is actually a negative quantity. Similarly, let w_{ab} and w_{bb} be the potential energies of A-B and B-B pairs at the equilibrium separation in the liquid state. For simplicity it will be supposed that the two kinds of molecules are very similar in size and shape so that the average number of nearest neighbours of any one molecule has the same value, z, in the pure liquids and in the mixture. The transfer of one molecule of A from the pure liquid A to the pure liquid B, and the transfer of one molecule of B in the reverse direction, will increase the potential energy of the system by†

$$z(2w_{ab} - w_{aa} - w_{bb}). \qquad (8·1)$$

It will be shown in more detail in Chapter 14 that the same factor is significant however many molecules are interchanged.

Now this change of potential energy may be expected to be the major part of the ΔH of mixing and will account for it entirely if the thermal motion of the molecules may be supposed to be unaffected by the nature of their neighbours. It follows that whenever w_{ab} is a *smaller* negative quantity than the mean of w_{aa} and w_{bb} there is an *increase* in enthalpy on mixing; this works in opposition to the entropy increase in regard to the minimization of G. If the relative magnitude of w_{ab} is sufficiently small it may occur that the least value of G is attained at some degree of mixing which does not correspond to complete mutual solubility of the two liquids. This frequently occurs, for example, when a polar and a non-polar substance are mixed, due to the large interaction energy w_{aa} of the polar molecules between themselves as compared to the smaller interaction energy w_{ab} of the polar with the non-polar molecules.

(b) The origin of the ideal solution laws. In considering the origin of Raoult's and Henry's laws at the molecular level it is instructive first of all to discuss a very rough *kinetic interpretation*. Consider the pure liquid A in equilibrium with its vapour. The rate of condensation of vapour molecules on unit area of the liquid surface is proportional to the vapour pressure and may be written as kp_a^*. Let r be the intrinsic evaporation rate of molecules per unit area. At equilibrium these rates are equal and therefore

$$r = kp_a^*. \qquad (8·2)$$

† For simplicity it has been supposed that only the nearest-neighbour interactions are appreciable; however, an allowance for the next nearest-neighbour interactions, w'_{aa}, etc., will clearly give a term of the same form as (8·1).

Suppose now that the substance A is dissolved in another liquid B; it is reasonable to expect that a gaseous A molecule can enter any part of the surface, provided that A and B form a true solution. The condensation rate of A molecules per unit area of the surface can therefore be taken as kp_a, where p_a is the partial pressure of A. On the other hand, evaporation of A molecules can only take place from the fraction of the surface which is occupied by these molecules. This fraction may be taken as approximately equal to x_a, the mole fraction of A in the solution, the two types of molecules being supposed approximately equal in size and the surface composition the same as the internal composition. The evaporation rate per unit area may thus be written as $r'x_a$ and at equilibrium

$$r'x_a = kp_a. \tag{8·3}$$

Eliminating the coefficient k between (8·2) and (8·3) we finally obtain

$$p_a = \frac{r'}{r}p_a^* x_a. \tag{8·4}$$

Now the quantities r and r' depend on the molecular forces which oppose the escape of an A molecule from pure A and from the solution respectively. If these forces remain unchanged over the whole range of composition, then $r = r'$ and (8·4) reduces to Raoult's law. On the other hand, even if r' varies with the composition it may be expected to be constant when component A is sufficiently dilute. For under these conditions all of the neighbours of a given A molecule are B molecules, and the force field which opposes its evaporation is constant. Thus in dilute solution r' will be independent of composition and (8·4) may be written

$$p_a = \text{constant} \times x_a,$$

which is Henry's law.

An interpretation of the ideal laws may also be developed on a statistical basis. For this purpose it is more convenient to discuss a solid solution than a liquid one—or, alternatively, to suppose that a liquid solution has a quasi-crystalline structure, which is to say that there exists something equivalent to the countable lattice sites. As in § 1·17, we consider the intermixing of two crystals, one consisting originally of N_a molecules of A and the other of N_b molecules of B. The two kinds of molecules will be supposed to be very closely similar in regard to their volumes and forces. In particular, it will be assumed (a) that the original A and B crystals and the mixed crystal all have the same lattice structure; (b) that the molecules are interchangeable between the lattice sites without causing any change in the molecular energy states or the total volume of the system. Under these conditions the expression (8·1) is zero and in the mixing process

$$\Delta U = \Delta H = 0. \tag{8·5}$$

With the same assumptions it may be expected that the A and B molecules are distributed at random over the lattice sites in the equilibrium mixture. This state can therefore be realized in

$$\Omega = \frac{(N_a + N_b)!}{N_a!\, N_b!} \tag{8·6}$$

different ways, as regards the position of the molecules on the $N_a + N_b$ lattice sites. The original unmixed crystal can be realized in only *one* such way. Therefore for the entropy of mixing we have

$$\Delta S = k \ln \Omega / 1$$

$$= k\{\ln (N_a + N_b)! - \ln N_a! - \ln N_b!\}, \tag{8·7}$$

as in equation (1·22). Using Stirling's theorem†

$$\Delta S = k\{(N_a + N_b) \ln (N_a + N_b) - N_a \ln N_a - N_b \ln N_b\}$$

$$= -k\left(N_a \ln \frac{N_a}{N_a + N_b} + N_b \ln \frac{N_b}{N_a + N_b}\right). \tag{8·8}$$

It was indicated in § 1·17 (and will be proved in § 12·7) that Boltzmann's constant k is equal to R/L, where L is the Avogadro constant. Therefore (8·8) may be written

$$\Delta S = -R\left(n_a \ln \frac{n_a}{n_a + n_b} + n_b \ln \frac{n_b}{n_a + n_b}\right), \tag{8·9}$$

where n_a and n_b are the numbers of moles of A and B. Using (8·9) and (8·5) we obtain the change in the Gibbs function on isothermal mixing

$$\Delta G = \Delta H - T\,\Delta S$$

$$= RT\left(n_a \ln \frac{n_a}{n_a + n_b} + n_b \ln \frac{n_b}{n_a + n_b}\right). \tag{8·10}$$

Let μ_a and μ_a^* be the chemical potentials of A in the mixture and in the original pure A crystal respectively. Then from the definition of μ,

$$\mu_a - \mu_a^* = \left(\frac{\partial \Delta G}{\partial n_a}\right)_{T,\, p,\, n_b}. \tag{8·11}$$

By differentiation of (8·10) we therefore obtain

$$\mu_a - \mu_a^* = RT \ln x_a, \tag{8·12}$$

where $x_a = n_a/(n_a + n_b)$. Now μ_a and μ_a^* may be equated to the chemical potentials of the vapour of A above the mixture and above the pure A crystal respectively. Using (3·19) we obtain

$$\mu_a - \mu_a^* = RT \ln \frac{p_a}{p_a^*}, \tag{8·13}$$

† When x is a very large number $\ln x!$ may be approximated by
$$\ln x! \doteq x \ln x - x.$$

where p_a and p_a^* are the vapour pressures of A above the mixture and above pure A respectively. Thus from (8·12) and (8·13) we finally obtain

$$x_a = p_a/p_a^*,$$

which is Raoult's law.

It is evident that this simple statistical derivation may be expected to correspond to reality only if we know the true number of particles which are 'randomized' in the mixing process. The result (8·9) for the entropy of mixing would clearly be too large if the B molecules were actually present as a dimer B_2 and would be too small if they dissociated to form ions. Therefore, as indicated previously in § 7·4b, the ideal laws may be expected to approximate most closely to observed behaviour when the mole fraction which appears in the expression of these laws is that which would be computed allowing for the existence of all particles which are able to make independent movements in the solution.

The statistical derivation depends also on the assumptions which were made in the lines preceding equation (8·5). Deviations from the ideal laws may therefore be expected to be due to either of the following factors:

(a) inequalities in the volumes of the molecules,

(b) disparities in the molecular forces.

Both factors can result in either positive or negative deviations from Raoult's law. With polymer solutions, where inequalities in the molecular volumes may be extreme, the factor (a) is very important and here it leads to negative deviations; with more normal solutions it usually leads to positive deviations, but for such solutions this factor tends often to be rather less significant than factor (b).†

If there are strong specific attractions between the species A and B, the potential energy w_{ab} in equation (8·1) is more negative than the mean of w_{aa} and w_{bb}. This gives rise to a ΔH of mixing which is negative. Therefore ΔG is more negative‡ than is indicated by equation (8·10), and the difference between μ_a and μ_a^* is larger than appears from equation (8·12). Hence $p_a < p_a^* x_a$ and the deviations from Raoult's law are negative.

The converse case, where w_{ab} is less negative than the mean of w_{aa} and w_{bb}, seems to occur more frequently in organic mixtures, especially when the molecules are not strongly polar. In this case the molecular attractions are mainly due to the London dispersion forces and these seem to be rather smaller between unlike molecules than between like ones. This situation leads to a ΔH of mixing which is positive and to partial pressures which are larger than would be expected from Raoult's law.

These results are in general agreement with observation. For example, Hildebrand has shown that negative deviations are usually accompanied by exothermic mixing and also by contraction in volume. Conversely,

† See Rowlinson, *Liquids and Liquid Mixtures*.

‡ It is here assumed that ΔS remains unchanged. This cannot be strictly true whenever the expression (8·1) is not zero, since the solution can then no longer be supposed to be completely random. The rough arguments which are used in this section will be examined in more detail in Chapter 14.

positive deviations are accompanied by absorption of heat and by expansion.

8·2. Definition of the ideal solution

A solution will be said to be ideal if the chemical potential of *every* component is a linear function of the logarithm of its mole fraction according to the relation

$$\mu_i = \mu_i^* + RT \ln x_i, \qquad (8·14)$$

where μ_i^* is function only of temperature and pressure. (Its significance will be discussed in § 8·3 below.)

The advantage of using (8·14) as the definition of ideality, rather than Raoult's or Henry's laws, is that the present chapter is brought more closely into relation with Chapter 3 on gaseous mixtures. In fact an equation of the same form as (8·14) has already been used in § 3·9 for the definition of the *ideal gaseous* solution. All mixtures which are called ideal—and also the special type of gaseous mixture which is called *perfect*—show the same dependence of the chemical potentials on the composition.

A solution is ideal only if (8·14) applies to every component in a given region of composition. However, it is not necessary that (8·14) shall apply over the whole range of composition. A solution may approximate to ideal behaviour in one region and not in others; as shown in the last section, there are grounds for expecting that a solution will approach closer and closer to ideality the more dilute it becomes in all but one of its components.

The limited number of solutions which are approximately ideal over the whole composition range are sometimes called *perfect solutions*. In many practical applications of thermodynamics it occurs that the behaviour of a mixture of interest has not been investigated experimentally; in such cases the assumption that the solution is perfect may be necessary in order to obtain an approximate answer to the problem in hand. It is to be emphasized, however, that deviations of partial pressures, etc., from the values predicted by the ideal laws are often very large, as may be seen from Figures 26 and 29 in § 7·4.

8·3. Raoult's and Henry's laws

These laws follow at once from the defining equation (8·14) together with the supposition that the gaseous phase above the solution is perfect. For the ith component the condition of equilibrium is

$$\mu_i^{\text{soln.}} = \mu_i^{\text{vap.}}.$$

Substituting from (8·14) and (3·19) we obtain

$$\mu_i^* + RT \ln x_i = \mu_i^0 + RT \ln p_i, \qquad (8·15)$$

or after rearrangement $p_i/x_i = K_i,$ (8·16)

where $K_i \equiv \exp\{(\mu_i^* - \mu_i^0)/RT\}.$ (8·17)

From the nature of μ_i^* and μ_i^0 it follows that K_i is *independent of composition*. In the case where (8·16) holds up to $x_i = 1$, it is evident that K_i is the same as p_i^*, the vapour pressure of pure component i.†
Thus (8·16) is the same as Raoult's law. Otherwise, this equation merely asserts a proportionality between p_i and x_i and is the same as Henry's law.

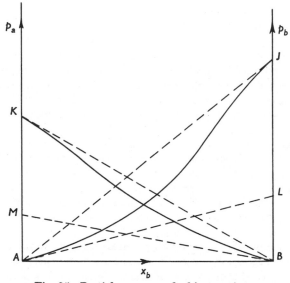

Fig. 35. Partial pressures of a binary mixture.

The situation can be made clearer by means of Fig. 35, which shows typical partial pressure curves for a binary liquid mixture of components A and B such as was discussed in §7·4. In the extreme left-hand region of the diagram, where x_b is quite small, we are concerned with an approximately ideal dilute solution of B in A. Here we have

$$\left.\begin{array}{l} p_a = p_a^* x_a, \\ p_b = K_b x_b, \end{array}\right\}$$ (8·18)

† Strictly speaking p_i^* refers to the vapour pressure of pure component at the *same total pressure* as the solution under discussion. This follows from equation (8·17), where μ_i^* must refer to this pressure. The difference between the value of p_i^* when the total pressure is p and the value of p_i^* when the total pressure is also p_i^* (as is the case for the pure component under its own pressure) is obtainable from equation (6·24) and is usually quite trivial. These remarks are equivalent to those made previously in a footnote on p. 224.

where p_a^* and K_b are represented by the lengths AK and BL respectively. In the relation $\mu_i = \mu_i^* + RT \ln x_i,$

μ_i^* is the value of μ_i when $x_i = 1$. Thus as regards component A, μ_a^* is simply the chemical potential of pure A, at the same temperature and pressure as the solution under discussion. As regards component B, μ_b^* signifies the chemical potential of pure B in a physically unattainable state corresponding to extrapolation from the actual solution along the line AL up to $x_b = 1$. (K_b may be similarly interpreted as the vapour pressure of pure B in this non-attainable state.)

It is to be emphasized that it is the tangent line AL which defines the ideal dilute solution of B in A. How far along this line the actual solution obeys the ideal laws has to be determined by experiment. But every point along this line can be thought of as representing the partial pressure of a *hypothetical ideal* solution of B in A.

Similar considerations apply to the extreme right-hand side of the diagram where we are concerned with an ideal dilute solution of A in B. In equation (8·14) μ_b^* will now stand for the real chemical potential of pure B and μ_a^* will stand for the chemical potential of A in a non-realizable state—and, of course, is *not* the same as the value of μ_a^* for the ideal solution on the left of the diagram.

The quantities μ_i^* which occur in the defining equation are thus properly to be regarded as constants, which may or may not correspond to a physically realizable state. In the latter case their values (and also the values of K_i) are strongly dependent on the nature of the other component (i.e. the 'solvent') which determines the position of the tangent lines AL and BM.

The above derivation of Raoult's and Henry's laws is based on the condition of equilibrium (equation (8·15)), and this in its turn is based on the conservation of mass in a virtual transfer of species i from one phase to the other.† It follows that the symbols p_i and x_i which occur in these laws *must refer to the same species*. For example, if a substance occurs both as a monomer A and as a dimer A_2, the laws may be applied to each species separately,

$$\left.\begin{array}{l} p_a = K_a x_a, \\ p_{a2} = K_{a2} x_{a2}, \end{array}\right\} \tag{8·19}$$

but there is no proportionality between the total amount of A in the one phase and the total amount in the other. Similar considerations arise in equilibria such as

$$NH_3(g) \rightleftharpoons NH_3(\text{soln.}) \rightleftharpoons NH_4OH \rightleftharpoons NH_4^+ + OH^-.$$

† Referring back to §2·9b $dn_{i\beta}$ was put equal to $-dn_{i\alpha}$ in the virtual transfer between the α and β phases.

In conclusion to this section, it may be noted that Raoult's law may be expressed in an alternative form: when there is a single solute (subscript s) the relative lowering of the vapour pressure of the solvent (subscript 0) is given by

$$\frac{p_0^* - p_0}{p_0^*} = 1 - x_0 = x_s. \tag{8·20}$$

Another useful variant is known as von Babo's law. Consider two ideal solutions at different temperatures but having the same mole fraction of solvent. At the one temperature $p_0 = p_0^* x_0$ and at the other temperature $p_0' = p_0^{*'} x_0$. Therefore

$$\frac{p_0}{p_0'} = \frac{p_0^*}{p_0^{*'}}, \tag{8·21}$$

and the ratio of the partial pressures of the solvent at the two temperatures is equal to the ratio of the vapour pressures of the pure solvent.

8·4. Imperfect vapour phase

When the vapour phase deviates appreciably from the perfect gas laws it follows from the treatment of the last section that Raoult's and Henry's laws must be expressed in terms of fugacities. Thus

$$f_i = K_i x_i, \tag{8·22}$$

and if the liquid phase remains ideal up to $x_i = 1$,

$$f_i = f_i^* x_i, \tag{8·23}$$

where f_i^* is the fugacity of the pure *liquid* component at the same temperature and total pressure as the solution under discussion.

This result may be compared with equation (3·72) which refers to an ideal gaseous solution and where f_i' is the fugacity of the pure vapour component at the temperature and total pressure under discussion. If it occurs that the liquid and vapour phases in equilibrium are both ideal solutions, and remain ideal up to $x_i = y_i = 1$, we obtain

$$f_i^* x_i = f_i' y_i, \tag{8·24}$$

since f_i is the same in (8·23) and in (3·72).

8·5. The mixing properties of ideal solutions

The defining equation for the ideal solution may be written

$$\frac{\mu_i}{T} = \frac{\mu_i^*}{T} + R \ln x_i,$$

and since this is an identity the partial differential coefficients of the left-hand and right-hand sides may be equated. Thus

$$\left(\frac{\partial \mu_i/T}{\partial T}\right)_{p,\,n_i,\,n_j} = \left(\frac{\partial \mu_i^*/T}{\partial T}\right)_p ,$$

and similarly

$$\left(\frac{\partial \mu_i}{\partial p}\right)_{T,\,n_i,\,n_j} = \left(\frac{\partial \mu_i^*}{\partial p}\right)_T .$$

The left-hand sides of these equations are given in terms of enthalpies and volumes by (2·113) and (2·111) respectively. Substituting from these relations we obtain

$$-\frac{H_i}{T^2} = \left(\frac{\partial \mu_i^*/T}{\partial T}\right)_p , \tag{8·25}$$

$$V_i = \left(\frac{\partial \mu_i^*}{\partial p}\right)_T . \tag{8·26}$$

Now μ_i^*, and therefore also its partial derivatives, is independent of composition. The same must therefore be true of H_i and V_i; *in an ideal solution the partial molar enthalpy and volume are independent of composition.*

In the case of liquid mixtures which are ideal over the complete range of composition, from $x_i = 0$ to $x_i = 1$, H_i and V_i are therefore the same as h_i and v_i, the enthalpy and volume respectively, per mole of the pure component. It follows that the total enthalpy $H(=\Sigma n_i H_i)$, and the total volume $V(=\Sigma n_i V_i)$, of any such mixture, are the same as those of the components before mixing. Such mixtures can therefore be prepared, at constant temperature and pressure, without heat effect or volume change.

The great majority of solutions, however, are those which approach ideality only when one of the species, the *solvent*, is in great excess and the remainder, the *solutes*, are very dilute. In such cases H_i and V_i can be interpreted as h_i and v_i respectively only in the case of the solvent, whose mole fraction approaches unity whilst the solution remains ideal. As regards the solutes, the partial molar enthalpy and volume are constant, in the region of ideality, but are not equal to the enthalpy and volume respectively per mole of the pure solutes in their normal states.† Moreover, the magnitude of these partial molar quantities is strongly dependent on the nature of the solvent, exactly as in the case of μ_i^* and K_i discussed previously. Suppose, for example,

---

† Cf. the discussion of μ_i^* in §8·3. The line AL in Fig. 35 defines a hypo-thetical ideal solution of B in A; such a solution has constant values of H_a, H_b, etc., at *all* compositions, but these values correspond to those of the actual solution only as B becomes very dilute.

that the substance i forms ideal solutions in two solvents A and B; either solution can be diluted with more of the *same* solvent without heat effect or volume change, but such effects will in general occur if the solution is diluted with the *other* solvent.

The change in the free energy and entropy on mixing may be obtained as follows. The total Gibbs free energy of an ideal solution is

$$G = \Sigma n_i \mu_i$$

$$= \Sigma n_i \mu_i^* + RT\Sigma n_i \ln x_i. \tag{8·27}$$

Now in the case of liquid mixtures which are ideal over the complete range of composition the μ_i^* are the free energies per mole of the pure liquids, as indicated in § 8·3. It follows that the first term on the right-hand side of (8·27) is the total free energy of the liquids before mixing. The free-energy change of mixing, at constant temperature and pressure, is therefore

$$\Delta_m G = RT\Sigma n_i \ln x_i. \tag{8·28}$$

and this is a negative quantity.

Since it has been shown already that the corresponding enthalpy change is zero, the entropy of mixing is

$$\Delta_m S = - R\Sigma n_i \ln x_i. \tag{8·29}$$

and this is correspondingly positive.

It may be noted that (8·28) and (8·29) are the same as for the perfect gas mixture (equation (3·37)).

Although constant values of H_i and V_i are a consequence of the solution being ideal it does not follow that, when a mixture is known from experiment to have these properties, it is necessarily ideal. Consider μ_i/T as a function of temperature, pressure and the $n - 1$ independent mole fractions:

$$\frac{\mu_i}{T} = f(T, p, x_1, x_2, \ldots, x_{n-1});$$

therefore

$$d\left(\frac{\mu_i}{T}\right) = \frac{\partial \mu_i/T}{\partial T} dT + \frac{\partial \mu_i/T}{\partial p} dp + \Sigma \frac{\partial \mu_i/T}{\partial x_i} dx_i$$

$$= -\frac{H_i}{T^2} dT + \frac{V_i}{T} dp + \frac{1}{T} \Sigma \frac{\partial \mu_i}{\partial x_i} dx_i. \tag{8·30}$$

If it is known that H_i and V_i are independent of composition the integral of this expression will be of the form

$$\frac{\mu_i}{T} = f(T, p) + \frac{1}{T} \Sigma \int \frac{\partial \mu_i}{\partial x_i} dx_i. \tag{8·31}$$

However, this is clearly not the same as equation (8·14), which gives μ_i as an explicit function of x_i. Some third experimental criterion is needed and this must reflect, in some way, the degree of randomness in the solution, which is to say the entropy change on mixing.

On the other hand, deviations from (8·29) are usually fairly small, and if the heat and volume change of mixing are zero, or almost zero, there is usually a fairly good presumption that the solution is at least approximately ideal.

8·6. The dependence of vapour-solution equilibria on temperature and pressure

The condition of equilibrium between a solution phase α and a vapour phase β for a component i may be written

$$\frac{\mu_i^\alpha}{T} = \frac{\mu_i^\beta}{T},$$

and therefore in any variation which maintains equilibrium we have

$$\mathrm{d}\left(\frac{\mu_i^\alpha}{T}\right) = \mathrm{d}\left(\frac{\mu_i^\beta}{T}\right). \tag{8·32}$$

If the solution is ideal μ_i^α depends only on the variables T, p and x_i according to (8·14). Similarly, if the vapour phase is perfect μ_i^β depends only on T and p_i in accordance with (3·19):

$$\mu_i^\beta = \mu_i^0(T) + RT \ln p_i. \tag{8·33}$$

The relation (8·32) may therefore be expanded in terms of the independent variables:

$$\frac{\partial \mu_i^\alpha/T}{\partial T}\mathrm{d}T + \frac{1}{T}\frac{\partial \mu_i^\alpha}{\partial p}\mathrm{d}p + \frac{1}{T}\frac{\partial \mu_i^\alpha}{\partial x_i}\mathrm{d}x_i = \frac{\partial \mu_i^\beta/T}{\partial T}\mathrm{d}T + \frac{1}{T}\frac{\partial \mu_i^\beta}{\partial p_i}\mathrm{d}p_i.$$

The partials with respect to temperature and pressure will by now be very familiar, and the partials with respect to x_i and p_i are obtained immediately from (8·14) and (8·33) respectively. Hence

$$-\frac{H_i}{T^2}\mathrm{d}T + \frac{V_i}{T}\mathrm{d}p + \frac{R}{x_i}\mathrm{d}x_i = -\frac{h_i^g}{T^2}\mathrm{d}T + \frac{R}{p_i}\mathrm{d}p_i,$$

where H_i and V_i refer to the partial molar quantities in solution and h_i^g is the enthalpy per mole of the gaseous component (§3·3c). Rearranging this equation we obtain

$$\mathrm{d}\ln\left(\frac{p_i}{x_i}\right) = \frac{h_i^g - H_i}{RT^2}\mathrm{d}T + \frac{V_i}{RT}\mathrm{d}p. \tag{8·34}$$

If the vapour phase is not perfect the partial pressure must be replaced by the fugacity, and h_i^g then refers to the enthalpy per mole of the gas at a pressure sufficiently low for the vapour phase to be perfect.†

The ratio p_i/x_i has been defined in (8·16) as the coefficient K_i of Henry's law. The temperature and pressure-dependence of this coefficient is therefore obtained from (8·34) as

$$\left(\frac{\partial \ln K_i}{\partial T}\right)_p = \frac{h_i^g - H_i}{RT^2}, \tag{8·35}$$

$$\left(\frac{\partial \ln K_i}{\partial p}\right)_T = \frac{V_i}{RT}. \tag{8·36}$$

In the first of these equations the quantity $h_i^g - H_i$ is, of course, the heat absorbed in the evaporation of 1 mole of the component from the ideal solution at constant temperature and pressure. The equation may be used for calculating the change of solubility with temperature.

In general, the effect of pressure on K_i, as on other properties of condensed phases, is very small. For the same reason equation (8·35), which is strictly correct only when the total pressure p is held constant (e.g. by use of an additional gaseous component which is insoluble in the liquid phase), is still a very good approximation when the left hand side is written $d \ln K_i/dT$.

In the special case where the component behaves ideally up to $x_i = 1$, K_i is the same as p^*, the vapour pressure of the pure liquid. In this instance equations (8·35) and (8·36) reduce to equations (6·25) and (6·24) respectively.

8·7. Nernst's law

From §7·4e it will be clear that if two liquids α and β are partially immiscible their behaviour towards each other is necessarily non-ideal. In the present section we consider, however, the behaviour of a third component (subscript i) which is present in each of the two liquid layers. If this substance is sufficiently dilute in each layer it may behave *individually* as an ideal solute in both of them, even though the system as a whole is non-ideal.‡

† This may perhaps be seen more clearly by carrying out an alternative derivation as follows. The condition of equilibrium $\mu_i^\alpha = \mu_i^\beta$ is written in the form
$$\mu_i^* + RT \ln x_i = \mu_i^0 + RT \ln f_i,$$
or
$$\ln \frac{f_i}{x_i} = \frac{\mu_i^*}{RT} - \frac{\mu_i^0}{RT}.$$
Differentiation followed by the application of (8·25), (8·26) and (3·64) gives (8·34) with p_i replaced by f_i.

‡ The effect of increasing the concentration of i in each solvent is usually to increase their mutual solubility and this often leads to a homogeneous mixture at a sufficiently high concentration. On the ternary diagram this occurs at the 'Plait point'.

When this condition is satisfied the condition of equilibrium,

$$\mu_i^\alpha = \mu_i^\beta,$$

may be replaced by

$$\mu_i^{*\alpha} + RT \ln x_i^\alpha = \mu_i^{*\beta} + RT \ln x_i^\beta,$$

and therefore

$$\ln \frac{x_i^\alpha}{x_i^\beta} = \frac{\mu_i^{*\beta} - \mu_i^{*\alpha}}{RT}, \tag{8·37}$$

which is independent of composition. The ratio x_i^α / x_i^β, which may be denoted N,

$$x_i^\alpha / x_i^\beta \equiv N, \tag{8·38}$$

is therefore independent of the individual values of x_i^α and x_i^β in the region where each solution is ideal. This is the Nernst distribution law.

The ratio N is known as the *partition coefficient*, and a little consideration will show that it is equal to the ratio K_i^β / K_i^α of the Henry law coefficients in the two solvents. Its temperature and pressure-dependence is obtained either by direct differentiation of (8·37) followed by the application of (8·25) and (8·26), or from (8·35) and (8·36):

$$\left.\begin{aligned}
\left(\frac{\partial \ln N}{\partial T}\right)_p &= \frac{H_i^\alpha - H_i^\beta}{RT^2}, \\
\left(\frac{\partial \ln N}{\partial p}\right)_T &= \frac{V_i^\beta - V_i^\alpha}{RT},
\end{aligned}\right\} \tag{8·39}$$

where H_i^α, etc., refer to the partial molar quantities in the two solvents.

8·8. Equilibrium between an ideal solution and a pure crystalline component

As an example, consider a solution in water of the crystalline substance A, and it will be supposed that these two components do not form solid solutions with each other. When plotted against the mole fraction x_a, the temperature at which the solution is in equilibrium with one or other of the pure solid phases will be of the type shown in Fig. 36.

On cooling a solution of the composition R, ice will separate out at the temperature P, and on cooling a solution of the composition S, the substance A will separate out at the temperature Q. The curve PT is conventionally called the freezing-point curve of the solvent, and the curve QT is called the solubility curve of the solute. However, there is no thermodynamic necessity to distinguish between the two components, and the general purpose of the present discussion is to obtain an equation giving the tangent to either of the two curves. As applied to the solute, this equation gives the temperature coefficient of the solubility. As applied to the solvent the same equation, after integration, gives the depression of the freezing-point.

The student will have observed that there are two main methods by which the temperature and pressure coefficients of the various equilibria may be derived. These are mathematically equivalent but somewhat different in the details of the manipulation.

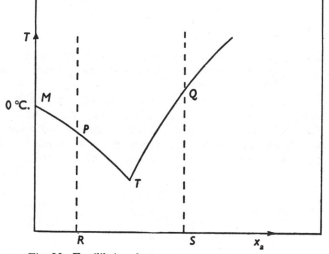

Fig. 36. Equilibrium between a pure solid and its solution.

(1) The method by which the equilibrium relation

$$\mu_i^\alpha = \mu_i^\beta, \tag{8·40}$$

is expressed in the differential form

$$d\mu_i^\alpha = d\mu_i^\beta,$$

or
$$d\left(\frac{\mu_i^\alpha}{T}\right) = d\left(\frac{\mu_i^\beta}{T}\right).$$

which is then expanded in terms of the chosen independent variables. This method was adopted in the derivation of (7·8), (8·34) and many other equations.

(2) The method adopted in the footnote on p. 254, and also in the derivation of (6·28) and (6·29).

The first of these has been our usual procedure, and for sake of variety we shall henceforth make more extensive use of the second.

Let substance i be present as a pure solid and also as a component of an ideal solution. The condition of equilibrium (8·40) may therefore be rewritten, in view of (8·14), as

$$\mu_i^s = \mu_i^* + RT \ln x_i,$$

where μ_i^s is the chemical potential of the pure solid and x_i is the mole fraction in the solution. Rearranging, we obtain

$$\ln x_i = -\frac{\mu_i^*}{RT} + \frac{\mu_i^s}{RT}. \tag{8·41}$$

In accordance with the phase rule, the two-component, two-phase system has two degrees of freedom, and the temperature and pressure may therefore be varied independently. Consider the temperature-dependence of $\ln x_i$ when the pressure is held constant; from (8·41) we obtain

$$\left(\frac{\partial \ln x_i}{\partial T}\right)_p = -\frac{1}{R}\frac{\partial \mu_i^*/T}{\partial T} + \frac{1}{R}\frac{\partial \mu_i^s/T}{\partial T}$$

$$= \frac{H_i - h_i^s}{RT^2} \tag{8·42}$$

by (8·25) and (2·113b). H_i is the partial molar enthalpy of the component in the ideal solution and h_i^s is its enthalpy per mole as the pure solid, both referring to the temperature T. The equation may therefore be rewritten

$$\left(\frac{\partial \ln x_i}{\partial T}\right)_p = \frac{L_i}{RT^2}, \tag{8·43}$$

where L_i is the heat absorbed, at constant temperature and pressure, when 1 mol of the component dissolves in the ideal solution.

As shown previously, H_i is independent of composition in the region of ideality. Therefore, in the case where the solution remains ideal up to $x_i = 1$, H_i is the same as the enthalpy per mole of the pure *liquid* component and L_i is its latent heat of melting. It is to be noted, however, that these quantities refer to the temperature T at which the solution of mole fraction x_i is in equilibrium with the pure solid. Therefore L_i is not quite equal to the latent heat at the melting-point.

Equation (8·43) gives the tangent $\partial x_i/\partial T$ to either of the curves of Fig. 36 at any particular mole fraction x_i. When applied to what is conventionally called the solute it gives the temperature coefficient of the solubility in terms of the heat of solution at the appropriate temperature. When applied to the solvent it gives, after integration, the depression of the freezing-point due to the solute. This can be seen most clearly by inverting the relation; thus

$$\frac{\partial T}{\partial \ln x_i} = \frac{RT^2}{L_i},$$

which shows how the equilibrium temperature T varies with the mole fraction of the *solvent*.

8·9. Depression of the freezing-point

In the second application discussed above, the concentration of the solutes is normally small and x_i is close to unity. If the solution is ideal at the particular mole fraction x_i, it will continue to be so *a fortiori* over the range of mole fractions up to the pure solvent. This makes possible the integration of (8·43) to give the depression of the freezing-point, allowing only for the temperature variation of L_i over the curve MP of Fig. 36.

It is to be noted that (8·43) applies strictly under conditions of constant total pressure, as when the solution is under the pressure of the atmosphere. Under the same conditions the temperature coefficient of L_i is given by

$$\left(\frac{\partial L_i}{\partial T}\right)_p = \left(\frac{\partial H_i}{\partial T}\right)_p - \left(\frac{\partial h_i^s}{\partial T}\right)_p$$

$$= c_{p_i} - c_{p_i}^s$$

$$\equiv \Delta c_p, \tag{8·44}$$

where c_{p_i} is the partial molar heat capacity of i in solution, equal to its value as a pure liquid (since H_i is independent of composition in the region of ideality) and $c_{p_i}^s$ is the molar heat capacity of solid i.

Over the small ranges of temperature which are usually encountered in the measurement of freezing-point depression, Δc_p may usually be assumed constant. The integration of (8·44) therefore gives

$$L_M - L = \Delta c_p (T_M - T), \tag{8·45}$$

where L_M is the enthalpy of melting at the melting-point T_M (the subscript i will henceforth be deleted). Substitution of this result in (8·43) gives

$$\frac{\partial \ln x}{\partial T} = \frac{L_M - \Delta c_p (T_M - T)}{RT^2}. \tag{8·46}$$

Integrating from $x = 1$ to the particular mole fraction x at which the freezing-point depression is to be calculated,

$$\ln(1/x) = \frac{1}{R} \int_T^{T_M} \frac{L_M - \Delta c_p (T_M - T)}{T^2} \, dT$$

$$= \frac{(L_M - \Delta c_p T_M)}{R} \int_T^{T_M} \frac{dT}{T^2} + \frac{\Delta c_p}{R} \int_T^{T_M} \frac{dT}{T}$$

$$= \frac{(L_M - \Delta c_p T_M)}{R} \left(\frac{1}{T} - \frac{1}{T_M}\right) + \frac{\Delta c_p}{R} \ln\left(\frac{T_M}{T}\right). \tag{8·47}$$

In using this equation for the purpose of calculating molecular weights, measured values of the quantities on the right-hand side are used for the purpose of calculating x, the mole fraction of the *solvent*. If there is only a single solute its mole fraction is obtained as $1-x$. The weight concentration of the solution being known, the molecular weight of the solute is immediately calculable.

A less accurate integration can be carried out by neglecting the temperature coefficient of the enthalpy of melting, equivalent to putting $\Delta c_p = 0$. In place of (8·47) we obtain

$$\ln \frac{1}{x} \doteqdot \frac{L_M}{R}\left(\frac{1}{T} - \frac{1}{T_M}\right)$$

$$\doteqdot \frac{L_M \theta}{RT_M^2}, \tag{8·48}$$

where $\theta = T_M - T$ is the depression of the freezing-point. This equation may also be written

$$-\ln\left(1 - \Sigma x_j\right) \doteqdot \frac{L_M \theta}{RT_M^2}, \tag{8·49}$$

where Σx_j is the sum of the mole fractions of all the dissolved species. Provided this quantity is small—which is to say the solution is very dilute—the expansion of the logarithm in (8·49) gives the approximate expression

$$\Sigma x_j \doteqdot \frac{L_M \theta}{RT_M^2}. \tag{8·50}$$

8·10. Elevation of the boiling-point

The last two sections were concerned with the equilibrium between an ideal solution and one of its components as a pure solid phase. An exactly similar theory will apply to the equilibrium between an ideal solution and one of its components as a pure vapour. When one or more involatile solutes are dissolved in a volatile solvent, and when the solution is sufficiently dilute to be ideal, it is readily confirmed that the elevation of the boiling-point of the solvent is given by equations closely similar to (8·43), (8·47) and (8·50).

Let T be the equilibrium temperature when the mole fraction of the solvent is x. Then the equations in question are

$$\left(\frac{\partial \ln x}{\partial T}\right)_p = -\frac{L}{RT^2}, \tag{8·51}$$

$$\ln(1/x) = \frac{(L_B - \Delta c_p T_B)}{R}\left(\frac{1}{T_B} - \frac{1}{T}\right) + \frac{\Delta c_p}{R}\ln\frac{T}{T_B}, \tag{8·52}$$

$$\Sigma x_j \doteqdot \frac{L_B \theta}{RT_B^2}, \tag{8·53}$$

where L is the enthalpy increase in the evaporation of 1 mol of the solvent at the temperature T, L_B is the corresponding latent heat at the boiling-point T_B of the pure solvent, Δc_p is equal to the molar heat capacity of the solvent as a vapour minus its value as a pure liquid, and $\theta = T - T_B$ is the elevation of the boiling-point.

8·11. The osmotic pressure of an ideal solution

In a solution, such as that of sugar in water, the *solvent* is the component whose mole fraction can be varied up to unity. Let it be supposed that such a solution is separated from a quantity of the pure solvent, at the same temperature, by means of a membrane permeable only to the solvent molecules. Then what is called the *osmotic pressure* of the solution is the excess pressure which must be placed on it in order to prevent any diffusion of solvent through the membrane.

The nature of osmotic pressure has been discussed very clearly by Guggenheim.† The osmotic pressure of a solution which is contained in a beaker open to the atmosphere is *not* a pressure which it actually exerts; it is rather to be regarded as being one of the thermodynamic properties of this solution in a manner closely similar to, say, its freezing-point. When a solution is said to have a freezing-point of $-5\,^{\circ}\mathrm{C}$ this does not imply that the solution *is* necessarily at this temperature but rather that $-5\,^{\circ}\mathrm{C}$ is the temperature at which the solution *would be* in equilibrium with one of its components as a solid phase. Similarly, when a solution is said to have an osmotic pressure of 10 atm this does not mean that the solution necessarily exerts this pressure but only that the solution *would be* in equilibrium with pure solvent, through the semi-permeable membrane, if it were given an excess pressure of this amount.

The cause of osmosis is simply diffusion; the solvent is able to diffuse through the membrane but the solute is not. It is only when there is a membrane which has this property that the phenomenon can occur. As discussed in §2·9b, this diffusion itself arises from a difference in chemical potential; at the same temperature and pressure the solvent substance is at a *lower* chemical potential in the solution than in its own pure liquid (on account of its lower mole fraction), and there is therefore a tendency for it to pass through the membrane in the direction pure solvent→solution. The osmotic pressure is the excess pressure which will just prevent this flow, and if a pressure greater than this were applied to the solution the solvent would diffuse in the reverse direction and the solution would become *more* concentrated.

† *Modern Thermodynamics by the Methods of Willard Gibbs* (London, Methuen, 1933).

In a perfect gas osmotic system, as shown in §3·3 b, equilibrium is attained when the partial pressure of the permeating gas is the same on each side of the semi-permeable membrane. This implies equal volume concentrations. On the other hand, in the case of solutions, the volume concentration is not an adequate criterion of the diffusion tendency; and it is possible for a substance to diffuse spontaneously from a region of *lower* to a region of *higher* concentration.† In such systems the only proper criterion of equilibrium, which is to say of the absence of a diffusion flow, is the equality of the chemical potential in the two regions.

Consider an osmotic system consisting of pure solvent together with a solution and an intervening membrane which is permeable to the solvent. Let p and p' be the pressures on the pure solvent and on the solution respectively when there is a condition of osmotic and thermal equilibrium. The difference will be called Π, the osmotic pressure:

$$\Pi \equiv p' - p. \tag{8·54}$$

Now the pure solvent, since its mole fraction is unity, has a chemical potential of μ_p^*, in the notation of this chapter, and the subscript denotes that this is the value at the pressure p. The same substance in the solution, if this is ideal, has a chemical potential given by

$$\mu_{p'}^* + RT \ln x,$$

where the subscript denotes that $\mu_{p'}^*$ refers to the pressure p' and x is the mole fraction of the solvent in the solution. The condition of equilibrium is‡

$$\mu_p^* = \mu_{p'}^* + RT \ln x.$$

Rearranging this we obtain

$$RT \ln x = \mu_p^* - \mu_{p'}^*$$
$$= \int_{p'}^{p} v \, dp, \tag{8·55}$$

by equation (2·111 b), where v is the molar volume of the pure solvent. This quantity does not vary much with pressure§ and therefore, to a fairly high degree of accuracy, the last equation can be written

$$RT \ln x = v_m (p - p'), \tag{8·56}$$

---

† See also pp. 86–7.

‡ Note that $RT \ln x$ is a *negative* quantity, and it is for this reason that $\mu_p^* < \mu_{p'}^*$.

§ v cannot be taken as being entirely constant, since quite large values of $p' - p$ are often encountered in practice.

where v_m is the mean value of the molar volume over the pressure range. This is easily evaluated if the compressibility of the solvent has been measured. Using the definition (8·54), we can rearrange (8·56) to obtain the following expression for the osmotic pressure of the ideal solution

$$\varPi = -\frac{RT}{v_m}\ln x. \tag{8·57}$$

Let Σx_j be the sum of the mole fractions of the various solutes. Then (8·57) may be written

$$\varPi = -\frac{RT}{v_m}\ln(1 - \Sigma x_j), \tag{8·58}$$

and if Σx_j is very small, $\varPi \doteqdot \dfrac{RT}{v_m}\Sigma x_j.$ \hfill (8·59)

By further approximation, this may be reduced to the form originally put forward in the early days of physical chemistry on the erroneous assumption that osmotic pressure is analogous to the 'bombardment' pressure in a gas. Now Σx_j is equal to $\Sigma n_j/n$, where Σn_j is the total number of solute moles and n is the total number of moles of all kinds including the solvent. Therefore (8·59) can be written

$$\varPi \doteqdot \frac{RT}{nv_m}\Sigma n_j. \tag{8·60}$$

If the solution is very dilute, n will not appreciably exceed the number of moles of the solvent only and nv_m will not differ appreciably from V, the total volume of the solution. Therefore an approximation to (8·60) is

$$\varPi \doteqdot RT\frac{\Sigma n_j}{V}$$

or $\varPi \doteqdot RT\,\Sigma c_j,$ \hfill (8·61)

where Σc_j is the sum of the volume concentrations of the solutes. This result is accurate only for very dilute solutions even when the solution continues to be ideal to much higher concentrations.

8·12. The ideal solubility of gases in liquids

Let it be supposed that a solution of a gas in a liquid obeys Raoult's law over the whole range of composition and in particular up to the mole fraction of unity which corresponds to the liquefied gas at the particular temperature. Under such conditions the solubility of the gas can be calculated from a knowledge of its vapour pressure. This may be illustrated by means of an example.

At 20 °C liquid carbon dioxide has a vapour pressure of 56.3 atm. If carbon dioxide dissolves in a solvent and obeys Raoult's law its mole fraction in the solution at 20 °C will be given by

$$x = p/p^*,$$

where p^* has the value 56.3 atm and p is the partial pressure of CO_2 above the solution. In particular, the solubility at 1 atm is given by

$$x = 1/p^* = 0.0178 \qquad (8·62)$$

This value may be called the *ideal solubility* at atmospheric pressure and is evidently independent of the nature of the solvent when it is expressed in the above form as a mole fraction. The table† shows some of the observed solubilities, and it is seen that the ideal solubility is of the right order of magnitude in most of the solvents.

Mole fraction of CO_2 *in various solvents at* 20 °C *and* 1 atm

Amyl acetate	0.027 0
Pyridine	0.012 9
Ethylene chloride	0.012 5
Chloroform	0.012 3
Toluene	0.010 7
Carbon tetrachloride	0.010 0
Ethyl alcohol	0.007 0
Carbon disulphide	0.002 2

In some cases, in order to apply the theory, it is necessary to extrapolate the vapour pressure of the liquefied gas beyond the critical point. For example, suppose that it is desired to estimate the ideal solubility of methane at a temperature of 25 °C, which is far above critical. If the observed vapour pressures are extrapolated by means of the Clausius–Clapeyron equation, the estimated value of p^* at 25 °C is found to be 289 atm—but of course this does not correspond to a stable state of gas-liquid equilibrium. The ideal solubility of methane at 25 °C is therefore $1/289 = 0.0035$. Some of the observed solubilities, as quoted by Hildebrand and Scott, are given in the table.

Mole fraction of CH_4 *in various solvents at* 25 °C *and* 1 atm

Ethyl ether	0.004 5
Cyclohexane	0.002 8
Carbon tetrachloride	0.002 9
Acetone	0.002 2
Benzene	0.002 1
Methyl alcohol	0.000 7
Water	0.000 02

† From Hildebrand and Scott, *Solubility of Non Electrolytes* (New York, Reinhold, 1950), Chapter xv, Table 7. It appears that this table refers to 20 °C and not 0 °C as stated (cf. Hildebrand, *Solubility* (New York, Reinhold, 1936), 2nd ed., p. 137).

According to the theory, the solubility of a gas may be expected to decrease with rising temperature, since this causes an increase in p^*; this is in general agreement with experience. The most soluble gases may also be expected to be those which are the most condensible, corresponding to low values of p^*.

8·13. The ideal solubility of solids in liquids

We have previously obtained (equation (8·43))

$$\left(\frac{\partial \ln x}{\partial T}\right)_P = \frac{L}{RT^2},$$

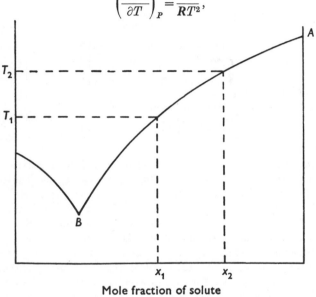

Mole fraction of solute

Fig. 37. Ideal solubility of a solid.

which gives the temperature coefficient of the solubility of a solute in terms of its heat of solution. If the equation is integrated at constant pressure on the assumption that L is independent of temperature over a small enough range, we obtain

$$\ln x_1/x_2 = \frac{L}{R}\left(\frac{1}{T_2} - \frac{1}{T_1}\right),$$

where x_1 and x_2 refer to the solubilities, expressed as mole fractions, at temperatures T_1 and T_2 respectively.

Let it be supposed that this equation continues to hold, at least approximately, up to a mole fraction of unity. This corresponds to point A in Fig. 37, where the pure liquid solute is in equilibrium with

its own solid at the melting-point. The previous equation may therefore be written

$$\ln x = \frac{L}{R}\left(\frac{1}{T_M} - \frac{1}{T}\right). \tag{8·63}$$

In this equation x is the solubility at the temperature T and T_M is the
melting-point of the solute. L is the enthalpy of solution, but, by the
nature of the assumptions which have been made, it is constant
along the curve BA and is therefore equal to the enthalpy of melting
of pure solute.

Equation (8·63), which was first put forward by Schroeder in 1893,
may be used to estimate the *ideal solubility* of solids in liquids, from
a knowledge of their melting-points and latent heats. It will be noted
that the equation is entirely analogous to (8·48) which gives the
depression of the freezing-point of the solvent; a more accurate form
of (8·63) could have been obtained by allowing for the temperature
dependence of L, as in § 8·9.

As an example, the enthalpy of melting of naphthalene is
18 580 J mol^{-1} and its melting-point is 80.05 °C. Its ideal solubility
at 20 °C may therefore be calculated from the above equation to be
$x = 0.273$ and is the same in every solvent. Some of the experimental
data for solvents of low polarity are given in the table.†

Mole fraction of naphthalene in various solvents at 20 °C.

Chlorobenzene	0.256
Benzene	0.241
Toluene	0.224
Carbon tetrachloride	0.205
Hexane	0.090

As deductions from equation (8·63) we have: (a) the solubility of
a solid may be expected to increase with rise of temperature; (b) the
solubility of a solid may be expected to be the greater the lower is its
melting-point and the smaller is its enthalpy of melting. These
results, although they are based on the supposition of ideal solutions,
are in fairly general agreement with experience.

PROBLEMS

1. It was shown by Zawidski that mixtures of benzene and ethylene
chloride obey Raoult's law quite accurately. At 50.0 °C their vapour
pressures as pure liquids are 268.0 and 236.2 mmHg respectively. At
this temperature calculate the total pressure and the composition of the
vapour which is in equilibrium with the liquid containing mole fractions
of 0.25, 0.50 and 0.75 of benzene.

† From Hildebrand and Scott, *The Solubility of Non-Electrolytes*, Chapter
XVII.

2. The solubility of succinic acid in 100 g of water is 2.35 g at $0°$ C and 6.76 g at 24.8 °C. Estimate the heat of solution of succinic acid in water.

3. The following data relates to the concentration of benzoic acid in layers of benzene and water which are in equilibrium:

Conc. in water as g/100 cm³	0.289	0.195	0.150	0.098	0.079
Conc. in benzene as g/100 cm³	9.7	4.12	2.52	1.05	0.737

Examine whether either of the following hypotheses would account for the results: (a) the benzoic acid is almost completely dissociated in the water; (b) the benzoic acid is almost completely associated to a dimer in the benzene.

4. The boiling-point of liquid ethane at 1 atm pressure is -88.6 °C, and its critical temperature and pressure are 32.2 °C and 48.2 atm respectively. Estimate its ideal solubility in hexane at 20 °C and a partial pressure of 1 atm.

5. The solubility in water of CO_2, expressed as cm³ of gas (at 0 °C and 1 atm) per cm³ of water, is 0.878 at 20 °C and 0.665 at 30 °C. Estimate the heat of solution. Establish any formula used.

It is required to remove CO_2 from a less soluble gas by absorption of the CO_2 in water in a counter-current tower operating at 10 °C and at a pressure of 20 atm. The gas mixture contains 10% of CO_2 by volume. Estimate the minimum amount of water required per 1000 m³ (reduced to 0 °C and 1 atm) of the entering gas.   [C.U.C.E. Qualifying, 1951]

6. The solubility of CO_2 in rubber has been expressed in two different publications by the following relations:

(a)
$$\frac{d \ln s}{dT} = \frac{\Delta H}{RT^2},$$

where $s =$ cm³ of dissolved gas measured at s.t.p. per 100 cm³ of rubber under a partial pressure of 1 atm,

$\Delta H =$ heat of solution of the gas in the rubber

$= -1.38 \times 10^4$ J mol⁻¹ experimentally at 20 °C.

(b)
$$\frac{d}{dT}\left(\ln \frac{c_B}{c_A}\right) = \frac{B}{RT^2},$$

where $c_B =$ concentration of the gas in the rubber phase,

$c_A =$ concentration of the gas in the gas phase, and

$B =$ experimental constant.

Derive the first of the above equations from the thermodynamic theory of an ideal solution, indicating any additional assumptions which must be made. Hence, find the corresponding value at 20 °C of B in the second expression and state what thermodynamic quantity it represents.

[C.U.C.E. Tripos, 1950]

7. A liquid X consists of an equilibrium mixture of a monomer, A, and a dimer, B, $2A \rightleftharpoons B$. At a particular temperature T the vapour pressure of liquid X, which may be assumed due to the monomer only, is p^0 and the equilibrium constant expressed in mole fraction units is K. At the same temperature the vapour pressure of a solvent C is p_C^0.

A solution consists of N_C moles of solvent C and N moles of X computed as if it were entirely monomer. If the solution is assumed ideal, show that at temperature T the vapour pressure of the monomer over the solution is given by

$$p_A = \frac{2N_A}{N_A + 2N_C + N} \, p^0 \, \frac{2K}{\sqrt{(1 + 4K)} - 1},$$

and the vapour pressure of the solvent by

$$p_C = \frac{2N_C}{N_A + 2N_C + N} p_C^0,$$

where the quantity N_A is given by the equation

$$\frac{(N - N_A)(N_A + 2N_C + N)}{4N_A^2} = K.$$

[C.U.C.E. Tripos, 1951]

8. It is required to depress the freezing-point of water to $-10\ ^\circ\mathrm{C}$ in order to prevent freezing. Estimate the weight percentage of glycerol which might be adequate for this purpose. Is this estimate likely to be too high or too low?

The enthalpy of melting of water is $6008\ \mathrm{J\ mol^{-1}}$.

[C.U.C.E. Qualifying, 1955]

CHAPTER 9

NON-IDEAL SOLUTIONS

9·1. Conventions for the activity coefficient on the mole fraction scale

The ideal solution was defined as one for which the chemical potential of every component is related to its mole fraction by the relation

$$\mu_i = \mu_i^* + RT \ln x_i,$$

where μ_i^* is a function of temperature and pressure only. It has been shown that the properties commonly attributed to ideal solutions are all deducible from this relation. These properties form a very convenient standard against which to compare the properties of all real solutions.

The method usually adopted in dealing with real solutions is to find the magnitude of the pure number which, when multiplied by the mole fraction of the particular species, makes applicable a relation of the above form. That is to say, we define† an activity coefficient γ_i such that the equation

$$\mu_i = \mu_i^* + RT \ln \gamma_i x_i \qquad (9\cdot1)$$

is correct, however large is the deviation from ideality. In this identity μ_i^* is to be taken as a function of *temperature and pressure only*, but γ_i may be a function of these variables together with the mole fractions of all substances in the solution.

Equation (9·1) as it stands does not provide a complete definition of γ_i, since μ_i^* is also an unknown in this equation. The definition of both quantities becomes complete as soon as it is specified under what conditions γ_i becomes equal to unity. For this purpose it is convenient to adopt either of two conventions. These are based on the fact that a component i of a real solution is normally found to approach ideal behaviour *both* as $x_i \to 0$ and as $x_i \to 1$, as discussed on p. 225.

Convention I. This is usually applied to solutions in which all of the components, in their pure states, are *liquids* at the same temperature and pressure as the solution (e.g. a water-alcohol mixture). For each

---

† The definition of activity coefficient adopted in most American text-books is a little different from that used here and is discussed in §9·10.

component the activity coefficient is taken as approaching unity as the mole fraction approaches unity. Thus

$$\left.\begin{aligned} \mu_i &= \mu_i^* + RT \ln \gamma_i x_i, \\ \gamma_i &\to 1 \quad \text{as} \quad x_i \to 1. \end{aligned}\right\} \tag{9·2}$$

It will be seen that the logarithm vanishes under the limiting conditions, and therefore μ_i^* is equal to the Gibbs free energy per mole of the pure substance at the same temperature and pressure as the solution under discussion.

Convention II. This is usually applied to solutions in which some of the components are *gases* or *solids* at the temperature and pressure in question (e.g. aqueous solutions of oxygen or of sugar). For such solutions it is convenient to distinguish between the solvent and the solutes. The former, denoted by the subscript 0, is a component which is present in excess and whose mole fraction can be varied up to unity without change of phase. The convention is then as follows:

$$\left.\begin{aligned} \text{for solvent:} &\quad \mu_0 = \mu_0^* + RT \ln \gamma_0 x_0 \quad \text{and} \quad \gamma_0 \to 1 \quad \text{as} \quad x_0 \to 1, \\ \text{for solutes:} &\quad \mu_i = \mu_i^* + RT \ln \gamma_i x_i \quad \text{and} \quad \gamma_i \to 1 \quad \text{as} \quad x_i \to 0. \end{aligned}\right\} \tag{9·3}$$

Thus, as regards the solvent, the situation is the same as on Convention I. As regards the solutes their activity coefficients are taken as approaching unity at *infinite dilution*; the quantity μ_i^* therefore stands for the chemical potential of pure solute in a hypothetical liquid state corresponding to extrapolation from infinite dilution along the Henry law gradient. This will be made clearer in the next section.

It is entirely a matter of convenience which convention is adopted. The great advantage of the first is that it is symmetrical in all of the components. On the other hand, it is not convenient to use this convention in the case of components whose mole fraction cannot be varied up to unity without change of phase.

9·2. The activity coefficient in relation to Raoult's and Henry's laws

For a non-ideal liquid mixture in contact with a perfect vapour phase we obtain, as the condition of equilibrium, in place of (8·15),

$$\mu_i^* + RT \ln \gamma_i x_i = \mu_i^0 + RT \ln p_i,$$

or after rearrangement
$$\frac{p_i}{\gamma_i x_i} = K_i, \tag{9·4}$$

where
$$K_i \equiv \exp\{(\mu_i^* - \mu_i^0)/RT\}. \tag{9·5}$$

From the nature of μ_i^* and μ_i^0 it follows that K_i is independent of composition. (If the vapour phase is not perfect, p_i must be replaced by the fugacity in equation (9·4).)

If we choose γ_i to approach unity as x_i approaches unity—as on Convention I for all species and on Convention II for the solvent— then K_i must clearly be the same as p_i^*, the vapour pressure of the pure component at the temperature and total pressure of the solution. Thus (9·4) may be written

$$p_i = p_i^* \gamma_i x_i, \qquad (9·6)$$

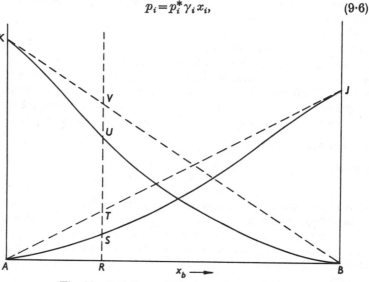

Fig. 38. Activity coefficients on Convention I.

and the extent to which γ_i differs from unity is seen to be a measure of the deviation from Raoult's law. It is greater than unity for positive deviations and less than unity for negative deviations.

The significance of γ_i may perhaps be seen more clearly from Fig. 38, which shows typical vapour-pressure curves of a binary mixture of components A and B. Consider the particular composition x_b corresponding to a point R. If the solution were ideal the partial pressure of B above the solution would have the value $p_b^* x_b$, which is represented by the length RT. The actual partial pressure is represented by the length RS. Therefore from equation (9·6) we have

$$\gamma_b = \frac{p_b}{p_b^* x_b} = \frac{RS}{RT}.$$

Similarly
$$\gamma_a = \frac{RU}{RV}.$$

In brief, whenever we choose $\gamma_i \to 1$ as $x_i \to 1$ the activity coefficients have a simple interpretation as the ratio of the actual partial pressure to the partial pressure which would occur if Raoult's law were obeyed. One of the most direct methods of determining these coefficients is by measurement of the partial pressures.

On the other hand, if we choose $\gamma_i \to 1$ as $x_i \to 0$, as on Convention II for the solutes, then K_i in equation (9·4) is clearly the tangent, at infinite dilution, of the curve of p_i against x_i. Therefore, according to Convention II, the extent to which the activity coefficient of the solute at any finite concentration differs from unity is a measure of its deviation from Henry's law.

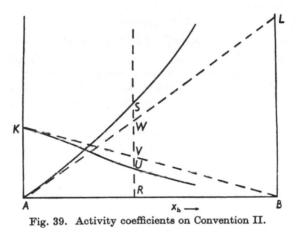

Fig. 39. Activity coefficients on Convention II.

This interpretation is made clearer in Fig. 39, which shows typical vapour-pressure curves for a type of binary system where the mole fraction of component B cannot be varied up to unity (e.g. oxygen in water). In such instances, instead of using Raoult's law as the criterion of ideality for the solute, it is more convenient to use Henry's law. Consider the solution whose composition corresponds to the point R. If the component B obeyed Henry's law at this composition its partial pressure would be $K_b x_b$, which is represented by the length RW.† The actual partial pressure is represented by the length RS. Therefore from equation (9·4) we have

$$\gamma_b = \frac{p_b}{K_b x_b} = \frac{RS}{RW}.$$

The activity coefficient is thus again equal to the ratio of the actual to the 'ideal' partial pressure, but the criterion of the latter is Henry's

† K_b itself is represented by the length BL.

rather than Raoult's law. (Of course as regards the solvent we continue, using Convention II, to regard Raoult's law as the measure of ideality and in Fig. 39 the activity coefficient γ_a is equal to the ratio RU/RV.)

If the solute is one such as sugar, whose vapour pressure is minute, it will be impossible to measure its partial pressure along the curve AS of Fig. 39. However, we can continue to regard the tangent AW as defining the ideal solution. Although this tangent now almost coincides with the axis AB, the activity coefficient of component B remains equal to the ratio RS/RW, and can be determined by methods other than the direct measurement of partial pressure, as will be described in §9·8.

The significance of μ_i^* in the equation

$$\mu_i = \mu_i^* + RT \ln \gamma_i x_i$$

also depends on the choice of convention. When we choose $\gamma_i \to 1$ as $x_i \to 1$, μ_i^* is simply the free energy per mole of the pure component at the same temperature and pressure as the solution under discussion. When we choose $\gamma_i \to 1$ as $x_i \to 0$ the situation can best be understood by imagining the existence of a hypothetical ideal solution whose partial pressure continues to obey Henry's law at all mole fractions up to $x_i = 1$. For example, in the case of Fig. 39 a solution is imagined for which the partial pressure of component B would lie everywhere along the line AL. γ_b would be unity at all compositions and μ_b^* could be interpreted as the free energy per mole of pure B in the hypothetical state corresponding to the extrapolation up to $x_b = 1$.

However, it is simplest to regard μ_i^* as being merely a constant in the above equation, a quantity independent of composition but dependent on the temperature and pressure of the solution.

9·3.   The use of molality and concentration scales

The quantity of a solid or a gas in solution is very often expressed as a molality rather than as a mole fraction. For example, most of the data on the free energy of formation of substances in aqueous solution refer to a reference state of unit molality. This is a matter of convenience, and the molality has less theoretical significance than the mole fraction.

The *molality* of a solute is the amount of it (mols, i.e. 'gram formula weights') per kg of solvent. Consider a solution containing n_i mols of solute and n_0 mols of solvent whose molar mass† is M_0 kg mol^{-1}. Then the molality of the solute is

† Note that 'molar mass' (kg mol $^{-1}$) differs by a factor of 1000 mol kg $^{-1}$ from the 'molecular weight', now to be known as the 'relative molecular mass', which is dimensionless. If, in place of the molar mass, the molecular weight had been included in (9·7), as in earlier editions, a factor of 1000 mol kg $^{-1}$ would have to be included in the denominator.

$$m_i \equiv \frac{n_i}{M_0 n_0} \tag{9·7}$$

and its mole fraction is $x_i = \dfrac{n_i}{n_0 + \Sigma n_i},$ (9·8)

where the summation is over all solutes. From (9·7) and (9·8)

$$\frac{m_i}{x_i} = \frac{n_0 + \Sigma n_i}{M_0 n_0} = \frac{1}{M_0 x_0} \tag{9·9}$$

and as $\Sigma n_i \to 0$, $m_i/x_i \to 1/M_0$. Therefore in very dilute solution

$$m_i \approx x_i/M_0, \tag{9·10}$$

and the molality and mole fraction are proportional to each other.

Occasionally a volume concentration scale is used. If V is the volume of the above solution in m^3, then the concentration (mol m^{-3}) of the particular solute is defined by[†]

$$c_i \equiv n_i/V. \tag{9·11}$$

Let ρ be the density of the solution (kg m^{-3}). Then the volume V weighs ρV kg and therefore

$$\rho V = M_0 n_0 + \Sigma M_i n_i,$$

where M_i is the molar mass (kg mol^{-1}) of species i. Hence

$$V = \frac{M_0 n_0 + \Sigma M_i n_i}{\rho}, \tag{9·12}$$

and from (9·11) we obtain

$$c_i - \frac{\rho n_i}{M_0 n_0 + \Sigma M_i n_i}. \tag{9·13}$$

Using (9·8) and (9·13) we obtain the ratio

$$\frac{c_i}{x_i} = \frac{\rho(n_0 + \Sigma n_i)}{M_0 n_0 + \Sigma M_i n_i}. \tag{9·14}$$

As $\Sigma n_i \to 0$, $c_i/x_i \to \rho/M_0$, and therefore in very dilute solution

$$c_i \approx \frac{\rho x_i}{M_0}, \tag{9·15}$$

and the concentration and mole fraction are proportional to each other.

It follows from (9·13) that the concentration of a given solution does not have the same value at two different temperatures on

† What was previously called the ' molality ' was the volume concentration expressed as mols per litre.

account of the change of density. The molality scale is therefore to be preferred. But the mole fraction scale has more theoretical significance than either. As shown in §8·1, there are reasons for expecting that the partial pressure of a component will be proportional to its mole fraction under certain especially simple molecular conditions, and this is one of the reasons why the mole fraction, rather than any alternative scale, is used for the *definition* of the ideal solution. In an ideal solution p_i is proportional to x_i, but is *not* proportional to m_i or c_i, except under the very dilute conditions where the mole fraction, molality and concentration are all proportional to each other.

9·4. Convention for the activity-coefficient on the molality scale

In the discussion of solutions containing solid or gaseous substances, the activity coefficient of the *solvent* is almost always defined on the mole fraction scale, but the activity coefficient of the solute is often chosen on the molality scale. This may be called Convention III.

Convention III:

$$\text{for solvent:} \quad \mu_0 = \mu_0^* + RT \ln \gamma_0 x_0 \quad \text{and} \quad \gamma_0 \to 1 \quad \text{as} \quad x_0 \to 1, \\ \text{for solutes:} \quad \mu_i = \mu_i^\square + RT \ln \gamma_i m_i \quad \text{and} \quad \gamma_i \to 1 \quad \text{as} \quad m_i \to 0. \tag{9·16}$$

This is closely related to Convention II and differs from it only in regard to the use of the molality scale for the solutes.†

As concerns the solvent the significance and value of μ_0^* and γ_0 are precisely the same as on Conventions I and II. As concerns the solute, $\mu_i^\square$ is the value of the chemical potential when m_i and γ_i are both equal to unity. For this reason it is often referred to, somewhat inexactly, as the chemical potential of the solute in *a hypothetical ideal solution of unit molality* (at the same temperature and pressure as the solution under discussion). This approximate interpretation of $\mu^\square$ can be seen more clearly with the aid of Fig. 40, which shows the partial pressure of component B plotted against its mole fraction. The Henry law coefficient K_b is again represented by the length BL. Let R be the point along the mole fraction axis at which component B has unit molality. If γ on the molality scale were unity when γ on

---

† In the case of electrolyte solutions, the molalities of *all* electrolytes must fall to zero for the activity coefficient of any of them to become unity (see equation (10·53) below). Thus in place of $\gamma_i \to 1$ as $m_i \to 0$ in the above convention, it is actually preferable to write $\gamma_i \to 1$ as $x_0 \to 1$.

As in the instances of equations (3·2) and (3·18), the quantity m_i in (9·16) is more correctly a ratio $m_i/m_i^\square$, where $m_i^\square$ is unit molality. The term of which the logarithm appears in (9·16) is therefore dimensionless.

the mole fraction scale is unity (which however is not exactly the case except in very dilute solution—see below) the 'hypothetical ideal' solution of unit molality would have a partial pressure of this component equal to the length of RT and $\mu_b^\square$ would refer to this solution. But because this interpretation is not exact, $\mu_b^\square$ is preferably to be regarded simply as a part of the chemical potential as given by (9·16).

The relationship between the activity coefficients defined in terms of Conventions II and III can be obtained as follows. Consider a solute whose mole fraction is x and whose molality is m. Its actual chemical potential is, of course, quite independent of the choice of scale and therefore

$$\mu = \mu^* + RT\ln\gamma_{II}x$$
$$= \mu^\square + RT\ln\gamma_{III}m, \qquad (9·17)$$

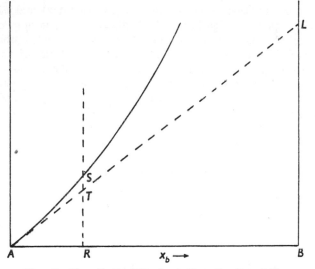

Fig. 40. Hypothetical ideal solution of unit molality.

where γ^{II} and γ^{III} are used, for the moment, to denote activity coefficients according to the Conventions II and III respectively. Rearranging we obtain

$$RT\ln\frac{\gamma^{III}m}{\gamma^{II}x} = \mu^* - \mu^\square. \qquad (9·18)$$

Under limiting conditions of very high dilution this reduces to

$$RT\ln\frac{m}{x} = \mu^* - \mu^\square,$$

or in view of equation (9·10)

$$RT \ln \frac{1}{M_0} = \mu^* - \mu^\square. \qquad (9\cdot19)$$

This equation determines the relation between μ^* and $\mu^\square$. Since these quantities are independent of composition, their difference in (9·19) has precisely the same numerical value as in (9·18). Hence, between these two equations we obtain

$$RT \ln \frac{\gamma^{\mathrm{III}} m}{\gamma^{\mathrm{II}} x} = RT \ln \frac{1}{M_0}$$

or

$$\frac{\gamma^{\mathrm{III}}}{\gamma^{\mathrm{II}}} = \frac{x}{M_0 m}, \qquad (9\cdot20)$$

which determines the relation between γ^{III} and γ^{II} for a solution which is not necessarily dilute.

The *general* relation between x and m is equation (9·9), and it follows from this equation, together with (9·20), that γ_{III} is not equal to γ_{II} except for a very dilute solution. Thus, even though a solution of appreciable concentration may be ideal, in the sense of the mole fraction scale (i.e. $\gamma_{\mathrm{II}} = 1$), it would not have a value of γ_{III} equal to unity. The activity coefficients defined on the molality scale evidently do not give a satisfactory measure of the deviations from ideality, except for a very dilute solution. This is equivalent to the remarks made at the end of the previous section.

Activity coefficients of a solute as calculated relative to the mole fraction and molality scale are sometimes called *rational* and *practical* respectively. Occasionally still another type of activity coefficient is used for the solute, defined by

$$\left. \begin{aligned} \mu_i &= \mu_i^\diamond + RT \ln y_i c_i, \\ y_i &\to 1 \quad \text{as} \quad c_i \to 0, \end{aligned} \right\} \qquad (9\cdot21)$$

where c_i is the concentration.

9·5. The effect of temperature and pressure

By rearranging equation (9·1) we obtain

$$\frac{\mu_i^*}{T} = \frac{\mu_i}{T} - R \ln \gamma_i - R \ln x_i.$$

Since this is an identity it may be partially differentiated at constant pressure and composition:

$$\frac{\partial \mu_i^*/T}{\partial T} = \frac{\partial \mu_i/T}{\partial T} - \frac{R \, \partial \ln \gamma_i}{\partial T}.$$

Therefore

$$\frac{\partial \mu_i^*/T}{\partial T} = \frac{-H_i}{T^2} - R \frac{\partial \ln \gamma_i}{\partial T}, \qquad (9\cdot22)$$

where H_i is the partial molar enthalpy of the particular component in the solution under discussion.

Let (9·22) be applied to the limiting composition where γ_i approaches unity according to the particular choice of convention. In this case the last term of (9·22) is zero and we obtain *either*

$$(a) \quad \frac{\partial \mu_i^*/T}{\partial T} = -\frac{h_i}{T^2}, \qquad (9·23)$$

where we have chosen $\gamma_i \to 1$ as $x_i \to 1$ and h_i is the enthalpy per mole of the pure liquid component, *or*

$$(b) \quad \frac{\partial \mu_i^*/T}{\partial T} = -\frac{H_i^0}{T^2}, \qquad (9·24)$$

where we have chosen $\gamma_i \to 1$ as $x_i \to 0$ and H_i^0 is the partial molar enthalpy of the component at infinite dilution.

Now μ_i^* is independent of composition and therefore $\dfrac{\partial \mu_i^*/T}{\partial T}$ has the same value whether we consider some particular solution or the limiting solution where $\gamma_i \to 1$. Comparing (9·22) with (9·23) or (9·24) we thus obtain the temperature dependence of the activity coefficient:

$$(a) \quad \frac{\partial \ln \gamma_i}{\partial T} = \frac{h_i - H_i}{RT^2}, \qquad (9·25)$$

for the case $\gamma_i \to 1$ as $x_i \to 1$ or

$$(b) \quad \frac{\partial \ln \gamma_i}{\partial T} = \frac{H_i^0 - H_i}{RT^2} \qquad (9·26)$$

for the case $\gamma_i \to 1$ as $x_i \to 0$.

By an exactly similar argument concerning the molality scale and Convention III we obtain

$$(c) \quad \frac{\partial \mu_i^\square/T}{\partial T} = -\frac{H_i^0}{T^2}, \qquad (9·27)$$

$$\frac{\partial \ln \gamma_i}{\partial T} = \frac{H_i^0 - H_i}{RT^2}, \qquad (9·28)$$

where $\gamma_i \to 1$ as $m_i \to 0$.

In these equations H_i^0 and H_i again refer to the partial molar enthalpy at infinite dilution and in the solution where the activity coefficient is γ_i respectively. These quantities can be determined by experiment as discussed in §2·14. It will be noted from (9·26) and (9·28) that the activity coefficients of a solute defined by Conventions II and III both have the same temperature coefficient.

The temperature-dependence of K_i in equation (9·5) is readily obtained by differentiation followed by application of equations

(3·10) together with (9·23) or (9·24) according to the choice of convention. For the case $\gamma_i \to 1$ as $x_i \to 0$ we obtain

$$\frac{\partial \ln K_i}{\partial T} = \frac{h_i^g - H_i^0}{RT^2}, \tag{9·29}$$

where h_i^g is the molar enthalpy in the gaseous phase. Thus $h_i^g - H_i^0$ is the enthalpy of evaporation from an infinitely dilute solution. For the case where $\gamma_i \to 1$ as $x_i \to 1$, K_i is the same as p_i^*, the vapour pressure of pure component, and its temperature-dependence is given by the Clausius–Clapeyron equation.

The pressure-dependence† of the activity coefficient may be obtained by similar reasoning. The results are

$$(a) \quad \frac{\partial \ln \gamma_i}{\partial p} = \frac{V_i - v_i}{RT}, \tag{9·30}$$

where $\gamma_i \to 1$ as $x_i \to 1$ and v_i is the volume per mole of the pure liquid component, and

$$(b) \quad \frac{\partial \ln \gamma_i}{\partial p} = \frac{V_i - V_i^0}{RT}, \tag{9·31}$$

where $\gamma_i \to 1$ as either the mole fraction or the molality approaches zero and V_i^0 is the partial molar volume at infinite dilution. V_i is the partial molar volume in the solution in question.

It may be remarked that it is possible for a solution to show zero values of $\partial \ln \gamma / \partial T$ and $\partial \ln \gamma / \partial p$ without the activity coefficients themselves being equal to unity. Solutions in which both of these conditions are satisfied have been called 'semi-ideal';‡ they behave like ideal solutions with regard to the enthalpy and volume, but they have a non-ideal free energy and entropy.

The dependence of γ on pressure, as given by (9·30) and (9·31) is, of course, extremely small. The dependence on temperature is much more significant, but for practical purposes it may often be necessary to neglect it, for lack of knowledge of the heat effects $(h_i - H_i)$ or $(H_i^0 - H_i)$ which occur in equations (9·25) and (9·26). Let it be supposed that this heat effect is actually 2000 J mol^{-1}. Then at 100 °C we obtain

$$\frac{\partial \ln \gamma}{\partial T} = \frac{2000}{8.314 \times 373^2} = 1.7 \times 10^{-3}$$

Therefore, for a variation of 30 °C, the change in γ would be about 5%. In studies on distillation it is frequently necessary to estimate partial pressures at one temperature from experimental measurements at another. If these two temperatures are not too far apart, the change in the activity coefficient will be rather small and may have to be neglected in the absence of the enthalpy data.

† See §9·10 for the pressure dependence of γ_i as it is defined in most American texts.

‡ Guggenheim, *Modern Thermodynamics by the Methods of Willard Gibbs*, Chapter VII (London, Methuen, 1933).

9·6. The determination of activity coefficients

In Chapter 8 various equations were derived relating to the partial pressures, freezing-point depression, osmotic pressure, etc., of an ideal solution. The corresponding expressions which are applicable to non-ideal solutions may be obtained simply by substituting $\gamma_i x_i$ in place of x_i.† The coefficient γ_i has, in fact, been constructed to have this property. It follows that an activity coefficient may be determined from experiment by application of these equations. As soon as a value of γ has been calculated from a measured property of the solution, for example a partial pressure, it may be used immediately to calculate the value of some other property of the same solution, e.g. its osmotic pressure, at the same temperature and pressure.

However, it must be borne in mind that the activity coefficients depend on temperature and pressure, as well as on composition, as discussed in the last section. Therefore a determination of the activity coefficient of the solvent in a particular solution by measurement of its freezing-point depression could not be used for the purpose of an accurate calculation of the boiling-point elevation of the same solution, without an additional knowledge of the enthalpies which appear in equation (9·25). Over appreciable ranges of temperature, the temperature-dependence of the enthalpies themselves must be allowed for, this dependence being expressible in terms of molar and partial molar heat capacities.

The various methods of determining activity coefficients will not be discussed in detail but may be summarized as follows.‡

(a) *Vapour-pressure measurements.* The principle of this method has already been discussed in § 9·2. For types of solution such as an alcohol-water mixture, where all of the components are liquids, the activity coefficient of each component is calculated using Convention I. That is to say, Raoult's law is taken as the criterion of ideality and the activity coefficients are evaluated by use of equation (9·6). On the other hand, for solutions such as salt in water or oxygen in water it is convenient to distinguish between the solvent and the solute. The activity coefficients are therefore calculated on the basis of Conventions II and III. For the solvent, Raoult's law is the criterion of ideality and the coefficients are again evaluated by use of equation

† The student may confirm that this is the case by carrying out a derivation *de novo* of equations (9·35), (9·36) and (9·37) below.

‡ For a more complete discussion see Lewis and Randall, *Thermodynamics* (New York, McGraw-Hill, 1923) or Glasstone, *Thermodynamics for Chemists* (New York, Van Nostrand, 1947).

(9·6); for the solute, Henry's law is the criterion and (provided that the partial pressure is measurable) the activity coefficient may be evaluated by use of equation (9·4),

$$p_i = K_i \gamma_i x_i.$$

In order to apply this equation it is necessary to know the value of K_i. Consider, for example, a solution of oxygen in water. By determining the value of the ratio p_{O_2}/x_{O_2} in a series of very dilute solutions and extrapolating this ratio to infinite dilution we obtain the value of K_{O_2}—since γ_{O_2} is unity at infinite dilution. The same value of K_{O_2} can then be used to calculate γ_{O_2} in solutions which are not very dilute, by substitution in the above equation.

Henry's law can also be expressed in terms of the molality scale and Convention III. Let the above equation, relating to Convention II, be written

$$p = K^{II} \gamma^{II} x. \tag{9·32}$$

Then the analogous equation relating to Convention III is

$$p = K^{III} \gamma^{III} m, \tag{9·33}$$

where

$$K^{III} \equiv \exp\{(\mu^\square - \mu^0)/RT\}.$$

Application of equation (9·5) and (9·19) shows that the relationship between K^{II} and K^{III} is

$$K^{III} = M_0 K^{II}. \tag{9·34}$$

The corresponding relationship between the activity coefficients γ^{II} and γ^{III} is given by equation (9·20).

(b) Application of Nernst's distribution law. If a solute is distributed between two solvents α and β, then in place of equation (8·38) we have

$$\frac{x^\alpha \gamma^\alpha}{x^\beta \gamma^\beta} = N, \tag{9·35}$$

where x^α and x^β refer to the mole fractions of the solute in the two solvents and γ^α and γ^β are the corresponding activity coefficients. The ratio N is a constant at a particular temperature and pressure, and its value can be determined experimentally by extrapolating measured values of the ratio x^α/x^β to infinite dilution.

Let it be supposed that γ^β is known by some independent method; then the above equation can clearly be used for the purpose of calculating γ^α. Alternatively, it may be that, by suitable choice of the solvent β, the value of γ^β is effectively unity even in solutions of appreciable concentration. In this case the equation can again be used for the calculation of values of γ^α.

(c) *Freezing-point depression.* In place of equation (8·47) we obtain for a non-ideal solution

$$\ln\left(\frac{1}{x\gamma}\right) = \frac{(L_M - \Delta c_p T_M)}{R}\left(\frac{1}{T} - \frac{1}{T_M}\right) + \frac{\Delta c_p}{R}\ln\left(\frac{T_M}{T}\right), \qquad (9·36)$$

where T is the freezing-point of a solution in which the mole fraction and activity coefficient of the solvent are x and γ respectively. L_M, Δc_p and T_M refer to the pure solvent. The measurement of the freezing-point depression can therefore be used for a determination of the activity coefficient.

It will be noted that it is the activity coefficient of the *solvent* which is obtained as a result of the measurement. The value also refers specifically to the temperature T at which melting takes place.

(d) *Osmotic-pressure measurement.* For a non-ideal solution in place of equation (8·57) we obtain

$$\Pi = -\frac{RT}{v_m}\ln\gamma x, \qquad (9·37)$$

and therefore γ may be determined from a measurement of the osmotic pressure. The activity coefficient and the mole fraction refer, of course, to the species which is able to pass through the semi-permeable membrane. Therefore it is the activity coefficient of the solvent, e.g. water, which is normally determined by this method.†

(e) *Solubility measurements.* Consider a solution of a substance in a certain solvent and let $x_i^{\text{sat.}}$ and $\gamma_i^{\text{sat.}}$ refer to the mole fraction and activity coefficient respectively of this solute when it is saturated. The value of $x_i^{\text{sat.}}$ can be varied somewhat, at constant temperature, by addition to the solution of other substances. On the other hand, in the relation $\mu_i^{\text{sat.}} = \mu_i^* + RT\ln\gamma_i^{\text{sat.}} x_i^{\text{sat.}}$,

the values of $\mu_i^{\text{sat.}}$ and μ_i^* are both unchanged by the addition of these other substances—the former because it is equal to the chemical potential of the solid solute‡ and the latter because it is in any case independent of composition.

It follows that for any saturated solution of the solute, at a fixed temperature, the quantity $\gamma_i^{\text{sat.}} x_i^{\text{sat.}}$ (9·38)

is also independent of the presence of other solutes. This principle can be used as a method of measuring $\gamma_i^{\text{sat.}}$, but it will not be discussed further as the method is of rather limited value.

† The value of γ, as obtained by this method, refers to the pressure p' of the solution, but may be corrected to any other pressure, e.g. atmospheric, by use of equation (9·30). (See Problem 5 below.)

‡ It is assumed that the added substances do not form solid solutions with the solute.

9·7. The Gibbs–Duhem equation applied to activity coefficients

In §7·1 we obtained the following form of the Gibbs–Duhem equation as applied to a two-component solution in which the mole fractions are x_a and x_b:

$$x_a \left(\frac{\partial \mu_a}{\partial x_a}\right)_{T,\,p} + x_b \left(\frac{\partial \mu_b}{\partial x_a}\right)_{T,\,p} = 0. \qquad (9\cdot39)$$

Now the activity coefficients are defined by the relations

$$\mu_a = \mu_a^* + RT \ln \gamma_a x_a,$$
$$\mu_b = \mu_b^* + RT \ln \gamma_b x_b.$$

Performing the necessary differentiations and substituting in (9·39) we obtain the following alternative form of the Gibbs–Duhem equation:

$$x_a \left(\frac{\partial \ln \gamma_a}{\partial x_a}\right)_{T,\,p} + x_b \left(\frac{\partial \ln \gamma_b}{\partial x_a}\right)_{T,\,p} = 0. \qquad (9\cdot40)$$

This relation† shows that the activity coefficients are not independent.

Any empirical or theoretical equations which purport to express the activity coefficients as functions of composition must be such as will satisfy (9·40). Typical of such equations are those of Margules and van Laar discussed in § 7·8.

9·8. The calculation of the activity coefficient of the solute

As described in § 9·6 the immediate result of measuring a freezing-point depression or an osmotic pressure is the activity coefficient of the *solvent*. Provided that these results are available over a range of concentrations which extend up to very high dilution it is possible to calculate the activity coefficient of the *solute* by integration of the Gibbs–Duhem equation.

Equation (9·40), referring to a binary solution, may be written

$$x_a \, d \ln \gamma_a + x_b \, d \ln \gamma_b = 0,$$

where it is understood that we are concerned with changes of composition at constant temperature and pressure. Therefore

$$d \ln \gamma_b = -\frac{x_a}{x_b} d \ln \gamma_a$$

$$= -\left(\frac{1-x_b}{x_b}\right) d \ln \gamma_a. \qquad (9\cdot41)$$

† The student should check the consistency of the various equations by carrying out an alternative derivation of (9·40) by use of the Duhem–Margules equation (7·42) together with (9·4).

We shall regard species A as being the solvent whose activity coefficients are already known.

Integrating the equation over the range $x_b = 0$ to $x_b = x_b'$ we obtain

$$\ln \gamma_b' = - \int_0^{x_b'} \left(\frac{1 - x_b}{x_b} \right) d \ln \gamma_a, \tag{9·42}$$

where γ_b' is the value of γ_b at the particular mole fraction of the solute which is the upper limit of the integration. The value of γ_b at the lower limit, $x_b = 0$, has been put equal to unity in accordance with Convention II.

Provided that experimental values of γ_a are available over the range of integration it is possible, at least in principle, to evaluate the integral by a graphical method and thus to obtain γ_b'. However, it will be seen that the term $(1 - x_b)/x_b$ in the integrand approaches infinity as $x_b \to 0$. Of course this does not imply that the area under the curve of $(1 - x_b)/x_b$ plotted against $\ln \gamma_a$ is infinitely large, but it does make it difficult to estimate the area with any accuracy.

A device for avoiding this difficulty, consisting in a change of variable, has been described by Lewis and Randall.† An alternative procedure‡ is to assume a simple algebraic relationship between $\ln \gamma_a$ and x_b in the region where x_b is *very small*. An expression which is suitable when the solute, species B, is a non-electrolyte is

$$\ln \gamma_a = \text{constant} \times x_b^2, \tag{9·43}$$

which is based on using only the first term on the right-hand side of equation (7·63). Now the integral in (9·42) can be separated into the sum of two integrals each extending over part of the range: (1) An integral over the range $x_b = 0$ up to a very low mole fraction $x_b = x_b''$, where (9·43) is still valid. The use of this equation then allows of a direct mathematical integration over this part. (2) An integral extending from x_b'' up to the particular mole fraction x_b' which is of interest. The latter can be evaluated graphically by using the experimental data. For more adequate discussion the reader is referred to the literature.

9·9. Excess functions of non-ideal solutions

In the literature the thermodynamic characteristics of solutions, especially of non-electrolyte mixtures, are frequently expressed by means of the excess functions. These are the amounts by which the free energy, entropy, etc., of the given solution exceed those of a hypothetical ideal

---

† Lewis and Randall, *Thermodynamics*, Chapter XXII. Also Glasstone, *Thermodynamics for Chemists*, Chapter XVI and Problem 6 below.

‡ Guggenheim, *Thermodynamics*, first edition, pp. 252–3. This discussion is in terms of the osmotic coefficient as defined in § 9·11 below.

solution of the same composition.

The excess Gibbs free energy is closely related to the activity coefficients. The total Gibbs free energy of a solution is

$$G = \Sigma n_i \mu_i.$$

In this equation we substitute from (9·1) and obtain

$$G = \Sigma n_i \mu_i^* + RT\Sigma n_i \ln x_i + RT\Sigma n_i \ln \gamma_i, \qquad (9·44)$$

and if the solution were ideal the last term would be zero. The excess free energy is therefore

$$G^E = RT\Sigma n_i \ln \gamma_i. \qquad (9·45)$$

Differentiating this expression at constant temperature we obtain

$$\mathrm{d}G^E = RT\Sigma n_i \,\mathrm{d}\ln \gamma_i + RT\Sigma \ln \gamma_i \,\mathrm{d}n_i.$$

Under conditions of constant temperature and pressure the first term on the right-hand side is zero, as follows from the Gibbs–Duhem equation. Therefore

$$\left(\frac{\partial G^E}{\partial n_i}\right)_{T,\,p,\,n_j} = RT \ln \gamma_i. \qquad (9·46)$$

If some molecular theory of the solution can be used to obtain an expression for G^E as a function of the composition, the activity coefficients of the various species can evidently be determined by differentiation of G^E with respect to the amounts of substances.

It is essential that any theoretical or empirical expressions for G^E shall be a homogeneous function of the first degree in the n_i. That is to say, the expression must be such that increasing each of the n_i k times shall increase G^E k times. This follows from the fact that the Gibbs free energy is an extensive property of a system. The expression must also, of course, approach zero under the limiting conditions where the solution approaches ideality.

Taking the case of a *binary* mixture, a type of power series expansion in the mole fractions x_1 and x_2 which satisfies these requirements is:

$$\frac{G^E}{n_1 + n_2} = RTx_1 x_2 \{A_0 + A_1(x_2 - x_1) + A_2(x_2 - x_1)^2 + A_3(x_2 - x_1)^3\},$$

where the left-hand side is now the excess free energy per mole of the mixture.

According to Guggenheim there is no case known where the accuracy required warrants the use of higher terms than are shown in the above equation. He further defines *symmetrical* mixtures as those for which all odd A's vanish, and *simple* mixtures as those for which only the A_0 term is significant.

The excess free energy may be readily calculated from experimental measurements of activity coefficients by use of equation (9·45). It will be seen that expressing it as a power series is a means of giving empirical description to deviations from ideality which is alternative to the power series expansions, referring to the individual activity coefficients, discussed in § 7·8. It is now preferred by many authors. One important reason is that it can be related more conveniently to the other global properties of the mixture, such as the enthalpy and volume change of mixing,† than can the individual activity coefficients which represent the deviations divided up, as it were, among the components.‡

9·10. The activity

The product of the mole fraction x_i of a component of a solution and its activity coefficient γ_i is often called the *activity* of this component

$$a_i = \gamma_i x_i. \tag{9·47}$$

Similarly on the molality scale

$$a_i = \gamma_i m_i \tag{9·48}$$

(and, of course, neither the a's nor the γ's in these two equations have the same numerical value).

In some of the American text-books definitions of the activity and activity coefficient are used which are not quite identical with those adopted in the present chapter. The difference depends on the effect of total pressure on the free energy of a liquid phase and is usually trivial unless the solution under discussion is at a pressure greatly in excess of one atmosphere.

The difference of usage can be illustrated by considering the solvent in a particular solution. Let μ be its chemical potential and let x be its mole fraction. According to the usage of the present chapter its activity coefficient γ and its activity a are defined by the relations

$$\mu = \mu^* + RT \ln \gamma x \tag{9·49}$$

$$= \mu^* + RT \ln a, \tag{9·50}$$

where μ^* is the chemical potential of the pure solvent at the same temperature T and *also at the same pressure* p as the solution under discussion.

In most American text-books, on the other hand, the activity coefficient and activity are defined by means of the relations

$$\mu = \mu^\wedge + RT \ln \gamma x \tag{9·51}$$

$$= \mu^\Delta + RT \ln a, \tag{9·52}$$

where μ^Δ is the chemical potential of the pure solvent at the same temperature T as the solution but at *a standard pressure of* 1 *atm*. §

† See for example Guggenheim, *Thermodynamics*, third edition, § 5·43.

‡ For a criticism of the use of excess functions see, however, Everett, Disc. Faraday Soc., No. 15 (1953), p. 126. Some typical curves for the excess functions are also given by this author on p. 177 of the same volume.

§ Alternatively $\qquad\qquad \gamma x = a = f/f^\Delta,$

where f is the fugacity of the solvent in the solution and f^Δ is its fugacity as a pure liquid at one atmosphere pressure.

These definitions are clearly not quite the same; μ^* is taken to vary with the pressure p of the solution whilst μ^Δ does not. However, provided that this pressure is not excessive the numerical values of the activity coefficients according to the two definitions will not differ significantly.

The dependence of γ, as defined in (9·51), on the pressure p, is given by a formula similar to equation (9·30) but not containing v_i, the volume per mole of the pure solvent. The theoretical disadvantage of this definition of the activity coefficient is that its numerical value is a measure not only of the deviations from ideality but also of the effect of the total pressure.

9·11. The osmotic coefficient

Still another function which is used, especially in connexion with electrolyte solutions, is the osmotic coefficient of the *solvent*. This is simply a logarithmic function of the activity coefficient, as already defined, but it is useful whenever the activity coefficient of the solvent differs from unity by only a very small amount. For example, in the case of dilute electrolyte solutions, the activity coefficient of the solvent may differ from unity by less than one part in 10^4, whilst the activity coefficient of the solute may differ from unity by several per cent. In such cases it is desirable to use a function which results in a larger numerical measure of the departure of the solvent from ideality.

The osmotic coefficient g of the solvent is defined by the relation†

$$\mu_0 = \mu_0^* + gRT \ln x_0,$$ (9·53)

where

$$g \to 1 \quad \text{as} \quad x_0 \to 1.$$

The relation between g and the activity coefficient γ_0 of the solvent is obtained from equations (9·3) and (9·53)

$$\mu_0^* + RT \ln \gamma_0 x_0 = \mu_0^* + gRT \ln x_0,$$

and therefore

$$\ln \gamma_0 = (g - 1) \ln x_0.$$ (9·54)

The osmotic coefficient g differs from unity by a much larger amount than does γ_0. The usefulness of g in a region of concentration where γ_0 has a value of, say, 0·9999 will be apparent.

PROBLEMS

1. The Bunsen absorption coefficient is the volume of a gas (reduced to 0 °C and 1 atm) which dissolves in unit volume of a solvent when the partial pressure of the gas is 1 atm. Show that the Bunsen coefficient is proportional to the reciprocal of K_i, as defined by equation (9·4), when the solution is dilute.

2. The partial pressure of mercury above an amalgam with thallium in which its mole fraction (i.e. the Hg) is 0.497 is 43.3% of its value over pure mercury at the same temperature (325 °C). Calculate the activity coefficient of the mercury in this solution. What is the change in free

† The symbol g is used here as in Guggenheim, *Thermodynamics*, first edition. C.f. third edition, §6·12. The IUPAC symbol is now ϕ.

energy in transferring 1 mol of mercury from an infinitely large quantity of this solution into pure mercury at the same temperature and pressure?

3. The melting-point of δ is 1808 K and the enthalpy of melting is 1.536×10^4 J mol^{-1}. The heat capacity of the liquid iron exceeds that of the solid by about 1.3 J K^{-1} mol^{-1}. Show that the difference of free energy of liquid and solid iron as a function of temperature at constant pressure is given in J mol^{-1} by

$$\Delta G^0 = 1.3 \times 10^4 - 1.3T \ln T + 2.6T.$$

At 1673 K a liquid mixture of iron and iron sulphide containing 0.870 mole fraction of iron is in equilibrium with almost pure solid δ iron. Estimate the activity coefficient of the iron in this liquid melt. State clearly the reference basis. [C.U.C.E. Qualifying, 1951]

4. Two partially miscible liquid substances A and B are in equilibrium with each other. If γ_a and γ_b are the activity coefficients (on Convention I) of the components in one of the saturated phases, in which the mole fraction of component A is x_a, and if γ'_a and γ'_b are the corresponding activity coefficients in the other phase, show that these are related to each other as follows:

$$x_a(\gamma_a \gamma'_b - \gamma'_a \gamma_b) = \gamma'_a(\gamma'_b - \gamma_b).$$

5. At 0 °C the osmotic pressure of a solution containing 141.0 g of cane sugar per 100 g of water is 1.365×10^7 Nm^{-2}. Calculate the ratio of the vapour pressure of this solution to the vapour pressure of pure water at the same temperature and compare with the observed ratio of 0.8988. The mean partial specific volume of water in the solution up to 1.365×10^7 Nm^{-2} is 0.983 21 cm^3 g^{-1}.

6. Transform equation (9·42) to the form

$$\ln\left(\frac{\gamma'_b x'_b}{r'}\right) = -h' - \int_0^{r'} \frac{h}{r}\, dr,$$

where $r = x_b/x_a$ and h is defined by

$$h = \frac{\ln \gamma_a x_a}{r} + 1.$$

r' and h' are the particular values of these variables at the mole fraction x'_b at which it is required to calculate γ_b. (This is the change of variable used by Lewis and Randall and referred to in § 9·8.)

7. If a binary solution has the following properties (a) the entropy is the same as that of a hypothetical ideal solution of the same composition, and (b) the enthalpy, per mole of the mixture, exceeds that of the corresponding ideal solution by an amount bx_1x_2, where b is a function of pressure only, show that its activity coefficients are given by

$$RT \ln \gamma_1 = bx_2^2,$$
$$RT \ln \gamma_2 = bx_1^2.$$

The vapour pressures of pure liquid benzene and pure liquid cyclohexane at 40 °C are respectively 182.6 mmHg and 184.5 mmHg, and at this temperature they form an azeotrope containing 49.4 mole % benzene, and exerting a total pressure of 206.2 mmHg.

Examine whether this azeotropic data is consistent with postulates (a) and (b). Estimate the total pressure of the vapour which would be in equilibrium at 40 °C with a liquid containing 12.8 mole % benzene.

[C.U.C.E. Qualifying, 1950]

8. At atmospheric pressure ethyl acetate and ethyl alcohol form an azeotropic mixture containing 53.9 mole % of the former component and boiling at 71.8 °C

Estimate: (a) the values of the constants A and B in the empirical equations of van Laar:

$$\log_{10}\gamma_1 = \frac{A}{\left(1 + \dfrac{Ax_1}{Bx_2}\right)^2},$$

$$\log_{10}\gamma_2 = \frac{B}{\left(1 + \dfrac{Bx_2}{Ax_1}\right)^2},$$

and (b) if A and B remain unchanged, the azeotropic composition and the corresponding total pressure for boiling at 56.3 °C.

The vapour pressures, in mmHg of the pure liquids are as follows:

	71.8 °C	56.3 °C
Ethyl alcohol	587	298
Ethyl acetate	636	360

[C.U.C.E. Tripos, 1950]

9. Both phases of a two-phase two-component liquid-vapour system are ideal solutions. The components are 1 and 2 respectively, and when the mole fraction of 2 in the liquid phase is x, that in the vapour phase is y and the total vapour pressure is p.

For pure component 1 at temperature T, the vapour pressure is p_1 and the liquid volume is v_1'. The corresponding quantities for pure component 2 at the same temperature are p_2 and v_2'. For a pressure p between p_1 and p_2 the compressibility factors of the pure vapours at T are $a_1 + b_1 p$ and $a_2 + b_2 p$ respectively where a_1, b_1, etc., are constants.

If p_2 is greater than p_1, which of these quantities must be estimated indirectly?

If the liquid is under the vapour pressure due to the two components only and v_1' and v_2' are not affected appreciably by pressure, show that for a given x, the values of y and p can be calculated from the simultaneous equations

$$(1-y) = (1-x)\left(\frac{p_1}{p}\right)^{a_1}\exp\left\{\left(\frac{v_1'}{RT} - b_1\right)(p - p_1)\right\},$$

$$y = x\left(\frac{p_2}{p}\right)^{a_2}\exp\left\{\left(\frac{v_2'}{RT} - b_2\right)(p - p_2)\right\}.$$

[C.U.C.E. Tripos, 1950]

10. The activity coefficients of the components of certain binary mixtures are given by the relations

$$RT \ln \gamma_1 = \alpha x_2^2,$$
$$RT \ln \gamma_2 = \alpha x_1^2,$$

where α is a function of pressure only. Obtain expressions for the increase in the Gibbs function and enthalpy in the process of mixing the pure components at constant temperature and pressure.

In a steady-flow process an equimolal mixture, for which $\alpha = 418$ J mol^{-1}, is separated into the pure components by a process of distillation. The inflow and outflow are at 20 °C and 1 atm. The only source of energy is a heat reservoir maintained at a steady temperature of 100 °C. Calculate the amount of heat which must be removed by cooling water at 20 °C, per mole of the mixture distilled, if the energy of the reservoir is used at maximum efficiency. [C.U.C.E. Qualifying, 1952]

11. Two components A and B, boiling-points T_A, T_B, form an approximately regular binary solution. Their activity coefficients in the mixture are given by

$$RT \ln \gamma_A = \alpha x_B^2,$$
$$RT \ln \gamma_B = \alpha x_A^2.$$

The solution forms an azeotrope boiling under atmospheric pressure at T_z.

By using Trouton's rule show that the mole fraction x of component A in the azeotropic mixture is given by the approximate relation

$$\left(\frac{1-x}{x}\right)^2 = \frac{T_A - T_z}{T_B - T_z}.$$

The variation of enthalpy of vaporization with temperature may be neglected. [C.U.C.E. Qualifying, 1953]

12. A binary solution is in equilibrium with its vapour and the latter may be assumed to be a perfect gaseous mixture.

If the enthalpies of vaporization L_1 and L_2 of the two components do not vary significantly over the temperature range, show that

$$\ln \frac{y_1}{x_1 \gamma_1} = \frac{L_1}{R}\left(\frac{1}{T_1} - \frac{1}{T}\right), \quad \ln \frac{y_2}{x_2 \gamma_2} = \frac{L_2}{R}\left(\frac{1}{T_2} - \frac{1}{T}\right),$$

where T_1, T_2 and T are the boiling-points of the two components and the solution respectively, at the same pressure.

If the solution may be assumed to be ideal, obtain an expression for its composition as a function of its boiling-point.

[C.U.C.E. Qualifying, 1955]

13. For the case of a binary solution at constant temperature, show that

$$\int_0^1 \ln\left(\frac{\gamma_1}{\gamma_2}\right) dx_1 = 0,$$

and therefore that the plot of $\ln(\gamma_1/\gamma_2)$ against x_1 must define two regions of equal area but of opposite sign.

CHAPTER 10

REACTION EQUILIBRIUM IN SOLUTION. ELECTROLYTES

10·1. Reaction equilibrium in solution

The condition of equilibrium for the generalized chemical reaction

$$\Sigma \nu_i M_i = 0$$

was shown in §4·4 to be $\qquad \Sigma \nu_i \mu_i = 0.$ $\hfill$ (10·1)

This expression is applicable to a reaction in any phase, or in any system of phases which is in equilibrium, and it contains the whole of the purely thermodynamic knowledge concerning the reaction equilibrium.

The form taken by the equilibrium constant depends on the type of expression which is substituted in the above equation for the purpose of expressing the chemical potentials in terms of the composition; this in its turn depends on additional physical knowledge concerning whether or not the real system in question may be approximately represented by means of a model, such as the perfect gas or the ideal solution. If the system does not approximate to either of these models it is still possible, of course, to formulate an equilibrium constant in terms of fugacities or in terms of mole fractions and activity coefficients. However, this is a purely formal process; the fugacities and activity coefficients are themselves defined in terms of the chemical potentials and therefore the knowledge contained in equation (10·1) is in no way increased, but is obtained in a more convenient form.

As mentioned in §7·3 it is always possible to discuss the equilibrium of a reaction in solution in terms of the partial pressures in the saturated vapour above the solution (provided that this vapour is a perfect mixture). However, for many purposes it is more useful to express the equilibrium constant of a liquid phase reaction directly in terms of the composition of the liquid. This is done by substituting in equation (10·1) any of the appropriate expressions for the chemical potential of a component of a solution which have been developed in the last two chapters. It will save space if the equations are developed in a general form applicable to a non-ideal solution. The limiting forms of these expressions applying to reaction equilibrium in an ideal solution may be obtained by putting the activity coefficients equal to unity.

If the composition and the activity coefficients of all species are expressed on the mole fraction scale (Conventions I and II of the last chapter) we obtain from equation (10·1) together with (9·2) or (9·3)

$$\Sigma \nu_i \mu_i^* + RT\Sigma \nu_i \ln \gamma_i x_i = 0.$$

Rearranging $\qquad -RT \ln \Pi(\gamma_i x_i)^{\nu_i} = \Sigma \nu_i \mu_i^*, \qquad$ (10·2)

or $\qquad\qquad -RT \ln K = \Delta G_T^*, \qquad$ (10·3)

where $\qquad\qquad K \equiv \Pi(\gamma_i x_i)^{\nu_i} \qquad$ (10·4)

and $\qquad\qquad \Delta G_T^* \equiv \Sigma \nu_i \mu_i^*. \qquad$ (10·5)

Similarly, if the composition and the activity coefficients are expressed on the molality scale (Convention III of the last chapter) we obtain from equations (10·1) and (9·16)

$$-RT \ln K = \Delta G_T^{\square}, \qquad (10·6)$$

where† $\qquad\qquad K \equiv \Pi(\gamma_i m_i)^{\nu_i} \qquad$ (10·7)

and $\qquad\qquad \Delta G_T^{\square} \equiv \Sigma \nu_i \mu_i^{\square}. \qquad$ (10·8)

The significance of the quantities μ_i^* and $\mu_i^{\square}$ were discussed in Chapters 8 and 9; they refer to certain definite composition states but otherwise to the same temperature and pressure as that of the reaction system under discussion. Let it be supposed for the moment that this system is at 1 atm and 25 °C Then the sums $\Sigma \nu_i \mu_i^*$ and $\Sigma \nu_i \mu_i^{\square}$, which appear in equations (10·5) and (10·8), will be equal to the corresponding sums of the free energies of formation of the various compounds from their elements,‡ as quoted in the literature for the standard pressure of 1 atm and a temperature normally chosen as 25 °C and in *the appropriate composition states*. The application of the above equations then allows of a calculation of K at 25 °C and 1 atm, and its value at other temperatures and pressures can be computed by use of the equations of § 10·4 below.

The appropriate composition states referred to above are usually either:

(a) the pure liquid. This corresponds to the use of μ^* and Convention I.

or (b) the hypothetical ideal solution of unit molality. This corresponds to the use of $\mu^{\square}$ and Convention III.

For example, the free energy of formation of sucrose in aqueous solution refers to 1 mol of this substance in a hypothetical ideal

† Although the same symbol has been used, the K's of equation (10·4) and (10·7) are not equal numerically. Notice that the K of the equation (10·7) *is dimensionless*. This follows from a footnote in §9·4; cf. also the discussion in §4·5.

‡ More simply we could take each μ^* or $\mu^{\square}$ as being *equal* to the free energy of formation of the particular compound in the same temperature, pressure and composition state. This would be equivalent to taking the free energies of the elements as zero. (See p. 149.)

solution of unit molality, formed from graphite, hydrogen and oxygen each at 1 atm pressure.

There is clearly no necessity that the composition should be expressed in the same terms—mole fractions or molalities—for all species taking part in the reaction. In considering a problem concerning reaction equilibrium in solution *it is always best to start out with the general relation* (10·1) and to substitute in this equation whatever expressions for the chemical potentials seem appropriate to the particular problem.

As an example consider the formation of urea in aqueous solution by the reaction

$$CO_2(g) + 2NH_3(g) = CO(NH_2)_2 + H_2O.$$

It will be supposed that it is required to calculate the equilibrium constant from known values of the free energies of formation of the various species. At 25 °C these are reported in the literature as follows (cal mol^{-1}):

$CO_2(g)$	$\Delta_f G^0 = -94\ 260$
$NH_3(g)$	$\Delta_f G^0 = -\ 3\ 976$
$H_2O(l)$	$\Delta_f G^0 = -56\ 690$
$CO(NH_2)_2$ (ideal one molal solution)	$\Delta_f G^\square = -48\ 720$

Now the general condition of equilibrium for the reaction is

$$\mu_{CO_2} + 2\mu_{NH_3} = \mu_u + \mu_{H_2O},$$

where the subscript u denotes urea. Consider first the CO_2 and the NH_3. If in the above equation we substitute the relations

$$\mu_{CO_2} = \mu_{CO_2}^0 + RT \ln f_{CO_2},$$

$$\mu_{NH_3} = \mu_{NH_3}^0 + RT \ln f_{NH_3},$$

the quantities $\mu_{CO_2}^0$ and $\mu_{NH_3}^0$ will clearly refer to CO_2 and NH_3 in the same physical states as for the given free energies of formation, namely, the gases at 1 atm pressure (strictly unit fugacity). Similarly for the water if we substitute

$$\mu_{H_2O} = \mu_{H_2O}^* + RT \ln \gamma_{H_2O} x_{H_2O},$$

the quantity $\mu_{H_2O}^*$ will refer to the same physical state† as for the given free energy of formation, namely, pure liquid water ($x_{H_2O} = 1$). Finally, with regard to the urea if we substitute

$$\mu_u = \mu_u^\square + RT \ln \gamma_u m_u,$$

the quantity $\mu_u^\square$ will refer to the same state† as for the given free

† Strictly speaking $\mu_{H_2O}^*$ and $\mu_u^\square$ refer to the total pressure p of the reaction system in question, whereas the standard free energies of formation refer to unit pressure. However, as shown in §§4·10–4·12, the effect of pressure on condensed phases may usually be neglected. Using the notation of those sections $\mu_{H_2O(l)}^0$ the chemical potential of water at 1 atm pressure is not significantly different from $\mu_{H_2O}^*$, the chemical potential at pressure p.

energy of formation, namely, the hypothetical ideal solution of unit molality.

It is the identity of the reference states for $\mu_{CO_2}^0$, etc., with those for the given values of the free energies of formation that makes the above equations (rather than any other formal alternatives) the most suitable choice for a substitution in the general condition of equilibrium. The result is to obtain a 'mixed' equilibrium constant given by

$$-RT\ln\frac{\gamma_u m_u \gamma_{H_2O} x_{H_2O}}{f_{CO_2} f_{NH_3}^2} = \mu_u^\square + \mu_{H_2O}^* - \mu_{CO_2}^0 - 2\mu_{NH_3}^0. \qquad (10\cdot9)$$

If $T = 298.15$ K the numerical value of the right-hand side of this equation is -3200 cal mol^{-1}, being equal to the corresponding sums and differences of the free energies of formation of the various substances in the same states.

If the gas pressure is not too high and if the solution is dilute this may be approximated by

$$-RT\ln\frac{m_u x_{H_2O}}{p_{CO_2} p_{NH_3}^2} = -3200 \text{ cal mol}^{-1} \text{ at } T = 298.15 \text{ K},$$

and under the same conditions a further approximation is to put $x_{H_2O} = 1$.

In brief the kind of equilibrium constant which is used in a practical problem is entirely a matter of choice and is most conveniently decided in relation to the standard states which are conventionally adopted for the various substances which take part in the reaction. In the case of dissolved substances in aqueous solution the free-energy data are almost invariably reported for the state of an ideal solution of unit molality, in accordance with Convention III of the last chapter. On the other hand, for a reaction of the type

$$CH_3OH + CH_3COOH = CH_3COOCH_3 + H_2O,$$

it would be normal to use an equilibrium constant containing mole fractions of all species, since all of them are obtainable as pure liquids and the free energies of formation are quoted for these states.

10·2. Free energy of formation in solution. Convention concerning hydrates

It is useful at this stage to describe one of the important methods of determining the free energy of formation of substances in solution. Consider the equilibrium between a gaseous substance A and its aqueous solution

$$\mu_a^{gas} = \mu_a^{soln}.$$

Using the molality scale for the dissolved substance this equilibrium relation may be written

$$\mu_a^0 + RT \ln f_a = \mu_a^\square + RT \ln \gamma_a m_a,$$

or rearranging

$$\mu_a^\square - \mu_a^0 = RT \ln \left(\frac{f_a}{\gamma_a m_a} \right). \tag{10·10}$$

Since $\mu_a^\square$ and μ_a^0 are both independent of the concentration of the particular solution the same applies also to the ratio $f_a/\gamma_a m_a$. The value of this ratio may be determined therefore by plotting measured values of p_a/m_a against m_a and extrapolating to infinite dilution where $\gamma_a \to 1$ and $f_a \to p_a$.[†] Now the difference $\mu_a^\square - \mu_a^0$ in equation (10·10) is equal to the difference of the free energies of formation of the substance in the hypothetical ideal solution in water of unit molality and as a gas at 1 atm (strictly at unit fugacity) respectively. If the latter is already known the equation can therefore be used to calculate the former, i.e. the standard free energy of formation in solution.

In certain instances hydrates are formed, and, if the proportions of hydrate and non-hydrate are unknown, it is convenient to adopt the convention [‡] that the standard free energy of formation refers to an ideal solution in which the *total* molality is unity.

As an example consider gaseous ammonia in equilibrium with its solution. Since ammonia is a weak base the concentration of the NH_4^+ ion may be neglected, relative to the concentrations of NH_3 and NH_4OH, provided that the solution is not extremely dilute. The relevant equilibria are therefore

$$NH_3(g) = NH_3(\text{soln.}), \tag{A}$$

$$NH_3(\text{soln.}) + H_2O = NH_4OH. \tag{B}$$

Considering the second of these, let m_{NH_4OH} and m_{NH_3} denote the actual molalities of hydrate and of dissolved NH_3 as such, respectively. The equilibrium constant takes the form $m_{NH_4OH}/m_{NH_3} = K$ for any solution which is sufficiently dilute for the mole fraction of the water, and also for the various activity coefficients, to be very nearly unity. Let m be the total molality

$$m \equiv m_{NH_3} + m_{NH_4OH}$$

$$= m_{NH_3} + K m_{NH_3}$$

$$= m_{NH_3}(1 + K).$$

This equation shows that m and m_{NH_3} are proportional to each other in dilute solution. It follows that if the ratio p_{NH_3}/m_{NH_3} approaches

† It is tacitly assumed that no other gas is present at appreciable pressure.

‡ See also Lewis and Randall, *Thermodynamics*, Chapter XXIV and XXXIX. Their procedure is equivalent to that used here.

constancy—as may be expected from Henry's law—the ratio p_{NH_3}/m will also approach constancy.

Let μ_{NH_3} be the chemical potential of dissolved NH_3 as such and let $\mu_{NH_3}^{\square}$ be defined by

$$\mu_{NH_3} = \mu_{NH_3}^{\square} + RT \ln \gamma m, \qquad (10\cdot11)$$

so that $\mu_{NH_3}^{\square}$ denotes the chemical potential of dissolved NH_3 in an ideal solution ($\gamma = 1$) in which the *total* molality m is unity. Considering the gas-liquid equilibrium ((A) above) we obtain, as in equation (10·10),

$$\mu_{NH_3}^{\square} - \mu_{NH_3}^{0} = RT \ln \frac{f_{NH_3}}{\gamma m}$$

$$= RT \ln \left(\frac{p_{NH_3}}{m}\right)^0,$$

where $(p_{NH_3}/m)^0$ is the limit of the measured ratio p_{NH_3}/m as extrapolated to $m = 0$.† This has the value $1/56.7$ at 25 °C. Hence

$$\mu_{NH_3}^{\square} - \mu_{NH_3}^{0} = -2384 \text{ cal mol}^{-1} \text{ at 25 °C.}$$

This difference of chemical potentials is equal to the difference in the free energy of formation of ammonia in a hypothetical ideal solution of unit *total* molality and the free energy of formation of gaseous ammonia at unit pressure. The latter quantity is known from measurements on gas equilibria to be -3976 cal mol^{-1} at 25 °C. Therefore the former quantity has the value $-3976 - 2384 = -6360$ cal mol^{-1}. Thus

$$\tfrac{1}{2}N_2 (g, 1 \text{ atm}) + \tfrac{3}{2}H_2 (g, 1 \text{ atm}) = NH_3 (\text{aq.}); \Delta_f G_{298} = -6360 \text{ cal mol}^{-1},$$

where the symbol NH_3 (aq.) denotes dissolved ammonia in the ideal solution of unit total molality. (And it is consistent with the same convention to take the free energy of formation of NH_4OH as being equal to the above figure plus the free energy of formation of water.)

The need for the convention arises simply on account of the difficulty in making an experimental distinction between the species NH_3 and NH_4OH. On the other hand, the ions NH_4^+ and OH^- are readily distinguishable from the NH_3 and NH_4OH and the equilibrium constant of the process

$$NH_4OH = NH_4^+ + OH^-$$

may be investigated by conductivity and other methods. This allows of a determination of the free-energy change in the ionization, and thereby the total free energy of formation of the NH_4^+ and OH^- ions.

† It is here assumed that the experimental data which are extrapolated refer to solutions which are not so dilute that there is any appreciable concentration of ions. If, on the contrary, the data refer to extremely dilute solutions where the ions predominate, it would be the ratio p_{NH_3}/m^2 which would approach constancy.

10·3. Equilibrium constants expressed on the molality and volume concentration scales

If a set of reactants and products form an ideal solution over the whole range of composition the equilibrium constant

$$K = \Pi x_i^{\nu_i},$$

expressed in terms of mole fractions, would be constant over this range. As shown in §9·3 there is a proportionality between mole fractions and molalities or volume concentrations only in very dilute solution. Therefore for the same reaction system, even though it is ideal, the equilibrium constant expressed in either of these latter units *would be constant only in very dilute solution*.

As an example we quote below the data of Cundall† on the equilibrium

$$N_2O_4 = 2NO_2$$

in chloroform solution at $0°$ C. The figures were obtained by measurement of the colour of the solution, and they extend over almost the complete range of composition.

The true thermodynamic equilibrium constant may be expressed either as

$$\frac{x_{NO_2}^2 \gamma_{NO_2}^2}{x_{N_2O_4} \gamma_{N_2O_4}}$$

or as

$$\frac{m_{NO_2}^2 \gamma_{NO_2}^2}{m_{N_2O_4} \gamma_{N_2O_4}},$$

where the activity coefficients in the first expression are on the mole fraction scale and in the second expression on the molality scale. If the solution were ideal the former set of activity coefficients would be unity and the quantity K_x defined by

$$K_x \equiv x_{NO_2}^2 / x_{N_2O_4},$$

would be found to be constant. On the other hand, the quantity K_m defined by

$$K_m \equiv m_{NO_2}^2 / m_{N_2O_4}$$

would not be expected to be constant except in very dilute solution. In fact, using equation (9·9), the relation between K_x and K_m is

$$K_m = K_x / M_0 x_0 \qquad (10·12)$$

where M_0 is the molar mass of the chloroform and x_0 is its mole fraction in a particular solution.

† Cundall, *J. Chem. Soc.* 59 (1891), 1076; 67 (1895), 794. The figures quoted in the first four colums of the table are as calculated by Lewis and Randall (*Thermodynamics*, Chapter XXIV) from Cundall's data.

The values of K_x and K_m worked out from Cundall's data are shown in Table 9. Over the whole range K_x varies over a two-fold range, due to the deviations from ideality. On the other hand, as is to be expected from equation (10·12), K_m is roughly constant only in the dilute solutions where the mole fraction x_0 of the solvent is approximately constant.

TABLE 9

Mole fractions at equilibrium			$K_x \times 10^8$	$K_m \times 10^8$
$CHCl_3$	N_2O_4	NO_2		
0.00	1.00	0.000 94	88	∞
0.27	0.73	0.000 80	87	2700
0.46	0.54	0.000 67	83	1500
0.70	0.30	0.000 45	67	800
0.875	0.125	0.000 29	66	630
0.934	0.066	0.000 19	52	470
0.950	0.050	0.000 15	43	380
0.963	0.037	0.000 12	35	310
0.982	0.018	0.000 10	49	420

10·4. Temperature and pressure dependence of the equilibrium constant

Consider the equilibrium constant expressed in terms of mole fractions together with the corresponding activity coefficients as in equation (10·4)

$$K = \Pi(\gamma_i x_i)^{\nu_i}.$$

Then from equation (10·2) after rearrangement

$$R \ln K = -\Sigma \nu_i \frac{\mu_i^*}{T}.$$

The μ_i^* are functions of temperature and also, although only to a small extent, of pressure. The same must therefore apply to the equilibrium constant. The total differential of $\ln K$ is therefore given by

$$R\, d \ln K = -\Sigma \nu_i \left(\frac{\partial \mu_i^*/T}{\partial T}\, dT + \frac{\partial \mu_i^*/T}{\partial p}\, dp \right)$$

$$= -\Sigma \nu_i \left(\frac{\partial \mu_i^*/T}{\partial T}\, dT + \frac{1}{T} \frac{\partial \mu_i^*}{\partial p}\, dp \right). \qquad (10·13)$$

The term $\dfrac{\partial \mu_i^*/T}{\partial T}$ was obtained in §9·5, and it was shown to be equal either to $-h_i/T^2$ or to $-H_i^0/T^2$ according to whether the activity coefficient of the particular species is chosen as approaching unity as $x_i \to 1$ or as $x_i \to 0$ respectively. The sum of the terms $\nu_i \dfrac{\partial \mu_i^*/T}{\partial T}$ in equation (10·13) is therefore the negative of the heat of reaction when each of the reactants and products is in the limiting state where it approaches ideal behaviour (according to the choice of convention) divided by T^2. This ideal heat of reaction will be denoted ΔH^0. Similarly, the sum of the terms $\nu_i \partial \mu_i^*/\partial p$ in equation (10·13) is equal to ΔV^0, the volume change in the reaction when each of the reactants and products is in the limiting state where it approaches ideal behaviour. Equation (10·13) may therefore be written

$$R\mathrm{d}\ln K = \frac{\Delta H^0}{T^2}\,\mathrm{d}T - \frac{\Delta V^0}{T}\,\mathrm{d}p \tag{10·14}$$

and the two partial differential coefficients are

$$\left(\frac{\partial \ln K}{\partial T}\right)_p = \frac{\Delta H^0}{RT^2} \tag{10·15}$$

and

$$\left(\frac{\partial \ln K}{\partial p}\right)_T = -\frac{\Delta V^0}{RT}. \tag{10·16}$$

In accordance with the usual state of affairs for condensed phases, the dependence of $\ln K$ on the pressure is very small and is significant only when very large changes of pressure are considered. For the same reason the temperature-dependence of $\ln K$ is usually given with sufficient exactness by expressing the left-hand side of (10·15) as a complete differential coefficient

$$\frac{\mathrm{d}\ln K}{\mathrm{d}T} = \frac{\Delta H^0}{RT^2}, \tag{10·17}$$

even if the total pressure is not quite constant. For example, if we were concerned with the reaction

$$N_2O_4 = 2NO_2$$

in a liquid phase under its own vapour pressure, a change in temperature would clearly result in a change in the total pressure of the system. In this case (10·17) would not be applicable as a *precise* relation, but only in so far as the dependence of $\ln K$ on the total pressure is negligible.

If the equilibrium constant is expressed in terms of molalities together with the corresponding activity coefficients as in equation (10·7),

$$K = \Pi(\gamma_i m_i)^{\nu_i},$$

the temperature and pressure-dependence will be the same as in equations (10·15) and (10·16), with ΔH^0 and ΔV^0 taken as the heat and volume change in the reaction at infinite dilution in the given solvent respectively.

The significance of ΔH^0 and ΔV^0 in cases where a 'mixed' equilibrium constant is in use will be clear in each particular case. Consider, for example, the formation of urea in aqueous solution by the reaction

$$CO_2(g) + 2NH_3(g) = CO(NH_2)_2 + H_2O.$$

The temperature coefficient of the logarithm of the equilibrium constant which appears on the left-hand side of equation (10·9) is

$$\frac{\partial \ln K}{\partial T} = \frac{\Delta H^0}{RT^2},$$

where
$$\Delta H^0 \equiv H_u^0 + h_{H_2O} - h_{CO_2} - 2h_{NH_3}.$$

In this equation H_u^0 is the partial molar enthalpy of urea at infinite dilution and h_{H_2O}, h_{CO_2} and h_{NH_3} are the enthalpies per mole of liquid water, of gaseous CO_2 and gaseous NH_3 respectively.

10·5. Ratio of an equilibrium constant in the gas phase and in solution

Let it be supposed that a liquid phase is in equilibrium with its saturated vapour at the total pressure p. If there is reaction equilibrium in the one phase there is also reaction equilibrium in the other. It will be supposed that the vapour behaves as a perfect mixture and the equilibrium constant is therefore

$$K_\nu = \Pi p_i^{\nu_i}.$$

The corresponding equilibrium constant for the liquid phase is as given by equation (10·4)
$$K = \Pi(\gamma_i x_i)^{\nu_i}.$$

Taking the ratio of these and applying equation (9·4) we obtain

$$\frac{K_p}{K} = \Pi K_i^{\nu_i}, \tag{10·18}$$

where the K_i have the significance discussed in §9·2. Alternatively, if we use an equilibrium constant K_y for the gas phase which is also expressed in terms of mole fractions, as in equation (4·38), we obtain

$$\frac{K_y}{K} = \Pi\left(\frac{K_i}{p}\right)^{\nu_i}. \tag{10·19}$$

10·6. Notation for electrolytes

Consider the electrolyte $M_{\nu_+} A_{\nu_-}$ which dissociates according to the equation

$$M_{\nu_+} A_{\nu_-} = \nu_+ M^{z+} + \nu_- A^{z-}, \tag{10·20}$$

where ν_+ and ν_- are the numbers of positive and negative ions respectively, obtained by the dissociation of one molecule of the parent electrolyte, and z_+ and z_- are the charges of these ions, measured in units of the charge of the proton.

For example, in the dissociation

$$H_2SO_4 = 2H^+ + SO_4^{2-},$$

we have $\nu_+ = 2$, $z_+ = 1$, $\nu_- = 1$, $z_- = -2$. In this instance the parent electrolyte is electrically neutral and the following relation holds between the ν's and the z's:

$$\nu_+ z_+ + \nu_- z_- = 0. \tag{10·21}$$

On the other hand, this equation does not apply to the dissociation of a species which is itself an ion. For example, it does not apply to the process

$$HSO_4^- = H^+ + SO_4^{2-},$$

where $\qquad \nu_+ = 1, \quad z_+ = 1, \quad \nu_- = 1, \quad z_- = -2.$

The dissociation of electrolytes is evidently a particular case of reaction equilibrium in solution. Electrolytes are conventionally divided into the classes *weak* and *strong*, according to whether the degree of dissociation is small or very large respectively. However, this distinction will not be made except where it is useful and most of the following discussion will be entirely general in character.

The chemical potentials, etc., of the positive and negative ions and the undissociated part of the electrolyte will be denoted by the subscripts $+$, $-$ and u respectively. The subscript 0 will be applied to the solvent. For simplicity the discussion will be limited to the case where there is only a single electrolyte in the solution, but the generalization to the case of mixed electrolytes may be readily carried out.

The molality scale is almost invariably used for electrolyte solutions, together with the convention that the activity coefficients of the various solute species approach unity at infinite dilution. This is in accordance with Convention III of §9·4. In general, we shall discuss a solution prepared by dissolving m mols of electrolyte in n_0 mols of solvent. In the special case where n_0 is chosen as equal to $1/M_0$ (M_0 being the molar mass, kg mol^{-1}, of the solvent), m is the same as the *stoichiometric molality* of the electrolyte.†

† Readers may wish to be reminded of what was said in §9·3, i.e. that ' molar mass ' differs by a factor of 1000 mol kg^{-1} from 'molecular weight '.

The components of the solution are

 n_0 moles of solvent,

 m_u moles of undissociated electrolyte $M_{\nu_+} A_{\nu_-}$,

 m_+ moles of positive ions M^{z+},

 m_- moles of negative ions A^{z-}.

From the stoichiometry of the dissociation as expressed by equation (10·20) we have

$$m_+ = \nu_+(m - m_u), \\ m_- = \nu_-(m - m_u). \tag{10·22}$$

10·7. Lack of significance of certain quantities

An infinitesimal change in the Gibbs free energy of the solution may be written in the usual way:

$$dG = -S\,dT + V\,dp + \mu_u dm_u + \mu_+ dm_+ + \mu_- dm_- + \mu_0 dn_0. \tag{10·23}$$

This equation is formally correct, but the quantities μ_+ and μ_- do not have any real experimental significance. They are defined by the relations (§2·7)

$$\mu_+ = \left(\frac{\partial G}{\partial m_+}\right)_{T, p, m_-, n_0, m_u}, \\ \mu_- = \left(\frac{\partial G}{\partial m_-}\right)_{T, p, m_+, n_0, m_u} \tag{10·24}$$

and in the first of these expressions the differential coefficient signifies the change in G of the system, divided by the increment dm_+, when the quantity of every other species, *including the ion of opposite charge*, is held constant. Now all measurements of μ, relative to the elements, such as have been described previously for uncharged species, are based essentially on an integration over a finite change of composition. On the other hand, in the case of an ion, the variation dm_+, at constant m_-, cannot be other than an infinitesimal and cannot cause more than an immeasurably small change in the Gibbs free energy of the whole solution. This arises from the necessity for electroneutrality—to cause more than a minute departure from equivalence in the number of ions would give rise to an impossibly large electric charge.†

Therefore the process implied in the definition of μ_+, and also of μ_-, cannot be carried out to more than an infinitesimal extent, and these quantities are not actually measurable.‡ Nevertheless, it is

† As noted in §5·4, the necessity for electroneutrality also has the effect of reducing by one the number of components in the sense of the phase rule.

‡ Further consideration will show that this difficulty cannot be avoided by considering mixed electrolytes or by studies on the galvanic cell. For further discussion see Taylor, *J. Phys. Chem.* **31** (1927), 1478; Guggenheim, *J. Phys. Chem.* **33** (1929), 842; Guggenheim, *Thermodynamics*.

entirely correct to use them in a formal treatment of the thermo-dynamic properties of the electrolyte, and it will be shown below that a certain linear combination of μ_+ and μ_- can be measured, relative to the elements.

10·8. Dissociation equilibrium and the chemical potential of the electrolyte

In the case where the solution contains only the single electrolyte $M_{\nu_+} A_{\nu_-}$ the molalities m_+ and m_- are related through the equations (10·22). Substituting the differentials of these equations in (10·23) and considering, for simplicity, changes at constant temperature and pressure, we obtain

$$dG = \mu_u \, dm_u + \nu_+ \mu_+ (dm - dm_u) + \nu_- \mu_- (dm - dm_u) + \mu_0 \, dn_0$$

$$= (\mu_u - \nu_+ \mu_+ - \nu_- \mu_-)\, dm_u + (\nu_+ \mu_+ + \nu_- \mu_-)\, dm + \mu_0 \, dn_0. \quad (10·25)$$

Now the amounts m and n_0 of electrolyte and of solvent respectively can be varied independently. The amount m_u of the undissociated electrolyte can also vary whilst the solution is in process of attaining dissociation equilibrium. All variations in (10·25) are therefore independent.

Two conclusions may be drawn from this. In the first place consider a closed system so that m and n_0 are actually constant. The equilibrium of the dissociation occurs when G is at a minimum and the condition of equilibrium is obtained by putting

$$\left(\frac{\partial G}{\partial m_u} \right)_{T, p, m, n_0} = 0.$$

From (10·25) this condition is seen to be

$$\mu_u = \nu_+ \mu_+ + \nu_- \mu_-, \quad (10·26)$$

and is clearly of the same form as for any other type of chemical reaction.

Secondly, if the amounts m and n_0 are now supposed to be changed sufficiently slowly for the dissociation equilibrium relation (10·26) to be maintained, the general equation (10·25) reduces to

$$dG = (\nu_+ \mu_+ + \nu_- \mu_-)\, dm + \mu_0 \, dn_0. \quad (10·27)$$

Let us define a quantity μ, which we shall call the chemical potential of the electrolyte as a whole, by the relation

$$\mu \equiv \left(\frac{\partial G}{\partial m} \right)_{T, p, n_0}, \quad (10·28)$$

and since there are no restrictions on a change in m, the quantity μ has a real experimental significance. Comparing (10·27) and (10·28) it is seen to be related to μ_+ and μ_- as follows:

$$\mu = \nu_+\mu_+ + \nu_-\mu_-. \tag{10·29}$$

This particular linear combination of μ_+ and μ_- therefore has a reality which they do not possess separately.

Comparing (10·29) and (10·26) it is also seen that at equilibrium

$$\mu = \mu_u,$$

where μ_u is the chemical potential of the undissociated part of the electrolyte.

10·9. Activity coefficients

Using Convention III activity coefficients may be defined by the relations†

$$\left.\begin{aligned}\mu_+ &= \mu_+^\square + RT \ln \gamma_+ m_+, \\ \mu_- &= \mu_-^\square + RT \ln \gamma_- m_-,\end{aligned}\right\} \tag{10·30}$$

$$\mu_u = \mu_u^\square + RT \ln \gamma_u m_u, \tag{10·31}$$

where the $\mu^\square$'s are functions only of temperature and pressure and the γ's approach unity at infinite dilution. Substituting (10·30) in (10·29) we obtain

$$\mu' = \nu_+\mu_+^\square + \nu_-\mu_-^\square + RT \ln \gamma_+^{\nu_+} \gamma_-^{\nu_-} m_+^{\nu_+} m_-^{\nu_-}. \tag{10·32}$$

Since μ_+ and μ_- have been shown to be quantities which are not measurable by any actual operation the same must apply to the derived quantities $\mu_+^\square$, $\mu_-^\square$, γ_+ and γ_-. On the other hand, the linear combination

$$\nu_1\mu_+^\square + \nu\ \mu_-^\square \tag{10·33}$$

which occurs in (10·32) will have physical significance and so also will the product $\gamma_+^{\nu_+} \gamma_-^{\nu_-}$.

If the total number of ions is

$$\nu \equiv \nu_+ + \nu_-, \tag{10·34}$$

the *mean ion activity coefficient* γ_+ of the electrolyte is defined by the equation

$$\mu = \nu_+\mu_+^\square + \nu_-\mu_-^\square + RT \ln \gamma_\pm^\nu m_+^{\nu_+} m_-^{\nu_-}, \tag{10·35}$$

from which it follows that $\gamma_\pm$ is measurable whenever m_+ and m_- are known. Comparing (10·32) and (10·35), the physically significant combination of γ_+ and γ_- is seen to be

$$\gamma_+^{\nu_+} \gamma_-^{\nu_-} = \gamma_\pm^\nu. \tag{10·36}$$

† The symbols m_+, etc., are here to be understood as referring to molalities, i.e. the amount of solvent is now fixed at $n_0 = 1/M_0$.

Although $\mu_+^\square$ and $\mu_-^\square$ are not measurable separately it is convenient to retain these quantities in (10·35). As will be seen from equation (10·30) $\mu_+^\square$ signifies the chemical potential of the positive ion in a hypothetical ideal solution in which the molality of this ion is unity. Similarly with regard to $\mu_-^\square$.† By adopting the convention that the standard free energy of formation of the hydrogen ion is zero it is then possible to assign numerical values to the free energy of formation of all other ions. This convention will be discussed in more detail in § 10·14.

The equation (10·35) may also be written

$$\mu = \nu_+\mu_+^\square + \nu_-\mu_-^\square + \nu\,RT\ln\gamma_\pm\,m_\pm, \tag{10·37}$$

where the *mean ionic molality* $m_\pm$ is defined by

$$m_\pm = (m_+^{\nu_+}\,m_-^{\nu_-})^{1/\nu}. \tag{10·38}$$

In the case of many strong electrolytes it is known that the ionization is essentially complete.‡ In this case we have

$$m_+ = \nu_+ m,$$
$$m_- = \nu_- m,$$

so that $m_\pm$ is immediately known from a knowledge of the total molality m of the electrolyte. This makes possible the calculation of $\gamma_\pm$ in equation (10·37). Even in cases where the ionization may not be quite complete it is conventional to adopt the same procedure, i.e. to put $m_+ = \nu_+ m$ and $m_- = \nu_- m$. Activity coefficients $\gamma_\pm$ calculated on this basis are sometimes called *stoichiometric activity coefficients*, since they are based on the total molality of the electrolyte. This does not involve any thermodynamic inexactitude, but it means that the stoichiometric activity coefficients include the effect of incomplete ionization or dissociation, as well as the deviations from ideality and the effects of any lack of proportionality between the molality and the mole fraction scales (§ 9·4).

Equation (10·36) implies that the measurable quantity $\gamma_\pm$ is determined by the properties γ_+ and γ_- of the individual ions. Although these properties are not measurable the equation does have

---

† In the case of an electrolyte such as $BaCl_2$, which is not of the 1:1 type, a solution which is one molal for, say, the Ba^{2+} ion is *two* molal for the Cl^- ion. This does not affect the significance of $\mu_+^\square$ and $\mu_-^\square$ which each refer to an ideal solution which is one molal for the particular ion in question.

‡ This is not quite the same as speaking of complete *dissociation*. In any solution of a strong electrolyte such as sodium chloride there may be a certain number of ion pairs, Na^+Cl^-, ion triplets $Na^+Cl_2^{2-}$, etc., and in this respect dissociation is not complete. As regards the randomness of the solution it is the degree of dissociation which is significant rather than the degree of ionization.

a significant result. Consider, for example, a solution containing the ions Na^+, K^+, Cl^-, Br^-. There are *four* mean ion activity coefficients involving these ions in the given solution, but the equation (10·36) shows that only *three* of them are independent. The relation between them is seen to be

$$\frac{\gamma_{Na^+,Cl^-}}{\gamma_{K^+,Cl^-}} = \frac{\gamma_{Na^+,Br^-}}{\gamma_{K^+,Br^-}},$$

where γ_{Na^+,Cl^-}, etc., denote the mean ion activity coefficients.

The temperature coefficients of $\mu_+^\square$, $\mu_-^\square$ and of γ_+ and γ_- are given by equations (9·27) and (9·28):

$$\left.\begin{array}{c}\dfrac{\partial \mu_i^\square/T}{\partial T} = -\dfrac{H_i^0}{T^2}, \\[3mm] \dfrac{\partial \ln \gamma_i}{\partial T} = \dfrac{H_i^0 - H_i}{RT^2}, \end{array}\right\} \tag{10·39}$$

where H_i^0 and H_i refer to the partial molar enthalpy of the ion at infinite dilution and in the particular solution where the activity coefficient is γ_i respectively. It follows that the temperature coefficients of the physically significant quantities $(\nu_+\mu_+^\square + \nu_-\mu_-^\square)$ and $\gamma_\pm$ are given by the relations

$$\frac{\partial}{\partial T}\left(\frac{\nu_+\mu_+^\square + \nu_-\mu_-^\square}{T}\right) = -\frac{H^0}{T^2}, \tag{10·40}$$

$$\frac{\nu \partial \ln \gamma_\pm}{\partial T} = \frac{H^0 - H}{RT^2}, \tag{10·41}$$

where

$$H = \nu_+ H_+ + \nu_- H_-, \tag{10·42}$$

and is the partial molar enthalpy of the electrolyte $M_{\nu_+}A_{\nu_-}$ in the particular solution. H^0 is the value of the same quantity at infinite dilution.

10·10. Phase equilibrium of an electrolyte. Solubility product

The fact that it is the linear combination

$$\mu = \nu_+\mu_+ + \nu_-\mu_-$$

which is the physically significant quantity for an electrolyte may be seen very clearly by considering phase equilibrium. Consider the equilibrium between the saturated electrolyte solution and the pure solid electrolyte. At constant temperature and pressure any variation in the composition of the solution gives rise to a change in the free energy of this phase by the amount

$$dG = \mu \, dm + \mu_0 \, dn_0,$$

as given by equations (10·27) and (10·29). The corresponding variation in the free energy of the solid phase may be written

$$dG' = \mu' \, dm',$$

where μ' and m' refer to the chemical potential and the amount respectively of the solid material. Since $dm = -dm'$ it is evident that the least value of $G+G'$, the total free energy, occurs when

$$\mu = \mu'. \tag{10·43}$$

Therefore it is the quantity μ, which has been called the chemical potential of the electrolyte as a whole, which provides the criterion of phase equilibrium in the usual way. Similar considerations relate to a solution of an electrolyte (e.g. HCl) in equilibrium with its vapour (§ 10·17).

Substituting from (10·35) into (10·43) we obtain

$$\nu_+\mu_+^\square + \nu_-\mu_-^\square + RT \ln \gamma_\pm^\nu \, m_+^{\nu_+} m_-^{\nu_-} = \mu'. \tag{10·44}$$

Now the addition of other substances to the saturated solution may alter the solubility of the electrolyte, and in this case will alter the values of m_+ and m_-. On the other hand, these additions do not alter the value of μ', the chemical potential of the solid electrolyte (provided that solid solutions are not formed) and also they do not alter the values of $\mu_+^\square$ and $\mu_-^\square$. (The activity coefficient has been defined in such a way that these quantities are independent of composition.) It follows from equation (10·44) that the product

$$(\text{S.P.}) \equiv \gamma_\pm^\nu \, m_+^{\nu_+} m_-^{\nu_-} \tag{10·45}$$

is constant (at fixed temperature and pressure) in any saturated solution in a particular solvent of the given electrolyte. This quantity is called the *solubility product*. It is only in very dilute solution that we can write as an approximation

$$m_+^{\nu_+} m_-^{\nu_-} \doteq \text{constant}. \tag{10·46}$$

The temperature coefficient of (S.P.) is obtained by dividing (10·44) by T followed by differentiation in the usual way and the application of equations (10·40) and (2·113b). We obtain

$$\frac{\partial \ln (\text{S.P.})}{\partial T} = \frac{H^0 - h'}{RT^2}, \tag{10·47}$$

where H^0 is the partial molar enthalpy of the electrolyte at infinite dilution and h' is the enthalpy per mole of the pure crystal. The numerator is thus equal to the heat absorbed in the process of solution at infinite dilution.

10·11. Equilibrium constant for ionic reactions

For the dissociation reaction

$$M_{\nu_+} A_{\nu_-} = \nu_+ M^{z+} + \nu_- A^{z-}$$

we have obtained already the condition of equilibrium (equation (10·26))

$$\mu_u = \nu_+ \mu_+ + \nu_- \mu_- (=\mu),$$

where μ_u is the chemical potential of the undissociated part of the electrolyte. Substituting in this equation from (10·31) and (10·35) we obtain

$$-RT \ln K = \Delta G_T^\square, \qquad (10\cdot48)$$

where

$$K \equiv \frac{\gamma_\pm^\nu \, m_+^{\nu_+} m_-^{\nu_-}}{\gamma_u m_u} \qquad (10\cdot49)$$

and

$$\Delta G_T^\square = \nu_+ \mu_+^\square + \nu_- \mu_-^\square - \mu_u^\square. \qquad (10\cdot50)$$

K is the true dissociation constant of the electrolyte and is independent of composition because of the characteristics of the $\mu^\square$. It is only equal to the apparent dissociation constant

$$K_m \equiv \frac{m_+^{\nu_+} m_-^{\nu_-}}{m_u} \qquad (10\cdot51)$$

under conditions where $\gamma_\pm$ and γ_u are unity, which is to say approaching infinite dilution. In all solutions which are not very dilute K_m is very far from being constant, due to the large deviations from ideality which occur in electrolyte solutions.

Similar considerations apply when the dissociating species is itself an ion. For example, in a solution of carbonic acid there are the equilibria

$$H_2CO_3 = H^+ + HCO_3^-,$$

$$HCO_3^- = H^+ + CO_3^{2-}.$$

The first and second dissociation constants are given by the relations

$$K_1 = \frac{\gamma_{H^+, HCO_3^-}^2 \, m_{H^+} m_{HCO_3^-}}{\gamma_{H_2CO_3} \, m_{H_2CO_3}},$$

$$K_2 = \frac{\gamma_{H^+} \gamma_{CO_3^{2-}} \, m_{H^+} m_{CO_3^{2-}}}{\gamma_{HCO_3^-} \, m_{HCO^-}}$$

$$= \frac{\gamma_{H^+}^2 \gamma_{CO_3^{2-}} \, m_{H^+} m_{CO_3^{2-}}}{\gamma_{H^+} \gamma_{HCO_3^-} \, m_{HCO_3^-}}$$

$$= \frac{\gamma_{2H^+, CO_3^{2-}}^3 \, m_{H^+} m_{CO_3^{2-}}}{\gamma_{H^+, HCO_3^{2-}}^2 \, m_{HCO_3^-}}.$$

where the symbols $\gamma_{2H^+, CO_3^{2-}}$, and γ_{H^+, HCO_3^-} denote the mean ion activity coefficients of pairs or trios of ions which together are electrically neutral. Similar equilibrium constants can be set up for reactions between ions, such as

$$2Fe^{3+} + Sn^{2+} = 2Fe^{2+} + Sn^{4+},$$

and also for reactions involving ions together with neutral species such as the solvent.

The temperature coefficient of $\ln K$ may be obtained by differentiation of equation (10·48) and application of equation (10·40). The results are equivalent to those already obtained in § 10·4. For example, the temperature coefficient of the dissociation equilibrium discussed above is given by

$$\frac{\partial \ln K}{\partial T} = \frac{\Delta H^0}{RT^2}, \tag{10·52}$$

where ΔH^0 is the enthalpy increase in the dissociation process at infinite dilution.

10·12. Magnitude of activity coefficients of charged and uncharged species

At this point it is convenient to discuss very briefly the relative magnitudes of the activity coefficients of the ions and of the uncharged parent electrolyte. In the region of low concentrations where molality and mole fraction are proportional to each other, the extent to which the activity coefficients differ from unity is a measure of the deviations from ideality in the solution, and this in its turn depends on the forces between the various components of the solution.

Consider in the first place the uncharged components. Between these the attractive force is that which is known by the name of van der Waals, and, according to London's theory, the potential energy of the interaction varies as the inverse sixth power of the separation of the attracting centres. Therefore it falls off extremely rapidly with distance. It follows that in any solution which is moderately dilute the solute molecules do not interact with each other at all appreciably. As mentioned already in §8·1b, each of them is present in an almost constant force field due to the solvent molecules. For this reason it is found that deviations from ideality of the order of 1 % are generally not encountered until the mole fraction of the solute is at least 0.001,† and in many cases the region of ideal behaviour may extend up to much higher concentrations and in exceptional cases over the whole range of mole fractions.

Consider now the coulombic forces acting between the various

† Guggenheim, *Modern Thermodynamics by the Methods of Willard Gibbs* (1933), p. 92.

ions. Between two ions of charges z_+ and z_- at a separation r the force is proportional to $z_+ z_-/r^2$. The potential energy of the interaction therefore varies inversely as the first power of the separation and may be expected to be appreciable at quite high degrees of dilution. For this reason the activity coefficient may be expected to differ appreciably from unity even in very dilute solutions and to a greater extent the larger the charge on the ions.

TABLE 10. *Mean ion activity coefficients at 25 °C*†

Molality	$\gamma_\pm HCl$	$\gamma_\pm H_2SO_4$	$\gamma_\pm MgSO_4$	$\gamma_\pm Al_2(SO_4)_3$
0.000 5	(0.975 2)	0.885	—	—
0.001	(0.965 6)	0.830	—	—
0.005	0.928 5	0.639	—	—
0.01	0.904 8	0.544	—	—
0.05	0.830 4	0.340	—	—
0.1	0.796 4	0.265	(0.150)	0.035 0
0.5	0.757 1	0.154	0.068	0.014 3
1.0	0.809 0	0.130	0.049	0.017 5
2.0	1.009	0.124	0.042	—
4.0	1.762	0.171	—	—
6.0	3.206	0.264	—	—
8.0	5.907	0.397	—	—
10.0	10.49	0.553	—	—
12.0	17.61	0.742	—	—
14.0	27.77	0.967	—	—
16.0	41.49	1.234	—	—

Some values of the mean ion activity coefficient $\gamma_\pm$ of hydrochloric and sulphuric acids at 25 °C are given in Table 10. (These results were obtained by a method, to be described shortly, based on the measurement of electromotive force.) Some values for magnesium sulphate and aluminium sulphate are also quoted, and the effect of the charge of the ions in increasing the deviations from ideality is shown very clearly. Even in the case of hydrochloric acid, where the ions are singly charged, it will be seen that there are appreciable deviations even at extreme dilution. Another feature of the activity coefficients of electrolytes is that the values often pass through a minimum and then, with increasing concentration, rise to values considerably greater than unity.

According to the theory developed by Debye, Hückel and others, the departure of electrolytes from the ideal solution laws may be

† This data is from Harned and Owen, *The Physical Chemistry of Electrolytic Solutions* (1943), Tables 11–4–1 A, 13–11–1 A and 14–9–1 A, and also from the extensive compilations of Robinson and Stokes, *Trans. Faraday Soc.* **45** (1949), 612.

accounted for, at least partially, in terms of the electrostatic attraction between the ions. Debye's limiting law which becomes valid as the concentration of all ions approaches zero is

$$\log_{10}\gamma_{\pm} = -\alpha z_+ \,|\, z_- \,|\, I^{\frac{1}{2}}, \tag{10·53}$$

where $z_+ \,|\, z_- \,|$ is the positive number which is the product of the ionic charges for the particular electrolyte and the quantity I, called the *ionic strength* of the solution, is defined by

$$I \equiv \tfrac{1}{2}\Sigma z_i^2 m_i, (\text{mol kg}^{-1}) \tag{10·54}$$

the summation being taken over *all* ions in the solution. For a solution in water the constant α has the values 0.509 and 0.488 mol$^{-1/2}$ kg$^{1/2}$ at 25 and 0 °C respectively.

The limiting law (10·53) agrees with experiment only at very high dilutions. Even in the case of a 1 : 1 electrolyte, such as HCl, deviations are significant when I is as small as 10^{-2}. An empirical formula, not containing adjustable parameters, which can be used for many electrolytes up to $I = 0·1$, is

$$\log_{10}\gamma_{\pm} = -\alpha z_+ \,|\, z_- \,|\, \frac{I^{\frac{1}{2}}}{1+I^{\frac{1}{2}}}, \tag{10·55}$$

where α has the same value as above.

10·13. Free energy of dissociation

Consider the dissociation of a weak electrolyte such as acetic acid. The dissociation constant is

$$K = \frac{m_{\text{H}^+}m_{\text{Ac}^-}\gamma_{\pm}^2}{m_{\text{HAc}}\gamma_{\text{HAc}}}. \tag{10·56}$$

According to the remarks made in the previous section the activity coefficient γ_{HAc} of the undissociated acid can probably be taken as being nearly unity up to a mole fraction of at least 10^{-3}, which corresponds to a molality in water of about 0.05. On the other hand, the application of equation (10·53) shows that $\gamma_{\pm}$ is within 1 % of unity only when the molality of the ions is less than 10^{-4}. It follows that the apparent dissociation constant defined by

$$K_m = \frac{m_{\text{H}^+}m_{\text{Ac}^-}}{m_{\text{HAc}}} = \frac{\alpha^2 m}{(1-\alpha)}, \tag{10·57}$$

where α is the degree of dissociation and m is the total molality, can be expected to be constant only under conditions of great dilution.

In the Ostwald treatment of dissociation it was assumed that α is given by Λ/Λ_{∞}, where Λ is the molar conductivity at the molality m and Λ_{∞} is the molar conductivity at infinite dilution. Some values of K_m at 25 °C worked out on the basis of this supposition, and using $\Lambda_{\infty} = 390.7$, are shown in Table 11.

TABLE 11

Normality	Λ	$K_m \times 10^5$
0.000 1	131.6	1.71
0.001	48.63	1.77
0.005	22.80	1.81
0.01	16.20	1.79
0.05	7.36	1.81
0.10	5.20	1.80

The approximate constancy of the K_m values may be attributed partly to the fact that the concentration of the free ions in the various solutions is very low, so that $\gamma_{\pm}$ is close to unity, and partly to the fact that the Ostwald assumption $\alpha = \Lambda/\Lambda_{\infty}$ is probably fairly near to the truth under the same conditions of high dilution of the ions. In general, the Ostwald method gives reasonably satisfactory results for dilute solutions of weak electrolytes for which the dissociation constant is of the order 10^{-5} or less.

In stronger solutions, or in the case of electrolytes which have a higher dissociation constant, $\gamma_{\pm}$ is no longer close to unity. The assumption $\alpha = \Lambda/\Lambda_{\infty}$ also becomes erroneous. Whenever the concentration of the ions is appreciable the ions of opposite charge attract each other, with the result that their speed of migration in the electric field is no longer independent of their concentration, as is supposed in the Ostwald treatment.

In order to make proper allowance for these factors it is necessary to estimate the value of $\gamma_{\pm}$ (e.g. by use of equations (10·53) or (10·55)) and also to use a more adequate theory of the dependence of α on the molar conductivity. For details the reader is referred to the literature on electrochemistry.[†] When these corrections are applied the thermodynamic dissociation constant of acetic acid at 25 °C is found to be

$$K = 1.758 \times 10^{-5}.$$

(Using an independent method based on the e.m.f. of a galvanic cell Harned and Ehlers obtained the figure 1.754×10^{-5} in close agreement.) The standard free energy of the dissociation process at 25° C may now be calculated by using equation (10·48), and the value is

$$\Delta G_{298}^{\square} \equiv \mu_{H^+}^{\square} + \mu_{Ac^-}^{\square} - \mu_{HAc}^{\square}$$
$$= 6485 \text{ cal mol}^{-1}.$$

† Harned and Owen, *The Physical Chemistry of Electrolytic Solutions* (New York, Reinhold, 1943), p. 206; Glasstone, *Introduction to Electrochemistry* (New York, Van Nostrand, 1942), p. 165.

The free energy of dissociation of other weak electrolytes such as water and ammonia may be obtained by similar methods. For example for the process

$$H_2O\,(l) = H^+\,(aq.) + OH^-\,(aq.),$$

where the symbol (aq.) denotes the hypothetical ideal solution of unit molality, we obtain

$$\mu_{H^+}^{\square} + \mu_{OH^-}^{\square} - \mu_{H_2O}^{*} = +19\ 095\ \text{cal mol}^{-1}\ \text{at 25 °C.}$$

Now the standard free energy of formation of liquid water at 25 °C is $-56\ 690$ cal mol^{-1}.† Combining this with the above figure for the free energy of ionization we obtain

$$H^+\,(aq.) + OH^-\,(aq.), \quad \Delta_f G_{298}^{\square} = -37\ 595\ \text{cal mol}^{-1}, \quad (10\cdot58)$$

this being the free energy of formation of 1 mol of hydrogen ion and 1 mol of hydroxyl ion, each in an ideal solution at unit molality, from gaseous hydrogen and oxygen each at 1 atm pressure.

Similarly for the process

$$NH_3 + H_2O = NH_4^+ + OH^-,$$

a measurement of the degree of ionization gives

$$\mu_{NH_4^+}^{\square} + \mu_{OH^-}^{\square} - \mu_{NH_3}^{\square} - \mu_{H_2O}^{*} = +6465\ \text{cal mol}^{-1}\ \text{at 25 °C.}$$

In §10·2 the figure -6360 cal mol^{-1} was worked out for the standard free energy of formation of dissolved ammonia. If this figure, together with the value $-56\ 690$ cal mol^{-1} for the free energy of formation of water, is combined with the above result for the free energy of ionization, we obtain

$$NH_4^+\,(aq.) + OH^-\,(aq.), \quad \Delta_f G_{298}^{\square} = -56\ 585\ \text{cal mol}^{-1}, \quad (10\cdot59)$$

and this is the free energy of formation of the pair of ions in their standard states in aqueous solution.

10·14. The hydrogen ion convention and the free energies and enthalpies of formation of individual ions

It follows from what was said in §10·7 that it is impossible to obtain absolute values of the free energies of formation of individual ions. For example, in the case of ammonia discussed above the figure which was obtained from experiment was the total free energy of formation of ammonium and hydroxyl ions when they are present in the solution in equivalent amounts.

† This is a more recent value than the figure of $-56\ 560$ cal mol^{-1} quoted in §4·10.

Nevertheless for tabulation† it is convenient to be able to quote values for each ion separately, and this can be done if an arbitrary value is chosen for the free energy of formation of any one ion. According to the convention which is generally adopted the free energy of formation of the hydrogen ion, in its hypothetical ideal solution of unit molality, from gaseous hydrogen at 1 atm pressure, is taken as zero. Moreover, this is taken to apply *at all temperatures*, with the result that the corresponding enthalpy and entropy of formation of the hydrogen ion are also zero, as follows from the relations

$$\partial(\Delta G/T)/\partial T = -\Delta H/T^2 \quad \text{and} \quad \partial \Delta G/\partial T = -\Delta S.$$

In brief the convention is

$$\tfrac{1}{2}H_2\ (g,\ 1\ \text{atm}) = H^+\ (\text{aq.}) + e, \quad \Delta_f G^{\ominus} = 0, \tag{10·60}$$

and on this basis it is possible to obtain numerical values for the free energies of formation of all other ions. For example, from equation (10·58) we obtain for the hydroxyl ion

$$OH^-\ (\text{aq.}), \quad \Delta_f G^{\ominus}_{298} = -37\ 595\ \text{cal mol}^{-1},$$

and combining this with the result (10·59) we obtain for the ammonium ion

$$NH_4^+\ (\text{aq.}), \quad \Delta_f G^{\ominus}_{298} = -18\ 990\ \text{cal mol}^{-1}$$

The procedure will now be clear by which a table of arbitrary but self-consistent free energies of the individual ions may be built up. An important experimental method for determining these quantities, based on the use of the galvanic cell, has still to be described.

The method of obtaining a corresponding table of enthalpies of formation of the individual ions is straightforward and need not be discussed in detail. As an example, the heat of formation of gaseous hydrogen chloride at 25 °C is $-22\ 063\ \text{cal mol}^{-1}$. The heat of solution in water at infinite dilution is $17\ 960\ \text{cal mol}^{-1}$, with the result that the heat of formation of hydrochloric acid at infinite dilution is $-40\ 023\ \text{cal mol}^{-1}$. On the basis of the above convention the heat of formation of the hydrogen ion is zero, and hence for the chloride ion we obtain

$$Cl^-\ (\text{aq.}), \quad \Delta_f H_{298} = -40\ 023\ \text{cal mol}^{-1}$$

It will be noted that this figure refers equally to infinite dilution and to the hypothetical ideal solution of unit molality. This is because the enthalpies in ideal solutions are independent of composition, as discussed in § 8·5.

† A very large compilation of free energy and enthalpy data has been published by the National Bureau of Standards under the title *Selected Values of Chemical Thermodynamic Properties* (Washington, U.S. Government Printing Office, 1952).

Another important experimental method for the determination of ionic enthalpies is based on the temperature coefficient of the e.m.f. of galvanic cells. From a knowledge of the temperature coefficient, the enthalpy change in the cell reaction may be determined by use of the Gibbs–Helmholtz equation.

10·15. Activity coefficients and free energies as measured by the use of the galvanic cell

The subject covered by this heading is a large one, and we shall aim only at describing it in outline by means of some examples. In particular, the discussion will be confined to *cells without transference*. This is the class of cells in which the electrolyte solution has essentially† the same composition at each electrode. The other class of cells, those *with transference*, contain two or more solutions in contact which differ appreciably in composition. Between these solutions there is an irreversible diffusion process and a corresponding *liquid-junction potential*. Because of the irreversible diffusion, the e.m.f.'s of such cells cannot be discussed in a really exact manner by the methods of classical thermodynamics.‡

The general principles of the reversible cell have already been discussed in §§4·14 and 4·15. Let it be supposed that the chemical reaction which takes place in the cell when there is a flow of current is

$$\Sigma \nu_i M_i = 0$$

(the stoichiometric coefficients ν_i being taken as negative for those substances on the left of the chemical equation as usually written and positive for those substances on the right). (For a cell to be useful for thermodynamic purposes it is necessary, of course, that the cell reaction shall be completely known.) The electrical work done by the cell is equal to the decrease of Gibbs free energy. The e.m.f. is therefore given by

$$\Sigma \nu_i \mu_i = -zFE, \qquad (10\cdot61)$$

where the μ_i are the chemical potentials of the various reactants and products in the states in which they are present in the cell and z is the number of Faradays which pass through the cell for the chemical reaction as written.

For any dissolved strong electrolyte $M_{\nu_+} A_{\nu_-}$ in the cell the chemical potential is related to the stoichiometric activity coefficient by equation (10·37)

$$\mu = \nu_+ \mu_+^\square + \nu_- \mu_-^\square + \nu RT \ln \gamma_\pm m_\pm, \qquad (10\cdot62)$$

† As Guggenheim has remarked (*Thermodynamics* (1949), pp. 342, 347) the solution is never quite identical at the two electrodes.

‡ See, for example, the author's *Thermodynamics of the Steady State* (London, Methuen, 1951), pp. 5 and 78.

where the mean ionic molality is related to the stoichiometric molality m of the electrolyte by the relation

$$m_{\pm} = (m_+^{\nu_+} m_-^{\nu_-})^{1/\nu}$$
$$= m(\nu_+^{\nu_+} \nu_-^{\nu_-})^{1/\nu}. \qquad (10\cdot63)$$

A measurement of the electromotive force of a suitable cell can therefore be used for the purpose of determining activity coefficients and also the standard free energy change in the cell reactions.

(a) *The activity coefficient of hydrochloric acid.* To measure the activity coefficient of the electrolyte $M_{\nu_+} A_{\nu_-}$ we require a cell in which one of the electrodes is reversible† with respect to the ion M^{z+} and the other is reversible with respect to the ion A^{z-}.

Consider, for example, the cell

$$\text{Pt, } \text{H}_2 \,|\, \text{HCl}\,(m) \,|\, \text{Hg}_2\text{Cl}_2 \,|\, \text{Hg,}$$

which is reversible to the hydrogen ion at the hydrogen electrode on account of the rapid process

$$\tfrac{1}{2}\text{H}_2 = \text{H}^+ + e,$$

and is reversible to the chloride ion at the calomel electrode on account of the rapid process

$$\tfrac{1}{2}\text{Hg}_2\text{Cl}_2 + e = \text{Hg}\,(l) + \text{Cl}^-.$$

This cell can be used for measuring the activity coefficient of hydrochloric acid and also the standard free energy change in the cell reaction.

† The cell as a whole is reversible if it is in a *state of stable equilibrium* when there is no flow of current (otherwise equation (10·61) would not be applicable). If the direction of a small current through the cell is reversed it will exactly reverse all the changes taking place in the cell.

A cell such as $\text{Zn}\,|\,\text{H}_2\text{SO}_4\,|\,\text{Cu}$ is *not* reversible. In addition to the process $\text{Zn} = \text{Zn}^{2+} + 2e$, the process $2\text{H}^+ + 2e = \text{H}_2\,(g)$ can also take place with appreciable speed at the zinc electrode and causes short-circuiting. Both processes continue to take place spontaneously and irreversibly even when no current is taken from the cell. Thus the system is not at equilibrium.

From a practical standpoint it is desirable also that the passage of a very small current through the cell, such as is involved in the use of a potentiometer, shall cause a negligible change in the measured p.d. (Where electrodes are easily polarizable a current of 10^{-6} A may produce a potential several tenths of a volt different from the reversible value.) Therefore it is desirable that the establishment of equilibrium at the electrodes shall be extremely rapid.

Consider the metal M in contact with a solution containing the ion M^+; the electrode is said to be reversible with respect to the ion M^+ if the process $\text{M} = \text{M}^+ + e$ takes place extremely rapidly in either direction. All other conceivable electron exchange processes at this electrode must be so slow that they may be regarded as non-existent.

This reaction is

$$\tfrac{1}{2}H_2 + \tfrac{1}{2}Hg_2Cl_2 = HCl + Hg\,(l), \tag{10·64}$$

and therefore the general equation (10·61) takes the particular form

$$\mu_{HCl} + \mu_{Hg} - \tfrac{1}{2}\mu_{H_2} - \tfrac{1}{2}\mu_{Hg_2Cl_2} = -FE. \tag{10·65}$$

In this equation we can substitute

$$\mu_{HCl} = \mu_{H^+}^{\square} + \mu_{Cl^-}^{\square} + 2RT\ln\gamma_{\pm}\,m,$$

as follows from (10·62) and (10·63), m being the molality of the hydro-chloric acid solution in the cell when the e.m.f. has the value E. We also substitute

$$\mu_{H_2} = \mu_{H_2}^0 + RT\ln p_{H_2},$$

where p_{H_2} is the partial pressure of the hydrogen at the hydrogen electrode. Provided that the cell is operating at a pressure near to that of the atmosphere we can also put $\mu_{Hg} = \mu_{Hg}^0$ and $\mu_{Hg_2Cl_2} = \mu_{Hg_2Cl_2}^0$, where the μ^0's refer to the chemical potentials of mercury and calomel respectively at 1 atm. (See p. 164.)

The result of these substitutions is

$$RT\ln\frac{\gamma_{\pm}^2\,m^2}{p_{H_2}^{\frac{1}{2}}} - FE^0 = -FE, \tag{10·66}$$

where E^0, called the *standard e.m.f.* of the cell, is defined by

$$-FE^0 \equiv \mu_{H^+}^{\square} + \mu_{Cl^-}^{\square} + \mu_{Hg}^0 - \tfrac{1}{2}\mu_{H_2}^0 - \tfrac{1}{2}\mu_{Hg_2Cl_2}^0, \tag{10·67}$$

which is equal to the standard free-energy change in the cell reaction. Under conditions where the partial pressure of the hydrogen is exactly 1 atm (10·66) simplifies to

$$RT\ln\gamma_{\pm}^2\,m^2 - FE^0 = -FE. \tag{10·68}$$

It is clear that E^0 would be equal to the measured e.m.f. if the hydro-chloric acid in the cell were an ideal solution of unit molality, since the logarithmic term would then vanish.

Rearranging (10·68) we obtain

$$2RT\ln m + FE = FE^0 - 2RT\ln\gamma_{\pm}, \tag{10·69}$$

and the two quantities on the left are measurable. It follows that if the sum of these two measured quantities is extrapolated to infinite dilution the limiting value will be equal to FE^0, since $\ln\gamma_{\pm}$ vanishes at the limit. Having determined FE^0 in this way, the equation can then be used for the calculation of $\gamma_{\pm}$, for any of the solutions for which the e.m.f. has been measured.

Considerable care needs to be exercised in the extrapolation to infinite dilution. One of the simplest methods is to use Debye's limiting law (equation (10·53)). In the case of a hydrochloric acid solution this takes the form

$$\log_{10} \gamma_\pm = -\alpha \sqrt{m}, \qquad (10·70)$$

and it follows that if the left-hand side of equation (10·69) is plotted, not against the molality but against the square root of the molality, the experimental points may be expected to fall on an approximately straight line, thereby facilitating an accurate extrapolation. On account of the deviations from the Debye limiting law when the molality is appreciable, more elaborate kinds of extrapolation are usually used, based for example on equation (10·55).

Some of the very accurate e.m.f. measurements made by Hills and Ives[†] on the cell in question at 25 °C are quoted in the table and the extrapolation gives the following value for the standard e.m.f.:

$$E^0_{298} = 0.267\ 96 \text{ V}.$$

The activity coefficients worked out by these authors are in very close agreement with those already quoted in Table 10 in § 10·12, which were based on the measurements of Harned and Ehlers using a silver-silver chloride electrode in place of the calomel electrode used by Hills and Ives:

Molality of HCl	0.119 304	0.051 645	0.010 947 4	0.005 040 3	0.001 607 7
E.m.f.	0.389 48	0.429 94	0.505 32	0.543 665	0.600 80

The value of E_0 can be used to obtain the standard free energy change of the cell reaction by application of equation (10·67). The result is -6195 cal mol^{-1} of HCl (or twice this value per mol of Hg_2Cl_2). Since the other two substances involved are both elements, this figure is equal to the free energy of formation of hydrochloric acid in its ideal solution of unit molality minus half the free energy of formation of solid Hg_2Cl_2. If one of these two quantities is known the other can be calculated. (See one of the problems at the end of the chapter.)

(b) Single electrode potentials. In § 2·9c it was shown that the potential difference between identical phases is a meaningful quantity, but not the potential difference between phases which differ in composition. Therefore, although the e.m.f. of a complete cell is quite definite, it is not possible to attribute *absolute* values to the potentials of the separate electrodes.

However, for tabulation it is convenient to regard the e.m.f. as being the sum of the individual electrode potentials and to adopt the convention that the potential of some particular electrode is zero.

† *J. Chem. Soc.* (1951), p. 318.

On this basis, it is possible to give numerical values to the potentials of all other electrodes. The fact that such potentials can be combined additively follows from the fact that the cell e.m.f. is related to the free-energy change of the cell reaction and the latter is a change in a function of state.

In § 10·14 the convention was adopted that the standard free energy of formation of the hydrogen ion is zero at all temperatures. Thus

$$\tfrac{1}{2}H_2 \ (g, \ 1 \ \text{atm}) = H^+ \ (\text{aq.}) + e, \quad \Delta_f G^\square = 0.$$

Since this is the process which takes place at the hydrogen electrode an equivalent and self-consistent convention is to take the potential of this electrode as zero. That is to say the potential of the standard hydrogen electrode,

$$\text{Pt}, \quad H_2 \ (1 \ \text{atm}) \,|\, H^+ \ (\text{aq.})$$

(where the symbol aq. refers to an ideal solution at unit molality), is taken as zero at all temperatures.

Now the standard e.m.f. of the cell

$$\text{Pt}, \quad H_2 \ (1 \ \text{atm}) \,|\, \text{HCl} \ (m) \,|\, \text{HgCl} \,|\, \text{Hg}$$

has been shown to be $E^0_{298} = 0.267 \ 96$ V. To the electrode

$$\text{Cl}^- \,|\, \text{HgCl} \,|\, \text{Hg},$$

when written in this sequence, we therefore attribute the standard potential

$$E^0_{298} = 0.267 \ 96 \ \text{V},$$

and this refers of course to the case where the chloride ion is in an ideal solution at unit molality.

By similar measurements on the cell

$$\text{Pt}, \quad H_2 \,|\, \text{HCl} \,(m) \,|\, \text{AgCl} \,|\, \text{Ag},$$

the standard e.m.f. is found to be

$$F^0_{298} = 0.222 \ 4 \ \text{V},$$

and this is therefore the standard potential of the silver-silver chloride electrode when written in the sequence

$$\text{Cl}^- \,|\, \text{AgCl} \,|\, \text{Ag}.$$

As an example of the additivity principle, it follows that the standard e.m.f. at 25 °C of the cell

$$\text{Ag} \,|\, \text{AgCl} \ldots \text{HgCl} \,|\, \text{Hg}$$

is equal to $-0.222\,4 + 0.267\,96 = 0.045\,6$ V,

which is in good agreement with a directly measured value of 0.0455 V.†

(c) The standard free energy of formation of the zinc ion. It will be useful to discuss an example concerning an electrolyte which is not of the 1:1 type. Consider the cell

$$\text{Zn} \mid \text{ZnCl}_2\,(m) \mid \text{AgCl}\,(s) \mid \text{Ag},$$

for which the cell reaction is

$$\text{Zn} + 2\text{AgCl}\,(s) = 2\text{Ag} + \text{Zn}^{2+} + 2\text{Cl}^-.$$

The general equation (10·61) takes the form

$$2\mu_{\text{Ag}} + \mu_{\text{Zn}^{2+}} + 2\mu_{\text{Cl}^-} - \mu_{\text{Zn}} - 2\mu_{\text{AgCl}} = -2FE, \tag{10·71}$$

where the factor 2 on the right-hand side arises from the fact that the cell reaction as written involves the passage of two Faradays. In this equation we can substitute for the chemical potential of the zinc chloride in solution by means of equation (10·62)‡

$$\mu_{\text{Zn}^{2+}} + 2\mu_{\text{Cl}^-} = \mu_{\text{Zn}^{2+}}^{\square} + 2\mu_{\text{Cl}^-}^{\square} + 3RT \ln \gamma_{\pm} m_{\pm}, \tag{10·72}$$

where the mean ionic molality is related to the stoichiometric molality by equation (10·63)

$$m_{\pm} = 4^{\frac{1}{3}} m. \tag{10·73}$$

The result of the substitution is

$$2\mu_{\text{Ag}} + \mu_{\text{Zn}^{2+}}^{\square} + 2\mu_{\text{Cl}^-}^{\square} + 3RT \ln \gamma_{\pm} m_{\pm} - \mu_{\text{Zn}} - 2\mu_{\text{AgCl}} = -2FE.$$

On rearranging we obtain

$$3RT \ln m_{\pm} + 2FE = 2FE^0 - 3RT \ln \gamma_{\pm}, \tag{10·74}$$

where

$$-2FE^0 \equiv 2\mu_{\text{Ag}}^0 + \mu_{\text{Zn}^{2+}}^{\square} + 2\mu_{\text{Cl}^-}^{\square} - \mu_{\text{Zn}}^0 - 2\mu_{\text{AgCl}}^0, \tag{10·75}$$

the chemical potentials of the various solid species having been written as μ^0, their values at atmospheric pressure.

The sum of the terms on the left-hand side of (10·74) can be determined by experiment. By carrying out an extrapolation to infinite dilution (e.g. by plotting against $\sqrt{m}$) it is therefore possible to determine the value of the term $2FE^0$ on the right-hand side of the equation. The value of E^0 which is obtained in this way is

$$E_{298}^0 = 0.9834 \text{ V}.$$

† Quoted by Guggenheim and Prue, *Trans. Faraday Soc.* **50** (1954), 231.

‡ In working out a problem concerning an electrolyte which is not of the 1:1 type the author actually finds it easier to start from equations such as (10·30). This avoids the need to memorize, or to look up, the definition of the mean ionic molality, equation (10·63).

Now the cell can be considered as the sum of two electrodes

$$\text{Zn} \,|\, \text{Zn}^{2+} \quad \text{and} \quad \text{Cl}^- \,|\, \text{AgCl} \,|\, \text{Ag}.$$

The standard potential of the latter has been shown already to be 0.2224 V at 25°C. The standard potential of the zinc electrode written in the above sequence is therefore 0.7610 V. It follows that the free-energy change in the process

$$\text{Zn} = \text{Zn}^{2+} + 2e$$

is $-2F \times 0.7610 = -35\,084$ cal mol^{-1}, and this is the standard free energy of formation of the zinc ion at 25 °C.

(d) *Redox Electrode*. This type of electrode consists of an inert metal, such as gold or platinum, dipping into a solution containing oxidized and reduced ions which are in equilibrium with each other. For example, when platinum is immersed in a solution containing the ferric and ferrous ions the process

$$\text{Fe}^{2+} = \text{Fe}^{3+} + e$$

can take place in either direction at the electrode, according to the direction of flow of current. Measurements of the potentials of such electrodes can therefore be used for the purpose of calculating the free-energy change in the oxidation-reduction process.

10·16. Activity coefficients by use of the Gibbs–Duhem equation

In a solution of an electrolyte the activity coefficient† of the *solvent* can be determined by measurement of its partial pressure, or from the freezing-point depression or the osmotic pressure. The relevant equations are the same as those developed in §9·6. Provided that values of the activity or osmotic coefficient have been determined over a wide range of concentrations, including some solutions which are very dilute, it is possible to calculate the activity coefficient of the *solute* in some particular solution by application of the Gibbs–Duhem equation. This procedure, as applied to solutions of non-electrolytes, was described in §9·7 and 9·8.

Consider a solution consisting of m mols of electrolyte and n_0 mols of solvent. It follows from equation (10·27) and (10·29) that any change in the Gibbs free energy of the solution at constant temperature and pressure is given by

$$dG = \mu\,dm + \mu_0\,dn_0, \tag{10·76}$$

† The deviations from ideality of the *solvent* are more frequently expressed in terms of the osmotic coefficient, rather than the activity coefficient, for the reasons discussed in §9·11.

where μ is the chemical potential of the electrolyte as a whole and μ_0 is the chemical potential of the solvent. The complete equation which allows for variations of temperature and pressure is

$$dG = -S\,dT + V\,dp + \mu\,dm + \mu_0\,dn_0. \qquad (10\cdot77)$$

Since

$$G = \mu m + \mu_0 n_0, \qquad (10\cdot78)$$

it follows that the Gibbs–Duhem equation for the electrolyte solution is

$$S\,dT - V\,dp + m\,d\mu + n_0\,d\mu_0 = 0. \qquad (10\cdot79)$$

(To these equations additional terms must be added, of course, if there are other substances present whose quantities can be varied independently.)

Applying (10·79) to conditions of constant temperature and pressure and rearranging we obtain

$$-m\,d\mu = n_0\,d\mu_0.$$

If we take the quantity of solvent n_0 as being equal to $1/M_0$, m is equal to the stoichiometric molality of the electrolyte. Thus

$$-m\,d\mu = d\mu_0/M_0 \qquad (10\cdot80)$$

The activity coefficient of the electrolyte is related to its chemical potential by equation (10·35)

$$\mu = f(T, p) + RT\ln\gamma_{\pm}^{\nu}\,m_{+}^{\nu_+}\,m_{-}^{\nu_-}.$$

In this equation, if we put $m_+ = \nu_+ m$ and $m_- = \nu_- m$, the quantity $\gamma_{\pm}$ becomes the *stoichiometric activity coefficient* and includes within itself the effect of any incompleteness of the dissociation as well as the deviation from ideality, as discussed in § 10·9. Adopting this convention we obtain

$$\mu = f(T, p) + RT\ln\gamma_{\pm}^{\nu}\,m^{\nu}\,\nu_{+}^{\nu_+}\,\nu_{-}^{\nu_-},$$

and therefore for any variation of composition at constant temperature and pressure

$$d\mu = \nu RT(d\ln\gamma_{\pm} + d\ln m). \qquad (10\cdot81)$$

The activity coefficient γ_0 of the solvent is related to its chemical potential by equation (9·16),

$$\mu_0 = f(T, p) + RT\ln\gamma_0 x_0,$$

where x_0 is the mole fraction of the solvent. Therefore for any variation of composition at constant temperature and pressure

$$d\mu_0 = RT\,d\ln\gamma_0 x_0. \qquad (10\cdot82)$$

Substituting (10·81) and (10·82) in (10·80) we obtain

$$-\nu m(\mathrm{d} \ln \gamma_\pm + \mathrm{d} \ln m) = \frac{1}{M_0} \mathrm{d} \ln \gamma_0 x_0,$$

or rearranging $\quad -\mathrm{d} \ln \gamma_\pm = \mathrm{d} \ln m + \dfrac{1}{M_0 \nu m} \mathrm{d} \ln \gamma_0 x_0.$ (10·83)

The integration of this equation, by graphical or other methods, makes it possible to determine the value of $\gamma_\pm$, at some particular molality, from a series of measured values of $\gamma_0 x_0$ which extend up to this molality from high dilution. For this purpose it is often convenient to express $\mathrm{d} \ln \gamma_0 x_0$ directly in terms of some measurable variable, such as the differential $\mathrm{d}T$ of the freezing-point depression.†

10·17. Partial pressure of a volatile electrolyte

In § 10·10 it was shown that the equilibrium of an electrolyte in a saturated solution with its pure solid phase depends on μ, the chemical potential of the electrolyte as a whole. Similar considerations apply to the equilibrium of a volatile electrolyte, such as hydrochloric acid, with its vapour.

Let μ' be the chemical potential of hydrogen chloride in the vapour phase. For equilibrium with the solution

$$\mu' = \mu$$
$$= \mu_{\mathrm{H}^+}^\square + \mu_{\mathrm{Cl}^-}^\square + RT \ln \gamma_\pm^2 \, m_{\mathrm{H}^+} m_{\mathrm{Cl}^-}$$

by equation (10·35). If it is supposed for simplicity that the vapour behaves as a perfect gas

$$\mu' = \mu_{\mathrm{HCl}}^0 + RT \ln p_{\mathrm{HCl}}$$

where p_{HCl} is the partial pressure and μ^0 refers to gaseous hydrogen chloride at unit pressure. Between the two equations

$$RT \ln \frac{p_{\mathrm{HCl}}}{\gamma_\pm^2 \, m_{\mathrm{H}^+} m_{\mathrm{Cl}^-}} = \mu_{\mathrm{H}^+}^\square + \mu_{\mathrm{Cl}^-}^\square - \mu_{\mathrm{HCl}}^0.$$

The quantities on the right-hand side of this equation are independent of composition. Hence

$$\frac{p_{\mathrm{HCl}}}{\gamma_\pm^2 \, m_{\mathrm{H}^+} m_{\mathrm{Cl}^-}} = K,$$ (10·84)

where K depends only on the temperature and pressure. Since hydrochloric acid is a strong electrolyte we can put

$$m_{\mathrm{H}^+} = m_{\mathrm{Cl}^-} = m,$$

† For details of the integration procedure see, for example, Glasstone, *Thermodynamics for Chemists*, Chapter XVI.

where m is the stoichiometric molality of the solution (alternatively $\gamma_{\pm}$ may be regarded as a stoichiometric activity coefficient). Hence finally

$$\frac{p_{HCl}}{(\gamma_{\pm} m)^2} = K, \qquad (10\cdot85)$$

and K is the limiting value of p_{HCl}/m^2 as the molality approaches zero.

In principle, equation (10·85) can be used for the calculation of $\gamma_{\pm}$ as soon as this limit has been determined. The difficulty is to determine K with any certainty because the partial pressures become almost immeasurably small in a region of concentration where $\gamma_{\pm}$ is still appreciably different from unity.† On the other hand, if values of $\gamma_{\pm}$ for dilute solutions are taken from some independent source (e.g. e.m.f. measurements at the same temperature), the equation can be used for the calculation of $\gamma_{\pm}$ in more concentrated solutions, using the measured partial pressures.

10·18. Limiting behaviour at high dilution

Consider a solution of a single strong electrolyte whose molality is m. From equations (10·53) and (10·54) it is evident that the Debye limiting law can be written in the form

$$\ln \gamma_{\pm} = -Am^{\frac{1}{2}},$$

where A is a constant and is positive. Hence differentiating with respect to m

$$\frac{\partial \ln \gamma_{\pm}}{\partial m} = -\frac{A}{2m^{\frac{1}{2}}}. \qquad (10\cdot86)$$

As m approaches zero the right-hand side of this equation approaches minus infinity. Therefore if the logarithm of the activity coefficient is plotted against the molality (rather than against its root) the gradient becomes infinitely steep as the molality approaches zero. In this respect electrolytes differ sharply in their behaviour from non-electrolytes, and this is due to the long-range forces between the ions.

Although the slope becomes infinite, the activity coefficient itself approaches unity. For this reason the result of equation (10·85) is in no way contrary to the fact that the behaviour of the *solvent* approaches Raoult's law as a limit. This may be shown in more detail by using the Gibbs–Duhem equation as developed in § 10·16.

In place of equation (10·82), the change in the chemical potential of the solvent may be expressed in terms of the change of its partial pressure p_0. Thus

$$d\mu_0 = RT \, d\ln p_0,$$

---

† As shown in § 7·4 b the experimental results, as far as they go, indicate that p_{HCl} plotted against the square of the molality (or the square of the mole fraction) probably has a finite slope at the origin, as is to be expected from equation (10·85).

and in place of equation (10·83) we obtain

$$-d\ln\gamma_\pm = d\ln m + \frac{1}{M_0 \nu m} d\ln p_0, \qquad (10·87)$$

or in view of the fact that the variations in question are due to the variation in the molality m of the electrolyte,

$$-\frac{\partial \ln\gamma_\pm}{\partial m} = \frac{d\ln m}{dm} + \frac{1}{M_0 \nu m}\frac{\partial \ln p_0}{\partial m}.$$

In the extremely dilute solution where the Debye limiting law may be applied, the term on the left-hand side of this relation is given by equation (10·86). Making this substitution we obtain

$$\frac{A}{2m^{\frac{1}{2}}} = \frac{1}{m} + \frac{1}{M_0 \nu m}\frac{\partial \ln p_0}{\partial m},$$

or rearranging
$$\frac{\partial \ln p_0}{\partial m} = M_0 \nu \left(\frac{Am^{\frac{1}{2}}}{2} - 1\right). \qquad (10·88)$$

Therefore the limiting gradient of $\ln p_0$ plotted against m is given by

$$\lim_{m\to 0} \frac{\partial \ln p_0}{\partial m} = -M_0 \nu, \qquad (10·89)$$

and is seen to depend on the number of ions ν which are formed on dissociation of the electrolyte.

Consider the mole fraction x_0 of the solvent. In the solution whose molality is m the amounts of all the ions together is νm and the amount of the solvent is $1/M_0$. Hence

$$x_0 = \frac{1/M_0}{1/M_0 + \nu m}.$$

Rearranging and differentiating we obtain

$$\nu\,dm = -\frac{1}{M_0}\frac{dx_0}{x_0^2},$$

and thus
$$\frac{dx_0}{dm} \to -M_0\nu \quad \text{as} \quad x_0 \to 1 \quad \text{(i.e. } m \to 0\text{)}. \qquad (10·90)$$

Now (10·89) may be written

$$\lim_{m\to 0} \frac{\partial \ln p_0}{\partial x_0}\frac{dx_0}{dm} = -M_0\nu,$$

and combining this with (10·90) we obtain

$$\lim_{m\to 0} \frac{\partial \ln p_0}{\partial x_0} = 1,$$

in agreement with Raoult's law.

PROBLEMS

1. At 25 °C the solubility of chlorine in water under a partial pressure at 1 atm is $0.061\ 8$ mol kg^{-1}. Calculate the standard-free energy of formation of an aqueous solution of chlorine and also the free energy of formation of a solution which is 0.01 mol kg^{-1}. Assume the solution to behave ideally.

2. Using the following data, estimate the number of moles of ethyl acetate which may be obtained at 25 °C by mixing 1 mole each of acetic acid and ethyl alcohol and allowing the mixture to come to equilibrium. Estimate the corresponding yield which would be obtained at 200 °C under a pressure sufficient for the system to remain liquid. What approximations are involved in these calculations?

	$\Delta_f G^0_{298}$	$\Delta_f H_{298}$
EtOH (l)	$-40\ 200$	$-\ 64\ 710$
AcH (l)	$-94\ 500$	$-117\ 200$
EtAc (l)	$-77\ 600$	$-111\ 800$
H$_2$O (l)	$-56\ 700$	$-\ 68\,320$

3. The standard free energy of formation of a gaseous substance A at 25 °C is $-25\ 000$ cal mol^{-1}. At 25 °C its vapour pressure as a liquid is 76 mm Hg, and at this pressure its solubility in water is 0.001 mol kg^{-1}. Calculate the standard free energy of its formation in aqueous solution according to the following conventions:

    (a) mole fraction scale and $\gamma_A \to 1$ as $x_A \to 1$;
    (b) mole fraction scale and $\gamma_A \to 1$ as $x_A \to 0$;
    (c) molality scale and $\gamma_A \to 1$ as $m_A \to 0$.

Sketch the probable form of the partial pressure curves of A and water and decide which of the above conventions it would be most natural to adopt.

4. At 25 °C the mean ion activity coefficient of a 4 mol kg^{-1} solution of hydrochloric acid in water is 1.762, and its partial pressure of hydrogen chloride is 0.2395×10^{-4} atm.

The standard free energy of formation of gaseous hydrogen chloride is $-22\ 770$ cal mol^{-1} and the standard potential of the calomel electrode is $0.268\ 0$ V.

Calculate the standard free energy of formation of solid mercurous chloride.

5. "The activity coefficient of a component of a solution is a numerical factor by use of which all the deviations of the solution from ideality can be correlated with each other."

The following table relates to aqueous solutions of HCl at 25 °C; the activity coefficients were obtained by measurement of e.m.f. and the figures for the partial pressures of HCl over the solutions were also obtained experimentally.

(*a*) Show that these results are consistent with each other, in agreement with the above statement.

(*b*) Calculate the partial pressures which would be expected for these solutions if they behaved in accordance with Raoult's law.

The vapour pressure of liquid HCl is 41.6 atm at 20 °C and 64.5 atm at 40 °C.

Molality	4	6	8
Activity coefficient, $\gamma_\pm$	1.762	3.206	5.907
p_{HCl} (atm) $\times 10^4$	0.239 5	1.842	11.10

6. The electrochemical cell

$$(Pt)\ H_2\ (1\ atm);\quad HCl\ (aq.),\ AgCl\ (s);\quad Ag$$

has a standard e.m.f. of 0.222 4 V at 25 °C (Pt negative), and an e.m.f. of 0.152 1 V when the concentration of HCl is 3 mol kg^{-1}. The standard free energies of formation of HCl (*g*, 1 atm) and AgCl (*s*) are respectively $-22\ 740$ and $-26\ 220$ cal mol^{-1} at 25 °C. Estimate the partial vapour pressure of HCl above a 3 mol kg^{-1} solution at 25 °C, and the mean ionic activity of HCl in a solution saturated with HCl at a partial pressure of 0.1 atm. [C.U.C.E. Tripos, 1952]

7. Chlorine at a partial pressure of 0.5 atm is bubbled through water at 25 °C. Using the data given below estimate the molalities in the solution at equilibrium of (*a*) dissolved chlorine, (*b*) hypochlorous acid, (*c*) chloride ions and (*d*) hypochlorite ions. Indicate clearly any assumptions which must be made in using the data below for this purpose.

The standard free energies of formation on a molality scale, in cal mol^{-1} at 25 °C, of the various dissolved species from the gaseous elements at 1 atm pressure are as under:

Cl_2	$+\ 1\ 650$
$H^+ + Cl^-$	$-31\ 370$
$HClO$	$-19\ 020$
$H^+ + ClO^-$	$-\ 6\ 500$
$H_2O\ (l)$	$-56\ 560$

[C.U.C.E. Tripos, 1950]

8. A waste liquor consists of a 0.5-mol kg^{-1} solution of $CaCl_2$ in water, and is at atmospheric temperature and pressure. It is proposed to separate the liquor into anhydrous $CaCl_2$ and pure water. Estimate the minimum work per mol $CaCl_2$ required for the process and the corresponding quantity of heat absorbed.

The free energy and enthalpy of formation of $CaCl_2$ are given below. The free energy figures for the ions refer to the standard state on the

molality scale. The mean ionic activity coefficient of 0.5-mol kg^{-1}
CaCl$_2$ is 0.448:

	$\Delta_f G^{\ominus}_{298}$ kcal mol^{-1}	$\Delta_f H_{298}$ kcal mol^{-1}
CaCl$_2$ (s)	-179.3	-190.0
Ca^{2+} (aq.)	-132.18	—
Cl^{-} (aq.)	-31.35	—
CaCl$_2$ (0.5 mol kg^{-1})	—	-209.1

[C.U.C.E. Tripos, 1953]

9. Nitrogen containing a trace of CO$_2$ is bubbled steadily, at 1 atm
pressure, through a suspension of lime in water at 25 °C. Use the data
below to calculate the percentage of CO$_2$ in the nitrogen leaving the
system on the assumption that equilibrium is reached. State clearly any
other necessary assumptions.

Discuss briefly the application of the phase rule to the system.

Solubility of calcium hydroxide	0.0211 mol dm^{-3}
Dissociation constant of water	10^{-14}
Solubility product of CaCO$_3$	0.87×10^{-8}
Standard free energies of formation	$\Delta_f G_{298}$ in cal mol^{-1}
CO$_2$ (g)	$-94\ 260$
H$_2$O (l)	$-56\ 700$
CO$_3^{2-}$ (ideal molal)	$-126\ 400$

[C.U.C.E. Tripos, 1951]

10. A continuous process for the manufacture of potassium chlorate
is based on the electrolysis of an acidified potassium chloride solution.
The feed liquor consists of a fresh 4.5 mol kg^{-1} solution of potassium
chloride at 25 °C, together with the recirculated liquor described below.

The hot liquor which leaves the cell contains a high concentration of
potassium chlorate and on cooling to 25 °C the bulk of it crystallizes out.
Solid potassium chloride is added to the mother liquor; this is then
recirculated to the cell, together with the 4.5 mol kg^{-1} feed solution.

Using the following data calculate the minimum amount of energy,
and the corresponding e.m.f., which is necessary for the production of
one mole of potassium chlorate from the given raw materials by the cell
reaction KCl + 3H$_2$O = KClO$_3$ + 3H$_2$. Mention causes of inefficiency in
the process as it is actually carried out.

	$\Delta_f G_{298}$ cal mol^{-1}
KCl(aq.)	$-98\ 820$
KCl(s)	$-97\ 590$
KClO$_3$(s)	$-69\ 290$
H$_2$O(l)	$-56\ 690$

Activity coefficient of KCl in 4.5 m solution $= 0.583$.
Faraday equivalent $= 23\ 050$ cal V^{-1} mol^{-1}.

[C.U.C.E. Qualifying, 1955]

11. It is proposed to produce benzaldehyde by the reaction of carbon monoxide with benzene in presence of a catalyst. The process is to be carried out at 50 °C and 500 atm.

Describe the calculations by which you would estimate an upper limit to the fraction of the benzene which could be converted to benzaldehyde if the available data were as given below. State clearly any assumptions which are necessary for this purpose.

(a) Heat capacities and standard free energies and enthalpies of formation of the compounds in question at 25 °C.

(b) The densities of benzene and benzaldehyde at 25 °C.

(c) $p-V$ data for carbon monoxide at 20 °C and 30 °C in the pressure range 0–500 atm.

[C.U.C.E. Tripos, 1955]

PART III

THERMODYNAMICS IN RELATION
TO THE EXISTENCE OF MOLECULES

CHAPTER 11

STATISTICAL ANALOGUES OF ENTROPY AND FREE ENERGY

11·1. Thermodynamics and molecular reality

The whole of classical thermodynamics is based on the empirical laws which lead to the notions of temperature, internal energy and entropy as functions of state. At no point in the development of these laws, or of their consequences, does any reference need to be made to the idea that matter consists of ultimate particles, i.e. the atomic theory. But thermodynamics gives an insight neither into the possible origin of its laws nor any means of making a direct calculation of the thermodynamic properties of a substance. It is the purpose of statistical mechanics to advance knowledge in both of these directions.

First of all it is shown that the laws of thermodynamics are a consequence of the postulates of the quantum theory, together with one other postulate which is of a statistical character. Secondly statistical mechanics makes it possible to obtain important new theorems not known in pure thermodynamics, including methods of calculating heat capacities, free energies, etc., from spectroscopic data.

Statistical mechanics is a large subject, and it is not possible to put forward an entirely rigorous treatment in a short compass. In the present chapter the main ideas will be developed in a fairly simple manner, and the formulae obtained will be applied in subsequent chapters to perfect gases, crystals and reaction kinetics.

11·2. The quantum states of macroscopic systems

The whole of this chapter is concerned with the quantum states of *macroscopic systems*. This is not a very familiar idea because the chemist or engineer becomes accustomed to thinking of the quantum states of *single molecules*, for which approximate solutions of Schrödinger's equation may be obtained. However, these solutions have significance only if the molecules are independent and this is not the case in liquids or solids. The form of statistical mechanics to be aimed at is one which will apply to liquids and solids, as well as to gases.

In a diatomic molecule, as is well known, it is not possible to ascribe sets of quantized energy levels to each of the two atoms; the potential energy of interaction is *mutual* and, therefore, a separate energy cannot be attributed to each atom. Similar considerations must apply, in any precise sense, to the molecules within a sample of liquid or solid. Here again there is a strong potential energy of interaction among the component molecules, and the system is a single quantum-mechanical whole. We cannot speak, except approximately, of the quantum states of the individual molecules; we can speak only of the quantum states of the whole macroscopic system.

To be more definite, let it be supposed that ϵ_0, ϵ_1, ϵ_2, etc., are the energy levels of a hydrogen molecule, as obtained by solving the Schrödinger equation for a *single* molecule contained in a vessel of a given volume. If there were actually N molecules in this volume (where N is not so large that the molecules are close together), then the energy levels of the total system could be taken quite correctly as being superpositions of the energy levels of the single molecules, i.e. as being $N\epsilon_0$, $(N-1)\epsilon_0+\epsilon_1$, etc. But it would *not* be correct to attribute this set of levels to a sample of hydrogen at high density, and still less to liquid or solid hydrogen. The calculated levels ϵ_0, ϵ_1, etc., refer specifically to a single molecule in the absence of any interaction with other bodies. In fact, it is only in the case of the perfect gas that it is really justifiable to think of each molecule as having its 'private' energy, to use Schrödinger's phrase.

For a mass of solid or liquid, containing a total of, say, 10^{25} electrons and nuclei, the possible quantum states could be obtained, in principle, by solving a Schrödinger differential equation containing 3×10^{25} terms, one for each of the necessary co-ordinates. However, for the present purposes it is not necessary to suppose that this equation has been solved; all that we need for the statistical theory is the idea that these quantum states exist.

11·3. Quantum states, energy states and thermodynamic states

It is evident that the thermodynamic specification of a system, for example the statement that it has fixed values of its energy, volume and composition, is a very incomplete specification from a molecular point of view. In the case of a crystal, for example, a great many distinguishable arrangements of different atoms on the lattice may all correspond to the same thermodynamic state, as discussed already in § 1·17.

According to wave mechanics the most complete statement which is possible about a system is a statement of its wave function, the quantity ψ which appears in Schrödinger's equation. A statement of ψ, as a function of the co-ordinates of the elementary particles, is a specification of the quantum state of the system. A great many of these quantum states, indeed, astronomically large numbers in the case of a macroscopic system, may all be compatible with the same total energy, volume and composition. *In brief, a great many quantum states are always comprised within a given thermodynamic state.*

For the purposes of illustration consider a system of four indistinguishable atoms which move about independently in a large container. For simplicity it will be supposed that they are free to move only along the x co-ordinate. Let ϵ_0, ϵ_1, etc., be the quantized translational energies of the separate atoms, counting upwards from the lowest level ϵ_0. It will be supposed that these levels are equally spaced; thus

$$(\epsilon_1 - \epsilon_0) = (\epsilon_2 - \epsilon_1) = (\epsilon_3 - \epsilon_2) = \ldots.$$

Provided that the total energy of the four atoms is at least equal to $2\epsilon_1 + 2\epsilon_0$, there is more than one quantum state of *the total system* which is compatible with the given energy. We could have either two atoms with energy ϵ_0 and the other two with energy ϵ_1, or we could have three with energy ϵ_0 and the fourth with energy ϵ_2. These two possibilities correspond to two different quantum states of the total system, but to the same thermodynamic state. It is easy to see that any increase in the total energy of the system will give rise to a very rapid increase in the number of quantum states which are compatible with this energy.

The number of quantum states of a closed system which are compatible with fixed values of its energy and volume will be denoted Ω. It has been noted that Ω increases rapidly with the total energy, for fixed volume. Even if the total energy and volume of the system are constant, Ω may still be able to change in a spontaneous process, such as a reaction, due, for example, to a change in the nature and spacing of the energy levels as reaction proceeds. These points will become clearer in later sections.

11·4. Fluctuations

What is called equilibrium is the state of a closed system in which its various macroscopic properties do not change with time. However, as soon as we wish to consider the interpretation of these properties at the molecular level, it must be borne in mind that the equilibrium state is of a dynamic rather than a static character, as a result of the motion and collisions of the molecules. Macroscopic properties such as pressure, temperature and entropy have a meaning therefore only as a result of averaging over a great many molecules. For example, the pressure exerted on a manometer is the mean change in the momentum of the molecules per unit time and surface area. The number of molecules striking the surface changes from moment to moment and the instantaneous pressure is therefore a fluctuating quantity. Of course on any large surface the probability of appreciable deviation from the mean is quite trivial, but on a body of small area, such as a colloidal particle, the fluctuations become observable, as shown in the Brownian movement.

Similar considerations apply to the other macroscopic properties. For example, the density of a fluid is not absolutely constant throughout its bulk, but undergoes slight variations about a mean. Under the conditions of very low pressure prevailing in the upper atmosphere these fluctuations are sufficiently appreciable to give rise to the scattering of the sun's light which is the cause of the blue colour of the sky. From measurements of the intensity of the blue it is possible to calculate the Avogadro constant. Another type of fluctuation which is important in the present chapter is that of energy. If a body is in contact with a heat bath its energy is not absolutely constant but fluctuates about a mean value, due to the passage of energy to and fro because of the molecular collisons at the interface.

11·5. Averaging and the statistical postulate

From what has been said it is clear that the calculation of the macroscopic properties of a body involves a process of averaging. Let it be supposed that we could determine the precise position and velocity of every molecule in the body. Then it would appear that we could, at least in principle, compute the future state of the system at any moment, and also its average properties over a long period, by using only the laws of mechanics. But this completely detailed knowledge of the instantaneous state of a system is clearly impossible in practice and also, according to the uncertainty principle, it is impossible even in theory. Any measurement of the position of a molecule disturbs the velocity and vice versa.

A knowledge of the state of a system in complete detail is therefore unattainable. This is the origin of the second law, as discussed already in §1·18, and it is for the same reason that it is impossible to calculate the average properties of a system by seeking to apply the laws of mechanics to the individual molecules. In order to calculate this average *it is necessary to use an extra postulate*, over and above the postulates of mechanics.† This postulate is of a statistical character, and it asserts that if there are Ω quantum states of an isolated system, all of them compatible with the fixed value of the energy, then the system is as likely to be found in any one of the states as in any other.

† For further discussion on this point see Tolman, *Principles of Statistical Mechanics* (Oxford, 1938), §25, and Born, *Natural Philosophy of Cause and Chance* (Oxford, 1949).

Thus each of these quantum states is to be given an equal statistical weight when we proceed to take the average over all these states. This postulate will be expressed in a more precise form at the end of the next section and again in §11·9.

11·6. Accessibility

It may be remarked that thermodynamics and statistical mechanics always deal with idealized conditions. For example, in a discussion on the properties of hydrogen it is tacitly assumed that this is a stable substance, whereas it may actually be changing into helium over immense periods of time. The same considerations apply to a mixture of hydrogen and oxygen at room temperature. In the absence of a catalyst, we are again dealing with a system which is not stable from the aspect of geological ages, but *is* quite stable relative to the duration of our measurements on the system. If formation of water took place it would greatly increase the number, Ω, of quantum states of the system, and it is entirely consistent with what was said in §1·17 to regard the 'driving force' of the reaction, under adiabatic conditions, as being the increase in Ω. However, in the absence of a catalyst, this increase does not take place at an appreciable speed. In other words, those additional quantum states which would occur if the catalyst were present *are not accessible* to the mixture of hydrogen and oxygen alone.

Thus, when we compute the average properties of the hydrogen and oxygen mixture, we include neither the quantum states which could arise from the formation of helium nor those which could arise from the formation of water. Certain quantum states are regarded as inaccessible, on the basis of our empirical knowledge of the system. In statistical mechanics we take averages only over those quantum states which are believed to be accessible, but each of these is regarded as being equally accessible (which is to say as having equal weight) when the system in question is of constant energy.

In the averaging process we thus idealize the system which is under discussion by making a sharp distinction between two classes of quantum states: (a) those which are taken as being totally inaccessible during the time taken to measure the property of interest; (b) those which are so readily accessible that a representative sample of them are passed through during this period. We average only over the latter and our postulate asserts that they are each reached with equal frequency. The idealization thus amounts to assuming very rapid rates for certain types of molecular process and absolutely zero rates for all other types.†

† For further discussion on accessibility see Fowler and Guggenheim, *Statistical Mechanics* (Cambridge, 1949), Chapter I and Mayer and Mayer, *Statistical Mechanics*, §3f (New York, Wiley, 1940).

The basic postulate may now be expressed as follows: *each of the accessible and distinguishable quantum states of a system of fixed energy†* *is equally probable.* This number of quantum states, or 'complexions', is denoted Ω and will be shown to be a function of the energy, volume and composition of the system. The postulate will be given in a slightly modified and more concrete form in § 11·9.

It may be remarked that the assumption that each quantum state has an *equal a priori probability* is less arbitrary than any alternative assumption, in the absence of any information to the contrary.‡

11·7. The equilibrium state

The statistical interpretation of the trend towards equilibrium has been discussed in §§ 1·17 and 1·18, and the significant points will be briefly summarized. Consider any closed macroscopic system of chosen energy and volume. In general, the statement of the values of these variables is sufficient to fix neither the thermodynamic state of the system nor the value of Ω. Thus we need to know in addition whether the volume is divided internally by means of impermeable or non-conducting partitions, and also what catalysts are present. In brief, we need to know what quantum states are accessible to the system.

The removal of a partition or the introduction of a catalyst is the lifting of a restraint. They may result in either of the two possibilities: either the system stays as it was, with the same value of Ω, or it undergoes an internal change, such as the flow of energy or of matter, or a chemical reaction. If a change takes place our detailed knowledge of the positions, etc., of the particles diminishes. This is

† On account of the uncertainty principle the energy itself cannot be known with complete precision. Thus Ω_i is really the number of quantum states of a system whose energy lies within a small range near the eigen energy E_i. Throughout the chapter, whenever we speak of an energy E_i it is to be understood in the same sense.

‡ In this connexion the following remarks from David Hume are very much to the point: 'as chance is nothing real in itself, and, properly speaking, is merely the negation of a cause, its influence on the mind is contrary to that of causation; and it is essential to it to leave the imagination perfectly indifferent, either to consider the existence or non-existence of the object (event) which is regarded as contingent.... Since, therefore, an entire indifference is essential to chance, no one chance can possibly be superior to another, otherwise than as it is composed of a superior number of equal chances. For if we affirm that one chance can, after any other manner, be superior to another, we must at the same time affirm, that there is something which gives it the superiority, and determines the event rather to that side than the other: that is, in other words, we must allow of a cause, and destroy the supposition of chance, which we had before established ' (*A Treatise of Human Nature* (1740), Book I, Part III, §XI).

equivalent to an increase in Ω, the number of quantum states which the system can take up, each of which is compatible with the given values of the energy and volume. An irreversible process, in a system of constant energy and volume, thus corresponds to an increase in the number of accessible quantum states. The equilibrium state of such a system, i.e. the state in which there is no further change, is therefore the state in which Ω has reached its maximum value.

It may be remarked, once again, that the need for the basic postulate arises because of our inability to predict which one of the Ω states the system will actually be in, at any given moment. Failing the possibility of prediction the next best thing is to assume that the system is as likely to be in any one of the quantum states as any other.

So much for the effect of lifting a restraint. In the converse case, where a partition is inserted, or a catalyst is removed, the value of Ω, or at any rate the value of $\ln \Omega$, is not significantly affected. Intuitively this is fairly plausible, and we shall not seek to prove it in detail;† if a gas is uniformly distributed throughout a vessel, the replacement of a partition between the two halves of the vessel does not cause any significant decrease in the logarithm of the number of accessible quantum states.

11·8. Statistical methods

In the setting up of the statistical theory there are several alternative procedures, and the newcomer to the subject often finds it difficult to see the connexion between them. The essential problem, of course, is that of averaging and the various methods differ principally in the following respects:

(1) Whether the averaging is carried out over
    (a) the quantum states of molecules, or
    (b) the quantum states of macroscopic systems.

(2) Whether the averages which are obtained are
    (a) true mean values, or
    (b) 'most probable values' based on picking out the largest term in a summation, together with the use of Stirling's theorem.

(3) Whether the system under discussion is regarded
    (a) as having fixed values of its energy, volume and composition (isolated system);
    (b) as having fixed values of its volume and composition and secondly as being in thermal equilibrium with a heat bath;

† See, for example, Mayer and Mayer, *Statistical Mechanics*, §§ 3f, 4b and 4e (New York, Wiley, 1940).

(c) as having a fixed value of its volume and secondly as being in equilibrium both with a heat bath and with reservoirs of the various substances which it contains, through semipermeable membranes (open system).

Most introductory accounts use the combination (1a), (2b) and (3a), which is one of the easiest to grasp. However, the disadvantage of averaging over the quantum states of *molecules*, as in (1a), is that the statistical formulae which are obtained are applicable only to systems in which the particles are independent, as in perfect gases. In many elementary accounts of this method it is also necessary to adopt a subterfuge in order to introduce an important term, $n!$, into the formulae. The method (2b) may also lead to an erroneous impression of the nature of equilibrium, for it might be taken to imply that the system stays always in the particular distribution known as the 'most probable' distribution.

For these reasons we shall put forward an elementary treatment based on the combination (1b), (2a) and either (3a) or (3b) according to circumstances.† It may be remarked that (3a), (3b) and (3c) signify three different choices of the independent variables in the corresponding thermodynamic analysis of the problem—these are (U, V, n_i), (T, V, n_i) and (T, V, μ_i) respectively.

11·9. The ensemble and the averaging process

Let it be supposed that we wish to calculate the value of some property χ of a macroscopic system which is in equilibrium. Now the experimental measurement of this property always requires a certain interval of time for its execution, and if the calculated value of χ is to be relevant the latter must clearly be some sort of average over all the quantum states through which the system is likely to pass during the period of measurement.

During a virtually infinite interval of time, Δt, the system may be expected to pass through each of its accessible quantum states a great many times. Let P_0, P_1, P_2, etc., be the fractions of this interval which the system spends in the quantum states 0, 1, 2, etc. P_i may be called the *probability* of the system being in the quantum state i. Let χ_0, χ_1, χ_2, etc., be the values of the property χ which occur when

† The methods used in this chapter are similar to those adopted by Tolman, *Principles of Statistical Mechanics* (Oxford, 1938) and also, at a more elementary level, by Slater in his *Introduction to Chemical Physics* (New York, McGraw Hill, 1939). Somewhat similar procedures are introduced in later chapters of the following books: Mayer and Mayer, *Statistical Mechanics*, Chapter x (New York, Wiley, 1940); Rushbrooke, *Introduction to Statistical Mechanics* (Oxford, 1949), Chapter xv; Gurney, *Introduction to Statistical Mechanics* (New York, McGraw-Hill, 1949) Chapter viii.

the.system is in the quantum states 0, 1, 2, etc. (We shall consider only those properties χ to which a meaning can be attached in each quantum state). Then the time average of χ over the interval Δt is

$$\bar{\chi} = \Sigma P_i \chi_i, \tag{11·1}$$

where
$$\Sigma P_i = 1. \tag{11·2}$$

In each equation the summations are to be taken over all of the quantum states which are accessible under the prescribed macroscopic conditions of the system. The next question to be discussed is the values which must be attributed to the P_i when these prescribed conditions are either (a) constancy of energy and volume or (b) constancy of volume together with equilibrium with a heat bath.

Case (a). The system is of fixed energy and volume. As discussed in § 11·7, the fixing of the energy and the volume is usually sufficient to determine the thermodynamic state of a closed system, provided that it has reached a condition of internal equilibrium with respect to all chemical reactions, diffusions, etc., which are possible under the prescribed restraints. The equilibrium of such a system corresponds to a maximum value of Ω, the number of accessible quantum states. Now because the energy of the system is fixed, each of these possible quantum states must correspond to this particular energy and no other. Therefore, according to our basic postulate, each of these states is equally probable—the system may be expected to spend an equal fraction of the time Δt in each of them. Thus all of the P_i are the same and each is therefore equal to the reciprocal of the number of quantum states:

$$P_i = \frac{1}{\Omega} \quad (\text{fixed } U \text{ and } V), \tag{11·3}$$

and therefore from (11·1) $\quad \bar{\chi} = \frac{1}{\Omega} \Sigma \chi_i.$

This is the *time average* over the interval Δt, chosen as virtually infinitely long so that the system may pass through each of its accessible quantum states a great many times. It is by no means obvious that the calculated average $\bar{\chi}$ will necessarily be the same as would be measured during an experiment, since only a small fraction of the assessible quantum states will be passed through during the duration of the experiment. On the other hand, the very fact that the results of measurement do not vary from one moment to the next (when the system is in equilibrium) indicates that the duration of such experiments is usually sufficient to correspond to a true average value of the property. Indeed this will be so if it is supposed that the quantum states which *are* passed through are a random, and therefore a representative sample of the total.

This supposition is somewhat distinct from our previous postulate
of equal *a priori* probabilities (§11·6) but may be combined with
it to form a single postulate, in a more concrete form, as follows:
*the equilibrium properties of a system of constant energy and volume are
obtained by averaging over all the accessible quantum states, each of
these being given equal weight.*

The totality of quantum states is referred to as an *ensemble*. In
the special case under discussion where each of the states is of equal
energy the ensemble is said to be *microcanonical*, and it is only in
this special case that the probabilities P_i may all be assumed equal.

Case (b). The system has fixed volume and is in a heat bath.

The averaging procedure as discussed above for the isolated system
is of limited value for several reasons. In the first place there is difficulty in calculating the value of Ω for such systems. Secondly, we
are usually much more interested in the properties of a thermostated
system than of an isolated one. Finally, as shown in Chapter 1, the
second law is bound up with conditions relating to an energy transfer,
and this involves at least *two* bodies. The case discussed above is
thus of limited value as a statistical model precisely because the
system in question is of fixed energy.

We consider now the case of a closed system of constant volume
which is at internal equilibrium and immersed in a large heat bath.
Due to fluctuations between itself and the bath the body can have
a range of energies about a mean value. These are all quantized and
the energy states will be denoted E_0, E_1, etc., counting upwards from
the state of lowest energy. Each energy state E_i will comprise a very
large number Ω_i of quantum states. However, there is only a single
thermodynamic state because the body has a chosen volume and is in
thermal equilibrium with the given heat bath. (Thermodynamically
it would be said to have a fixed temperature, but one of our aims is
to define temperature statistically.)

The question which arises is: how shall we do the averaging for
such a system? In other words, what values must be attributed to
the P_i in equations (11·1) and (11·2)? For quantum states of *equal*
energy our postulate asserts that the correct averaging is obtained
by giving equal values to each of the P_i. We now have to consider what
is the relative frequency with which states of *unequal* energy are
occupied.

In answer to this question it seems reasonable to assume that the
probability of the occurrence of any one of the quantum states
depends only on the energy of that state. This statement is actually *not*
an additional assumption in the theory, for it is shown in Appendix 1
to this chapter that it may be derived from the basic postulate.

However, in order to avoid breaking up the continuity of the discussion by giving the proof at this stage, the statement will be used as if it were an additional assumption. In any case it is clearly a reasonable one.

We assume therefore that the probability P_i of the ith quantum state of the body, *when it is at equilibrium with the given heat bath*, is a function only of E_i:
$$P_i = f(E_i). \tag{11·4}$$

Consider a second body in the same large heat bath. Let the quantum and energy states for the first body be denoted by Roman letters, as in (11·4), and for the second body by Greek letters. For the second body the corresponding assumption is that the probability of its being in the quantum state κ is a function only of ϵ_κ:
$$P_\kappa = f(\epsilon_\kappa). \tag{11·5}$$

Let $P_{i\kappa}$ be the probability that the first body is in state i and the second is simultaneously in the state κ. The same assumption used once again asserts that the compound probability $P_{i\kappa}$ is a function only of the total energy,† $E_i + \epsilon_\kappa$, of the two bodies
$$P_{i\kappa} = f(E_i + \epsilon_\kappa). \tag{11·6}$$

Now because the heat bath is assumed in the first place to be very large compared to either of the two bodies, and therefore to be capable of a large supply of energy, the fraction of the time that the first body is in the state i is quite independent of the fraction of the time that the second body is in the state κ. Hence the fraction of the time for the combined event is equal to the product of the two fractions separately, i.e.
$$P_{i\kappa} = P_i P_\kappa. \tag{11·7}$$

(This is the multiplication rule for the probabilities of events which are independent.) Thus, using (11·4), (11·5) and (11·6)
$$f(E_i + \epsilon_\kappa) = f(E_i) f(\epsilon_\kappa). \tag{11·8}$$

We ask, what kind of function is it which satisfies this relation? A function of the sum of two variables must be equal to the product of the functions of these two variables taken separately (and, of course,

---

† As pointed out in §11·2, the energies of *molecules* cannot, in general, be combined additively because of their interaction. However, the energies of macroscopic systems can be so combined to a high degree of approximation. The larger the number of molecules that they contain, the smaller is the relative contribution of the interfacial energy, which arises from the molecular forces acting across the interface. The surface energy becomes a trivial part of the total energy of the bodies, because it varies as the square of the linear dimensions whilst the volume varies as the cube.

the physical situation requires that the functions are all of the same form). This is only satisfied if the functions are of the exponential form†

$$
\left.
\begin{aligned}
P_i &= f(E_i) = C_1\, e^{\beta E_i}, \\
P_\kappa &= f(\epsilon_\kappa) = C_2\, e^{\beta \epsilon_\kappa}, \\
P_{i\kappa} &= f(E_i + \epsilon_\kappa) = C_1 C_2\, e^{\beta(E_i + \epsilon_\kappa)},
\end{aligned}
\right\}
\tag{11·9}
$$

which are in accordance with (11·7). In these equations, C_1, C_2 and β are constants. C_1 is characteristic of the first body and C_2 of the second, but it is to be noted especially that the *constant β must be the same for both bodies*, as otherwise, the functional relationship would not be satisfied. It may be anticipated therefore that the *statistical* quantity β is related to the *thermodynamic* quantity called temperature. This will be discussed in a later section, but it may be remarked that β must clearly be a negative quantity, as otherwise, according to (11·9), P_i would become infinite for infinite values of the energy of the quantum state. (See also Appendix 1.)

We thus conclude that for a body in a large heat bath the probability of one of its quantum states is an exponential function of the energy of this state.

---

† A proof that (11·9) is the *only* possible functional form is as follows. Writing (11·8) as

$$
f(x)\,f(y) = f(x+y),
$$

differentiate partially with respect to y

$$
f(x)\,\frac{df(y)}{dy} = \frac{\partial f(x+y)}{\partial y} = \frac{\partial f(x+y)}{\partial(x+y)}\,\frac{\partial(x+y)}{\partial y} = \frac{df(x+y)}{d(x+y)} \times 1.
$$

Similarly, by differentiating partially with respect to x

$$
f(y)\,\frac{df(x)}{dx} = \frac{df(x+y)}{d(x+y)}.
$$

Between these two equations we obtain

$$
f(x)\,\frac{df(y)}{dy} = f(y)\,\frac{df(x)}{dx}
$$

or

$$
\frac{1}{f(y)}\,\frac{df(y)}{dy} = \frac{1}{f(x)}\,\frac{df(x)}{dx}.
$$

Now x and y are independent variables. Hence the left-hand side of this equation does not depend on x and the right-hand side does not depend on y. Thus each side must be equal to a constant, which is the *same* constant, β. Thus

$$
\frac{1}{f(x)}\,\frac{df(x)}{dx} = \beta,
$$

and the integral is $f(x) = A e^{\beta x}$.

Instead of writing the constants outside the exponentials as C_1 and C_2, it is more convenient to write them as $1/Q_1$ and $1/Q_2$. Thus

$$P_i = \frac{1}{Q_1} e^{\beta E_i}, \qquad (11\cdot10)$$

$$P_\kappa = \frac{1}{Q_2} e^{\beta \epsilon_\kappa}. \qquad (11\cdot11)$$

However, the second body, which has served its purpose, can now be forgotten. We proceed to consider the meaning of Q_1, cutting out the subscript. Summing over all quantum states we have, as in (11·2),

$$\Sigma P_i = 1,$$

and therefore from (11·10) $Q = \Sigma e^{\beta E_i}.$ (11·12)

Combining this equation with (11·10) we obtain finally

$$P_i = \frac{e^{\beta E_i}}{\Sigma e^{\beta E_i}}, \qquad (11\cdot13)$$

where the summation is taken over all of the accessible *quantum states*. This sum, which is Q, is known as the *partition function* of the system for constant temperature and volume.

The procedure for obtaining the average value of a property χ by use of equation (11·1) is therefore to use the values of the P_i which are given by (11·13).

The totality of the quantum states which are accessible to a body at equilibrium with a heat bath is called the *canonical ensemble*, and the equation (11·13) describes the canonical distribution of probabilities. (The equation implies, of course, that quantum states of equal energy are all equally probable, in accordance with our basic postulate.)

11·10. Statistical analogues of the entropy and Helmholtz free energy

Consider the quantity S' defined by the relation

$$S' \equiv -k\Sigma P_i \ln P_i, \qquad (11\cdot14)$$

where the summation is over all the accessible quantum states of the system and k is a positive constant whose value will be chosen in Chapter 12. S' will shortly be shown to have all the properties of the thermodynamic entropy and may be called a *statistical analogue* of the entropy.

The relationship of S' to another type of statistical analogue,

$$S'' \equiv k \ln \Omega, \qquad (11\cdot15)$$

which was tentatively introduced in § 1·17, will also be discussed, but for the present purposes S' is the more useful quantity.

In equation (11·14) the P_i's are all positive fractions and therefore S' is always positive. Suppose that the system could exist in only a single quantum state, say state 0. Then for this state we should have $P_0 = 1$ and all other P's would be zero. Such a distribution would correspond to no randomness, and (11·14) shows that it would give rise also to zero S'. On the other hand, a distribution of the ensemble over a large number of states, for each of which P_i is very small, will clearly give rise to a large positive value of S'. The similarity of (11·14) to the thermodynamic expressions for the entropy increase of mixing in perfect gases and ideal solutions may also be noted. These various points provide some preliminary justification for the idea that S' is related to entropy.†

Case (a). The microcanonical ensemble‡. The differentiation of (11·14) gives
$$dS' = -k\Sigma(\ln P_i + 1)\,dP_i. \tag{11·16}$$

Consider a virtual variation in the ensemble which involves the j and k states only, their probabilities changing by dP_j and dP_k respectively. Since $\Sigma P_i = 0$ it follows that $dP_j = -dP_k$ and the last equation becomes
$$dS' = -k(\ln P_j + 1)\,dP_j - k(\ln P_k + 1)\,dP_k$$
$$= -k(\ln P_j - \ln P_k)\,dP_j. \tag{11·17}$$

It follows that $\partial S'/\partial P_j$ *is zero* whenever the conditions of the system are such that P_j and P_k *are equal*. The same applies to a virtual change affecting any other pair of accessible quantum states.§

---

† In E. T. Jaynes' treatment of statistical mechanics on the basis of information theory, the starting point is that the quantity S', as defined above in (11.14), has the important property of describing the distribution of probabilities P_1 which is maximally non-committal with regard to missing information. The quantity S' thus measures the ' amount of uncertainty ' and it is to this that Jaynes gives the name entropy (*Phys. Rev.* **106** (1957), 620; **108** (1957), 171).

‡ Cases (a) and (b) of the present section correspond to (a) and (b) of the previous section.

§ The discussion of the quantum states *in pairs* is adopted for reasons of mathematical simplicity. Alternatively, Lagrange's method of undetermined multipliers may be used, or the following method shown to me by Professor N. R. Amundson. Equation (11·16) may be written
$$\frac{\partial S'}{\partial P_j} = -k \sum_i (\ln P_i + 1) \frac{\partial P_i}{\partial P_j} = -k \sum_i \ln P_i \frac{\partial P_i}{\partial P_j}.$$
The latter equality follows since $\sum_i \dfrac{\partial P_i}{\partial P_j} = 0$. For the same reason the quantity

Let it be supposed that the body is isolated so that all of these states are of equal energy. Then, according to the basic postulate, all of the P's become equal when the body reaches equilibrium. It follows from the above that the function S' simultaneously reaches a maximum† value and, of course, this is also a property of the thermo-dynamic function called entropy, under the same physical conditions.

The same postulate gives also

$$P_i = 1/\Omega$$

as in equation (11·3). Substituting this relation in (11·14) we obtain

$$S' = -k\Sigma(1/\Omega) \ln (1/\Omega),$$

and since there are altogether just Ω terms in the summation this reduces to

$$S' = k \ln \Omega.$$

Thus, in the case of the microcanonical ensemble, the quantities S' of equation (11·14) and S'' of equation (11·15) are identical. Of course, any change in the system which increases Ω, the number of accessible states, will cause an increase in S' or S''.

Case (b). The canonical ensemble. In the ensemble discussed above the quantum states of the system were all of equal energy. Therefore there was no question of difference of energy, and for this reason the concept of temperature did not enter. We turn now to the more useful canonical ensemble, i.e. the totality of the quantum states of a body in a thermostat.

Let the function A' be defined by

$$A' \equiv \overline{U} + S'/k\beta, \tag{11·18}$$

$k \ln P_j \sum_i \dfrac{\partial P_i}{\partial P_j}$ is zero, and therefore it may be added to the last equation. We thus obtain

$$\frac{\partial S'}{\partial P_j} = -k \sum_i \ln P_i \frac{\partial P_i}{\partial P_j} + k \ln P_j \sum_i \frac{\partial P_i}{\partial P_j}$$

$$= -k \sum_i \ln \frac{P_i}{P_j} \frac{\partial P_i}{\partial P_j}, \quad \text{for all } j.$$

This is *zero* if all P's are equal. A second differentiation of the last equation gives

$$\frac{\partial^2 S'}{\partial P_j^2} = -k \sum_i \frac{1}{P_i} \left(\frac{\partial P_i}{\partial P_j}\right)^2 - k \sum_i \ln \frac{P_i}{P_j} \frac{\partial^2 P_i}{\partial P_j^2},$$

and the last term is again zero when the P's are equal. Thus $\partial^2 S'/\partial P_j^2$ is negative, and S' is a *maximum*, since k, P_i and $(\partial P_i/\partial P_j)^2$ are all positive.

† A second differentiation of (11·17) gives

$$\frac{d^2 S'}{dP_j^2} = -k\left(\frac{1}{P_j} - \frac{d \ln P_k}{dP_j}\right) = -k\left(\frac{1}{P_j} + \frac{1}{P_k}\right),$$

since $dP_j = -dP_k$. Thus, because k is positive, S' is a maximum and not a minimum when $P_i = P_k$.

where $\bar{U}$ is the mean internal energy of the system and β is the quantity which appears in equation (11·13). Now, in accordance with equation (11·1),

$$\bar{U} = \Sigma P_i E_i. \tag{11·19}$$

Substituting (11·14) and (11·19) into (11·18) we obtain

$$A' = \frac{1}{\beta} \Sigma P_i (\beta E_i - \ln P_i), \tag{11·20}$$

and it is to be understood that the summation is over the *accessible* quantum states, and for these the P_i are greater than zero. It will now be shown that A' is a minimum if there is a state of equilibrium in the system, and therefore A' may be expected to be an analogue of the Helmholtz free energy.

Consider the effect on A' of a virtual displacement of the P's about their equilibrium values, i.e. small changes in the fraction of the time that the system resides in the various quantum states whose energies E_i, E_j, etc., remain unchanged.† In this process β remains constant because the system is at equilibrium with the heat bath. For the variation in A' we therefore obtain from (11·20)

$$\mathrm{d}A' = \frac{1}{\beta} \Sigma (\beta E_i - \ln P_i - 1) \,\mathrm{d}P_i, \tag{11·21}$$

but this is subject to the condition‡

$$\Sigma P_i = 1 \quad \text{or} \quad \Sigma \mathrm{d}P_i = 0. \tag{11·22}$$

Assume for simplicity that the variations in question affect only two quantum states i and j. Then (11·21) and (11·22) may be written

$$\mathrm{d}A' = \frac{1}{\beta} \{ (\beta E_i - \ln P_i - 1) \,\mathrm{d}P_i + (\beta E_j - \ln P_j - 1) \,\mathrm{d}P_j \},$$

$$\mathrm{d}P_i + \mathrm{d}P_j = 0.$$

Eliminating $\mathrm{d}P_j$ from these two equations we obtain

$$\frac{\partial A'}{\partial P_i} = \frac{1}{\beta} (\beta E_i - \ln P_i - \beta E_j + \ln P_j). \tag{11·23}$$

---

† We are concerned here with an ensemble of states, all of which are of the same volume, and it is shown in wave mechanics that the values of the E_i depend only on the volume in which a fixed number of fundamental particles are contained. (In certain problems other geometrical parameters may become important. For example in problems of interfacial tension, the area of the system is significant in its effect on the E_i of the surface phase. In the above it is assumed that all such parameters are held constant.)

‡ There is no additional restriction $\mathrm{d}\bar{U} = \Sigma E_i \mathrm{d}P_i = 0$ because the system can exchange energy with the heat bath.

Now according to equation (11·13), if there is a state of equilibrium, the P's satisfy the relation

$$P_i = e^{\beta E_i}/Q, \atop P_j = e^{\beta E_j}/Q,} \tag{11·24}$$

and Q is, of course, the same in both equations since we are dealing with the same body. Substituting these relations in (11·23) we obtain

$$\frac{\partial A'}{\partial P_i} = 0. \tag{11·25}$$

The same result may be obtained for a variation between any other pair of quantum states. The canonical distribution (11·24), which we have derived on strictly statistical grounds with the help of our basic postulate, therefore results in the function A' being at a minimum† when a body of fixed volume is at equilibrium in a thermostat. The thermodynamic function which has this property is the Helmholtz free energy A. It seems probable, therefore, that A', as previously defined, is a statistical analogue of the free energy of thermodynamics. Further grounds for this will be discussed in the next section.

Substituting (11·24) in (11·20) we obtain the minimum or equilibrium value of A' for our body as

$$A'_{eq.} = \frac{1}{\beta} \Sigma P_i (\beta E_i - \beta E_i + \ln Q)$$

$$= \frac{\ln Q}{\beta} \Sigma P_i$$

$$= \frac{\ln Q}{\beta}. \tag{11·26}$$

The mean value at equilibrium of any property χ is obtained from (11·1) and (11·24):

$$\bar{\chi}_{eq.} = \frac{1}{Q} \Sigma \chi_i e^{\beta E_i}. \tag{11·27}$$

In particular, the mean internal energy is

$$\bar{U}_{eq.} = \frac{1}{Q} \Sigma E_i e^{\beta E_i}, \tag{11·28}$$

and this equation may also be written in the form

$$\bar{U}_{eq.} = \left(\frac{\partial \ln Q}{\partial \beta}\right)_{V, comp.}, \tag{11·29}$$

† A second differentiation of (11·23) gives

$$\frac{\partial^2 A'}{\partial P_i^2} = -\frac{1}{\beta}\left(\frac{1}{P_i} + \frac{1}{P_j}\right).$$

As shown in the lines following equation (11·10) β must be a negative quantity. Therefore A' is a minimum at equilibrium and not a maximum.

and may be confirmed by differentiation of the defining equation

$$Q \equiv \Sigma \, e^{\beta E_i}. \tag{11·30}$$

In (11·29) the subscripts denote constant volume and composition, as otherwise the E_i are not constant.

Finally, the value of S' at equilibrium is obtained from (11·18), (11·26) and (11·29):

$$S'_{eq.} = k\beta(A'_{eq.} - \overline{U}_{eq.})$$

$$= k \ln Q - k\beta \left(\frac{\partial \ln Q}{\partial \beta} \right)_{V, \, comp.} \tag{11·31}$$

11·11. Comparison of statistical analogues with thermodynamic functions

In the last section we defined S' and A' by means of the relations

$$S' = -k\Sigma P_i \ln P_i, \tag{11·32}$$

$$A' = \overline{U} + \frac{S'}{k\beta}, \tag{11·33}$$

where k is a positive constant. Both S' and A' are clearly functions of state, being determined by the P_i and the E_i. Also it has been shown that S' is at a maximum in a system of constant energy and volume and A' is at a minimum in a system of constant volume in a heat bath, whenever there is equilibrium. The same properties apply to the thermodynamic quantities, entropy and Helmholtz free energy respectively. Now the latter quantities are related by the equation

$$A = U - TS.$$

Therefore, if we identify S with S' and A with A', we must have

$$\frac{1}{k\beta} = -T. \tag{11·34}$$

That this is not unsatisfactory may be seen by returning to the discussion following equation (11·10). It was shown that β is a negative quantity and also that it has the same value for any two bodies which are in equilibrium in a heat bath. The statistical quantity β is therefore a function of the thermodynamic temperature. Equation (11·34) is clearly consistent with everything that has been established so far.

However, a more complete identification of the statistical with the thermodynamic quantities can be carried out if we make use of a result from wave mechanics, namely, that the energies E_i depend

only on the volume and other geometrical parameters of the system.†
That is to say, the E_i depend on those variables which measure *work*,
as distinct from heat.

Consider the differentiation of equation (11·19), $\bar{U} = \Sigma P_i E_i$, giving

$$\mathrm{d}\bar{U} = \Sigma E_i \,\mathrm{d}P_i + \Sigma P_i \,\mathrm{d}E_i. \tag{11·35}$$

The first term on the right-hand side represents change due to the P_i
when the E_i are constant. This term therefore represents a change in
the system due to a change in the relative frequency with which
various quantum states are occupied, but it implies no change in these
quantum states themselves and therefore no volume change of the
system. This term is clearly a change in the energy of the system in
the absence of any performance of work and therefore it must be
identified as *heat*. The absorption of heat by a system is interpreted
as a shift in the relative occupation frequency towards quantum
states of higher energy, the nature of the quantum states remaining
unchanged. Also, since $\mathrm{d}\bar{U} = \mathrm{d}q + \mathrm{d}w$, the second term on the right-
hand side of (11·35) must be identified as the work done on the
system. The quantum interpretation of work is a change in the E_i,
whilst the relative distribution over the quantum states, as measured
by the P's, remains unchanged.

For the present purposes we therefore have

$$\mathrm{d}q = \Sigma E_i \,\mathrm{d}P_i. \tag{11·36}$$

Consider now the quantity S' defined by

$$S' = -k\Sigma P_i \ln P_i.$$

Differentiating, we obtain

$$\mathrm{d}S' = -k\Sigma(\ln P_i + 1)\,\mathrm{d}P_i.$$

Now whenever there is a state of equilibrium with a heat bath, the
P_i are given by

$$P_i = \mathrm{e}^{\beta E_i}/Q.$$

Substituting this in the previous equation we obtain for the change of
S' at equilibrium

$$\mathrm{d}S'_{\text{eq.}} = -k\Sigma(\beta E_i - \ln Q + 1)\,\mathrm{d}P_i$$
$$= -k\beta\Sigma E_i \,\mathrm{d}P_i, \tag{11·37}$$

since $\Sigma \mathrm{d}P_i = 0$. Therefore between equations (11·36) and (11·37)

$$\mathrm{d}S'_{\text{eq.}} = -k\beta \,\mathrm{d}q. \tag{11·38}$$

Now S' is a function of state because of the way in which it has been
defined. The result of equation (11·38) is therefore to show that if

† See footnote † on p. 348. A proof of the statement for the special case of
a perfect gas will be given in the next chapter.

dq is multiplied by $-k\beta$ it becomes equal to the differential of a function of state. On the other hand, in pure thermodynamics, it is shown that the *only way* in which dq can be converted into the differential of a function of state is by multiplying by $1/T$, where T is the thermodynamic temperature. Therefore we can make the identification

$$-k\beta = \frac{1}{T}. \tag{11·39}$$

Combining this with (11·38) we obtain

$$dS'_{eq.} = \frac{dq}{T}. \tag{11·40}$$

We thus conclude that S' satisfies all the requirements of the thermodynamic entropy. Of course the latter is only defined as a *change* of entropy, and therefore any statistical analogue which purports to give the absolute value of entropy, as in the case of S', is satisfactory provided that its differential obeys (11·40).

Using (11·39) we can now replace β by $-1/kT$ in the important equations (11·12) and (11·13) and (11·26)–(11·31). We shall also delete the primes, etc., and adopt the normal thermodynamic notation. The resulting equations are

$$Q = \Sigma\, e^{-E_i/kT}, \tag{11·41}$$

$$P_i = e^{-E_i/kT}/Q, \tag{11·42}$$

$$A = -kT \ln Q, \tag{11·43}$$

$$U = kT^2\left(\frac{\partial \ln Q}{\partial T}\right)_{V,\,comp.}, \tag{11·44}$$

$$S = k \ln Q + kT\left(\frac{\partial \ln Q}{\partial T}\right)_{V,\,comp.}. \tag{11·45}$$

The equilibrium pressure of the system is given by the thermodynamic relation

$$p = -\left(\frac{\partial A}{\partial V}\right)_{T,\,comp.},$$

and is therefore

$$p = kT\left(\frac{\partial \ln Q}{\partial V}\right)_{T,\,comp.}. \tag{11·46}$$

The other thermodynamic functions G, H, etc., can all be obtained by linear combination of those above. Values of the heat capacities and expansion and compressibility coefficients may also be obtained by appropriate differentiation. Therefore all of the thermodynamic properties of a system of fixed composition may be calculated if it is possible to evaluate Q and to know its dependence on volume and temperature.†

This quantity Q, as given by (11·41), is a sum over all quantum states

$$Q = \sum_{\text{states}} e^{-E_i/kT}. \tag{11·47}$$

Let the number of states of energy E_0 be denoted Ω_0, etc. Ω_i is the 'degeneracy' of the energy state E_i. Therefore in (11·47) each term of the type $e^{-E_i/kT}$ repeats itself Ω_i times. The partition function can therefore be expressed in the alternative form

$$Q = \sum_{\text{levels}} \Omega_i e^{-E_i/kT}, \tag{11·48}$$

when the summation is over the distinct energy states only. Similarly, the probability P'_i of the occurrence of a particular *energy* state at equilibrium is given by

$$P'_i = \Omega_i P_i$$

$$= \Omega_i e^{-E_i/kT}/Q, \tag{11·49}$$

because there are Ω_i quantum states comprised within the given energy state. Finally, it remains to be proved in the next chapter that

$$k = R/L \tag{11·50}$$

where R is the gas constant per mole and L is the Avogadro number.

11·12.   Thermal and configurational entropy

The quantity Ω as previously defined (§ 11·6) is the number of accessible quantum states of a system. Therefore Ω has a clear meaning only in the case of a system of fixed energy. Now for such a system it was shown in § 11·10 that the two entropy analogues

$$S' = -k\Sigma P_i \ln P_i,$$

$$S'' = k \ln \Omega,$$

are identical (because all of the P's are equal and have the value $1/\Omega$). Consider now a system which is able to exchange energy with neighbouring systems. On account of the fluctuations the energy of the system in question is no longer definite and neither is Ω. On the other hand, in any macroscopic system the fluctuations are very minute, and for almost all of the time the system remains very close

† As noted previously Q is called the *partition function* at fixed temperature and volume and it refers to a *macroscopic system*. It is not to be confused with the molecular partition function which is discussed in the next chapter. In some text-books the symbol Z is used in place of Q and Rushbrooke denotes it as (P.F.).

to its mean energy $\bar{U}$. It can thus be shown that S' and S'' are entirely *equivalent* for such a system, although no longer quite *identical*. (See the second appendix to this chapter.)

Of the two formulae above the second has probably the greater intuitive significance, and it was for this reason that it was originally introduced in § 1·17. On the other hand, even if we consider systems of fixed energy rather than of fixed temperature, the value of Ω is usually quite unknown. The advantage of the formulation

$$S' = -k\Sigma P_i \ln P_i$$

is that the P's can be obtained in terms of the partition function and the latter can often be evaluated completely. From this point of view the Helmholtz free energy, which is given by equation (11·43), is more directly calculated from statistical mechanics than the entropy itself.

There are, however, certain problems in which it is possible to calculate the *ratio* of Ω, as between two states, and therefore the corresponding entropy change. This occurs in situations where the system can exist in a number of geometrical configurations each having the same energy levels. In such instances the use of the expression

$$S_2 - S_2 = k \ln \Omega_2/\Omega_1$$

can be very helpful. Applications occur mainly in connexion with the solid state, adsorption and the quasi-crystalline model of liquids.

Suppose, for example, that there are two distinguishable types of molecules, A and B, distributed over lattice sites. Such a system was discussed in § 1·17. Each geometrical arrangement which is physically distinct from all other arrangements gives rise to a fresh quantum state of the system. In addition to this there is the randomness of the energy distribution. The molecules can vibrate about their mean situations on the lattice with varying amounts of quantized energy, and there are also the vibrations within the molecules and states of electronic excitation, etc. All of these forms of kinetic energy may be called the *thermal* energy of the system, i.e. that part of U which is not potential energy on the molecular scale. Each alternative distribution of the thermal energy, for a given geometrical distribution of molecules on the lattice, clearly gives rise to a new quantum state. We can thus speak of a *configurational* randomness, as pertaining to the number of distinguishable arrangements of molecules on the lattice, and a *thermal* randomness as pertaining to the energy distribution. Changes in the former are often much more easily calculated than changes in the latter. However, in certain problems it may occur that the energy levels are the same for each of the geometrical states. In such instances the thermal randomness will not be affected by changes in the configurational randomness, and this makes it possible to calculate the entropy of mixing by quite simple methods.

For example, in § 1·17 we discussed the mixing of four A atoms with four B atoms on eight lattice sites. The number of configurational states is

$$\Omega_{\text{config.}} = \frac{8!}{4!\,4!} = 70.$$

Consider the system as having a constant amount of thermal energy, and let it be supposed that each of the seventy arrangements has the *same number* of quantum states which arise from the quantized vibration of the atoms and are accessible for the fixed amount of thermal energy. Let this number be $\Omega_{\text{th.}}$. Thus the first geometrical arrangement has $\Omega_{\text{th.}}$ accessible quantum states, the second has also $\Omega_{\text{th.}}$ quantum states and so on. Therefore the total number of accessible quantum states is $70 \times \Omega_{\text{th.}}$. It follows that the entropy of mixing when the system passes adiabatically from a first state, in which it is known that all the A atoms are to the left of a certain plane and all the B atoms to the right of this plane, into a second state in which the system may be in any of the seventy configurations, is given by

$$\Delta S = k \ln \frac{70 \times \Omega_{\text{th.}}}{1 \times \Omega_{\text{th.}}} = k \ln 70.$$

It is evident that this simple result would not have been obtained if $\Omega_{\text{th.}}$ were not the same for each of the configurations. However, whenever this condition is satisfied we can write

$$\Omega = \Omega_{\text{config.}}\, \Omega_{\text{th.}}. \tag{11·51}$$

It is useful also to define

$$S_{\text{config.}} \equiv k \ln \Omega_{\text{config.}}, \tag{11·52}$$

$$S_{\text{th.}} \equiv k \ln \Omega_{\text{th.}}, \tag{11·53}$$

and therefore the total entropy is given by

$$S = k \ln \Omega$$
$$= k \ln \Omega_{\text{config.}} + k \ln \Omega_{\text{th.}}$$
$$= S_{\text{config.}} + S_{\text{th.}}. \tag{11·54}$$

An application of the equation

$$S_2 - S_1 = k \ln \frac{\Omega_{\text{config. 2}}}{\Omega_{\text{config. 1}}}, \tag{11·55}$$

which may be used under the above conditions, has already been given in § 8·1, where it was used to derive the laws of ideal solutions.

For this purpose it was assumed that the molecules are located on a quasi-crystalline lattice and can be moved about on this lattice *without affecting the spacing of the energy levels of the system.* This will be discussed again in Chapter 14 in connexion with 'regular solutions', and the equations above will also be applied to the problem of adsorption.

11·13. Appendix 1. Origin of the canonical distribution

In this section we shall seek to derive equation (11·13) without the need for any extension to our basic postulate. The discussion will also give some additional insight into the nature of thermal equilibrium.

Consider a body in a very large heat bath. Let the energy and quantum states of the body be denoted by Roman letters and those of the heat bath by Greek letters. Let the energy of the total system, body plus heat bath, be E and let Ω be the number of quantum states of the total system which are accessible when the energy is fixed at this value. Because of our basic postulate, the probability $P_{i\kappa}$ that the body is in a quantum state i and the bath is simultaneously in the quantum state κ is

$$P_{i\kappa} = 1/\Omega, \tag{11·56}$$

as discussed in §11·9. Now if the body is in a quantum state i, whose energy is E_i, the quantum state κ of the bath may be *any one* of those quantum states which all have the same energy ϵ_κ given by

$$\epsilon_\kappa = E - E_i. \tag{11·57}$$

Let Ω_κ be the number of such quantum states of the bath, i.e. Ω_κ is the degeneracy of the bath level ϵ_κ. Then the probability that the body is in the quantum state i independently of the particular quantum state of the bath is obtained by multiplying (11·56) by Ω_κ:

$$P_i = \Omega_\kappa/\Omega. \tag{11·58}$$

For convenience the heat bath will be considered as being a very large amount of a monatomic gas. For such a substance the number Ω_κ of quantum states which are comprised within the energy state ϵ_κ is known. It will be shown in the next chapter (equation (12·93)) that this number is

$$\Omega_\kappa = \alpha \epsilon_\kappa^\mu, \tag{11·59}$$

where α depends on the volume of the gas and μ is equal to $\frac{3}{2}$ times the number of atoms in the gas. This equation is based on wave mechanics and is quite independent of our statistical postulate. For the present purpose the essential point is this: for the case of a macroscopic heat bath, where μ is of the order of 10^{23} or more, Ω_κ *varies as an extremely high power* of ϵ_κ. (The same is true of substances which are not perfect gases but for such substances equation (11·59) is no longer exact. It may be remarked that the choice of a monatomic gas for our heat bath, in order conveniently to use the above equation, does not affect the behaviour of the body which is immersed in this bath.)

Let $\bar{U}$ and $\bar{\epsilon}$ be the mean energies of the body and the bath respectively when they are in contact. Thus

$$\bar{\epsilon} + \bar{U} = E, \tag{11·60}$$

and combining this with (11·57)

$$\epsilon_\kappa = \bar{\epsilon} + \bar{U} - E_i$$
$$= \bar{\epsilon}\left(1 + \frac{\bar{U} - E_i}{\bar{\epsilon}}\right). \tag{11·61}$$

Using equations (11·58), (11·59) and (11·61) we obtain

$$P_i = \frac{\alpha}{\Omega}\bar{\epsilon}^\mu\left(1 + \frac{\bar{U} - E_i}{\bar{\epsilon}}\right)^\mu. \tag{11·62}$$

Now almost all energy states E_i of the body whose probabilities are at all appreciable lie very close to $\bar{U}$, the mean energy. This is because of the smallness of the energy fluctuations (see also below). Also the body has a much smaller mean energy than the heat bath, which is taken as being relatively large in heat capacity. That is to say

$$\frac{\bar{U} - E_i}{\bar{\epsilon}} \ll 1, \tag{11·63}$$

and therefore equation (11·62) can be closely approximated by the exponential form

$$P_i = \frac{\alpha}{\Omega}\bar{\epsilon}^\mu \, e^{\mu(\bar{U} - E_i)/\bar{\epsilon}}$$
$$= \frac{\alpha}{\Omega}\bar{\epsilon}^\mu e^{\mu\bar{U}/\bar{\epsilon}} e^{-\mu E_i/\bar{\epsilon}}. \tag{11·64}$$

Now if the body is at equilibrium with the heat bath, the number Ω of quantum states of the combined system has reached a maximum value and is constant. The last equation can therefore be written

$$P_i = \text{constant} \times e^{-\mu E_i/\bar{\epsilon}}. \tag{11·65}$$

It will be shown below that $\mu/\bar{\epsilon}$ is equivalent to $-\beta$, as introduced in equation (11·10) and (11·13). The equation (11·65) is therefore the same as these equations, and the apparent extension in the scope of the basic postulate which was used in § 11·9 was not actually needed.

Since $\mu/\bar{\epsilon}$ is obviously a positive quantity it follows from (11·65) that the probability P_i of a *quantum state* of the body decreases exponentially with increase in the energy of this state, for fixed mean energy of the heat bath. On the other hand, the probability of an *energy state* of the body has an exceedingly sharp maximum at a particular value of E. This probability, denoted P_i', is obtained by multiplying P_i by Ω_i, the number of quantum states of the body whose energy is E_i:

$$P_i' = \Omega_i P_i'. \tag{11·66}$$

In this equation there are two opposing effects: (a) P_i *decreases* exponentially with increase in E_i, and this is due essentially to the rapid increase of Ω_κ with ϵ_κ for the heat bath, as shown above; (b) on the other hand, Ω_i for the body, like the bath, also *increases* rapidly with E_i. Because of these opposing effects P'_i passes through a very sharp maximum.

This important conclusion is equivalent to saying that the body spends almost all of its time in energy states very close to its mean energy. This result may be proved in more detail if we assume a relation connecting Ω_i and E_i similar to equation (11·59):

$$\Omega_i = A E_i^m, \tag{11·67}$$

where m depends on the number of particles in the body. This relation is strictly accurate only if the body is a perfect gas, but for the present purposes is sufficiently accurate as applied to other states of aggregation as well. Combining equations (11·65), (11·66) and (11·67)

$$P'_i = \text{constant } E_i^m \times e^{-\mu E_i/\bar{\epsilon}}. \tag{11·68}$$

Treating P'_i and E_i as continuous variables, we obtain by differentiation,

$$\frac{\partial P'_i}{\partial E_i} = \text{constant } E_i^m \times e^{-\mu E_i/\bar{\epsilon}} \left(\frac{m}{E_i} - \frac{\mu}{\bar{\epsilon}} \right). \tag{11·69}$$

Therefore P'_i has its maximum value at a particular energy E_* given by

$$\frac{m}{E_*} = \frac{\mu}{\bar{\epsilon}}, \tag{11·70}$$

and this energy state is the one which is by far the most frequently occupied out of the whole range $E_0, E_1, ..., E_i, ...$

The extreme sharpness of the maximum may be shown as follows. From equation (11·68) the ratio of the probability P'_* that the system is in the energy state E_* to the probability P'_i that it is in any other energy state E_i is given by

$$\frac{P'_i}{P'_*} = \frac{E_i^m e^{-\mu E_i/\bar{\epsilon}}}{E_*^m e^{-\mu E_*/\bar{\epsilon}}}. \tag{11·71}$$

Putting $E_i = E_*(1 + \delta)$, this becomes

$$\frac{P'_i}{P'_*} = (1+\delta)^m e^{-\mu \delta E_*/\bar{\epsilon}}$$

$$= (1+\delta)^m e^{-\delta m}, \tag{11·72}$$

on account of equation (11·70). Taking logarithms and expanding we obtain

$$\ln \frac{P'_i}{P'_*} = m(\delta - \tfrac{1}{2}\delta^2 + ...) - \delta m$$

$$= -\tfrac{1}{2}m\delta^2, \tag{11·73}$$

for small values of δ. For example, if we choose m equal to 10^{24}, corresponding to about a mole of material, and δ equal to 10^{-6} we obtain

$$\frac{P'_i}{P'_*} = e^{-10^{12}}, \tag{11·74}$$

which is exceedingly minute. The chance that a system containing about a mole of material, when placed in a thermostat, should have a momentary energy differing from its most probable value by as little as one part in a million, is thus entirely negligible. This confirms what has been said previously about the smallness of the energy fluctuations.

For the same reason E_* is practically the same as $\bar{U}$, the mean energy of the body. Equation (11·70) can thus be inverted and written

$$\frac{\bar{U}}{m} = \frac{\bar{\epsilon}}{\mu}. \tag{11·75}$$

This equation shows that at equilibrium there is an equality between the quantity $\bar{U}/m$, which is a characteristic of the body, and the quantity $\bar{\epsilon}/\mu$ which is a characteristic of the heat bath. Either of these quantities is therefore a function of the thermodynamic variable T. In fact, the comparison of (11·65) and (11·10) shows that $\bar{\epsilon}/\mu$ is the same as $-1/\beta$, which was identified in equation (11·39) with kT. Thus

$$\frac{\bar{\epsilon}}{\mu} = -\frac{1}{\beta} = kT. \tag{11·76}$$

In the special case of a monatomic gas it will be shown in the next chapter that μ is $\frac{3}{2}$ times the number of atoms. It follows from (11·76) that the mean energy per atom is $\frac{3}{2}kT$, a well-known result.

11·14. Appendix 2. Entropy analogues

There are several statistical analogues, S', S'', etc., of the entropy, all of which have the required thermodynamic properties. Such functions are not always algebraically identical but are numerically identical to a very high degree of accuracy. One of the difficulties in studying statistical mechanics is to recognize that the various analogues, although they may appear to be quite different in form, are actually equivalent when they are worked out numerically. This arises for several reasons:

(1) The entropy is related to the *logarithm* of the statistical quantities P_i, Ω, etc. Now the product xy is certainly not the same as x, but the value of $\ln xy$ is almost the same as $\ln x$, whenever $x \gg y$.

(2) According to Stirling's theorem the following relation is a very good approximation, at large values of x, but is not an identity

$$\ln x! = x \ln x - x. \tag{11·77}$$

(3) Because of the smallness of the energy fluctuations of a macroscopic body in a thermostat, its thermodynamic behaviour is the same as if it were an isolated system having a constant energy equal to its mean energy $\bar{U}$ in the thermostat.

(4) For the same reason it may occur that only one term in a summation is significant. Thus, as shown in the last section, a body in a thermostat behaves as if it spends all of its time in the 'most probable state'. This is equivalent to neglecting terms corresponding to all other states in a summation over such states.

PROBLEM

1. (*a*) The Helmholtz free energy A of a system is known as a function of temperature and volume. Obtain expressions for S, P, U, H, C_v, C_p, α (expansivity coefficient) and κ (compressibility coefficient) in terms of A, T and V.

(*b*) Show that if A is known as a function of temperature and pressure, it is *not* possible to calculate the volume V.

[C.U.C.E. Tripos, 1951]

CHAPTER 12

PARTITION FUNCTION OF A PERFECT GAS

12·1. Distinguishable states of a gas and the molecular partition function

In the last chapter it was shown that the thermodynamic properties of a system may be calculated from a knowledge of its partition function. The latter is defined by equations (11·47) or (11·48):

$$Q = \sum_{\text{states}} e^{-E_i/kT} \tag{12·1}$$

$$= \sum_{\text{levels}} \Omega_i\, e^{-E_i/kT}, \tag{12·2}$$

where the E_i are the quantized energy states of the whole macroscopic system. In the present chapter we shall discuss perfect gases, and it will be shown how (12·1) may be expressed in terms of a function of the individual molecules—the molecular partition function—and how this may be evaluated from Schrödinger's equation.

Let it be supposed, to start with, that there are just two molecules in the given volume V and also that *these molecules are different in kind*. Let ϵ_0, ϵ_1, etc., be the quantized energy states of the first kind of molecule, as obtained by solving the Schrödinger equation for this molecule when it is alone in the given volume. Similarly, let ϵ_0', ϵ_1', etc., be the energy states of the other molecule. Because the molecules of a gas do not appreciably interact, the energy of the system when both molecules are present in the volume V is the sum of the energies of the individual molecules when each is present on its own. If the quantum states of the molecules are not degenerate, then (12·1) may be written†

$$Q = e^{\beta(\epsilon_0+\epsilon_0')} + e^{\beta(\epsilon_0+\epsilon_1')} + e^{\beta(\epsilon_0+\epsilon_2')} + \ldots$$
$$+ e^{\beta(\epsilon_1+\epsilon_0')} + e^{\beta(\epsilon_1+\epsilon_1')} + e^{\beta(\epsilon_1+\epsilon_2')} + \ldots$$
$$+ \ldots, \text{ etc.} \tag{12·3}$$

However, in general, each level ϵ_i of the first type of molecule will be ω_i-fold degenerate, i.e. the ϵ_i level comprises ω_i different quantum states (see § 12·2 below). Similarly, let ω_j' be the degeneracy of the jth level of the other molecule. When the first type of molecule has the

† For convenience $-1/kT$ is replaced by the symbol β, as used in the previous chapter.

energy ϵ_i and the second has the energy ϵ_j', there are therefore $\omega_i \omega_j'$ quantum states of the combined system each of energy $\epsilon_i + \epsilon_j'$. It follows that the term $e^{\beta(\epsilon_i + \epsilon_j')}$ occurs $\omega_i \omega_j'$ times in the summation for Q. Thus, in general, (12·3) must be replaced by

$$Q = \omega_0 \omega_0' e^{\beta(\epsilon_0 + \epsilon_0')} + \omega_0 \omega_1' e^{\beta(\epsilon_0 + \epsilon_1')} + \ldots + \omega_1 \omega_0' e^{\beta(\epsilon_1 + \epsilon_0')} + \ldots$$

$$= \omega_0 e^{\beta\epsilon_0} (\Sigma \omega_i' e^{\beta\epsilon_i'}) + \omega_1 e^{\beta\epsilon_1} (\Sigma \omega_i' e^{\beta\epsilon_i'}) + \ldots$$

$$= (\Sigma \omega_i e^{\beta\epsilon_i}) (\Sigma \omega_i' e^{\beta\epsilon_i'}), \tag{12·4}$$

where the summations are over the different energy states of the molecules.

In the case where there are not two but N molecules, all different in kind, in the volume V, it will be seen that the partition function again factorizes into a product of N terms. This property of factorization occurs whenever the energies are additive, due to the independence of the molecules, and is a consequence also of the exponential structure of the partition function:

$$e^{x+y+z} = e^x e^y e^z.$$

However, what we are actually interested in is the partition function of a gas all of whose molecules are the same. The above discussion has been included only for the sake of clarifying what follows. Now it is to be remembered that Q is a summation over quantum states which are independent, i.e. which are distinct from each other. On the other hand, if the two molecules in the vessel are identical, so that $\epsilon_0 = \epsilon_0'$, etc., then the quantum state of the system in which the one molecule has the energy ϵ_i and the other has the energy ϵ_j, is in no way distinguishable from the state in which the first molecule has the energy ϵ_j and the other the energy ϵ_i. In fact these two 'states' are not two at all, but only one. This arises from the fact that the particles are identical and because they are both free to move throughout the whole of the common volume V.

Returning to equations (12·3) or (12·4) it will now be seen that these equations are not applicable as they stand to a system containing two molecules of the same kind. Thus almost all of the terms in these equations occur *twice* too often. For example, the term $e^{\beta(\epsilon_0 + \epsilon_1')}$ is the same as the term $e^{\beta(\epsilon_1 + \epsilon_0')}$, and only one of these should be included. These considerations apply to every term in the matrix except those along the diagonal, e.g. $e^{\beta(\epsilon_0 + \epsilon_0')}$, $e^{\beta(\epsilon_1 + \epsilon_1')}$, etc. In brief, every term for which $i \neq j$ occurs twice too often. Now there are obviously an enormously larger number of terms of this type than those for which $i = j$, i.e. those along the diagonal. There will perhaps be a negligible error if we divide *all* of the terms in the summation by 2, and not merely those for which $i \neq j$, in order to allow for the identity of the two molecules.

Before proceeding we must consider rather more carefully whether the simplification which has just been proposed is a valid one. Now our purpose is to evaluate Q which is a summation over all of the possible quantum states of the system right up to virtually infinite energies. It is therefore entirely correct that the number of terms in the summation for which $i \neq j$ enormously exceeds—really infinitely exceeds—the number for which $i = j$. However, in evaluating Q we are also concerned with the *magnitude* of the terms. Each term is of the form

$$e^{-(\epsilon_i + \epsilon_j')/kT},$$

and therefore the higher is the energy of the particular quantum state the smaller is its numerical contribution to the partition function. Beyond a certain point in the series of terms, additional terms make an almost negligible contribution to Q, and this will clearly occur at a lower total energy, $\epsilon_i + \epsilon_j'$, the lower is the temperature T. Suppose, for example, that only the terms containing ϵ_0 and ϵ_1 are significant in equations (12·3) and (12·4). In this case it is evident that there will be appreciable error in dividing all of these terms by 2, in order to eliminate indistinguishable states, as we have proposed above. However it can be shown by numerical evaluation of the translational energies of a molecule in a container that this kind of situation only arises at exceedingly low temperatures, of the order of 1°K. Under almost all conditions which are met with in practice the procedure of dividing *all* of the terms in equations (12·3) or (12·4) by two leads to an entirely negligible error. Thus in place of (12·4) we obtain

$$Q = \tfrac{1}{2}(\Sigma \omega_i \, e^{\beta \epsilon_i})^2 \qquad (12·5)$$

as the partition function of a system containing two identical molecules.

Consider now the case where there are *three* molecules of the same kind in the enclosure. If, at any moment, these three molecules are in the energy states ϵ_i, ϵ_j and ϵ_k respectively, then the total energy of the system is

$$\epsilon_i + \epsilon_j + \epsilon_k.$$

Let it be supposed that the three molecules are actually distinguishable. Then the above energy state of the system could arise in 3! different ways. Thus the ϵ_i level could be filled in three ways, by choosing between the three molecules, and each of these could be combined with the two alternative ways of filling the ϵ_j level from amongst the two remaining molecules. Because the molecules are actually identical it is evident therefore that in the summation for Q all terms for which $i \neq j \neq k$ occur 3! times too often. Using the same argument as previously, there will be no significant error if we divide all terms in Q by 3!, in order to avoid including terms which are not independent. We thus obtain as the partition function for a perfect gas system containing three molecules of the same kind

$$Q = \frac{1}{3!}(\Sigma \omega_i \, e^{\beta \epsilon_i})^3, \tag{12·6}$$

and in general for N molecules of the same kind

$$Q = \frac{1}{N!}(\Sigma \omega_i \, e^{-\epsilon_i/kT})^N. \tag{12·7}$$

In this expression β has been replaced by $-1/kT$, and the summation is over the molecular energy levels ϵ_0, ϵ_1, etc., whose degeneracies are ω_0, ω_1, etc.

The whole basis of the simplification is that there are always far more quantum states per molecule, of energies such that $e^{-\epsilon_i/kT}$ is not insignificant, than there are molecules in the vessel.† It follows that *almost all the terms which make appreciable contribution to Q correspond to every molecule being in a different quantum or energy state.* When this condition is satisfied, the kind of molecular statistics which are obtained, as given in the present chapter, are known as *Maxwell–Boltzmann.* When it is not satisfied, the procedure for evaluating Q is more difficult but leads to the two kinds of molecular statistics known as *Bose–Einstein* and *Fermi–Dirac* respectively. (The latter includes the application of the Pauli exclusion principle and the former does not.) However, the Maxwell–Boltzmann statistics are satisfied by all molecular gases at all temperatures and pressures at which they are effectively perfect, i.e. at which these gases consist of essentially independent molecules, which is one of the necessary assumptions of the present chapter. The only types of perfect 'gas' to which the formulae of the present chapter may not be applied are radiation and the free electrons in metals.

Equation (12·7) may be rewritten

$$Q = \frac{1}{N!}f^N, \tag{12·8}$$

where

$$f \equiv \Sigma \omega_i \, e^{-\epsilon_i/kT}, \tag{12·9}$$

which is called the *molecular partition function.*

In the case of a mixture of two gases, consisting of N_1 molecules of type 1 and N_2 molecules of type 2, it is readily seen that the partition function for the whole system is

$$Q = \frac{1}{N_1! N_2!}f_1^{N_1} f_2^{N_2}, \tag{12·10}$$

where f_1 and f_2 are the molecular partition functions for the two types of molecules.

† See Problem 1 at the end of the chapter.

12·2. Schrödinger's equation

The discussion so far has been almost entirely statistical and has been based on the postulate concerning the equal probability of quantum states of equal energy. The quantum aspects must now be considered in more detail, and in particular we shall determine the value of the ω_i and ϵ_i which appear in equation (12·9).† This information makes it possible to calculate f and Q and therefore the thermodynamic functions of the gas. For this purpose it is necessary to introduce an extra postulate—Schrödinger's equation.

As is well known, it is not possible to make a direct application of Newtonian mechanics to phenomena which are on a very fine scale. This shows itself in the first place in a certain ambiguity in the conventional mental images of the particle and the wave. It would have been obliging on the part of nature if it were possible to attribute to the smallest pieces of matter (electrons and protons) the same Newtonian properties—in particular, definite positions and velocities —as are normally associated with the notion of a particle. Experiment shows, however, that these minute objects do not have such clear-cut properties. To be sure, in certain kinds of experiment they behave in the compact manner of a particle, but in other types of experiment, such as electron and proton diffraction, they seem to have the extended character of a wave.

In a closely analogous manner radiation is found to have both a particle and a wave aspect, the former being shown, for example, in the photoelectric effect. In short, the things of the senses, matter and radiation, do not correspond in a simple one-to-one relationship with the mental images of the particle and wave respectively.

For our present purposes we shall take as our additional postulate the supposition that the mechanical behaviour of matter on the atomic scale is in accordance with the Schrödinger wave equation. Its numerical solution, for appropriate conditions, expresses the observable properties without contravening the principle that it is impossible to make an exact and simultaneous specification of position and velocities. It may be remarked that it is because of this principle of uncertainty that wave mechanics seek to describe the state of a system by means of the function ψ, whose purpose is to describe *probabilities* and not certainties.

Consider a system containing n fundamental particles, of masses m_1, m_2, ..., m_n, some or all of which may be present as atoms or molecules. The system would be described classically by assigning

---

† The chapter will actually be concerned mainly with translational states and not with vibration or rotation.

values to the Cartesian co-ordinates x_i, y_i, z_i, of the various particles, together with corresponding values of the momenta.

Wave mechanically, the system is described by means of a wave function ψ, which has the following property: the probability that the various particles have their co-ordinates in the range x_1 to x_1+dx_1, ..., z_n to z_n+dz_n, is given by†

$$\psi^2\,dx_1\,dy_1\ldots dz_n. \tag{12·11}$$

ψ^2 is thus a probability density.

We shall be interested in systems of the above type which are in a time-independent state and where the particles move in a field of force which can be described by a potential energy V which is a function of the co-ordinates $x_1, y_1, ..., z_n$. For such systems we shall now take as our postulate that ψ may be obtained by solving the equation

$$\sum_i \frac{1}{m_i}\left(\frac{\partial^2\psi}{\partial x_i^2}+\frac{\partial^2\psi}{\partial y_i^2}+\frac{\partial^2\psi}{\partial z_i^2}\right) = -\frac{8\pi^2}{h^2}(E-V)\,\psi, \tag{12·12}$$

together with any boundary conditions which are appropriate to the particular problem. The summation in (12·12) is over all the particles. h is Planck's constant and E is the total energy. It is an ordinary differential equation, and complications arise only on account of the large number of co-ordinates. Similar equations are used in classical wave theory and ψ may therefore be regarded as being the amplitude of a probability wave. However, it is not a wave in physical space because the equation is concerned with $3n$ dimensions.

The essential requirement of ψ, if it is to fulfil its wave-mechanical purpose, is that it shall describe a probability, in accordance with equation (12·11). For this purpose it must be *single-valued, continuous and finite* throughout the range of the co-ordinates. For example, if it is known that all of the particles are contained within a certain range of the co-ordinates (e.g. if they are in a box), then ψ^2 must have the above properties within this range but must be zero at all points outside it. Otherwise ψ^2 does not have the required properties as a probability density.

In brief, not all possible solutions of the differential equation (12·12) necessarily lead to values of ψ which correspond to physical reality; *it may occur that significant solutions are obtained only for particular values E_0, E_1, etc., of the energy.*‡ These are called *eigenvalues*

---

† If ψ is complex, the product of ψ and its conjugate ψ^* is to be used in place of ψ^2.

‡ It should be noted that discrete energy states occur in most problems, but not all. For example the translational energies of a particle not confined in a container are not quantized. Quantization arises from the requirement that ψ shall be single-valued, etc., together with the boundary conditions of the particular problem.

and the corresponding values of ψ are ψ_0, ψ_1, etc., and are called *eigensolutions*.†

Each eigensolution specifies a possible quantum state of the system, and their successive numbering gives rise to a *quantum number*. Instead of numbering the quantum states, an alternative procedure is to number the energy levels (eigenvalues) successively. If it occurs that there are Ω_i different quantum states of the system, all of which correspond to the same energy E_i, this level is said to be Ω_i-fold degenerate. This is the same notation as was used in equations (12·1) and (12·2). For molecules, on the other hand, the symbols ϵ_i and ω_i are used in place of E_i and Ω_i respectively.

12·3. Separability of the wave equation

The quantum states of samples of solids, liquids or gases, of which we have spoken in the statistical discussion, are obtained in principle by solving equation (12·12). However, the equation can only be solved in actuality under especially simple conditions, particularly where there are *separable energy states*. The particular significance of separability—and this applies whether we carry out the separation as between different particles (as in the case of a perfect gas) or as between the different kinds of energy (translational, rotational, etc.) of any single molecule—is that it allows equation (12·12) to be broken down into a number of simpler differential equations, each containing a smaller number of variables.

In order to illustrate this point consider the form of equation (12·12) when it contains only two independent variables x and y. We shall set about the problem of separating this equation into two others, one containing x only and the other containing y only. For simplicity it will be supposed that the physical situation corresponds to one of constant potential energy, i.e. V is not a function of x or y

---

† The student should not be put off by the seemingly abstract notion of eigenvalues and eigensolutions. As an example of their occurrence in ordinary algebraic problems Margenau and Murphy (*The Mathematics of Physics and Chemistry* (New York, Van Nostrand, 1943), §8·1) quote the pair of simultaneous equations

$$(1-E)\,x + 2y = 0,$$

$$2x + (1-E)\,y = 0.$$

Solution of these equations (other than the trivial solution $x = 0$, $y = 0$) can be obtained only when E has the values 3 or -1. For any other values of E the equations are inconsistent and thus the two equations could not refer simultaneously to the same physical problem involving the variables x and y. The eigenvalue $E = 3$ has as its solution, or 'eigenfunction', $x = y$ and the eigenvalue $E = -1$ has as its eigenfunction $x = -y$.

and can be taken as zero by appropriate choice of the basis for computing energies. Equation (12·12) therefore reads

$$\frac{\partial^2 \psi}{\partial x^2} + \frac{\partial^2 \psi}{\partial y^2} = -\frac{8\pi^2 m}{h^2} E\psi. \tag{12·13}$$

This would be the wave equation of a particle of mass m which moves only in an x-y plane of constant potential energy. Alternatively, it could refer to a system of two particles, of equal mass, one confined to motion along the x axis and the other to motion along the y axis.

Let it be assumed, for purposes of trial, that an eigenfunction ψ, corresponding to a permitted value of E, may be expressed as a product of a function ψ_x which depends only on x and a function ψ_y which depends only on y:

$$\psi = \psi_x \psi_y. \tag{12·14}$$

The correctness of this assumption depends on whether (12·14) satisfies (12·13), and this remains to be seen. From (12·14) we obtain

$$\left. \begin{aligned} \frac{\partial^2 \psi}{\partial x^2} &= \psi_y \frac{\mathrm{d}^2 \psi_x}{\mathrm{d}x^2}, \\ \frac{\partial^2 \psi}{\partial y^2} &= \psi_x \frac{\mathrm{d}^2 \psi_y}{\mathrm{d}y^2}. \end{aligned} \right\} \tag{12·15}$$

Adding these equations

$$\frac{\partial^2 \psi}{\partial x^2} + \frac{\partial^2 \psi}{\partial y^2} = \psi_y \frac{\mathrm{d}^2 \psi_x}{\mathrm{d}x^2} + \psi_x \frac{\mathrm{d}^2 \psi_y}{\mathrm{d}y^2}. \tag{12·16}$$

Comparing this with (12·13) it is evident that (12·14) will meet all requirements provided that we can satisfy the relationship

$$\psi_y \frac{\mathrm{d}^2 \psi_x}{\mathrm{d}x^2} + \psi_x \frac{\mathrm{d}^2 \psi_y}{\mathrm{d}y^2} = -\frac{8\pi^2 m}{h^2} E\psi.$$

Dividing through by $\psi = \psi_x \psi_y$ we must therefore be able to satisfy

$$\frac{1}{\psi_x} \frac{\mathrm{d}^2 \psi_x}{\mathrm{d}x^2} + \frac{1}{\psi_y} \frac{\mathrm{d}^2 \psi_y}{\mathrm{d}y^2} = -\frac{8\pi^2 m}{h^2} E. \tag{12·17}$$

What does this condition imply? Now E, which is a particular eigenvalue, is a constant of the motion and does not vary with x or y. Therefore each of the left-hand terms in the equation must be separately constant. (The first term is a function only of x and is thus not affected by any change in y and similarly the second term is a function only of y and is independent of x.) Provided that we impose these conditions (12·14) is a satisfactory solution of (12·13). ψ_x and ψ_y are therefore obtained by solving the equations

$$\frac{1}{\psi_x}\frac{d^2\psi_x}{dx^2} = -\frac{8\pi^2 m}{h^2}E_x, \tag{12·18}$$

$$\frac{1}{\psi_y}\frac{d^2\psi_y}{dy^2} = -\frac{8\pi^2 m}{h^2}E_y, \tag{12·19}$$

where E_x and E_y are constants which satisfy

$$E_x + E_y = E. \tag{12·20}$$

The original differential equation has thus been separated into two equations, (12·18) and (12·19), each involving only a single independent variable.

In brief, in the example discussed, the property of separability implies that the wave function ψ can be expressed as a product of two wave functions ψ_x and ψ_y, each dependent on only a single co-ordinate. It implies also that the total kinetic energy E (or $E - V$) can be expressed as the sum of kinetic energies, E_x and E_y, which are characteristic of motion along the x and y co-ordinates respectively. For example, if (12·13) refers to a single particle moving in the x, y plane then E_x and E_y may be interpreted as the kinetic energies for motion along the x and y co-ordinates respectively. Alternatively, if (12·13) refers to two particles, one moving along x and the other along y, then E_x and E_y refer to the individual kinetic energies of the two particles respectively. At any rate these are what may be called the 'classical' interpretations.

Similar considerations apply when the wave equation contains more than two independent variables. However, it is to be emphasized that separability is not a *necessary* property of the wave equation, and it can be achieved only when the potential energy is constant or is a particularly simple function of the co-ordinates. In the present connexion three important types of separability are as follows:

(a) The wave equation for a macroscopic quantity of a perfect gas can be separated into wave equations characteristic of each separate molecule in the same volume. This follows immediately as a generalization from the discussion above, together with the fact that in a perfect gas the potential energy of interaction between the molecules is zero. This result has already been tacitly assumed in § 12·1, where the total energy of the gas was taken as being the sum of the energies of the individual molecules. A similar assumption is not true of liquids or solids because the molecules are not independent.

(b) The wave equation for each individual molecule of a gas can be separated into a part characteristic of its translational motion and a part characteristic of its internal motion. This is because the potential energy for the internal states does not depend on the position of the molecule in space. It follows that any molecular energy ϵ_i, such as appears in equation (12·7), may be taken as the sum of a translational part and an internal part. Thus

$$\epsilon_i = \epsilon_j^{\text{tr.}} + \epsilon_k^{\text{int.}}, \tag{12·21}$$

where $\epsilon_j^{\text{tr.}}$ is the jth level of translational energy and $\epsilon_k^{\text{int.}}$ is the kth level of internal energy (this usage of the term internal energy is not to be confused with the usage in thermodynamics).

(c) The wave equation for the internal state of the molecules may be further factorized, but not with quite the same degree of accuracy as in the two cases above. A detailed investigation, which cannot be given here, shows that $\epsilon_k^{\text{int.}}$ may justifiably be regarded as the sum of two terms, one due to the state of electronic excitation† of the molecule and the other due to the combined rotational and vibrational motion.‡ At a rather poorer degree of approximation the latter may also be regarded as separable. From a classical point of view, the inaccuracy in this separation is due to the following reciprocal effect: the centrifugal force due to the rotation causes a stretching of the molecule, and thus affects its vibration, whilst the change of internuclear distance during vibration causes a variation in the moment of inertia and thus modifies the rotation. However, to the extent to which this approximation is valid, we can write

$$\epsilon^{\text{int.}} = \epsilon^{\text{rot.}} + \epsilon^{\text{vib.}} + \epsilon^{\text{elec.}}. \tag{12·22}$$

(Subscripts to denote the particular rotational, vibrational or electronic quantum states have not been included.)

In summary, separability is attained whenever there is negligible interaction between two or more degrees of freedom of the system. Under these conditions the total energy can be expressed as a *sum* of the energies which are attributable to each degree of freedom and the complete eigenfunction ψ can be expressed as a corresponding *product*. In the next section it will be shown that, under the same conditions, the partition function can also be expressed as a product

† At normal temperatures the electronic states of most molecules are not 'excited'. That is to say the separation between the lowest electronic state and the one above is large compared to kT so that only the lowest state contributes a significant term to the partition function and almost all molecules are in the lowest state. An important exception is nitric oxide.

‡ In the present simplified treatment no account is taken of states internal to the nucleus, nor of spin variables.

of partition functions, one for each separable degree of freedom. The fact that independent eigenfunctions and partition functions combine as *products* (and not as sums) is related, of course, to the fact that they both determine independent probabilities.

12·4. Factorization of the molecular partition function

In equation (12·8) the partition function Q was expressed as a product of the partition functions f, of each of the N molecules, together with the factor $N!$ which allows for indistinguishability. As noted in the last section, the justification for this procedure is the separability of the wave equation in the case of a perfect gas. In addition, there is also the separability which is expressed by equations (12·21) and (12·22), and this permits a factorization of f itself. Let $\omega_j^{tr.}$ and $\omega_k^{int.}$ be the degeneracies of the jth translational level and the kth internal level respectively. The degeneracy of the energy state

$$\epsilon_i = \epsilon_j^{tr.} + \epsilon_k^{int.}$$

is

$$\omega_i = \omega_j^{tr.} \cdot \omega_k^{int.}. \tag{12·23}$$

This is because any of the $\omega_j^{tr.}$ translational quantum states, all of energy $\epsilon_j^{tr.}$, may be combined with any of the $\omega_k^{int.}$ internal quantum states, all of energy $\epsilon_k^{int.}$, to give a state of a given total energy. Substituting the above in (12·9) we obtain

$$f = \sum_{j,\,k} \omega_j^{tr.} \cdot \omega_k^{int.} \cdot \exp\left[-(\epsilon_j^{tr.} + \epsilon_k^{int.})/kT\right]. \tag{12·24}$$

This may be factorized in the same way as in equation (12·4). This gives

$$f = f^{tr.} \cdot f^{int.}, \tag{12·25}$$

where

$$f^{tr.} = \Sigma \omega_j^{tr.} \exp\left[-\epsilon_j^{tr.}/kT\right], \tag{12·26}$$

$$f^{int.} = \Sigma \omega_k^{int.} \exp\left[-\epsilon_k^{int.}/kT\right], \tag{12·27}$$

and the summations are over the translational and internal states respectively. To the extent to which the further separation of equation (12·22) is valid, we can proceed a step further and write

$$f = f^{tr.} \cdot f^{rot.} \cdot f^{vib.} \cdot f^{elec.}, \tag{12·28}$$

where $f^{rot.}$, $f^{vib.}$ and $f^{elec.}$ are rotational, vibrational and electronic partition functions respectively.

In speaking of, say, the translational and internal parts of the energy as being independent forms of energy, it is to be borne in mind that whenever we have an *isolated* system the total energy is constant. Thus any decrease in the translational energy of the molecules would result in an increase in the energy of the internal motions and vice versa. For the isolated system Ω tends to a maximum and the equilibrium distribution between the translational and internal parts of the energy is determined by the maximum of the product

$$\Omega = \Omega_{\text{trans.}} \cdot \Omega_{\text{int.}} \cdot$$

The allocation of a fixed quantity of energy between the various forms of motion thus depends on the 'spreading' of this energy in such a way as to maximize Ω, due allowance being made for the different spacing of the quantum levels. This point was mentioned previously in § 1·17.

The equilibration of the various forms of energy implies also that the statistical parameter T, the temperature, has the same value in all of the various parts, such as (12·26) and (12·27), of the total partition function. This follows from the same kind of argument as was used in § 11·9, where it was shown that two bodies at equilibrium have the same value of β. To demonstrate the equality of T or β between, say, translational and internal states, we should set up equations similar to (11·4)–(11·6), noting on the one hand that the probability of a chosen translational state and a chosen internal state is multiplicative and, on the other hand, that the energies are additive.

It may be remarked that the equilibration between the different forms of energy of a system of molecules which are not undergoing reaction is usually attained quite rapidly due to the collisional process. It is only under rather exceptional conditions that the equilibration is not attained, for example, in flames or in the rapid adiabatic compressions due to sound waves. In the latter instance the vibrational energy does not attain equilibrium with the translational (and rotational) energy within the period of the wave. Under such conditions it may occur that the various translational states are at approximate equilibrium with each other and have a statistical parameter $T_{\text{tr.}}$, and also that the vibrational states are at equilibrium amongst themselves with a characteristic temperature $T_{\text{vib.}}$. However, if there is only a relatively slow equilibration between the translational and vibrational states $T_{\text{tr.}}$ is not equal to $T_{\text{vib.}}$ during the time required to attain this equilibrium.

12·5. The translational partition function

From equation (12·25) we have

$$f = f^{\text{tr.}} \cdot f^{\text{int.}},$$

and within the scope of the present chapter we shall deal in detail only with the translational partition function. Now, as shown in § 12·3, the wave function of a gaseous molecule has the property of being separable into parts which are characteristic of the translational and internal motions,

$$\psi = \psi^{\text{tr.}} \psi^{\text{int.}},$$

where $\psi^{\text{tr.}}$ is an eigenfunction of the equation

$$\frac{\partial^2 \psi^{\text{tr.}}}{\partial x^2} + \frac{\partial^2 \psi^{\text{tr.}}}{\partial y^2} + \frac{\partial^2 \psi^{\text{tr.}}}{\partial z^2} = -\frac{8\pi^2 m}{h^2} \epsilon \psi^{\text{tr.}}. \tag{12·29}$$

ϵ is the total translational energy of the molecule, and it has been assumed that the potential energy is constant and zero throughout the containing vessel. The permissible values of ϵ are those we require in order to evaluate $f^{\text{tr.}}$ by means of equation (12·26).

The last equation is again separable, and we obtain

$$\psi^{\text{tr.}} = \psi_x \psi_y \psi_z, \tag{12·30}$$

where ψ_x, ψ_y and ψ_z depend only on x, y and z respectively. This follows in exactly the same manner as in the treatment of equation (12·13) where, however, the discussion was concerned with the case of two dimensions only. In this way we obtain the three simpler equations

$$\frac{1}{\psi_x} \frac{d^2 \psi_x}{dx^2} = -\frac{8\pi^2 m}{h^2} \epsilon_x, \tag{12·31}$$

$$\frac{1}{\psi_y} \frac{d^2 \psi_y}{dy^2} = -\frac{8\pi^2 m}{h^2} \epsilon_y, \tag{12·32}$$

$$\frac{1}{\psi_z} \frac{d^2 \psi_z}{dz^2} = -\frac{8\pi^2 m}{h^2} \epsilon_z, \tag{12·33}$$

where

$$\epsilon_x + \epsilon_y + \epsilon_z = \epsilon. \tag{12·34}$$

The solution to (12·31) is

$$\psi_x = A \sin\left(\frac{\pi x}{h} \sqrt{(8m\epsilon_x)} + B\right), \tag{12·35}$$

where A and B are constants, as is readily verified. Similar solutions are obtained for (12·32) and (12·33).

Let it be supposed that the gas is confined in a rectangular box whose sides have lengths a, b and c parallel to the x, y and z axes respectively. In accordance with the probability interpretation (equation 12·11), ψ_x must therefore be zero at all values of x which lie outside the range of length a. This is because we specify that the molecules are *certainly* inside the box. Thus ψ_x is zero for all values of x less than zero and greater than a, and it must approach these zero values quite smoothly within the box, as otherwise ψ_x would not be a continuous function and the probability interpretation would become ambiguous. The application to (12·35) of the boundary conditions $\psi_x = 0$ at $x = 0$, $x = a$, gives

$$B = 0, \tag{12·36}$$

$$\frac{\pi a}{h}\sqrt{(8m\epsilon_x)} = q_x\pi,$$

where q_x is an integer (not including zero as otherwise ψ_x would be zero everywhere). The last equation may be rewritten

$$\epsilon_x = \frac{h^2}{8m}\left(\frac{q_x}{a}\right)^2. \tag{12·37}$$

Similarly

$$\epsilon_y = \frac{h^2}{8m}\left(\frac{q_y}{b}\right)^2, \tag{12·38}$$

$$\epsilon_z = \frac{h^2}{8m}\left(\frac{q_z}{c}\right)^2, \tag{12·39}$$

where q_y and q_z are also integers. These are the *quantum numbers* and they determine the permissible energies or eigenvalues. It will be noticed that the origin of these translational energy levels is the imposition of the boundary conditions; in the case of unconfined motion there is no quantization and the kinetic energy is continuous. Moreover, the larger are the dimensions a, b and c, the smaller is the separation of adjacent levels, reaching zero when the box is infinitely large (cf. the discussion following equation (12·7)).

It will be noticed from (12·37) that to each possible value of the energy ϵ_x there is only a single choice of the quantum number q_x. The levels ϵ_x are therefore not degenerate and the same applies also to ϵ_y and ϵ_z. The corresponding eigenfunctions are obtained by substituting (12·36) and (12·37) in (12·35):

$$\psi_x = A\sin\frac{q_x\pi x}{a}, \quad \text{etc.} \tag{12·40}$$

On the other hand, the total translational energy is given by

$$\epsilon = \epsilon_x + \epsilon_y + \epsilon_z$$
$$= \frac{h^2}{8m}\left(\frac{q_x^2}{a^2} + \frac{q_y^2}{b^2} + \frac{q_z^2}{c^2}\right), \tag{12·41}$$

and the complete eigenfunction for translation is

$$\psi = \text{constant} \times \frac{\sin q_x\pi x}{a}\frac{\sin q_y\pi y}{b}\frac{\sin q_z\pi z}{c}. \tag{12·42}$$

Now x, y and z are present in (12·42) but not in (12·41). It follows that each choice of the integers q_x, etc., corresponds to a *different* eigenfunction, but possibly to the *same* energy ϵ. For example, if the box is such that $a = b$, then the choice of quantum numbers $q_x = 7$, $q_y = 1$, $q_z = 3$ will give the same energy as the choice $q_x = 5$, $q_y = 5$, $q_z = 3$. Thus if a, b and c are commensurate, it will occur that many of the energy levels ϵ are degenerate.

The translational part of the molecular partition function is given by (12·26)
$$f^{\text{tr.}} = \Sigma \omega_j^{\text{tr.}} \exp[-\epsilon_j^{\text{tr.}}/kT],$$

where the summation is over the energy levels. Alternatively, we can sum over the quantum states
$$f^{\text{tr.}} = \sum_{\text{states}} \exp[-\epsilon_j^{\text{tr.}}/kT],$$

and substituting from (12·41)
$$f^{\text{tr.}} = \sum_{q_x, q_y, q_z} \exp\left[-\frac{h^2}{8mkT}\left(\frac{q_x^2}{a^2} + \frac{q_y^2}{b^2} + \frac{q_z^2}{c^2}\right)\right]. \tag{12·43}$$

This may be factorized
$$f^{\text{tr.}} = f_x f_y f_z, \tag{12·44}$$

where
$$f_x = \sum_{q_x} \exp\left(-\frac{h^2 q_x^2}{8mkTa^2}\right)$$
$$= \sum_{q_x} e^{-gq_x^2}, \quad \text{if} \quad g = \frac{h^2}{8mkTa^2}, \tag{12·45}$$

and similarly for f_y and f_z.

The Boltzmann constant k, which originated in (11·14), has not yet been finally identified with R/L (see equation 11·50). However if it is assumed for the moment that this is the case, it is readily shown, by inserting numerical values, that the coefficient g in (12·45) is exceedingly small for all reasonable values of T and a. It follows that the first term (i.e. $q_x = 1$) in the summation is practically unity, the second term is also practically unity and so on. In fact, the various terms diminish so slowly that the summation may be replaced by an integral.†

A plot of $e^{-gq_x^2}$ as a function of q_x is shown in Fig. 41, where, however, the decrease in the function is much magnified. The value of f_x is the sum of all the vertical lines up to $q_x = \infty$. Bearing in mind that the horizontal separation of the vertical lines is unity, the value of f_x is therefore equal to the area of the rectangles. This again is equal, to a very high degree of approximation, to the area under the dotted curve of the continuous function $e^{-gq_x^2}$, where q_x is no longer limited to integral values. Thus
$$f_x = \sum_{q_x=1}^{\infty} e^{-gq_x^2} = \int_0^{\infty} e^{-gq_x^2} dq_x. \tag{12·46}$$

† Degrees of freedom for which this property holds are usually called *classical*.

The integral is a standard one and we obtain

$$f_x = \tfrac{1}{2}(\pi/g)^{\frac{1}{2}},\qquad(12\cdot47)$$

or inserting the value of g

$$f_x = \frac{a}{h}(2\pi m k T)^{\frac{1}{2}}.\qquad(12\cdot48)$$

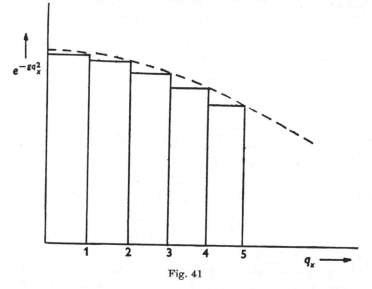

Fig. 41

The same result is obtained for f_y and f_z. Thus finally we have

$$f^{\text{tr.}} = f_x f_y f_z$$
$$= abc\left(\frac{2\pi m k T}{h^2}\right)^{\frac{3}{2}}$$
$$= V\left(\frac{2\pi m k T}{h^2}\right)^{\frac{3}{2}},\qquad(12\cdot49)$$

where V is the volume of the vessel. This result is not limited to rectangular boxes.

12·6. The internal partition function

For methods of determining the internal part of the molecular partition function the reader is referred to more comprehensive accounts in the literature.† Suffice it to say that it is a sum of ex-

† For example, Fowler and Guggenheim, *Statistical Thermodynamics* (Cambridge, 1949); Mayer and Mayer, *Statistical Mechanics* (New York, Wiley, 1940).

ponentials, as in equation (12·27), and the most accurate method of calculating $f^{\text{int.}}$ is to analyse the spectrum of the particular molecule and thereby to obtain the value of each energy level relative to the lowest. If the degeneracies, ω_i, are known from theory, it is therefore possible to calculate the numerical value of each term, $\omega_i e^{-\epsilon_i/kT}$, which contributes significantly to the summation, and these terms can then be added.

An alternative method is based on the approximation discussed in § 12·4, where $f^{\text{int.}}$ was expressed as a product of rotational, vibrational and electronic factors. Analytical expressions for $f^{\text{rot.}}$ and $f^{\text{vib.}}$ may be obtained by the methods of wave mechanics; these expressions are rather complicated, being expressed as series expansions, and will not be quoted here.

The convention is usually adopted of using an energy scale in which the lowest level of each type is taken as having zero energy of that particular kind. The use of (11·43) and (11·44) to calculate free energy and internal energy will therefore give values of these functions relative to the same zero.

Consider the vibrational part of the molecular partition function; according to the above convention the first term, $e^{-\epsilon_0/kT}$, in the summation is just unity. The second, and all succeeding terms, are very much smaller, and this is because the separation of the vibrational levels is usually rather larger than kT. The vibrational partition function is therefore not much larger than unity at room temperature, and the same usually applies *a fortiori* to the electronic part. By contrast, the rotational partition function, for which the spacing of the levels is much closer, is usually of the order of magnitude 10–100 at room temperature, and the corresponding value of $f^{\text{tr.}}$, as given by equation (12·49), is enormously larger still. This does not necessarily imply that it is the translational partition function which always makes the largest contribution to the thermodynamic functions of a gas. For example, from (11·44) it is evident that $C_v = \partial U/\partial T$ is strongly dependent on the rate at which the partition function increases with temperature. In certain regions of temperature this may be larger for the vibrational part than for the translational part, even though the latter is much greater in magnitude.

12·7. Thermodynamic properties of the perfect gas

Using (12·25) and (12·49) we obtain for the molecular partition function of a perfect gas whose volume is V,

$$f = V\left(\frac{2\pi m kT}{h^2}\right)^{\frac{3}{2}} f^{\text{int.}}, \tag{12·50}$$

where m is the mass of each molecule. Hence from (12·8) the partition function Q is

$$Q = \frac{1}{N!} V^N \left(\frac{2\pi m kT}{h^2}\right)^{\frac{3}{2}N} (f^{\text{int.}})^N, \qquad (12·51)$$

where N is the number of molecules in the volume V. We are now in a position to apply the important equations (11·42)–(11·46) which give the macroscopic properties of the gas in terms of T, V and N as the independent variables. Consider first of all the gas pressure, which is given by (11·46):

$$p = kT \left(\frac{\partial \ln Q}{\partial V}\right)_{T, N}.$$

Now in (12·51) the only quantity which is a function of V is the term V^N. In particular, $f^{\text{int.}}$ does not depend on V, and this was implicit in the separability discussed in §§ 12·3 and 12·4. (It is because the potential energy for the internal motions of the molecule does not depend on the distance between one molecule and another in a perfect gas phase.)

Taking the logarithm of (12·51) and differentiating we obtain

$$p = kT \frac{\partial \ln V^N}{\partial V}$$

$$= \frac{NkT}{V}. \qquad (12·52)$$

If there are n moles of gas in the system then $N = Ln$, where L is the Avogadro number. Thus (12·52) may be written

$$pV = n LkT, \qquad (12·53)$$

and this is therefore a derivation of the gas law from the principles of statistical and quantum mechanics. The discussion of § 12·3 shows that it arises because of the negligible interaction between one molecule and another whenever they are far apart, and it was this condition which allowed the wave equation for the whole macroscopic system to be separated into a number of equations, one for each individual molecule.

The last equation may be expressed as $pV \propto nT$. Now the coefficient of proportionality is usually denoted by the symbol R. Hence

$$R = Lk. \qquad (12·54)$$

This serves to identify the Boltzmann constant k as equal to the gas constant per molecule, as was foreshadowed in § 11·11.

Further application of equations (11·42)–(11·46) can only be made if the value of $f^{\text{int.}}$ is known. Methods of calculating this quantity were briefly outlined in the previous section, and it follows from what was said that $f^{\text{int.}}$ can usually be taken as unity in the case of monatomic gases.† In the remainder of this section we shall develop the equations which are applicable to this special case. First of all (12·51) reduces to

$$Q = \frac{1}{N!} V^N \left(\frac{2\pi m k T}{h^2}\right)^{\frac{3}{2}N},\qquad (12\cdot55)$$

and therefore, applying (11·44), we obtain for the internal energy

$$U = kT^2 \left(\frac{\partial \ln Q}{\partial T}\right)_{V,N}$$

$$= kT^2 \frac{\partial \ln T^{\frac{3}{2}N}}{\partial T}$$

$$= \tfrac{3}{2} N k T = \tfrac{3}{2} n R T. \qquad (12\cdot56)$$

This familiar result is also obtainable from the kinetic theory, and it corresponds to the fact that in a monatomic gas the energy is entirely translational. The enthalpy is

$$H = U + pV$$

$$= \tfrac{5}{2} n R T, \qquad (12\cdot57)$$

and the heat capacities per mole are obtained by differentiation and by putting $n - 1$:

$$\left.\begin{array}{l} c_V = \tfrac{3}{2} R, \\ c_p = \tfrac{5}{2} R. \end{array}\right\} \qquad (12\cdot58)$$

The Helmholtz free energy is obtained by substituting (12·55) in (11·43), $\qquad A = -kT \ln Q.$

This may be expressed in the following form, which is obtained by using the Stirling theorem (equation 11·77) for $N!$ and by putting $m = M/L$, where M is the molecular weight of the gas:

$$A = -nRT \left[\ln\left(\frac{V M^{\frac{3}{2}} T^{\frac{3}{2}}}{n}\right) + 1 + \ln \frac{(2\pi k)^{\frac{3}{2}}}{h^3 L^{5/2}} \right]. \qquad (12\cdot59)$$

The corresponding Gibbs free energy is

$$G = A + PV$$

$$= A + nRT$$

$$= -nRT \left[\ln\left(\frac{V M^{\frac{3}{2}} T^{\frac{3}{2}}}{n}\right) + \ln \frac{(2\pi k)^{\frac{3}{2}}}{h^3 L^{5/2}} \right]. \qquad (12\cdot60)$$

† If there is appreciable electronic excitation $f^{\text{int.}}$ cannot be taken as unity, even in the case of a monatomic gas. This occurs in the case of the alkali metal vapours where there are two electronic states of equal energy. In this instance $f^{\text{int.}} = f^{\text{elec.}} = 2$ very nearly.

This may be expressed more conveniently by using p, T and n as variables in place of V, T and n, by means of (12·53):

$$G = -nRT\left[\ln\left(\frac{M^{\frac{3}{2}}T^{\frac{5}{2}}}{p}\right) + \ln\left(\frac{2\pi}{L}\right)^{\frac{3}{2}}\frac{k^{\frac{5}{2}}}{h^3}\right]. \qquad (12·61)$$

The chemical potential, which in the case of a single component gas is simply G/n, is obtained by putting $n=1$ in the last equation. If p is expressed in atmospheres, the insertion of the numerical values of k, h and L gives

$$\mu = -RT\left[\ln\left(\frac{M^{\frac{3}{2}}T^{\frac{5}{2}}}{p}\right) - 3·66\right]. \qquad (12·62)$$

Finally, for the entropy of the gas, we obtain from (12·57) and (12·61)

$$S = (H - G)/T$$

$$= nR\left[\ln\left(\frac{M^{\frac{3}{2}}T^{\frac{5}{2}}}{p}\right) + \frac{5}{2} + \ln\left(\frac{2\pi}{L}\right)^{\frac{3}{2}}\frac{k^{\frac{5}{2}}}{h^3}\right] \qquad (12·63)$$

$$= nR\left[\ln\left(\frac{M^{\frac{3}{2}}T^{\frac{5}{2}}}{p}\right) - 1·16\right], \qquad (12·64)$$

where the units of pressure are again in atmospheres. This is known as the Sackur–Tetrode equation and was first obtained in 1912, in the early days of the quantum theory.

Although the above equations are applicable only when $f^{\text{int.}}$ is unity, they may be regarded as giving those parts of the internal energy, entropy, etc., of a polyatomic gas which are due to the translational motion. Equations analogous to those above may be readily obtained for the more general case. For example, the equation for the chemical potential is

$$\mu = -RT\left[\ln\left(\frac{M^{\frac{3}{2}}T^{\frac{5}{2}}}{p}\right) + \ln\left(\frac{2\pi}{L}\right)^{\frac{3}{2}}\frac{k^{\frac{5}{2}}}{h^3} + \ln f^{\text{int.}}\right], \qquad (12·65)$$

and this can be evaluated if $f^{\text{int.}}$ is known, for example, from band spectra. It is one of the great achievements of statistical mechanics that it has made possible a method of determining the thermo-dynamic functions of polyatomic molecules, a method which is at least as accurate as that which is based on calorimetry and the measurement of equilibrium constants. For example, the equilibrium constant of the reaction

$$CO + H_2O = CO_2 + H_2$$

at 1000 K is 1.37 as calculated† from spectroscopic data‡ on the various substances and is 1·5 according to the direct measurements of Haber and Hahn. The latter figure is based on gas analysis and is perhaps less accurate than the former.

It may be remarked that the above equations lead to perfectly definite numerical values for that part of the energy, entropy, etc., which is due to the translational motion. For example, from (12·56) it may be calculated that the kinetic energy of translation at 300 K is 3.74×10^3 J mol^{-1}, and of course, this is the value of the internal energy relative to the energy of the same molecules when they are at rest. Similarly, if equation (12·64) is evaluated for a gas of molecular weight 44 (CO_2) at 300 K and 1 atm the result is an entropy of 159 J K^{-1} mol^{-1}.

The fact that this is an *absolute* value arises from the statistical definition of entropy in equation (11·14). According to this equation the entropy would be zero if the system were known to be in a single quantum state. This point will be discussed in more detail in the next chapter in connexion with the third law. For the moment it may be noted that equation (12·64) leads to an apparent paradox—as T approaches zero it appears that S approaches an infinitely negative value, whereas the least value of S should be just zero, as occurs when the system is known to be in the single quantum state. This difficulty is due to the fact that equation (12·8), on which the equations of the present section are based, becomes invalid at very low temperature. Under such conditions the Boltzmann statistics must be replaced by Einstein–Bose or Fermi–Dirac statistics.

In conclusion to this section it is of interest to obtain an expression for the equilibrium constant of a gas reaction in terms of molecular partition functions. In equation (12·10) we obtained an expression for the partition function of a mixture of *two* gases, and the corresponding expression for the partition function of a mixture of *three* gases A, B and C is

$$Q = \frac{1}{N_a! N_b! N_c!} f_a^{N_a} f_b^{N_b} f_c^{N_c},$$

where N_a, etc., are the numbers of molecules of each species in the volume V of the system and f_a, etc., are the molecular partition functions. The Helmholtz free energy of the system is obtained by substituting the above expression in (11·43):

$$A = -kT \ln Q$$
$$= -kT(N_a \ln f_a + N_b \ln f_b + N_c \ln f_c - \ln N_a! - \ln N_b! - \ln N_c!).$$

† Guggenheim, *Thermodynamics*, §7·10.

‡ Together with a value for the heat of reaction. This is needed in order to fix the energy zero of the products relative to those of the reactants, i.e. the relative energies of the lowest levels or ground states.

Let it be supposed that there is a reaction

$$A + B = C,$$

so that $dN_a = dN_b = -dN_c$. Differentiating the above equation at constant temperature and volume and using these relations, together with Stirling's theorem, we obtain

$$\left(\frac{\partial A}{\partial N_c}\right)_{T,V} = -kT(-\ln f_a - \ln f_b + \ln f_c + \ln N_a + \ln N_b - \ln N_c).$$

Now the condition of equilibrium is $(\partial A/\partial N_c)_{T,V} = 0$, as follows from §2·8. Hence it follows that

$$\frac{N_c}{N_a N_b} = \frac{f_c}{f_a f_b}.$$

Let $[C]$, etc., denote the number of *molecules* per unit volume. Then $N_c = V[C]$, etc. From (12·50) we also have the result that the partition functions f_c, etc., are proportional to the volume V of the system. Let ϕ'_c, etc., denote partition functions per unit volume, thus $\phi'_c \equiv f_c/V$, etc. Then we readily obtain

$$\frac{[C]}{[A][B]} = \frac{\phi'_c}{\phi'_a \phi'_b}.$$

In this expression the partition functions ϕ'_c, etc., must all be calculated relative to the same zero. On the other hand, it is more convenient to calculate the partition function of each molecule relative to its own lowest level taken as having zero energy. Relative to any arbitrary zero let the lowest level of the A, B and C molecules have energies ϵ_a, ϵ_b and ϵ_c respectively. From equation (12·9) it is evident that the terms $e^{-\epsilon_a/kT}$, etc., may be factorized out of each of the molecular partition functions. The last equation can therefore be rewritten

$$\frac{[C]}{[A][B]} = \frac{\phi_c}{\phi_a \phi_b} e^{-(\epsilon_c - \epsilon_a - \epsilon_b)/kT},$$

where ϕ_c, etc., are now calculated relative to the lowest level of each molecule as having zero energy. This relation may also be expressed in the form

$$\frac{[C]}{[A][B]} = \frac{\phi_c}{\phi_a \phi_b} e^{-E_0/RT}, \qquad (12·66)$$

when E_0 is the difference in the zero-level energy *per mole* of the reactants and products.

Finally it may be remarked that, on account of the development of spectroscopic methods of calculating thermodynamic quantities, and also the extension of calorimetry to very low temperatures, it has become customary to tabulate these quantities in the form of values of $(H^0 - H_0^0)/T$ and $(G^0 - H_0^0)/T$, or preferably the dimensionless ones $(H^0 - H_0^0)/RT$ and $(G^0 - H_0^0)/RT$, whose meaning will be briefly described.†

As noted already in §12·6, the convention is usually adopted of evaluating the partition function as if the lowest levels of each type had zero energy. This is equivalent to the factoring out of a term ϵ_0/kT from the partition function and this is quite satisfactory. But when it is a question of chemical reaction the existence of the factored out term must very definitely be allowed for, since the ϵ_0 values of different molecules are in general not the same. In other words there exists a heat of reaction at absolute zero, this being essentially the algebraic sum of the ϵ_0's for reactants and products.

This may be formally taken into account on the following basis. Let ΔG_T^0 and ΔH_T^0 be the changes in the standard Gibbs free energy and enthalpy respectively at the temperature T. These may be expanded in the forms:

$$\Delta G_T^0 = \Delta H_0^0 + \Delta \left(\frac{G_T^0 - H_0^0}{RT} \right). \; RT$$

and

$$\Delta H_T^0 = \Delta H_0^0 + \Delta \left(\frac{H_T^0 - H_0^0}{RT} \right). \; RT$$

The term ΔH_0^0 may be taken as being the energy change in the reaction at the absolute zero and can be determined by applying the second of the above equations using a measured value of ΔH_T^0 together with the statistically calculated value of the second term on the right-hand side. It follows that the tabulated values of $(G^0 - H_0^0)/RT$ and $(H^0 - H_0^0)/RT$ (where the subscript $_T$ has now been omitted) can be used as effectively, in conjunction with the equations of previous chapters, for the calculation of equilibrium constants as any of the other equivalent forms of tabulation.

12·8. The Maxwell–Boltzmann distribution

Under this heading we are concerned with the question: what is the average number of molecules in a sample of a perfect gas which are in some particular energy state ϵ_i? To start with, it will perhaps be clearest if we deal with the *quantum* states of the molecules, rather than with the energy states, since the latter may be degenerate. Let the quantum states be numbered $0, 1, 2, \ldots, i, j, \ldots$.

Over any large period of time let p_{i1} be the fraction of the time that there is just one molecule in the sample of gas in the ith quantum state, let p_{i2} be the fraction of the time that there are two molecules in the ith

† Sometimes E_0^0 is used in place of H_0^0. This is of no significance since the energy and enthalpy of a perfect gas become equal at the absolute zero.

state, etc. The mean number of molecules in the ith state is therefore

$$\bar{n}_i = 1 p_{i1} + 2 p_{i2} + 3 p_{i3} + \dots . \tag{12·67}$$

Now, as discussed already in § 12·1,† the number of quantum states per molecule is enormously larger than the number of molecules in the sample. (This is the basis of the Maxwell–Boltzmann statistics and applies under all conditions of temperature and pressure at which a gas is effectively perfect.) In any sample of gas it is therefore very improbable that there will be more than one molecule in a particular quantum state —in fact, the *mean* number, $\bar{n}_i$, will be very much less than unity. In the above equation we can therefore neglect all p's except p_{i1} and write

$$\bar{n}_i = p_{i1}.$$

Alternatively—and this makes our discussion easier—we can put

$$\bar{n}_i = p_i, \tag{12·68}$$

where p_i is the probability that there is *at least one* molecule in the ith quantum state, this being effectively the same as the probability p_{i1}, that there is *only one* molecule in this state.

For simplicity consider a gas sample containing just three identical molecules. A quantum state of the whole system will be specified by saying how many molecules there are in each of the molecular quantum states, but without seeking to distinguish between these molecules, which is impossible. An expression for the probability of this state has already been obtained in equation (11·42). For example, the probability that one molecule is the zeroth molecular quantum state, a second also in the zeroth, and the third in the ith is

$$P = e^{\beta(\epsilon_0 + \epsilon_0 + \epsilon_i)}/Q,$$

where β has been written in place of $-1/kT$. Similarly, the probability that there is one molecule in the zeroth state, one in the first and the third again in the ith is

$$P = e^{\beta(\epsilon_0 + \epsilon_1 + \epsilon_i)}/Q.$$

The total probability that there is *at least one* molecule in the ith state is obtained by summing ‡ all expressions of the above kind in which at least one of the ϵ's is an ϵ_i. Thus

$$p_i = \frac{e^{\beta \epsilon_i}}{Q} \left[e^{\beta(\epsilon_0 + \epsilon_0)} + e^{\beta(\epsilon_0 + \epsilon_1)} + e^{\beta(\epsilon_0 + \epsilon_2)} + \dots \right.$$

$$+ e^{\beta(\epsilon_1 + \epsilon_1)} + e^{\beta(\epsilon_1 + \epsilon_2)} + \dots$$

$$+ e^{\beta(\epsilon_2 + \epsilon_2)} + \dots$$

$$\left. + \dots \right]. \tag{12·69}$$

Consider the expression in brackets. Going back to § 12·1 it is seen that

† See also Problem 1 at the end of the chapter.

‡ Note that we are concerned here with a *summation* of probabilities. The events represented, for example, by the two P's quoted above *exclude* each other and do not occur simultaneously.

this expression is a correct form for the partition function, Q, of a system consisting of only *two* identical molecules. In fact, this expression is the same as equation (12·3) with the identical terms in the latter left out. Therefore (12·69) can be written

$$p_i = e^{\beta \epsilon_i} \frac{Q_2}{Q_3},$$

where Q_2 and Q_3 are the partition functions for systems of two and three identical molecules respectively. In general, for a system which actually contains N molecules, the probability that at least one of them is in the ith state is

$$p_i = e^{-\epsilon_i/kT} \frac{Q_{N-1}}{Q_N}.$$

Substituting the values of Q_{N-1} and Q_N as given by (12·7)

$$p_i = e^{-\epsilon_i/kT} \frac{N!}{(N-1)!} \frac{(\Sigma \omega_i e^{-\epsilon_i/kT})^{N-1}}{(\Sigma \omega_i e^{-\epsilon_i/kT})^N}$$

$$= \frac{N e^{-\epsilon_i/kT}}{\Sigma \omega_i e^{-\epsilon_i/kT}}. \tag{12·70}$$

The denominator is a summation over molecular energy levels and is the same as the molecular partition function, f, of equation (12·9).

Combining (12·68) and (12·70) we thus obtain

$$\bar{n}_i = \frac{N e^{-\epsilon_i/kT}}{\Sigma \omega_i e^{-\epsilon_i/kT}} \tag{12·71}$$

as an expression for the mean number of molecules in the ith quantum state. Let $\bar{N}_i$ be the mean number of molecules in the ith energy level. If this is ω_i-fold degenerate then—since each of the ω_i quantum states is equally probable—we have

$$\bar{N}_i = \omega_i \bar{n}_i$$

$$= \frac{N \omega_i e^{-\epsilon_i/kT}}{\Sigma \omega_i e^{-\epsilon_i/kT}}. \tag{12·72}$$

These equations express the quantal analogue of the Maxwell–Boltzmann distribution law, and they are more general in their scope than the Maxwell expression for the distribution of velocities (the latter will be derived from (12·72) in § 12·11).

Equation (12·71) can be used to give a verbal interpretation to f, which is its denominator. If, in accordance with the usual convention, we take the lowest level as having zero energy, then from (12·71)

$$\frac{\bar{n}_0}{N} = \frac{1}{f}. \tag{12·73}$$

The molecular partition function may thus be interpreted as equal to $N/\bar{n}_0$, the ratio of the total number of molecules to the number in the lowest quantum state. Equation (12·72) can similarly be used to give a meaning to each term which occurs in f; for example, the term $\omega_i e^{-\epsilon_i/kT}$ is proportional to the mean population $\bar{N}_i$ of the particular level ϵ_i. Thus each term in the molecular partition function is proportional to the probability of the energy level to which it refers. The ratio of the population numbers of two levels, ϵ_i and ϵ_j, is obtained from (12·72)

$$\frac{\bar{N}_i}{\bar{N}_j} = \frac{\omega_i \, \mathrm{e}^{-\epsilon_i/kT}}{\omega_j \, \mathrm{e}^{-\epsilon_j/kT}}, \tag{12·74}$$

and similarly the mean total population of any *two* levels taken together, for example, the ith and jth levels, is

$$\bar{N}_i + \bar{N}_j = \frac{N}{f} (\omega_i \, \mathrm{e}^{-\epsilon_i/kT} + \omega_j \, \mathrm{e}^{-\epsilon_j/kT}). \tag{12·75}$$

12·9. Distribution over translational and internal states

The ideas of the last section may be extended to the various parts of the factorized partition function. As discussed previously, any molecular energy level ϵ_i may also be specified by stating that the molecule is in the jth translational level and the kth internal level. From equations (12·21) and (12·23)

$$\epsilon_i = \epsilon_j^{\mathrm{tr.}} + \epsilon_k^{\mathrm{int.}},$$

$$\omega_i = \omega_j^{\mathrm{tr.}} \, \omega_k^{\mathrm{int.}}.$$

Substituting in (12·72)

$$\bar{N}_{j,k} = \frac{N\omega_j^{\mathrm{tr.}} \omega_k^{\mathrm{int.}} \exp\left[-(\epsilon_j^{\mathrm{tr.}} + \epsilon_k^{\mathrm{int.}})/kT\right]}{f}, \tag{12·76}$$

where $\bar{N}_{j,k}$ is the mean number of molecules which are in the jth translational level and at the same time in the kth internal level. Let $\bar{N}_j^{\mathrm{tr.}}$ be the number of molecules in the jth translational level, irrespective of the internal level of the molecules. This is obtained by summing (12·76) over all internal levels. Thus

$$\bar{N}_j^{\mathrm{tr.}} = \sum_k \bar{N}_{j,k}$$

$$= \frac{N\omega_j^{\mathrm{tr.}} \exp\left[-\epsilon_j^{\mathrm{tr.}}/kT\right]}{f} \sum_k \omega_k^{\mathrm{int.}} \exp\left[-\epsilon_k^{\mathrm{int.}}/kT\right].$$

Here the summation is just the internal part of the molecular partition function, as given by equation (12·27). Therefore, using (12·25) and (12·27)

$$\bar{N}_j^{\mathrm{tr.}} = \frac{N\omega_j^{\mathrm{tr.}} \exp\left[-\epsilon_j^{\mathrm{tr.}}/kT\right]}{f^{\mathrm{tr.}}}. \tag{12·77}$$

Similarly, the number of molecules, $\bar{N}_k^{\mathrm{int.}}$, in the kth internal level, irrespective of their translational states, is

$$\bar{N}_k^{\mathrm{int.}} = \frac{N\omega_k^{\mathrm{int.}} \exp\left[-\epsilon_k^{\mathrm{int.}}/kT\right]}{f^{\mathrm{int.}}}. \tag{12·78}$$

The same property of factorization applied to the internal energy states —in so far as there is approximate separability of energies—allows similar equations to be obtained relating to rotational, vibrational and electronic distribution functions. For example, the number of molecules in the ith rotational level, irrespective of the particular translational, vibrational and electronic level, is

$$\bar{N}_i^{\text{rot.}} = \frac{N\omega_i^{\text{rot.}}\exp\left[-\epsilon_i^{\text{rot.}}/kT\right]}{f^{\text{rot.}}}. \qquad (12\cdot79)$$

In brief, the distribution of a particular type of energy, if it is separable, can be discussed without reference to the other types of energy; each distribution function is determined only by its own partition function. It will be apparent also that the larger is the value of, say, $f^{\text{rot.}}$, the smaller is the fraction of all the molecules which populate any *single* rotational level. The magnitude of a molecular partition function, or any factorized part of it, is a measure of the total number of levels which can be appreciably populated at the given temperature.

12·10. Number of translational states of a given energy

Before proceeding to the Maxwell distribution of velocities it will be useful to derive equation (11·59). This was assumed without proof in the appendix to the last chapter. It is

$$\Omega_k = \alpha E_k^{\frac{3}{2}N},$$

where Ω_k is the number of quantum states of a macroscopic sample consisting of N atoms of a monatomic gas of total energy E_k. (The expression will also give the number of *translational* quantum states of a polyatomic gas.)

From (12·37) the quantum number q_x is given by

$$q_x = \frac{a}{h}\sqrt{(8m\epsilon_x)}, \qquad (12\cdot80)$$

and this is therefore the number of translational quantum states of a single molecule corresponding to an x component of energy which lies between zero and a value ϵ_x. The number of quantum states which correspond to an energy which lies between ϵ_x and $\epsilon_x + \Delta\epsilon_x$ is therefore

$$\Delta q_x = \frac{a}{h}(8m)^{\frac{1}{2}}\left[(\epsilon_x + \Delta\epsilon_x)^{\frac{1}{2}} - \epsilon_x^{\frac{1}{2}}\right]$$

$$= \frac{a}{h}(8m)^{\frac{1}{2}}\epsilon_x^{\frac{1}{2}}\left[\left(1 + \frac{\Delta\epsilon_x}{\epsilon_x}\right)^{\frac{1}{2}} - 1\right],$$

and if $\Delta\epsilon_x$ is small compared to ϵ_x this expression can be approximated by

$$\Delta q_x = \frac{a}{h}(8m)^{\frac{1}{2}}\epsilon_x^{\frac{1}{2}}\left(1 + \frac{1}{2}\frac{\Delta\epsilon_x}{\epsilon_x} - 1\right)$$

$$= \frac{a}{h}\left(\frac{2m}{\epsilon_x}\right)^{\frac{1}{2}}\Delta\epsilon_x. \qquad (12\cdot81)$$

In fact, for large values of q_x and ϵ_x these variables can be regarded as continuous, and we obtain

$$dq_x = \frac{a}{h}\left(\frac{2m}{\epsilon_x}\right)^{\frac{1}{2}}d\epsilon_x, \qquad (12\cdot82)$$

which is the result of a direct differentiation of (12·80). Similarly

$$dq_y = \frac{b}{h} \left(\frac{2m}{\epsilon_y}\right)^{\frac{1}{2}} d\epsilon_y,$$

$$dq_z = \frac{c}{h} \left(\frac{2m}{\epsilon_z}\right)^{\frac{1}{2}} d\epsilon_z.$$

The total number of translational quantum states for which the energy lies in the range ϵ_x to $\epsilon_x + d\epsilon_x$, ϵ_y to $\epsilon_y + d\epsilon_y$ and ϵ_z to $\epsilon_z + d\epsilon_z$ is therefore

$$dq_x dq_y dq_z = \frac{V}{h^3} (2m)^{\frac{3}{2}} \frac{d\epsilon_x d\epsilon_y d\epsilon_z}{\epsilon_x^{\frac{1}{2}} \epsilon_y^{\frac{1}{2}} \epsilon_z^{\frac{1}{2}}}, \tag{12·83}$$

where $V = abc$ is the volume of the gas.

We now ask, what is the number g of translational quantum states of a single molecule when its total energy $\epsilon_x + \epsilon_y + \epsilon_z$ lies between zero and some particular value ϵ? This is the integral of (12·83):

$$g = \frac{V(2m)^{\frac{3}{2}}}{h^3} \iiint \frac{d\epsilon_x d\epsilon_y d\epsilon_z}{\epsilon_x^{\frac{1}{2}} \epsilon_y^{\frac{1}{2}} \epsilon_z^{\frac{1}{2}}}, \tag{12·84}$$

where the integration is to be taken over all positive values of ϵ_x, ϵ_y and ϵ_z subject to the condition

$$\epsilon_x + \epsilon_y + \epsilon_z \leqslant \epsilon. \tag{12·85}$$

This is the type of integral known by the name of Dirichlet. The integral

$$I = \int \cdots \iint \epsilon_1^{p-1} \epsilon_2^{p-1} \ldots \epsilon_N^{p-1} d\epsilon_1 d\epsilon_2 \ldots d\epsilon_N, \tag{12·86}$$

subject to the condition $\Sigma \epsilon_i \leqslant \epsilon$,

has the value

$$I = \frac{[\Gamma(p)]^N \, \epsilon^{Np}}{\Gamma(Np + 1)}, \tag{12·87}$$

where $\Gamma(p)$ is the gamma function,[†] which is a generalization of the factorial. (For example, if p is an integer $\Gamma(p) = (p-1)!$. In general, $\Gamma(p)$ is the 'smoothest' function which takes on the values $(p-1)!$ at the integers.) Numerical values of Γ are to be found in mathematical tables[‡] and two values we shall need are

$$\Gamma(\tfrac{1}{2}) = \pi^{\frac{1}{2}} \quad \text{and} \quad \Gamma(\tfrac{5}{2}) = \tfrac{3}{4}\pi^{\frac{1}{2}}.$$

In the case of (12·84) we have $p = \tfrac{1}{2}$ and $N = 3$. Hence

$$g = \frac{V(2m)^{\frac{3}{2}}}{h^3} \frac{(\pi^{\frac{1}{2}})^3}{\tfrac{3}{4}\pi^{\frac{1}{2}}} \epsilon^{\frac{3}{2}}$$

$$= \tfrac{4}{3}\pi V \frac{(2m)^{\frac{3}{2}}}{h^3} \epsilon^{\frac{3}{2}}. \tag{12·88}$$

[†] See, for example, F. S. Woods, *Advanced Calculus* (Boston, Ginn and Co., 1939), Chapter VII; and Margenau and Murphy, *The Mathematics of Physics and Chemistry* (New York, Van Nostrand, 1943), §3·2.

[‡] For example, Milne-Thomson and Comrie, *Standard Four Figure Mathematical Tables* (London, Macmillan, 1948).

This result could actually have been obtained more simply,† but it was useful to employ the method based on the gamma function because this method has an important application in reaction kinetics, especially for calculating the number of molecules having total energy greater than or less than a given value in S degrees of freedom.

The differential of (12·88) gives the number of translational quantum states of a single molecule whose energy lies in the range ϵ to $\epsilon + d\epsilon$:

$$dg = 2\pi V \frac{(2m)^{\frac{3}{2}}}{h^3} \epsilon^{\frac{1}{2}} d\epsilon. \qquad (12\cdot89)$$

If there are N molecules in the system—and let it be supposed for the moment that they are all distinguishable from each other—then the total number of translational states for the whole system when the translational energy of the first molecule lies in the range $d\epsilon_1$, of the second in the range $d\epsilon_2$, etc., is obtained from the last equation and is

$$dg_1 dg_2 \ldots dg_N = \left[2\pi V \frac{(2m)^{\frac{3}{2}}}{h^3} \right]^N \epsilon_1^{\frac{1}{2}} \epsilon_2^{\frac{1}{2}} \ldots \epsilon_N^{\frac{1}{2}} d\epsilon_1 d\epsilon_2 \ldots d\epsilon_N.$$

As before, we ask what is the number G of translational quantum states of the whole system of N molecules when their total energy is less than some particular value E? This is the integral of the last expression, i.e.

$$G = \left[2\pi V \frac{(2m)^{\frac{3}{2}}}{h^3} \right]^N \int \ldots \int \epsilon_1^{\frac{1}{2}} \epsilon_2^{\frac{1}{2}} \ldots \epsilon_N^{\frac{1}{2}} d\epsilon_1 d\epsilon_2 \ldots d\epsilon_N,$$

subject to the condition $\epsilon_1 + \epsilon_2 + \ldots + \epsilon_N \leqslant E$.

Evaluating the Dirichlet integral as before, we obtain

$$G = \left[2\pi V \frac{(2m)^{\frac{3}{2}}}{h^3} \right]^N \frac{[\Gamma(\frac{3}{2})]^N}{\Gamma(\frac{3}{2}N + 1)} E^{\frac{3}{2}N}. \qquad (12\cdot90)$$

If the molecules are actually identical this must be divided‡ by $N!$ because an interchange of particles does not give rise to a new quantum state of the system. However, for the present purposes, this is not important; the essential result is

$$G = CE^{\frac{3}{2}N}, \qquad (12\cdot91)$$

where C depends on V and N. The number of translational quantum states of the system whose energy lies in the small range between E and $E + \delta E$ is thus

$$\delta G = \frac{3}{2} NCE^{\frac{3}{2}N-1}\delta E,$$

or since N is very large

$$\delta G = C' E^{\frac{3}{2}N}\delta E, \qquad (12\cdot92)$$

where C' again depends on V and N.

Now throughout the above discussion E has been treated as a continuous function, whereas actually it is quantized. If we take δE as being the energy separation between one level and the next, then the number

† See, for example, Mayer and Mayer, *Statistical Mechanics* (New York, Wiley, 1940), §2i.

‡ It is assumed, as in §12·1, that the molecular quantum states which are occupied by more than one molecule may be neglected.

Ω_k of quantum states having the energy E_k is the increase in G in passing from the jth level to the adjacent kth level about it. Therefore

$$\Omega_k = C' E_k^{\frac{3}{2}N} \delta E$$

$$= \alpha E_k^{\frac{3}{2}N}, \qquad (12\cdot93)$$

where $\alpha = C'\delta E$. This is equation (11·59) which was used in §11·13; for the purposes of that section the meaning of α was of no importance, and the proof of the canonical distribution depended only on the fact that Ω varies as a very high power of E.

12·11. The Maxwell velocity distribution

According to (12·77) the mean number of molecules whose translational energy level is ϵ_j, irrespective of their particular internal level, is

$$\bar{N}_j = \frac{N\omega_j \mathrm{e}^{-\epsilon_j/kT}}{f^{\mathrm{tr.}}}.$$

The number $\bar{n}_j$ of molecules in each of the quantum states which is comprised within ϵ_j is obtained by dividing by ω_j:

$$\bar{n}_j = \frac{N\mathrm{e}^{-\epsilon_j/kT}}{f^{\mathrm{tr.}}}$$

$$= \frac{N\mathrm{e}^{-(\epsilon_x+\epsilon_y+\epsilon_z)/kT}}{f^{\mathrm{tr.}}}, \qquad (12\cdot94)$$

where ϵ_x, ϵ_y and ϵ_z correspond to the notation of §12·5. Let $\bar{n}_x$ be the mean number of molecules whose x component of translational energy is ϵ_x, irrespective of the y and z components. This is obtained by summing the last expression over all values of ϵ_y and ϵ_z, in the manner of §12·9. We thus obtain

$$\bar{n}_x = \frac{N\mathrm{e}^{-\epsilon_x/kT}}{f_x}, \qquad (12\cdot95)$$

where f_x is the part of the molecular partition function for translation in the x direction and is given by (12·48). Equation (12·95) thus refers to the molecules which have a unique value of the quantum number q_x, but any values whatever of q_y and q_z. We now ask, What is the number of molecules whose x component of translational energy lies in the small range ϵ_x to $\epsilon_x + \mathrm{d}\epsilon_x$? The number of quantum states which lie in this range is given by (12·82):

$$\mathrm{d}q_x = \frac{a}{h}\left(\frac{2m}{\epsilon_x}\right)^{\frac{1}{2}} \mathrm{d}\epsilon_x.$$

Multiplying together the last two expressions and inserting the value of f_x from (12·48) we find the required number as

$$\frac{N}{(\pi kT\epsilon_x)^{\frac{1}{2}}} \mathrm{e}^{-\epsilon_x/kT} \mathrm{d}\epsilon_x. \qquad (12\cdot96)$$

Now if v_x is the x component of velocity of these molecules and if $|v_x|$

is its scalar magnitude, we have

$$\epsilon_x = \frac{m}{2} \, |v_x|^2,$$

and thus
$$d\epsilon_x = (2m\epsilon_x)^{\frac{1}{2}} d\,|v_x|. \qquad (12·97)$$

Eliminating $d\epsilon_x$ between (12·96) and (12·97) and dividing the resulting expression by N, we find that the *fraction* of all the N molecules whose magnitude· of the x component of velocity lies in the range $|v_x|$ to $|v_x| + d\,|v_x|$ is

$$\left(\frac{2m}{\pi kT}\right)^{\frac{1}{2}} \exp[-mv_x^2/2kT]\,d\,|v_x|. \qquad (12·98)$$

In this expression $|v_x|$ can only be positive, since it refers to a scalar magnitude. By using the standard integrals tabulated at the end of the chapter it is readily confirmed that the integral of (12·98) is unity when taken over the range 0 to ∞. On the other hand, if, in place of $|v_x|$, we consider v_x itself, its values can range from $-\infty$ to $+\infty$. The fraction of the N molecules for which the x component of velocity lies between v_x and $v_x + dv_x$ is therefore obtained by dividing (12·98) by 2 and is†

$$\left(\frac{m}{2\pi kT}\right)^{\frac{1}{2}} \exp[-mv_x^2/2kT]\,dv_x, \qquad (12·99)$$

whose integral is unity over the range $-\infty$ to $+\infty$.

Entirely analogous expressions may be obtained relating to the y and z components. The product of all three fractions is the fraction of the N molecules for which the components of velocity lie simultaneously in the ranges dv_x, dv_y and dv_z. This is

$$\left(\frac{m}{2\pi kT}\right)^{\frac{3}{2}} e^{-mv^2/2kT}\,dv_x dv_y dv_z, \qquad (12·100)$$

where
$$v^2 = v_x^2 + v_y^2 + v_z^2. \qquad (12·101)$$

Finally, we may ask, what is the fraction of the N molecules for which the scalar magnitude of velocity lies between v and $v + dv$, without specifying the range of the component velocities? From (12·94), the fraction of the molecules in the *quantum state* for which the translational eigenvalue is ϵ is

$$\frac{e^{-\epsilon/kT}}{f^{\text{tr.}}}.$$

The number of such quantum states within the range ϵ to $\epsilon + d\epsilon$ is given by (12·89)

$$dg = 2\pi V \frac{(2m)^{\frac{3}{2}}}{h^3} \epsilon^{\frac{1}{2}} d\epsilon.$$

Multiplying together the last two expressions and inserting the value of $f^{\text{tr.}}$ from (12·49), we find that the fraction of the molecules for which the translational energy lies between ϵ and $\epsilon + d\epsilon$ is

† Half the molecules comprised in (12·98) have positive values of v_x and the other half have negative values.

$$\frac{2\pi}{(\pi kT)^{\frac{3}{2}}} \epsilon^{\frac{1}{2}} e^{-\epsilon/kT} d\epsilon. \tag{12·102}$$

If the corresponding velocity lies between v and $v + dv$, then

$$\epsilon = \tfrac{1}{2}mv^2$$

and
$$d\epsilon = (2m\epsilon)^{\frac{1}{2}} dv. \tag{12·103}$$

Therefore the fraction of the N molecules for which the scalar magnitude of velocity lies between v and $v + dv$ is

$$4\pi \left(\frac{m}{2\pi kT} \right)^{\frac{3}{2}} e^{-mv^2/2kT} v^2 dv. \tag{12·104}$$

This expression may also be obtained directly from (12·100) by transformation to spherical polar co-ordinates, followed by integration over the surface of a sphere. (The same process is implicit in the relationship between (12·82) and (12·89) which has been used in the above derivation.)

In the various forms of the Maxwell equation it will be noticed that Planck's constant h has cancelled between numerator and denominator. It appears therefore as if the equations might be independent of the quantum theory and they were, of course, obtained by Maxwell before this theory was developed. However, it is to be remembered that the results of the present chapter depend on the particles being very sparsely distributed over the quantum states.† Under the same conditions the separation of the translational states is very small compared to kT, and therefore the translational energy is virtually continuous, as was supposed by Maxwell. This is an aspect of the fact that quantum behaviour converges towards classical behaviour at high values of the quantum numbers, in accordance with Bohr's correspondence principle.

12·12. Principle of equipartition

This principle has a very limited applicability, because it is really a classical rather than a quantum theorem. Its derivation from the quantum point of view which has been adopted throughout this chapter may be outlined rather roughly as follows. Let it be supposed that the energy of a molecule is separable into a number of parts, ϵ', ϵ'', etc., and consider any of these, say ϵ', which depends on only a single variable or degree of freedom. (This applies, for example, to the x component of kinetic energy whose eigenvalues are determined by (12·31).) The fraction of the molecules in each of the quantum states whose energy is ϵ'_k, in the class of energies denoted by a prime, is given by an equation of the type of (12·95)

$$\frac{e^{-\epsilon'_k/kT}}{f'}, \tag{12·105}$$

where f' is the appropriate part of the molecular partition function. The molecules in the system have a mean energy of this class which is obtained by multiplying each of the quantized values, ϵ'_i, ϵ'_k, etc., by the corre-

† Otherwise the Fermi–Dirac or Einstein–Bose statistics must be used, as in the case of the electrons in metals.

sponding fraction (12·105) and summing over all quantum states. Thus the mean energy is

$$\bar{\epsilon}' = \frac{\Sigma \epsilon_k' \, e^{-\epsilon_k'/kT}}{\Sigma e^{-\epsilon_k'/kT}}.$$
(12·106)

If the separation of successive levels is small compared to kT, these summations may be expressed as integrals, as discussed in § 12·5. We obtain

$$\bar{\epsilon}' = \int_0^\infty \epsilon' \, e^{-\epsilon'/kT} \, dq' \Big/ \int_0^\infty e^{-\epsilon'/kT} \, dq',$$
(12·107)

where q' is the appropriate quantum number, of which ϵ' is now regarded as a continuous function. Now in certain instances ϵ' varies as the square of q':

$$\epsilon' = Cq'^2.$$
(12·108)

This has already been shown to be the case with regard to the components of translational energy, equations (12·37)–(12·39), and is also approximately true of the energy of rotation about an axis. Substituting (12·108) in (12·107) and evaluating the integrals, which are now of a standard form,† we obtain

$$\bar{\epsilon}' = \tfrac{1}{2}kT.$$
(12·109)

Therefore a molecule has an average energy, of the type in question, equal to $\tfrac{1}{2}kT$ and the corresponding contribution to the heat capacity of the gas is $\tfrac{1}{2}k$ per molecule or 4.157 J K^{-1} mol^{-1}. This result is in accordance with equations (12·56) and (12·58) which were concerned with the *three* degrees of translational freedom.

Equation (12·109), which expresses the principle of equipartition, is only applicable under the conditions stated, in particular that the separation of the levels is small and that the energy can be expressed as a 'square term', as in (12·108). These conditions certainly apply to the translational motion and also, *at room temperatures and above*, to the rotational motion. A linear molecule requires two co-ordinates for the specification of its rotation, and for each of these the contribution to c_V is $\tfrac{1}{2}R$ per mole. A non-linear molecule, on the other hand, requires three co-ordinates for the specification of its rotation and the rotational contribution to c_V is $\tfrac{3}{2}R$ per mole. The overall value of c_V would thus be 24.943 J K^{-1} mol^{-1}, and the corresponding value of c_p 33.257 J K^{-1} mol^{-1}.

Turning to the case of vibration, the above conditions are not obeyed in either respect; the separation of levels is not small compared to kT, at normal temperatures, and the eigenvalues are not proportional to the square of the quantum number. In fact, for a harmonic oscillator, the permissible energies are a *linear* function of the quantum number v.

$$\epsilon_i = (v + \tfrac{1}{2}) h\nu \quad (v = 0, 1, 2, \ldots),$$
(12·110)

where ν is the classical frequency of the oscillator. If this relation is substituted in (12·107), then, at *a sufficiently high temperature*, we find that the mean energy of the oscillators is

$$\bar{\epsilon} = kT,$$
(12·111)

† Integrals of this type are listed in § 12·13.

which is just twice the value given by (12·109). In classical terms this may be explained as follows: each harmonic oscillator has a kinetic energy proportional to the *square* of the momentum and a potential energy proportional to the *square* of the displacement. If each of these has a mean value of $\frac{1}{2}kT$ per molecule, the total is kT per molecule. This corresponds to a contribution to c_V of R per mole.

However, it is to be emphasized that (12·111) applies only at high temperatures such that kT is large compared to the separation of the vibrational levels. At room temperature the vibrational motion makes a much smaller contribution to the heat capacity of gases than would be expected on the basis of (12·111). The main value of this equation is for the estimation of lower and upper limits to the vibrational contribution to c_V, if this has not been measured. The lower limit, corresponding to no vibrational excitation, is zero and the upper limit is R per mole for each vibrational mode.

The number of these modes is $3n - 5$ in the case of a linear molecule and $3n - 6$ in the case of a non-linear one, where n is the number of atoms in the molecule. This may be seen as follows. The position of the n atoms may be completely specified by means of $3n$ co-ordinates. These may be chosen as the Cartesian co-ordinates of each atom or in any alternative way. For example, three co-ordinates can be used to describe the position of the centre of gravity of the molecule (corresponding to its translational motion), and its orientation in space can be specified by using either two or three angles, according to whether it is linear or non-linear respectively. These angles correspond to the rotational degrees of freedom. This leaves either $3n - 5$ or $3n - 6$ co-ordinates still to be specified, and these can be chosen as the internuclear distances. It follows that oither $3n - 5$ or $3n - 6$ of these distances are independent variables, and each of these describes an independent mode of vibration.

The experimental values of c_p for some common gases are shown in Fig. 42. The reader may find it instructive to compare these values with those estimated on the basis of the foregoing discussion and to note the temperatures at which the rotational and vibrational contributions are partially or completely effective. It may be remarked that electronic contributions to the heat capacity only become appreciable at very high temperatures, except in one or two exceptional cases such as nitric oxide.

12·13. Appendix. Some definite integrals

$$\int_0^\infty x^n e^{-ax} \, dx = \frac{n!}{a^{n+1}} \quad (n = 0, 1, 2, \ldots) \quad [\text{N.B. } 0! = 1],$$

$$\int_0^\infty e^{-ax^2} \, dx = \frac{1}{2} \left(\frac{\pi}{a}\right)^{\frac{1}{2}},$$

$$\int_0^\infty x^2 e^{-ax^2} \, dx = \frac{1}{4a} \left(\frac{\pi}{a}\right)^{\frac{1}{2}},$$

$$\int_0^\infty x^4 e^{-ax^2} \, dx = \frac{3}{8a^2} \left(\frac{\pi}{a}\right)^{\frac{1}{2}},$$

$$\int_0^\infty x^6 e^{-ax^2} \, dx = \frac{15}{16a^3} \left(\frac{\pi}{a}\right)^{\frac{1}{2}},$$

$$\int_0^\infty x^{2n+1} e^{-ax^2} \, dx = \frac{n!}{2a^{n+1}} \quad (n = 0, 1, 2, \ldots).$$

Values of the gamma function, defined by

$$\Gamma(n) = \int_0^\infty x^{n-1} e^{-x} \, dx \quad (n \text{ positive}),$$

and the error function, $\quad \mathrm{erf}\, x = \dfrac{2}{\sqrt{\pi}} \displaystyle\int_0^x e^{-x^2} \, dx,$

are given in Milne-Thomson and Comrie, *Standard Four-Figure Mathematical Tables.*

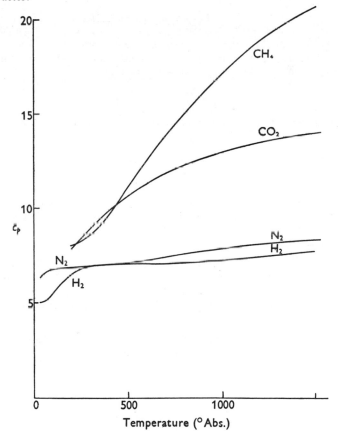

Fig. 42. Values of c_p cal K^{-1} mol^{-1}, at 1 atm pressure.

PROBLEMS

1. Investigate the simplification used in §12·1 which leads to the Maxwell–Boltzmann statistics. For this purpose use equation (12·88) to show that the number of translational energy states per molecule, whose energy ϵ_i is such that $e^{-\epsilon_i/kT}$ is not insignificant, is much larger than the number of molecules in the system under any conditions at which a molecular gas is perfect. Show, however, that this condition is not satisfied in the case of free electrons, if their concentration is of the order 10^{23} per cm³, as in metals.

2. Show that the number of molecules which cross a plane of unit area in a gas in unit time, and which have a component of velocity normal to this plane greater than a value v_{x0}, is given by

$$\frac{N}{V}\left(\frac{kT}{2\pi m}\right)^{\frac{1}{2}} \exp\left[-mv_{x0}^2/2kT\right], \qquad (12\cdot112)$$

where N is the number of molecules in the volume V.

Show also that the number crossing unit area in unit time whose total kinetic energy exceeds the value ϵ_0 is given by

$$\frac{N}{V}\left(\frac{kT}{2\pi m}\right)^{\frac{1}{2}}\left(1+\frac{\epsilon_0}{kT}\right)e^{-\epsilon_0/kT}. \qquad (12\cdot113)$$

Note that if no lower limit is given to v_x or to ϵ, the above expressions both reduce to

$$\frac{N}{V}\left(\frac{kT}{2\pi m}\right)^{\frac{1}{2}} = \frac{p}{(2\pi mkT)^{\frac{1}{2}}}, \qquad (12\cdot114)$$

where p is the pressure.

(The above expressions are usually taken as being applicable also to the number of gas molecules striking a plane solid surface. This neglects the fact that, close to the surface, the potential energy in the gas phase is no longer quite constant, on account of the attractive forces due to the solid.)

3. Consider the molecules which cross a given plane in a gas in unit time. Show that their mean kinetic energy in the direction normal to this plane is kT, and also that their mean total kinetic energy is $2kT$. Why is this larger than the value $\frac{3}{2}kT$, which is the mean kinetic energy of all molecules in the gas?

4. According to equation (12·64) the translational entropy of a gas is larger the greater is the molecular weight. Why is this?

5. From equation (12·104) obtain expressions for the following means: $\bar{v}, \bar{v}^2, \overline{v^2}$.

CHAPTER 13

PERFECT CRYSTALS AND THE THIRD LAW

13·1. Normal co-ordinates

The thermodynamic properties of a system, as shown already, may be calculated from its partition function

$$Q = \Sigma\, e^{-E_j/kT},$$

which is a summation over all of the quantum states. This can be evaluated only if the Schrödinger equation for the system can be simplified, in particular only if the system can be regarded as consisting of independent entities for each of which the Schrödinger equation takes a sufficiently simple form. Now any sample of a crystal (or liquid) is really a single quantum-mechanical whole and does not consist of independent molecules. This is because of the strong internal forces; the state of any one molecule depends on the state of all the others and the system is said to be *co-operative*. In this situation Q cannot be evaluated as simply as in the case of a gas.

Despite what has been said, it is possible to attribute approximately independent energies, not to the molecules of a crystal lattice, but to the *normal modes* of its vibration. At any rate this is the case when the atoms are arranged in a fixed and regular pattern and when the forces between them may be approximated by Hooke's law. This will become apparent in § 13·2 below, but it is necessary first of all to discuss normal co-ordinates.

The thermal energy of a crystal containing N atoms consists in the small vibrations of these atoms about their equilibrium positions in the lattice. To specify the momentary positions of all the atoms, a total of $3N$ co-ordinates are required, and these may be numbered $x_1, x_2, \ldots, x_{3N}$. They may be chosen conveniently as the displacements of the various atoms from their equilibrium positions. Thus x_1, x_2 and x_3 will denote the momentary displacements, in three directions at right angles, of the atom occupying the first lattice site; x_4, x_5 and x_6 will denote the displacements of the atom occupying the second numbered site, and so on. The total kinetic energy of all the atoms is therefore

$$T = \frac{1}{2} \sum_{i=1}^{3N} m_i \dot{x}_i^2, \tag{13·1}$$

where m_i is the mass which is appropriate to the ith co-ordinate. (Of course if all of the atoms are the same, all of the m_i are equal.)

The potential energy of the system is of more immediate importance and is a function of all of the displacements,

$$V = f(x_1, x_2, \ldots, x_{3N}). \tag{13·2}$$

For example, if there were just two significant displacements, x_1 and x_2, we should have $\quad V = f(x_1, x_2), \tag{13·3}$

and this may be expanded by means of Taylor's theorem

$$V = E_0 + \frac{\partial V}{\partial x_1} x_1 + \frac{\partial V}{\partial x_2} x_2 + \frac{1}{2} \frac{\partial^2 V}{\partial x_1^2} x_1^2 + \frac{\partial^2 V}{\partial x_1 \partial x_2} x_1 x_2$$

$$+ \frac{1}{2} \frac{\partial^2 V}{\partial x_2^2} x_2^2 + \frac{1}{3!} \frac{\partial^3 V}{\partial x_1^3} x_1^3 + \ldots. \tag{13·4}$$

In this expression E_0 is the potential energy when the displacements are zero. Under the latter conditions V is at a *minimum* and therefore the first differential coefficients in (13·4) are zero. If we make the approximation of neglecting the terms which involve the cubes and higher powers of the displacements, then (13·4) may be written

$$V = E_0 + \frac{1}{2} \frac{\partial^2 V}{\partial x_1^2} x_1^2 + \frac{\partial^2 V}{\partial x_1 \partial x_2} x_1 x_2 + \frac{1}{2} \frac{\partial^2 V}{\partial x_2^2} x_2^2 \tag{13·5}$$

or $\quad V - E_0 = \frac{1}{2}(c_{11} x_1^2 + c_{12} x_1 x_2 + c_{21} x_2 x_1 + c_{22} x_2^2), \tag{13·6}$

where c_{11}, etc., denote the second differential coefficients ('force constants') and $c_{12} = c_{21}$. (The use of both of the latter symbols, instead of a single one equal to $\partial^2 V / \partial x_1 \partial x_2$, enables the $\frac{1}{2}$ to be brought outside the bracket.)

A similar Taylor expansion of (13·2), in the general case, gives

$$V - E_0 = \frac{1}{2} \Sigma c_{ij} x_i x_j, \tag{13·7}$$

where the summation is over all values of i and j from 1 to $3N$. E_0, the potential energy of the crystal when none of the oscillations are excited, will be taken as being computed relative to the same atoms at infinite separation. It is the negative of the work which would have to be done in order to separate the atoms against their attractive forces.

The expressions (13·6) and (13·7) contain the *cross-terms* $x_1 x_2$, etc. For the purpose of attaining separability in the Schrödinger equation for the crystal it is necessary to define new co-ordinates such that these cross-terms are no longer present. For this purpose consider the simple example, as discussed above, of only two degrees of freedom x_1 and x_2. We define new co-ordinates q_1 and q_2 by linear relations

$$\left. \begin{aligned} x_1 &= q_1 + q_2, \\ x_2 &= a q_1 + b q_2, \end{aligned} \right\} \tag{13·8}$$

where a and b are arbitrary constants. Equations (13·1) and (13·6), when expressed in terms of q_1 and q_2, become

$$2T = m_1 \dot{x}_1^2 + m_2 \dot{x}_2^2$$

$$= \dot{q}_1^2(m_1 + a^2 m_2) + 2\dot{q}_1 \dot{q}_2(m_1 + abm_2) + \dot{q}_2^2(m_1 + b^2 m_2), \quad (13·9)$$

$$2(V - E_0) = q_1^2[c_{11} + a(c_{12} + c_{21}) + a^2 c_{22}]$$

$$+ q_1 q_2[2c_{11} + (a + b)(c_{12} + c_{21}) + 2abc_{22}]$$

$$+ q_2^2[c_{11} + b(c_{12} + c_{21}) + b^2 c_{22}]. \quad (13·10)$$

The cross-terms $\dot{q}_1 \dot{q}_2$ and $q_1 q_2$ vanish from these equations, if a and b are chosen to satisfy the relations

$$\left. \begin{array}{r} m_1 + abm_2 = 0, \\ 2c_{11} + (a + b)(c_{12} + c_{21}) + 2abc_{22} = 0. \end{array} \right\} \quad (13·11)$$

By means of these two equations the two constants a and b may be expressed in terms of the masses, m_1 and m_2, together with the force constants c_{11}, etc., which are physical properties of the system.

In the general case involving $3N$ Cartesian co-ordinates it is again possible to choose an alternative set of co-ordinates, $q_1, q_2, ..., q_{3N}$. These are related to the x_i by independent linear relations,

$$x_i = a_{i1} q_1 + a_{i2} q_2 + ... + a_{i3N} q_{3N}, \quad (13·12)$$

and they have the property that neither the kinetic nor the potential energy involves cross-terms between different co-ordinates. However, this can be achieved only as a result of the important physical assumption that terms higher than the square in the Taylor expansion may be neglected. This is equivalent to the assumption that the inter-atomic forces obey Hooke's law, i.e. the force is proportional to the displacement or the potential energy is proportional to the square of the displacement.

Co-ordinates having the above property, namely, that the energy can be expressed as a sum of squares, are called *normal* or *principal* co-ordinates. Their significance in regard to the Schrödinger equation will be discussed in the next section. From a classical viewpoint their importance is that when a system of harmonic oscillators (e.g. pendulums coupled with springs) is given an initial displacement in one of these co-ordinates, the motion continues as a simple harmonic motion in this co-ordinate only. The energy is not transferred to the other normal 'modes' which therefore remain unexcited. Any more complex oscillation can be conveniently analysed as a superposition of the simple sinusoidal motion in the normal modes. At the same time it may be remarked that the transformation of the mathematics of a physical problem by use of normal co-ordinates, or normal modes,† is a purely formal process—they have neither more nor less physical significance than the original Cartesian co-ordinates.

13·2. The Schrödinger equation for the crystal

Using the normal co-ordinates the potential energy of the crystal may be expressed, analogously to (13·10), as

$$V = E_0 + \tfrac{1}{2}\Sigma\lambda_i q_i^2, \tag{13·13}$$

which contains no cross-terms. The generalized force constants, λ_i, are related to the c_{ij}, the force constants in Cartesian co-ordinates, in a manner analogous to the two-dimensional example of equation (13·10). On this basis the Schrödinger equation for the crystal may be written as follows (see equation 12·12)

$$\Sigma \frac{1}{b_i}\frac{\partial^2\psi}{\partial q_i^2} = -\frac{8\pi^2}{h^2}(E - E_0 - \tfrac{1}{2}\Sigma\lambda_i q_i^2)\,\psi, \tag{13·14}$$

where the b_i's are constants, related to the masses m_i, arising from the transformation from the Cartesian to the normal co-ordinates. The very important result of using these co-ordinates is that (13·14) is separable, in the sense of § 12·3, and this is because there are no cross-terms. Therefore ψ can be written as a product

$$\psi = \psi_1\psi_2\dots\psi_i\dots\psi_{3N}, \tag{13·15}$$

where each ψ_i is a function only of the co-ordinate q_i. This means that equation (13·14) can be separated into $3N$ equations each involving only a single independent variable q_i,

$$\frac{d^2\psi_i}{dq_i^2} = -\frac{8\pi^2 b_i}{h^2}(\epsilon_i - \tfrac{1}{2}\lambda_i q_i^2)\,\psi_i, \tag{13·16}$$

where the ϵ_i satisfy the relation

$$\Sigma\epsilon_i = E - E_0. \tag{13·17}$$

In order to see the significance of this result let it be supposed that, instead of $3N$ oscillations, we are concerned with only one. For a single particle making a harmonic oscillation in a particular co-ordinate let the displacement from the position of equilibrium be q. The potential energy is given by

$$V = \tfrac{1}{2}\lambda q^2, \tag{13·18}$$

† Corresponding to the $3N$ normal co-ordinates there are actually only $3N - 6$ normal vibrational modes of the crystal. This is because six co-ordinates are needed to specify the centre of gravity and orientation of the crystal as a whole (see § 12·12). However 6 may be neglected relative to $3N$.

where λ is the force constant. The Schrödinger equation (12·12) can therefore be written

$$\frac{d^2\psi}{dq^2} = \frac{-8\pi^2 m}{h^2}(\epsilon - \tfrac{1}{2}\lambda q^2)\,\psi, \tag{13·19}$$

where m is the mass of the particle and ϵ is an energy eigenvalue. It is seen that (13·16) and (13·19) are of precisely the same form, and *the whole crystal may therefore be regarded as a system of $3N$ independent harmonic oscillators each obeying* (13·16). The sum of their energies is equal to $E - E_0$, which is the total energy of the crystal relative to the chosen zero point E_0. These oscillators, which should usually be spoken of as 'normal modes', are not to be confused with the atoms themselves. It is true that if the crystal were to vibrate in a single mode only, all of the N atoms would vibrate with the same frequency (although with different amplitudes). But in general the motion of the atoms is a superposition of the $3N$ (actually $3N-6$) normal modes, just as the vibration of a violin string is a superposition of a fundamental and its harmonics.

13·3. The energy levels of the harmonic oscillator

In solving the differential equation (13·19) it has to be borne in mind that ψ^2 must have the physical significance of a probability density and ψ must therefore be single-valued, continuous and finite over the range of the co-ordinate q. When these conditions are applied it is found that (13·19) only has a solution† when ϵ takes on certain eigenvalues as given by

$$\epsilon = (v + \tfrac{1}{2})\frac{h}{2\pi}\left(\frac{\lambda}{m}\right)^{\tfrac{1}{2}}, \tag{13·20}$$

where $v = 0, 1, 2, \ldots$.

In short, only certain discrete values of the energy are permissible if ψ is to 'make sense' and fulfil its quantum-mechanical purposes.

In (13·20) the quantity $(\lambda/m)^{\tfrac{1}{2}}/2\pi$ has the dimension of a frequency and will be denoted by ν,

$$\frac{1}{2\pi}\left(\frac{\lambda}{m}\right)^{\tfrac{1}{2}} \equiv \nu. \tag{13·21}$$

The energy levels of the oscillator may therefore be written

$$\epsilon = (v + \tfrac{1}{2})h\nu, \tag{13·22}$$

but the introduction of the symbol ν has no particular significance in quantum mechanics. Equation (13·22) is conventionally written in

† For more detailed discussion see, for example, Margenau and Murphy, *The Mathematics of Physics and Chemistry* (New York, Van Nostrand, 1943), §11·11.

this form because $(\lambda/m)^{\frac{1}{2}}/2\pi$ is the frequency with which the oscillator *would* vibrate if it obeyed classical theory. Its classical equation of motion is

$$m\ddot{q} = -\lambda q, \qquad (13\cdot23)$$

and therefore

$$q = A \sin\left[\left(\frac{\lambda}{m}\right)^{\frac{1}{2}} t + \delta\right], \qquad (13\cdot24)$$

where A and δ are constants having the significance of amplitude and phase respectively. If ν is the frequency of this classical motion then q must have the same value at t and at $t+1/\nu$. Since $\sin\theta = \sin(\theta+2\pi)$ it follows that the classical oscillation frequency has the value given by (13·21). On the other hand, in the quantum theory a particle cannot be regarded as returning to a given point after a precise time, on account of the uncertainty principle, and therefore the quantity $(\lambda/m)^{\frac{1}{2}}/2\pi$ which appears in (13·20) is only to be regarded as a frequency by convention.

It is for the same reason that the oscillator has a non-zero amount of energy, $\frac{1}{2}h\nu$, even when it is in the lowest vibrational state, as follows from (13·22). If an oscillator could have zero energy it would correspond to a precise location, $q=0$. The least allowable energy $\frac{1}{2}h\nu$ is called the *zero-point energy* of the oscillator.

13·4. The partition function

We first collect together the important results. The partition function for the crystal is

$$Q = \Sigma\, e^{-E_j/kT}, \qquad (13\cdot25)$$

where the summation is over all the independent quantum states of the crystal. The energies E_j corresponding to these quantum states can be expressed by means of (13·17) in terms of the energies of the normal modes

$$E_j = E_0 + \Sigma\epsilon_i, \qquad (13\cdot26)$$

where the summation is over the $3N$ normal modes. Finally, each ϵ_i can have the values

$$\epsilon_i = (v_i + \tfrac{1}{2})\, h\nu_i \quad (v_i = 0, 1, 2, \ldots), \qquad (13\cdot27)$$

where ν_i is the classical frequency of the ith oscillator. In general, the various modes are not to be expected to have the same frequency, and the main difficulty in calculating the thermodynamic properties of the crystal lies in finding the distribution function for the ν_i. This will be discussed in later sections.

As mentioned above, the summation (13·25) is over the independent quantum states. Now a state of the crystal in which a particular oscillator has an energy of, say, 2 units and another particular oscillator has an energy of 3 units, is distinguishable from the state in which the first oscillator has the energy of 3 and the second of 2. This is because the oscillators, or normal modes, are specified in terms of the normal co-ordinates and these in turn are functions of $3N$ Cartesian co-ordinates which specify the displacements of the atoms from the numbered lattice sites. Since the latter are physically distinguishable so also are the normal modes. For this reason the kind of indistinguishability which was discussed in §12·1 does not arise and the factor $N!$ will not make any appearance in the formulae to be obtained.

Substituting (13·26) in (13·25)

$$Q = \Sigma\, \mathrm{e}^{-E_0/kT}\, \mathrm{e}^{-\Sigma\epsilon_i/kT}. \tag{13·28}$$

In order to make clear the nature of the indicated summations, suppose that there are just two normal modes, denoted A and B, of conventional frequencies ν_a and ν_b. Let the eigenvalues for these oscillators be $\epsilon_{a0}, \epsilon_{a1}, \ldots;\ \epsilon_{b0}, \epsilon_{b1}, \ldots,$ as given by (13·27). Then (13·28) can be written
$$Q = \sum_{\text{states}}\, \mathrm{e}^{-E_0/kT}\, \mathrm{e}^{-\epsilon_{ai}/kT}\, \mathrm{e}^{-\epsilon_{bj}/kT}.$$

Let it be supposed that when the two oscillators are each in specified states, it is still possible for the crystal to have Ω_0 different complexions. This degeneracy of the crystal, if it occurs, is due to factors, such as different possibilities of orientation of the molecules in the lattice,† which are quite distinct from the oscillators—the latter are not degenerate, as may be seen from (13·22). The last expression can therefore be written
$$Q = \Omega_0\, \mathrm{e}^{-E_0/kT} \Sigma\, \mathrm{e}^{-\epsilon_{ai}/kT}\, \mathrm{e}^{-\epsilon_{bj}/kT},$$

and this is now a summation over all combinations of the permitted energies of the two oscillators. Writing it out in full

$$\begin{aligned}
Q &= \Omega_0\, \mathrm{e}^{-E_0/kT}\{\mathrm{e}^{-\epsilon_{a0}/kT}\,(\mathrm{e}^{-\epsilon_{b0}/kT} + \mathrm{e}^{-\epsilon_{b1}/kT} + \ldots) \\
&\qquad + \mathrm{e}^{-\epsilon_{a1}/kT}\,(\mathrm{e}^{-\epsilon_{b0}/kT} + \ldots) \\
&\qquad + \ldots\} \\
&= \Omega_0\, \mathrm{e}^{-E_0/kT} \sum_i \mathrm{e}^{-\epsilon_{ai}/kT} \sum_j \mathrm{e}^{-\epsilon_{bj}/kT}.
\end{aligned}$$

Therefore in the general case of $3N$ oscillators, (13·28) becomes

$$Q = \Omega_0\, \mathrm{e}^{-E_0/kT} \Sigma\, \mathrm{e}^{-\epsilon_a/kT} \Sigma\, \mathrm{e}^{-\epsilon_b/kT} \ldots \Sigma\, \mathrm{e}^{-\epsilon_{3N}/kT}, \tag{13·29}$$

where each sum refers to one of the $3N$ oscillators and is a summation over all of its energy levels. The equation may be written

$$Q = \Omega_0\, \mathrm{e}^{-E_0/kT} \prod_{i=1}^{i=3N} f_i, \tag{13·30}$$

where
$$f_i = \Sigma\, \mathrm{e}^{-\epsilon_i/kT}, \tag{13·31}$$

† The significance of Ω_0 will become clearer in §13·11.

and may be called the partition function of the ith oscillator. Substituting (13·22) in (13·31)

$$f_i = e^{-h\nu_i/2kT} (1 + e^{-h\nu_i/kT} + e^{-2h\nu_i/kT} + \dots),$$

the series extending over all of the vibrational quantum numbers from zero to infinity. The quantity in brackets is a geometrical progression and can readily be evaluated in closed form. We thus obtain

$$f_i = \frac{e^{-h\nu_i/2kT}}{1 - e^{-h\nu_i/kT}}. \tag{13·32}$$

Under conditions of high temperature where $h\nu_i \ll kT$, this expression approaches the value

$$f_i = \frac{kT}{h\nu_i}. \tag{13·33}$$

Substituting (13·32) in (13·30) we obtain an expression for the partition function Q of the crystal in terms of the frequencies. Its logarithm is more useful and is

$$\ln Q = \ln \Omega_0 - \frac{E_0}{kT} - \frac{1}{kT} \sum_i \frac{h\nu_i}{2} - \sum_i \ln (1 - e^{-h\nu_i/kT}). \tag{13·34}$$

The second and third terms on the right-hand side may be conveniently grouped together. $\Sigma \frac{1}{2} h\nu_i$ is the sum of the zero point energies of all the oscillators, and therefore the quantity

$$U_0 \equiv E_0 + \Sigma \tfrac{1}{2} h\nu_i \tag{13·35}$$

is the value taken by the internal energy† of the crystal at the absolute zero of temperature where all of the oscillators are in their lowest quantum states. Substituting (13·35) in (13·34) we obtain

$$\ln Q = \ln \Omega_0 - \frac{U_0}{kT} - \sum_i \ln (1 - e^{-h\nu_i kT}). \tag{13·36}$$

From (11·43) and (11·44) we now obtain the following expressions for the Helmholtz free energy and internal energy of the crystal:

$$A = - kT \ln \Omega_0 + U_0 + kT \Sigma \ln (1 - e^{-h\nu_i/kT}), \tag{13·37}$$

$$U = U_0 + h \Sigma \frac{\nu_i}{e^{h\nu_i/kT} - 1}, \tag{13·38}$$

where the summations are over the $3N$ values of ν_i. It may be remarked that these frequencies are a function of the volume of the crystal and, if this function were known, it would be possible to apply equation (11·46) to obtain an expression for the pressure (which varies with the volume when the crystal is compressed.)

† Relative to the component atoms at infinite separation.

13·5 The Maxwell–Boltzmann distribution

Before going on to discuss the possible values of ν_i it is useful at this stage to consider the analogue of the Maxwell–Boltzmann distribution as applied to the oscillators. Let it be supposed that N' of the oscillators have an equal, or approximately equal, frequency ν'. Equation (13·38) may be applied to this group of oscillators so that their total internal energy is

$$U' = U'_0 + N'h\nu'/(e^{h\nu'/kT} - 1)$$

$$= E'_0 + \frac{N'h\nu'}{2} + N'h\nu'/(e^{h\nu'/kT} - 1)$$

by (13·35). The total vibrational energy of the group of oscillators is the sum of the last two terms in this equation and is

$$N'h\nu'\left(\frac{1}{2} + \frac{1}{e^{h\nu'/kT} - 1}\right). \tag{13·39}$$

Now although the oscillators in this group all have the same frequency their individual energies are not necessarily equal. Let N'_i be the number whose vibrational energy is ϵ_i. $N'_i\epsilon_i$ is therefore the total vibrational energy of this subgroup, and if this quantity is summed over all values of ϵ_i the result must be the same as (13·39). Thus

$$\Sigma N'_i\epsilon_i = N'h\nu'\left(\frac{1}{2} + \frac{1}{e^{h\nu'/kT} - 1}\right). \tag{13·40}$$

It will be shown that this relation is satisfied if we put†

$$\frac{N'_i}{N'} = \frac{e^{-\epsilon_i/kT}}{\Sigma e^{-\epsilon_i/kT}}. \tag{13·41}$$

The denominator in this expression is the partition function for any oscillator of the particular frequency ν' and may be summed as in the case of (13·31). Substituting (13·27) in (13·41) we obtain

$$\frac{N'_i}{N'} = (1 - e^{-h\nu'/kT})\,e^{-h\nu_i\nu'/kT}, \tag{13·42}$$

where v_i is the vibrational quantum number corresponding to ϵ_i. The left-hand side of (13·40) may thus be written

$$\Sigma N'_i\epsilon_i = N'(1 - e^{-h\nu'/kT)}) \Sigma e^{-h v_i\nu'/kT}\,(v_i + \tfrac{1}{2})\,h\nu',$$

and the summation is an arithmetico-geometric series. If this is evaluated it is readily seen that (13·40) is satisfied.

Equation (13·41) is closely analogous to (12·72) and expresses the Maxwell–Boltzmann distribution for a group of normal modes of approximately equal frequency. Of course if *all* of the $3N$ normal modes were of

---

† A more rigorous derivation of this equation is given by Tolman, *Principles of Statistical Mechanics* (Oxford, 1938), §113.

equal frequency—as is supposed in the Einstein approximation shortly to be discussed—then (13·41) or (13·42) would give the fraction of all oscillators in the crystal for which the vibrational energy is ϵ_i.

The fraction of the N' equal oscillators whose energy is equal to or exceeds a specified value ϵ_i is readily calculated from (13·41), and the result is of considerable interest in connexion with the reaction rate or evaporation rate of solids. This fraction, which will be denoted $\phi_{\epsilon i}$, is obtained by adding to (13·41) the corresponding expression for the next higher level ϵ_j, and so on up to infinity. We obtain

$$\phi_{\epsilon_i} = \frac{e^{-\epsilon_i/kT} + e^{-\epsilon_j/kT} + \cdots}{e^{-\epsilon_0/kT} + e^{-\epsilon_1/kT} + \cdots}.$$

Since the energies of the adjacent levels of a harmonic oscillator all differ by the constant amount $h\nu$, as shown in equation (13·22), the above expression may be written

$$\phi_{\epsilon_i} = \frac{e^{-\epsilon_i/kT}(1 + e^{-h\nu'/kT} + e^{-2h\nu'/kT} + \cdots)}{e^{-h\nu'/2kT}(1 + e^{-h\nu'/kT} + \cdots)}$$

$$= e^{(\frac{1}{2}h\nu' - \epsilon_i)/kT}. \tag{13·43}$$

If ϵ_i is large compared to the zero-point energy this expression can be approximated by
$$\phi_{\epsilon_i} = e^{-\epsilon_i/kT}. \tag{13·44}$$

13·6. The high temperature approximation

Of the $3N$ normal modes of vibration of the crystal, those with the *lowest* frequencies are the ordinary acoustic vibrations which appear in the theory of sound. On the other hand, the *highest* possible frequency is determined by the smallest possible independent wavelength, and this can be shown[†] to be of the order of 10^{-7}–10^{-8} cm, the distance between atoms in the lattice. If the wave velocity may be assumed approximately constant over the whole range of frequencies, it will be equal to the velocity of sound; the highest value of ν_i may therefore be calculated to be of the order 10^{12}–10^{13}/sec.

Consider the largest possible value of $h\nu/kT$. Using the known values of h and k, this ratio may be calculated to be about $50/T$ or $500/T$, according to whether the largest frequency of the crystal is of the order 10^{12} or 10^{13} respectively. Clearly this ratio will be small compared to unity at sufficiently high temperatures and for certain crystals, those for which the highest frequency is about 10^{12}, even at room temperature.

Under conditions of temperature where $h\nu/kT \ll 1$, the terms beyond the third in the expansion

$$e^{h\nu/kT} = 1 + \frac{h\nu}{kT} + \frac{1}{2!}\left(\frac{h\nu}{kT}\right)^2 + \cdots$$

† Seitz, *Modern Theory of Solids* (New York, McGraw-Hill, 1940), §19.

may be neglected. With this approximation (13·38) becomes

$$U = U_0 + 3NkT - \Sigma \frac{h\nu_i}{2}$$
$$= E_0 + 3NkT, \tag{13·45}$$

in view of (13·35). At sufficiently high temperatures the internal energy is therefore independent of the frequencies. The corresponding heat capacity *at constant volume*, $\partial U/\partial T$, is $3Nk$ or $3R$ per mole, in close agreement with the empirical rule of Dulong and Petit.† As is well known, this is a good approximation for many elements at room temperature, but in the case of diamond, beryllium and silicon the value of c_V approaches $3R$ only at high temperatures. This is presumably because the highest frequency in these crystals is larger than 10^{12} due to the strength of the bonding.

It may be remarked that transition metals have values of c_V which considerably exceed $3R$ at high temperatures; for example, γ iron has a value of about 38 J K^{-1} mol^{-1}. This is believed to be due to electronic excitation from the unfilled d shells, a factor which is not allowed for in the theory as developed above. The applicability of the Dulong and Petit rule to ionic and molecular lattices will be discussed in a later section.

The same result as (13·45) could also be obtained by applying the principle of equipartition (§12·12) which will be valid under conditions where $h\nu/kT \ll 1$. Each oscillator will have a mean energy of kT, as shown in equation (12·111), and the total thermal energy of the crystal is therefore $3NkT$ in agreement with (13·45).

The form taken by the Helmholtz free energy at high temperatures does not have quite the same simplicity as in the case of the internal energy. Under conditions where $h\nu_i/kT \ll 1$, the expansion of the exponential allows (13·37) to be written in the approximate form

$$A = -kT \ln \Omega_0 + U_0 + kT \Sigma \ln \frac{h\nu_i}{kT}$$
$$= -kT \ln \Omega_0 + U_0 + kT \ln \left(\frac{h\nu_1}{kT}\right) \left(\frac{h\nu_2}{kT}\right) \cdots \left(\frac{h\nu_{3N}}{kT}\right),$$

where ν_1, etc., are the frequencies. If we define a geometric mean frequency $\bar{\nu}$ by means of the relation

$$\bar{\nu}^{3N} = \nu_1 \nu_2 \ldots \nu_{3N},$$

† According to this rule the heat capacity of atomic crystals tends to the value 6.4 cal K^{-1} mol^{-1} (i.e. 26.8 J K^{-1} mol^{-1}). This actually refers to the value of the heat capacity at *constant pressure* which is the heat capacity most easy to measure experimentally. The difference between c_p and c_V may be calculated by means of equation (2·91) and is usually about 2 J K^{-1} mol^{-1} at temperatures in the region of 300 K.

then the high-temperature value of F may be expressed as

$$A = -kT \ln \Omega_0 + U_0 + 3NkT \ln \left(\frac{h\bar{\nu}}{kT}\right). \tag{13·46}$$

Combining this with (13·45) and neglecting the zero-point energy, we obtain the high-temperature value of the entropy as

$$S = \frac{U-A}{T} = k \ln \Omega_0 + 3Nk\left(1 - \ln \frac{h\bar{\nu}}{kT}\right). \tag{13·47}$$

It will be remembered from § 12·1 that the crystal under discussion contains N atoms. Thus the values of U, A and S per mole are obtained from the preceding equations by replacing N by L, the Avogadro constant, or by replacing Lk by R, the gas constant per mole.

13·7. The Einstein approximation

As shown above, the heat capacity c_V (or at any rate that part of it which is due to the vibrations) may be expected to have a value of $3R$ whenever $h\nu/kT \ll 1$. This would be so, even at the lowest temperatures, if Planck's constant h were zero, and this is the case in the classical or pre-quantum mechanics. In fact, classical theory leads to the expectation that, for any crystalline substance, c_V has the constant value of $3R$ per mole. This is contrary to experiment, and it is known that c_V usually diminishes below $3R$, with fall of temperature, and seems to approach zero at the absolute zero. One of the early successes of the quantum theory consisted in finding the reason for this decrease in c_V which is quite inexplicable in classical theory. The explanation is implicit in the previous equations and is due to the fact that the oscillators can only take up finite increments of energy. When a system of oscillators is held at low temperature, most of them are in their lowest energy level, and a small rise of temperature is insufficient to excite them to the next higher level. Therefore c_V, which measures the intake of energy per unit increase of temperature, is smaller than at higher temperature.

In order to make use of equations such as (13·37) and (13·38) some assumption must be made with regard to the frequencies. As early as 1907 Einstein used the approximation of assuming that all of the $3N$ frequencies are equal. This is equivalent to supposing that each of the N atoms in the lattice makes quantized harmonic oscillations in three dimensions, and these oscillations are quite unaffected by the motion of the neighbouring atoms. This supposition of atomic independence cannot be correct, but nevertheless it leads to an expression for c_v which is fair agreement with experiment except at very low temperatures.

If all the frequencies are put equal to a constant value ν_E, equations (13·37) and (13·38) become

$$A = - kT \ln \Omega_0 + U_0 + 3NkT \ln (1 - e^{-h\nu_E/kT}), \qquad (13\cdot48)$$

$$U = U_0 + 3Nh\nu_E/(e^{h\nu_E/kT} - 1), \qquad (13\cdot49)$$

and therefore

$$C_V = 3Nk\left(\frac{h\nu_E}{kT}\right)^2 \frac{e^{h\nu_E/kT}}{(e^{h\nu_E/kT} - 1)^2}. \qquad (13\cdot50)$$

The entropy is obtained from (13·48) and (13·49) and is

$$S = k \ln \Omega_0 + 3Nk\left\{\frac{h\nu_E}{kT(e^{h\nu_E/kT} - 1)} - \ln (1 - e^{-h\nu_E/kT})\right\}. \qquad (13\cdot51)$$

If a suitable value of ν_E is used, (13·50) fits the experimental data quite well at temperatures at which c_V is more than half of its maximum value, $3R$. But at low temperatures there is a serious discrepancy, and the Einstein equation predicts a value of c_V which approaches zero too rapidly. This is due to the error in treating all the normal modes as if they have the same frequency ν_E. In the actual crystal there are no doubt a large number of modes which have quite a small frequency and therefore a small separation of their energy levels. At low temperatures these modes will have a much greater probability of becoming excited, with absorption of heat, than would be expected on the Einstein model.

13·8. The Debye approximation

At a rather later date Debye made the assumption that the frequency spectrum of the crystal—consisting as it does of discrete particles—can be adequately represented by the frequency spectrum of a *continuous* elastic medium. Let such a medium be in the form of a rectangular block of sides a, b and c. Then a standing wave can be developed within the material parallel to the side of length a if this length is an integral number of half-waves. Similarly with regard to the other sides. These conditions fix the allowable wavelengths within the elastic solid and thereby also the allowable frequencies.

Proceeding in this way it can be shown† that the number of modes whose frequencies lie in the range ν to $\nu + d\nu$ is

$$dN = A V \nu^2 d\nu, \qquad (13\cdot52)$$

† For further details see Fowler and Guggenheim, *Statistical Thermodynamics* (Cambridge, 1949) or Seitz, *Modern Theory of Solids* (New York, McGraw-Hill, 1940).

where V is the volume of the elastic medium and A is a function of the wave velocity. Now in such a medium there is no lower limit to the wavelength and thus no upper limit to the frequency. On the other hand, in an atomic solid there is such an upper limit, $\nu_{max.}$, as discussed already in §13·6. There are also a finite number, $3N$, of normal modes. The essential feature of the Debye theory is to assume that (13·52) is applicable to the atomic solid right up to the frequency $\nu_{max.}$, but beyond it dN falls abruptly to zero. The value of $\nu_{max.}$ is determined by the condition that the integral of (13·52) must be equal to $3N$. Thus

$$3N = A V \int_0^{\nu_{max.}} \nu^2 d\nu = \frac{A V}{3} \nu^3_{max.}, \tag{13·53}$$

and eliminating A and V between (13·52) and (13·53)

$$dN = \frac{9N}{\nu_D^3} \nu^2 d\nu, \quad (\nu \leqslant \nu_D), \tag{13·54}$$

where ν_D has been written in place of $\nu_{max.}$.

Within the scope of the present section we shall apply (13·54) to the internal energy and heat capacity only. The free energy and entropy on the Debye model may be worked out by similar methods. From (13·35) and (13·38) we have

$$U = E_0 + \frac{h}{2} \Sigma \nu_i + h \Sigma \frac{\nu_i}{e^{h\nu_i/kT} - 1},$$

where the summations are over the $3N$ frequencies. By using (13·54) the summations may be replaced by integrations (cf. the discussion in § 12·5) and thus

$$U = E_0 + \frac{h}{2} \int_0^{\nu_D} \frac{9N}{\nu_D^3} \nu^3 d\nu + h \int_0^{\nu_D} \frac{9N}{\nu_D^3} \frac{\nu^3 d\nu}{e^{h\nu/kT} - 1}$$

$$= E_0 + \frac{9Nh}{8} \nu_D + \frac{9Nh}{\nu_D^3} \int_0^{\nu_D} \frac{\nu^3 d\nu}{e^{h\nu/kT} - 1}. \tag{13·55}$$

It is convenient to define

$$\theta_D \equiv h\nu_D/k, \tag{13·56}$$

$$x \equiv h\nu/kT, \tag{13·57}$$

and therefore (13·55) may be written

$$U = E_0 + \frac{9Nk\theta_D}{8} + 9NkT \left(\frac{T}{\theta_D}\right)^3 \int_0^{\theta_D/T} \frac{x^3 dx}{e^x - 1}, \tag{13·58}$$

in which the second term on the right-hand side is the total zero-point energy of the crystal. The value of C_V is obtained by differentiating with respect to temperature and is

$$C_V = 9Nk \left(\frac{4T^3 I}{\theta_D^3} + \frac{T^4}{\theta_D^3} \frac{dI}{dT}\right),$$

where I is the integral which occurs in (13·58). The upper limit of this integral is a function of T. When the differentiation is carried out† we obtain

$$C_V = 9Nk\left\{4\left(\frac{T}{\theta_D}\right)^3 \int_0^{\theta_D/T} \frac{x^3}{e^x-1}\,dx - \left(\frac{\theta_D}{T}\right)\frac{1}{e^{\theta_D/T}-1}\right\}. \qquad (13\cdot59)$$

This expression takes a much simpler form at very low temperatures where $\theta_D/T \gg 1$. Under such conditions the second term in (13·59) is negligible, because of the exponential, and the upper limit in the integral can be replaced by infinity with only a small error. The integral is therefore independent of temperature, and we obtain the result that C_V is proportional to T^3—the 'Debye T^3 law'. The actual result of the integration is

$$C_V = \frac{12\pi^4}{5}Nk\left(\frac{T}{\theta_D}\right)^3. \qquad (13\cdot60)$$

13·9. Comparison with experiment

The quantity θ_D has the dimensions of a temperature and is called the 'Debye characteristic temperature' of the particular substance. According to (13·59) C_V is the same function of the ratio T/θ_D for all substances, and the heat capacity per mole may therefore be written

$$c_V = 3Rf(T/\theta_D), \qquad (13\cdot61)$$

where the function is three times the value of the bracket in (13·59). The ratio T/θ_D is evidently a 'reduced temperature' such that two crystals having the same value of T/θ_D are in a 'corresponding state' as regards their heat capacity.

The same considerations apply to the Einstein theory. An Einstein characteristic temperature θ_E may be defined by the relation

$$\theta_E \equiv h\nu_E/k, \qquad (13\cdot62)$$

where ν_E is the Einstein frequency. For one mole of a substance, (13·50) may therefore be written

$$c_V = 3R\left(\frac{\theta_E}{T}\right)^2 \frac{e^{\theta_E/T}}{(e^{\theta_E/T}-1)^2}$$

$$= 3Rf(T/\theta_E). \qquad (13\cdot63)$$

It will be seen from (13·62) that θ_E is the temperature at which kT becomes equal to $h\nu_E$, the separation of the vibrational energy levels according to the Einstein model. It is in the temperature region between $0.1\theta_E$ and θ_E that c_V increases most rapidly because it is in

† See, for example, Aston's chapter in Taylor and Glasstone's, *Treatise on Physical Chemistry* (New York, Van Nostrand, 1942), vol. I, p. 622.

this region that the number of normal modes which are excited above the lowest level increases most rapidly, with absorption of heat. (See Problem 1 at end of chapter.)

In Fig. 43 the lower curve represents the value of c_V as a function of T/θ_E, as calculated from equation (13·63) of the Einstein theory. The upper curve represents c_V as a function of T/θ_D, according to equation (13·59) of the Debye theory. (Tabulated values of the

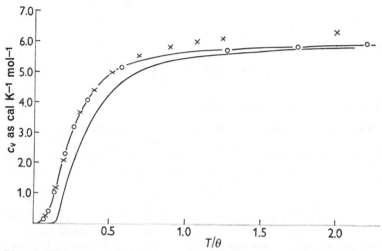

Fig. 43. Upper curve, Debye approximation; lower curve, Einstein approximation. ⊙, c_V for silver, $\theta_D = 215\mathrm{K}$; $\times$, c_p for KCl per gram atom, $\theta_D = 230$ K. N.B. The experimental points have been worked out with the best value of θ to fit the Debye curve; corresponding points to show the best agreement which can be obtained with the Einstein curve have not been included.

integral which occurs in this equation, and also tabulated values of c_V and the other thermodynamic functions as predicted by the Debye theory, are available in the literature.†)

The limiting values of c_V are the same on both theories, namely, $3R$ at high temperature and approaching zero at the absolute zero. But at intermediate values of T/θ the Einstein theory predicts lower values of c_V, and as mentioned previously, these values approach zero too rapidly, when compared with experimental results.

This comparison with experiment is carried out by finding the value of θ_E, or θ_D, for the particular substance which brings the experimental data into as close agreement as possible with the theoretical

† See, for example, Appendix IV to Aston's chapter in Taylor and Glasstone's *Treatise on Physical Chemistry* (New York, Van Nostrand, 1942), vol. I.

equations over a wide temperature range. Now in the case of a large number of solid elements, and also many inorganic compounds which have an ionic structure, a very close agreement to the Debye curve can be obtained over the whole temperature range. Fig. 43 shows the values of c_V for silver, as calculated from the experimental data on c_p by use of equation (2·91). With θ_D chosen as 215 K, these results fall almost exactly on the Debye curve. Good agreement with the Einstein curve could also be obtained in the region above $T = 0·4\theta_E$, by appropriate choice of θ_E, but this choice would result in very poor agreement in the low-temperature region.

The figure also shows the experimental values of the heat capacity *per gram-atom* of potassium chloride (half the value per mole). These experimental figures refer to the heat capacity at constant pressure, whereas the Debye theory predicts the value of the heat capacity at constant volume. In the low-temperature region, where the difference between c_p and c_V is very small, the agreement is excellent, and in the higher temperature region the experimental figures lie above the Debye curve by an amount such as would be expected for this difference. (In the case of potassium chloride the data are not available for the direct application of equation (2·91) which gives $c_p - c_V$.)

The values of θ_D which bring the experimental data into agreement with the Debye theory are usually about 100–400 K, corresponding to a frequency ν_D, as calculated from (13·56), of 2×10^{12} to 8×10^{12}. On the other hand, those substances which approach the Dulong and Petit value only at elevated temperatures have a much larger value of θ_D. Diamond, for example, has a value of 1860 K, corresponding to a classical frequency of 4×10^{13}.

As mentioned in § 13·6, the transition metals have heat capacities which rise considerably above $3R$ per mole, and the same applies to their ionic salts. Germanium and hafnium also have heat capacities which are not in agreement with the Debye theory. Both metals have peaks in their curves at about 70 K, and in the case of hafnium the heat capacity at the peak is as much as 46 J K⁻¹ mol⁻¹. At higher temperatures the heat capacity settles down to the Dulong and Petit value.

At a sufficient temperature the heat capacity at constant pressure of many ionic lattices approaches the value $26.8n$ (J K⁻¹ per g formula weight), where n is the number of atoms per gram formula weight. This is equivalent to the Kopp rule, as used in §4·13. However, this rule is not even approximately correct in the case of a molecular lattice, such as solid benzene, or where there are molecular ions such as NO_3^-, SO_4^{2-}, etc. In such systems there are internal vibrations within the molecules, in addition to the vibrations of the lattice. The force constants for the former are usually much larger than for the lattice vibrations, and therefore they become appreciably excited only at considerably higher temperatures. The heat capacities of such crystals can often be represented quite adequately by applying the Debye theory to the lattice vibrations and treating the remainder of the heat capacity as in the case of a gas molecule, using spectroscopically observed frequencies.†

Although the Debye theory predicts a heat-capacity curve which is often in good agreement with experiment, the assumptions in the theory with regard to the frequency spectrum are not necessarily correct. For certain simple types of lattice, Blackman (1937) was able to make a detailed calculation of the frequencies of the normal modes, allowing for the atomic structure of the system. It was found, as is assumed in the Debye theory, that there is an upper limit, ν_D, to the possible frequencies, and it was also confirmed that the T^3 law should hold at very low temperatures. On the other hand, Blackman found that the frequency distribution does not have such a simple form (equation (13·54)) as was assumed by Debye and, in fact, it may have two or more peaks. In view of these results it seems that the agreement of the Debye theory with experiment is better than might reasonably have been expected.

Blackman's calculations also give an indication why it is that the Einstein approximation works as well as it does—almost as well as the Debye in the high-temperature range. It seems that a large fraction of the actual frequencies are closely packed in the vicinity of a peak which is just below ν_D. Therefore it is a fairly good approximation, at temperatures which are not too low, to regard *all* of the normal modes as having a single frequency ν_E which is a little smaller than the maximum frequency ν_D of the Debye theory.‡

13·10. Vapour pressure at high temperature

Provided that the temperature is high enough for the relation $h\nu_i \ll kT$ to be satisfied, even for the highest lattice frequencies, the Helmholtz free energy of the crystal is given by equation (13·46):

$$A = -kT \ln \Omega_0 + U_0 + 3NkT \ln \left(\frac{h\bar{\nu}}{kT}\right),$$

where $\bar{\nu}$ is a geometric mean frequency. This expression is not dependent on the assumptions of the Einstein or Debye theories. The chemical potential of the crystal when it is under its vapour pressure p is therefore

$$\mu_c = pv_c - kT \ln \Omega_0 + u_0 + 3RT \ln \left(\frac{h\bar{\nu}}{kT}\right), \qquad (13·64)$$

† See, for example, §62 of Aston's chapter in Taylor and Glasstone, *Treatise on Physical Chemistry* (New York, Van Nostrand, 1942), vol. I.

‡ For further discussion see, for example, Fowler and Guggenheim, *Statistical Thermodynamics* (Cambridge, 1949), Chapter IV.

where v_c is the volume per mole of the crystal and u_0 is the internal energy per mole of the crystal relative to the gas at the absolute zero. For convenience we define ω_0 by the relation

$$\Omega_0 = \omega N_A \qquad (13·65)$$

and in view of the meaning of Ω_0 (§ 13·4), ω_0 is the degeneracy of the lattice *per atom*. Therefore

$$\mu_c = p \, v_c + u_0 - RT \ln \omega_0 + 3RT \ln \left(\frac{h\bar{\nu}}{kT} \right). \qquad (13·66)$$

It was shown in equation (12·65) that the chemical potential of a perfect gas may be expressed in the form

$$\mu_g = - RT \left[\ln \left(\frac{M^{\frac{3}{2}} T^{\frac{5}{2}}}{p} \right) + \ln \left(\frac{2\pi}{L} \right)^{\frac{3}{2}} \frac{k^{\frac{5}{2}}}{h^3} + \ln f^{\text{int.}} \right],$$

where p is the pressure and M is the molecular weight. For equilibrium between the crystal and its vapour, $\mu_c = \mu_g$, and therefore from the last two equations,

$$\ln p = \frac{p v_c + u_0}{RT} - \tfrac{1}{2} \ln T + \ln \left(\frac{h\bar{\nu}}{k} \right)^3 M^{\frac{3}{2}} \frac{f^{\text{int.}}}{\omega_0} + \ln \left(\frac{2\pi}{L} \right)^{\frac{3}{2}} \frac{k^{\frac{5}{2}}}{h^3}.$$

The quantity $p v_c + u_0$ is very nearly the negative of the latent heat of evaporation L_0 at the absolute zero.† The last equation can therefore be rearranged to give

$$\ln p = - \frac{L_0}{RT} - \tfrac{1}{2} \ln T + \ln \left(\frac{\bar{\nu}^3 M^{\frac{3}{2}} f^{\text{int.}}}{\omega_0} \right) + \ln \left(\frac{2\pi}{L} \right)^{\frac{3}{2}} \frac{1}{k^{\frac{1}{2}}}. \qquad (13·67)$$

In the special case where the substance is monatomic the factor $f^{\text{int.}}/\omega_0$ may be expected to be unity. Therefore

$$\ln p = - \frac{L_0}{RT} - \tfrac{1}{2} \ln T + 3 \ln \bar{\nu} M^{\frac{1}{2}} + \ln \left(\frac{2\pi}{L} \right)^{\frac{3}{2}} \frac{1}{k^{\frac{1}{2}}}. \qquad (13·68)$$

Comparison with equation (6·20) shows that (13·67) is a thermodynamically correct expression for the vapour pressure of a monatomic substance at a high enough temperature where Δc_p has a value of about $-\tfrac{1}{2} R$. In fact what has been achieved in (13·67), by means of the statistical theory, is a definite value for the integration constant in the thermodynamic equation (6·20).

The vapour pressures of a large number of substances are represented in the literature by an empirical equation

$$\log_{10} p = - A/T + B. \qquad (13·69)$$

For the purpose of comparing the experimental values of B with the terms on the right-hand side of (13·68) which do not involve $1/T$ we shall

---

† $\Delta H = \Delta U + \Delta(pV) = - u + RT - p v_c.$ Hence, when $T = 0$, $L_0 = - u - p v$ (neglecting the temperature coefficient of v_c).

put $\frac{1}{2}\ln T = 3$, which will be a good approximation when T is of the order of 1000 K. After inserting numerical values for k, L, etc., we obtain

$$\log_{10} \bar{\nu} M^{\frac{1}{2}} = \frac{1}{3}(B+31), \qquad (13·70)$$

when p is expressed in mmHg. From this equation it is possible to estimate an approximate value of $\bar{\nu}$ from the empirical values of B. The results are given in the fourth column of Table 12 and are seen to be in the range 10^{12}–10^{13}. Values of the Debye frequency ν_D, as calculated from the heat capacity, are given in the fifth column. A precise agreement is not to be expected because the empirical equation (13·69) is not thermo-dynamically exact and the values of the constant B could no doubt have been chosen rather differently without appreciably affecting the corre-lation of the vapour-pressure data. In any case, $\bar{\nu}$ and ν_D do not have the same significance and the former would be expected to be rather smaller than the latter.

TABLE 12

Substance	Temp. range (°C)	B	$\bar{\nu}$	ν_D
Ca	500– 700	9.69	5×10^{12}	4.8×10^{12}
Cd	150– 320	8.56	2×10^{12}	3.5×10^{12}
Mo	1800–2240	10.84	8×10^{12}	8.0×10^{12}
Pt	1425–1765	7.79	6×10^{11}	4.7×10^{12}
Tl	2230–2770	9.92	3×10^{12}	2.0×10^{12}
Zn	250– 420	9.20	3×10^{12}	4.9×10^{12}
FeCl$_2$	700– 930	8.33	1×10^{12}	—
As$_2$O$_3$	100– 310	12.13	2×10^{13}	—

Data from *International Critical Tables*, vol. III.

A more general discussion of the integration constant in the vapour pressure equation will be found in the literature.† Our purpose in developing equation (13·67), applicable to high temperatures, is partly because of its simplicity and also because it has application in the theory of the evaporation rate and reaction kinetics of solids. If equation (12·114) may be taken as a correct expression for the condensation rate of a gas on its own crystal, then this equation may be combined with (13·67) to give an equation for the evaporation rate of the solid.

13·11. The third law—preliminary

In § 11·10 two statistical analogues of the entropy were introduced

$$S' = -k\Sigma P_i \ln P_i, \qquad (13·71)$$

$$S'' = k \ln \Omega. \qquad (13·72)$$

† See, for example, Fowler and Guggenheim, *Statistical Thermodynamics*, Chapter v.

Now quantum theory assigns definite values to the P's and also to Ω. Therefore, for any system, S' and S'' have perfectly definite values which are positive or zero. (They cannot be negative because the probabilities P_i are all fractions and the number of complexions Ω of the system cannot be less than one.)

It was shown subsequently that S' and S'' possess all of the properties of entropy and the primes were therefore deleted. On the other hand, since thermodynamics deals only with *changes* of entropy, it is clear that the statistical analogues would have been equally satisfactory if they had contained an additive constant. Let us write such a constant in the form $-k \ln \Omega_n$. Then the entropy analogue

$$S''' = k \ln \Omega - k \ln \Omega_n = k \ln \Omega/\Omega_n \qquad (13\cdot73)$$

will be entirely satisfactory for all thermodynamic purposes, and it will be zero when the system has a number of complexions equal to Ω_n. Therefore it would be a matter of convention if (13·71) or (13·72), containing no additive constant, were to be taken as the choice of the statistical definition. In actual fact our procedure in this chapter, and the previous one, has been based on taking the entropy as zero when the system may still possess a large number of possible complexions. This is because we have no means of computing the number of complexions due to alternative quantum states within the nuclei of the atoms.

In Chapter 12 we discussed randomization over translational and configurational states and also, to some extent, over rotational, vibrational and electronic states. Two other factors which are known to contribute small amounts to the entropy are (a) the number of possible orientations of the nuclear spins and (b) the entropy of isotope mixing. (The latter factor arises if the substance in question contains two or more isotopes.) However, this leaves quite untouched the possibility of *randomness within the nucleus*, about which nothing whatever is known. It is evident that even when we have allowed for all the known factors our computed entropy may still be incomplete.

In brief, *it is not possible to calculate absolute entropies*. Instead it is necessary to adopt some convention concerning what factors are to be included in Ω, in view of the present state of knowledge. The convention which is usually adopted in physical chemistry is that the entropy is taken as zero when the substance is in a physical state such that translational, configurational, rotational, vibrational and electronic contributions to the entropy are all zero. Contributions due to the nucleus, including its spin, are ignored and also the effect of isotope mixing. The justification for omitting these factors lies in the fact that nuclei are conserved in chemical processes and also because the isotopic composition usually remains almost constant.

Effects of this kind therefore cancel when we consider *changes* of entropy.

Returning to (13·73), the convention which is adopted is equivalent to taking Ω, not as the total number of complexions of all types, but as including only the number due to translation, etc., and at the same time arbitrarily putting Ω_n, the number of complexions due to the nucleus and isotope mixing, equal to unity. (This is what was done implicitly in Chapter 12 when $f^{\text{int.}}$ for a monatomic gas was put equal to unity.) If this convention is used whenever statistical mechanics is applied to chemical problems, it may be expected that the results will be self-consistent. The entropies as calculated on this basis are referred to as *practical or conventional entropy values*.

Having adopted this convention we may ask, Does there exist a physical state of a substance for which the conventional entropy is actually zero? Now perfect crystals are known to have a very orderly structure, and at very low temperatures the lattice vibrations will all be in their lowest states which correspond to the zero-point energy. Therefore it may be expected that a crystal will have a very low entropy at temperatures approaching the absolute zero, and in one of the original forms (Planck's version) of the third law it was asserted that the entropy of a pure substance is actually zero under such conditions. On the other hand, from (13·51), based on the Einstein approximation, it is seen that

$$\lim_{T \to 0} S = k \ln \Omega_0, \qquad (13·74)$$

and the same result is readily obtainable from the Debye equations. It seems therefore that, even after we have adopted the above conventions, the entropy of the crystal approaches zero only if Ω_0 is unity. This quantity, which was first introduced in equation (13·29), was tentatively identified with orientational factors in the crystal.

Before discussing this question it may be remarked, that *imperfect* crystals would *not* be expected to have zero entropy. Also it might be very difficult to determine whether or not a crystal *is* perfect, at very low temperature, except by the investigation of its entropy. Therefore there is some danger of circularity in the argument. Moreover there is no evidence that the absolute zero can ever be reached; on the contrary it seems quite unattainable, as if there is really a kind of infinity of temperature between 0 K and, say, 1 K. What we have to discuss, therefore, is really an *extrapolated entropy*, namely its apparent value at $T = 0$, as extrapolated from the lowest temperatures attainable in calorimetric measurements. This is normally a temperature of a few K upwards.

Suppose we have available a set of c_p values on a crystalline substance at closely spaced intervals from a temperature T' (say 10 K) up to its sublimation temperature T_s, and also a set of c_p values for the same substance as a gas from T_s up to some temperature T'' which is of interest. These values will all be taken to refer to the same pressure P. The entropy increase of the substance in passing from the solid state at (T', P) to the gaseous state at (T'', P) is therefore

$$S_{T''} - S_{T'} = \int_{T'}^{T_s} \frac{c_p}{T} \, dT + \frac{L_s}{T_s} + \int_{T_s}^{T''} \frac{c_p}{T} \, dT, \qquad (13\cdot75)$$

where L_s is the latent heat of vaporization at the sublimation temperature. The integrals may be evaluated by plotting c_p against $\ln T$ and taking the area under the curve.† Now below T' the Debye T^3 law may be assumed to hold with sufficient accuracy. Thus from $(13\cdot60)$‡

$$c_p = aT^3, \qquad (13\cdot76)$$

and in accordance with what was said above *we shall use this law for the purpose of extrapolating downwards to* $T = 0$. The estimated entropy change in the process $(T = 0, P) \rightarrow (T', P)$ is§

$$\begin{aligned}
S_{T'} - S_0 &= \int_0^{T'} \frac{c_p}{T} \, dT \\
&= \frac{aT'^3}{3} \\
&= \frac{c_p'}{3}, \qquad (13\cdot77)
\end{aligned}$$

where c_p' is the lowest measured value of c_p at the temperature T'.

Adding $(13\cdot75)$ and $(13\cdot77)$ we obtain a quantity which is usually denoted $S_{\text{calor.}}$ and is called the *calorimetric entropy* of the gas at the temperature T'':

$$S_{T'} - S_0 = \frac{c_p'}{3} + \int_{T'}^{T_s} \frac{c_p}{T} \, dT + \frac{L_s}{T_s} + \int_{T_s}^{T''} \frac{c_p}{T} \, dT$$
$$\equiv S_{\text{calor.}} \qquad (13\cdot78)$$

The significance of $S_{T'} - S_0$ in equation $(13\cdot77)$ is an entropy change based on the use of the Debye theory for the purpose of a smooth extrapolation from T' to $T = 0$. *It is therefore tacitly assumed*

† Additional integrals and latent heat terms must, of course, be included if the process passes through the liquid state or through more than one solid state.
‡ At low temperatures c_p and c_V are practically equal.
§ Note that the integral is convergent and $S_{T'} - S_0$ is finite. This is because of the T^3 law which shows that c_p/T approaches zero as $T \rightarrow 0$.

that the only cause of a decrease of entropy in passing from T' to $T = 0$ is a damping out of the lattice vibrations.

In accordance with (13·74) let us now substitute

$$S_0 = k \ln \Omega_0. \tag{13·79}$$

Then it follows that Ω_0 must similarly be interpreted not necessarily as the true degeneracy at $T = 0$ (even when nuclear factors are not included), but rather as its value at T' and upwards. This will become clearer in what follows.

The substitution of (13·79) in (13·78) gives

$$S_{T''} = k \ln \Omega_0 + S_{\text{calor.}}$$

$$= k \ln \Omega_0 + \frac{c_p'}{3} + \int_{T'}^{T_s} \frac{c_p}{T} \, dT + \frac{L_s}{T_s} + \int_{T_s}^{T''} \frac{c_p}{T} \, dT. \tag{13·80}$$

In summary, this is an expression for the entropy of the gas at the temperature T'' on the basis of (a) the foregoing conventions with regard to nuclear and isotope factors not being included; (b) the smooth extrapolation from T' to zero by use of the statistical theory of perfect crystals. Using just the same conventions as in (a) above an alternative value for $S_{T''}$ may be calculated by the methods of § 12·7. If the gas is monatomic the entropy in question is simply the translational entropy and is given by the Sackur–Tetrode equation (12·64). If the gas is polyatomic it is necessary to know the value of $f^{\text{int.}}$, e.g. by determination of the energy levels by spectroscopy. In either case the entropy, as computed in this way, is known as $S_{\text{spec.}}$. Thus (13·80) can be written

$$S_{\text{spec.}} = k \ln \Omega_0 + S_{\text{calor.}}. \tag{13·81}$$

A number of experimental values of $S_{\text{spec.}}$ and $S_{\text{calor.}}$ are given in Table 13,† and it is seen that there is an appreciable difference between them only in the case of H_2, CO and N_2O. The difference in the case of hydrogen is believed to be due to the *ortho-para* effect and will not be discussed here. With CO and N_2O the difference is about 1.1 cal K^{-1} mol^{-1}, and this may be accounted for on the following lines. In carbon monoxide the oxygen and carbon atoms probably do not differ much in size or in their force fields (and similar considerations apply to N_2O which is also a linear molecule). Therefore it may be expected that an arrangement of the crystal in which all of the CO (or N_2O) molecules are arranged with their dipoles pointing in the *same* direction, e.g. CO, CO, CO, etc., will not differ much in energy

† The figures refer to the ideal gas state at 25 °C and 1 atm and are in units of cal K^{-1} mol^{-1}. A much larger table is given by Aston in Taylor and Glasstone's *Treatise on Physical Chemistry*, vol. 1, p. 588.

TABLE 13

Substance	$S_{spec.}$	$S_{calor.}$
HCl	44.64	44.5
HBr	47.48	47.6
HI	49.4	49.5
N_2	45.78	45.9
O_2	49.03	49.1
H_2	31.23	29.74
CO	47.31	46.2
H_2S	49.10	49.15
CO_2	51.07	51.11
N_2O	52.58	51.44
NH_3	45.94	45.91
C_2H_4	52.47	52.48
CH_3Br	58.74	58.61

from an arrangement in which the molecules are turned end to end, e.g. CO, OC, CO, etc., at random. At high temperatures the random arrangement will certainly be the more stable and at the lowest temperatures, $c.$ 10 K, which can be reached conveniently in the heat-capacity measurements, the crystal may still be expected to have the random structure. (Either because the random arrangement continues to be the more stable or because the transition to a more orderly structure is exceedingly slow compared to the time in which the published heat capacity measurements were made.) We therefore postulate that each molecule can have two arrangements in the lattice, CO or OC, and thus, per mole of the crystal,

$$k \ln \Omega_0 = k \ln 2^L = 1.38 \qquad (13\cdot82)$$

This rather more than accounts for the discrepancy between $S_{spec.}$ and $S_{calor.}$, and it may be that some loss of randomness has taken place at the lowest temperatures attained.

13·12. Statement of the third law

As is well known, Thomsen and Berthelot believed that the tendency of reactions to take place is determined by the heat of reaction, and this view was subsequently modified in favour of the free energy as the correct criterion. Nevertheless it remains true that the majority of reactions take place in the direction in which heat is evolved, especially when the temperature is low, and when none of the substances are gaseous.

Evidently ΔF (or ΔG) is not very different from ΔU (or ΔH) for such processes, and this was substantiated by Richards who showed, in 1902, that the e.m.f. of galvanic cells becomes more and more nearly proportional to the internal energy change of the cell reaction the lower the temperature.

On the basis of such results Nernst† in 1906 put forward a new hypothesis: the curves of the changes in free energy and total energy of a chemical reaction between pure solid or liquid bodies become tangential to each other at the absolute zero. The shapes of these curves, according to Nernst's hypothesis, are shown in Fig. 44, and it is instructive to consider them in relation to the Gibbs–Helmholtz equation (2·68). Nernst's hypothesis is clearly equivalent to the supposition that the entropy change in a reaction between pure solids or liquids approaches zero at $T=0$.

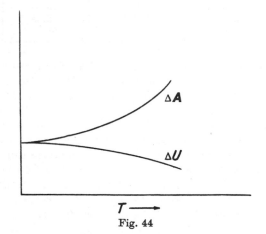

Fig. 44

Instead of considering *differences* of entropy, as was done by Nernst, Planck subsequently adopted‡ the stronger statement 'as the temperature diminishes indefinitely, the entropy of a chemically homogeneous body of finite density approaches indefinitely near to the value zero'. The discussion in §13·11 shows that this statement is not at all satisfactory. In the first place the absolute zero is not attainable, and we must discuss instead the properties of matter as smoothly extrapolated to $T = 0$ from the lowest attainable temperature T'. Secondly, the possibility of giving any value including zero, to the entropy is entirely conventional because nothing is known about the entropy of the nucleus. Finally, we have seen that even the 'conventional entropy' does not necessarily fall to a value of

---

† *Nachr. Ges. Wiss. Gottingen*, Klasse Math. Phys. (1906), p. 1.

‡ Planck, *Treatise on Thermodynamics*, transl. Ogg (London, Longmans, 1927), 3rd ed., p. 274.

zero, due to the presence of configurational randomness at the lowest attainable temperature.

To be sure, in the great majority of cases the comparison of $S_{spec.}$ and $S_{calor.}$ indicates that Ω_0 is unity, which means that the conventional entropy does approach zero. But it is precisely because there are exceptions that it is difficult to put forward an unambiguous law of nature on the lines of Planck's formulation. Modern versions of the law are much closer to Nernst's original statement, but contain safeguards against unstable states such as supercooled liquids and glasses.

Guidance with regard to the best formulation of the law may be obtained by considering the case of crystals, for which we have the limiting equation (13·74), of the Debye theory. From this equation

$$\lim_{T\to 0} (S_\beta - S_\alpha) = k \ln \Omega_{0\beta} - k \ln \Omega_{0\alpha}, \qquad (13\cdot83)$$

where α and β refer to two states of the crystalline substance or substances. Therefore, if Ω_0 is the same in both states, we have

$$\lim_{T\to 0} (S_\beta - S_\alpha) = 0, \qquad (13\cdot84)$$

and this will apply even if Ω_0 is not unity in the two states. For example, any isothermal process on solid carbon monoxide, such as a small change of pressure, may be expected to have zero entropy change at a low enough temperature, provided that it does not change the randomness of the molecular arrangement in the crystal. On the other hand, the reaction $CO + \frac{1}{2}O_2 = CO_2$ would not be expected to have a limiting entropy change of zero; in this process the randomness of the CO lattice is lost and there is no evidence that it is compensated by any corresponding factor in the CO_2 lattice (cf. the table in § 13·11).

A statement of Fowler and Guggenheim[†] which meets the situation very adequately is as follows: For any isothermal process involving only phases in internal equilibrium, or, alternatively, if any phase is in frozen metastable equilibrium, provided the process does not disturb this frozen equilibrium,

$$\lim_{T\to 0} \Delta S = 0. \qquad (13\cdot85)$$

[†] Fowler and Guggenheim, *Statistical Thermodynamics*, Chapter v. These authors prefer to deduce the above statement from a second principle which at first sight appears to be quite different. This is the principle of the unattainability of the absolute zero: 'it is impossible to reduce a system to the absolute zero in a finite number of operations.' The equivalence of the two statements is discussed in the well-known books by Epstein, Zemansky and Roberts and Miller and also by Simon in *Four Lectures on Low Temperature Physics* (1952).

For example, a change in the pressure of solid CO might be expected to fulfil this condition, whereas any reaction involving CO does not. 'Internal equilibrium' implies that the state of the phase is determined entirely by its temperature, pressure and composition, and this excludes glassy states or supercooled liquids whose state may be expected to be at least partially determined by their previous history, such as the rate of cooling.

13·13. Tests and applications of the third law

(*a*) *Liquid helium.* The above statement of the law is not limited to crystals, and it is actually the transition between liquid and solid helium which provides one of the best confirmations. This transition can be carried out reversibly, by suitable change of pressure, at temperatures far below 1 K. The equilibrium is determined by the Clausius–Clapeyron equation, (6·8), which can be expressed as

$$\frac{dp}{dT} = \frac{\Delta S}{\Delta V}.$$

It was shown experimentally by Simon and Swensen† that dp/dT is extremely small or zero below 1.4 K. The differential coefficient was actually found to be proportional to the seventh power of T,

$$\frac{dp}{dT} = 0.425 T^7,$$

and is thus quite minute at $T = 0.5$ K. It follows from the Clausius–Clapeyron equation that ΔS, the entropy difference between liquid and solid, must also approach zero very rapidly, in accordance with the third law. (It is of interest that in these low regions of temperature there is no latent heat, and the phase change does not take place by intake of heat but only by change of pressure.)

(*b*) *Coefficients of expansion.* Consider a small increase δp, in the pressure of a stable phase at a constant temperature. The entropy increase is

$$\delta S = \left(\frac{\partial S}{\partial p}\right)_T \delta p.$$

According to the third law δS tends to zero and therefore so also must $(\partial S/\partial p)_T$. Using one of the Maxwell relations it follows that $(\partial V/\partial T)_p$ must also tend to zero as $T \to 0$. This has been confirmed experimentally in the case of copper, aluminium, silver and other substances.

† Simon and Swensen, *Nature, Lond.*, **165** (1950), 829.

(c) Crystal-crystal and crystal-glass transitions. Let α and β be two crystalline forms of a substance both of which are in a state of internal equilibrium at very low temperature, as previously discussed. This does not necessarily imply that they are at equilibrium with respect to each other, when they are in contact. Let T_e be their normal equilibrium temperature, where their chemical potentials are equal. Let $S^\alpha_{\text{calor.}}$ and $S^\beta_{\text{calor.}}$ be the calorimetric entropies of the two phases at the temperature T_e, as calculated by use of equations such as (13·78):

$$S^\alpha_{\text{calor.}} = \frac{c'_{p\alpha}}{3} + \int_{T'}^{T_e} \frac{c'_{p\alpha}}{T}\, dT.$$

Finally, let $(S_\alpha - S_\beta)_{T_e}$ be the entropy increase in the process $\beta \to \alpha$ at the equilibrium temperature T_e, as calculated from the latent heat, and let $(S_\alpha - S_\beta)_0$ denote the extrapolated entropy difference as $T \to 0$. Since entropy is a function of state its change round a cycle is zero. Hence

$$(S_\alpha - S_\beta)_0 = S^\beta_{\text{calor.}} + (S_\alpha - S_\beta)_{T_e} - S^\alpha_{\text{calor.}}, \qquad (13·86)$$

and therefore, according to the third law, the sum of the three terms on the right-hand side should be approximately zero. Results quoted in Table 14, where the entropies are in cal K^{-1} mol^{-1}, show that this is the case.

TABLE 14

Substance	β form	α form	T_e/K	$S^\beta_{\text{calor.}}$	$(S_\alpha - S_\beta)_{Te}$	$S^\alpha_{\text{calor.}}$	$(S_\alpha - S_\beta)_0$
Sulphur	Rhombic	Monoclinic	368.6	8.69	0.24	8.91	0.02
Tin	Grey	White	286.4†	9.23	1.87	11.17	−0.07
Phosphine	β	α (and γ)	49.43	4.38	3.76	8.13	0.01
Cyclohexanol	β	α	263.5	33.5	7.4	41.2	−0.3

† The entropy data refer to 25 °C, although the transition temperature is 13.2 °C

The same agreement is not to be expected if we compare the entropy of a crystal with that of the same substance in a supercooled liquid or glassy state. For example, the entropy, extrapolated to $T = 0$, of glassy glycerol exceeds that of crystalline glycerol by 4.6 cal K^{-1} mol^{-1}, and the corresponding difference for glassy and crystalline alcohol is 2.6 cal K^{-1} mol^{-1}. These glassy states evidently have quite large values of Ω_0.

Despite the good agreement in the table above it must be admitted that any application of the third law to metastable states is somewhat precarious, and is perhaps justified only in cases where we seem to have a fairly clear statistical insight.

(d) Chemical reaction. A much more certain application of the third law is to chemical reactions, as discussed originally by Nernst. Consider a reaction

$$A + B = C,$$

in which the three substances all exist as perfect and stable crystals at a temperature less than, say, 10 K. The value of $S_{calor.}$ for each substance at any temperature T can be calculated from experimental measurements by use of equation (13·78). It follows from the third law (as in the derivation of equation (13·86) above) that

$$S^C_{calor.} - S^A_{calor.} - S^B_{calor.} \tag{13·87}$$

is equal to the entropy increase in the reaction at temperature T. This third law value of ΔS can be compared with a value obtained in the usual way, either from measured values of ΔH and ΔG, or from the temperature coefficient of the e.m.f. when the reaction is carried out in a cell.

TABLE 15

Reaction	T/K	ΔS third law cal K^{-1} mol^{-1}	ΔS direct (cal K^{-1} mol^{-1})
$Ag(s) + \frac{1}{2}Br_2(l) = AgBr(s)$	265.9	$- 3.01 \pm 0.40$	$\begin{cases} - 3.13 \pm 0.10 \\ - 3.02 \pm 0.10 \end{cases}$
$Ag(s) + \frac{1}{2}Cl_2(g) = AgCl(s)$	298.16	-13.85 ± 0.25	-13.73 ± 0.10
$Zn(s) + \frac{1}{2}O_2(g) = ZnO(s)$	298.16	-24.07 ± 0.25	-24.24 ± 0.05
$Mg(OH)_2(s) = MgO(s) + H_2O(g)$	298.16	35.85 ± 0.08	36.67 ± 0.10

A comparison quoted by Aston† is given in Table 15. It will be seen that the agreement is within the experimental error except in the fourth reaction which involves water. This discrepancy is attributed to lack of perfection in the ice crystal at very low temperatures. In fact a residual entropy of 0.806 cal K^{-1} mol^{-1} can be accounted for statistically and this almost completely removes the discrepancy.

The great value of the third law in physical chemistry is this application to chemical reactions. In the case of simple molecules, which exist as perfect or almost perfect gases, the standard free energy can be evaluated from the band spectrum, as discussed in § 12·7. This method is not applicable to the more complex organic substances, and it is for this reason that the third law is extremely valuable.

Let it be supposed that the substance in question passes into a crystalline form (and not a glass) on cooling. Its calorimetric entropy can be calculated at any temperature T by use of equation (13·78) provided that heat-capacity measurements are available at close

† Taylor-Glasstone, *Treatise on Physical Chemistry*, vol. I, p. 518.

intervals from about 10 K upwards. By subtracting from this value the corresponding calorimetric entropy of the elements of which the substance is formed we obtain the entropy of formation at the temperature T. A determination of the heat of combustion gives the enthalpy of formation, and a combination of the two results gives the free energy of formation.

By this method the equilibrium constants of reactions can be estimated by purely calorimetric methods. The main point where care must be exercised is in the measurement of the heat capacities—if, below a certain temperature, the substance were to pass into a glassy state, instead of a crystal, the computed entropy of formation might be considerably too low, as may be seen from the example of glycerol as previously discussed.

In conclusion, it may be remarked that the above method of applying the third law has now largely replaced the method of the 'Chemical Constants' as described in the older textbooks.

PROBLEMS

1. Using the Einstein model show that the total number of modes which are in the first or higher excited levels increases most rapidly with temperature when $T = \frac{1}{2}\theta_E$. Why is this not quite the same as the temperature at which c_V increases most rapidly? The figure in §13·9 shows that the latter occurs at about $T = 0.2\theta_E$.

2. Show quite generally that the application of the statistical theory determines the value of the integration constant in equation (6·19) for the vapour pressure of a crystal.

3. Obtain the following expression for the evaporation rate of a crystal at high temperature

$$\text{Molecules cm}^{-2} \text{ sec}^{-1} = \frac{2\pi M \bar{\nu}^3 f^{\text{int.}}}{RT} \frac{}{\omega_0} e^{-L_0/RT} \qquad (13\cdot88)$$

(This expression is usually multiplied by a factor α, the condensation coefficient, to allow for the possibility that molecules strike the surface without condensing. In any case it seems that α is usually close to unity.)

4. Verify the shape of the curves shown in Fig. 11 of Chapter 2 for the equilibrium of a crystal with its monatomic vapour. For this purpose plot the total energy and entropy of the crystal-vapour system as a function of the fraction present as vapour when suitable values of $\bar{\nu}$, M, etc., are substituted in the relevant equations. For simplicity, use the equations which are valid for high temperature.

5. Using the following data, which are from Landolt–Bornstein, compare the calorimetric entropy of mercury vapour at 343.9 K with a value calculated from the Sackur–Tetrode equation, assuming that the vapour is entirely monatomic.

P

Atomic weight 200.6.

Vapour pressure at 343.9 K $= 5.16 \times 10^{-2}$ mmHg.

Latent heat of evaporation of liquid at 343.9 K $= 14\,460$ cal mol^{-1}

Latent heat of melting at m.p. (234.3 K) $= 555$ cal mol^{-1},

Heat capacity of solid (cal K^{-1} mol^{-1}):

T/K	10	20	30	40	50	60	70
c_p	1.11	2.51	3.69	4.45	4.94	5.26	5.48

T/K	80	100	120	150	200	234.3
c-	5.63	5.86	6.01	6.20	6.47	6.65

Heat capacity of liquid (cal K^{-1} mol^{-1}):

T/K	234.3	240	250	270	300	323.2	353.2
c_p	7.00	6.95	6.90	6.76	6.65	6.63	6.60

6. Using the limited amount of data provided below, investigate the thermodynamic conditions which might be needed for the conversion of graphite into diamond. What additional data are necessary for the purpose of a more exact calculation? What other conditions may need to be satisfied if the conversion is to take place?

The first column refers to the standard enthalpy of formation (cal mol^{-1}), and the second column to the standard entropy (cal K^{-1} mol^{-1}), as obtained by third law methods. The third and fourth columns give the heat capacity (cal K^{-1} mol^{-1}) and density (g cm^{-3}) respectively, at atmospheric temperature and pressure.

	$\Delta_f H_{298}$	S°_{298}	C_p	d
Diamond	453	0.583	1.449	3.51
Graphite	0	1.361	2.066	2.25

[C.U.C.E. Tripos, 1955]

[For further discussion of the conditions necessary for diamond formation see Hall, *J. Chem. Educ.*, **38** (1961), 484.]

CHAPTER 14

REGULAR SOLUTIONS
AND ADSORPTION

14·1. Configurational energy and entropy

In §11·12 it was shown that, in certain types of problem, the difference in entropy between two states of a system is given by

$$S_2 - S_1 = k \ln \frac{\Omega_{\text{config.}2}}{\Omega_{\text{config.}1}}, \qquad (14 \cdot 1)$$

where the Ω's are the numbers of configurational arrangements in the states 1 and 2. The purpose of the present chapter is to make some simple applications of this equation by assuming, as a rough approximation, that certain phenomena of solutions, and of adsorption, are due to purely configurational effects.

Before proceeding, it is desirable to show in rather greater detail how the above equation may be derived from the more fundamental relation
$$A = -kT \ln Q. \qquad (14 \cdot 2)$$

Equation (14·1) was actually derived for certain types of process at constant energy; it must now be shown that the equation may also apply when the process is isothermal.

Consider a change from state 1 to state 2 which is at the same temperature. Then from (14·2)

$$A_2 - A_1 = -kT \ln \left(\frac{\sum_i e^{-E_{2i}/kT}}{\sum_i e^{-E_{1i}/kT}} \right), \qquad (14 \cdot 3)$$

where the explicit form of the partition functions has been inserted and the summations are taken over all of the quantum states i which are comprised within the macroscopic or thermodynamic states 1 and 2. In these summations it may occur, of course, that a particular term, $e^{-E_i/kT}$, repeats itself a great many times on account of the degeneracy of the particular level.

Let it be supposed that, as we pass from state 1 to state 2, the only change which occurs is a change in the number of geometrical configurations in which the system can exist. In particular, we shall suppose that each configuration has the same set of energy levels rising above the same zero. A typical example was that discussed in §1·17, namely, the mixing of very similar atoms or molecules between lattice sites. If the numbers of particles of type A and B are N_a and N_b

respectively, there are $3(N_a + N_b)$ normal modes of the lattice and the present assumption is that these modes are quite unaffected by the mixing process. (If A and B are actually molecules, rather than atoms, it must also be assumed that their internal states are unchanged.)

The second macroscopic state of the system will therefore be taken as having Ω_c times as many configurations as the first, the symbol Ω_c being used for brevity in place of the ratio $\Omega_{\text{config. 2}}/\Omega_{\text{config. 1}}$. Thus each term $e^{-E_i/kT}$ repeats itself Ω_c times more often in the numerator of (14·3) than in the denominator. This quantity can therefore be factorized out and we obtain

$$A_2 - A_1 = -kT \ln \left(\Omega_c \frac{\sum e^{-E_{2i}/kT}}{\sum e^{-E_{1i}/kT}} \right),$$

where each summation now refers to any single configuration only. *Also these two summations are equal.* This is because, as said already, the kind of isothermal process which is being discussed is one which involves a change in the number of configurations but for each of these configurations the quantum states are the same.

Hence
$$A_2 - A_1 = -kT \ln \Omega_c. \tag{14·4}$$

For the same reason
$$U_2 - U_1 = 0, \tag{14·5}$$

and therefore
$$S_2 - S_1 = k \ln \Omega_c, \tag{14·6}$$

in agreement with equation (14·1).

On the other hand, it may occur that the *spacing* of the energy levels is the same for each configuration of the system, but not their energy zeros. This kind of situation is indicated in Fig. 45. For example, in the mixing of A and B atoms it may occur that the A-B interaction energy is greater than the mean of A-A and B-B interaction energies. Those configurations of the mixture in which there are a large number of A-B nearest neighbours will therefore have a more negative potential energy than those in which there are a small number. If the physical conditions are such that the spacing is unaffected, then be-

Fig. 45

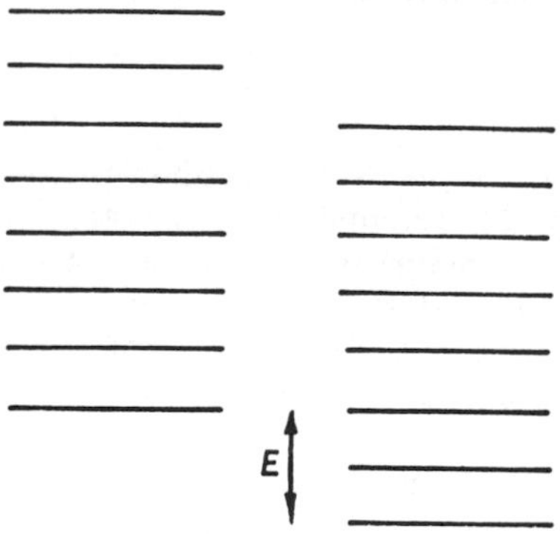

tween any two configurations there is merely a displacement of the levels relative to each other by some quantity E. However, the magnitude of E *will vary from one pair of configurations to another*, according to the relative numbers of A-B, A-A and B-B nearest neighbours. Under such conditions it is evident that the two summations do not cancel and no simple result is obtainable. However, a further approximation† is frequently adopted which is difficult to justify in a brief discussion. It amounts to the supposition that the energy zeros of the *vast majority* of configurations in state 2 of the system differ from those of state 1 by a constant amount E.

Consider the example of the mixing of the A and B particles. As mentioned above it may be that there is a tendency for a particle of a given type to be surrounded by an excess of particles of the other type. Conversely there may be a tendency for like particles to collect together. However, provided that the disparities in the A-A, A-B and B-B interaction energies are not very large, it seems that these tendencies may often be neglected. This is the approximation which is now being made, and it amounts to the assumption that those quantum states of the mixed system in which the particles are distributed at random over the lattice sites are so preponderant in numbers that all other states (which should really be included in the summations) may be neglected.

We thus assume that the energy zero of *every* configuration of the final state of the system differs from that of the initial state by the same amount E. Thus

$$E_{2i} = E_{1i} + E. \tag{14·7}$$

If, as before, the ratio of the number of geometrical configurations is Ω_c, it follows that the term $\Omega_c \, e^{-E/kT}$ can be factorized out of the numerator of (14·9). We thus obtain

$$A_2 - A_1 = -kT \ln \Omega_c \, e^{-E/kT}$$
$$= -kT \ln \Omega_c + E, \tag{14·8}$$

the remaining parts of the summations having cancelled.

By application of the Gibbs–Helmholtz equation, or equation (11·44), it is readily established that

$$U_2 - U_1 = E, \tag{14·9}$$

provided that E and Ω_c are independent of temperature. Thus we again obtain

$$S_2 - S_1 = k \ln \Omega_c, \tag{14·10}$$

whenever the above approximations are valid.

† Equivalent to the so-called 'zeroth' approximation or the Bragg and Williams approximation.

14·2. Regular solutions†

In §8·1 it was shown that equations (14·4)–(14·6) lead to a statistical deduction of the laws of ideal solutions. For this purpose it was supposed that a liquid solution can be approximated by a quasi-crystalline lattice, and also that the A and B molecules are of roughly the same shape and size, so that they can be interchanged between the lattice sites without change of lattice structure and without change in the lattice vibrations or the internal states of the molecules. Before mixing there is only *one* geometrical arrangement and after mixing there are

$$\Omega_{\text{config. 2}} = \frac{(N_a + N_b)!}{N_a! \, N_b!} \tag{14·11}$$

such arrangements.

It was also supposed that there are no preferential interaction energies between the molecules. Let w_{aa} be the increase in potential energy when a pair of A molecules are brought together from infinite distance to their equilibrium separation in the pure or mixed lattice. Similarly, let w_{ab} and w_{bb} be the potential energies of A-B and B-B pairs. Assuming for simplicity that only nearest-neighbour interactions are appreciable and that the average number of nearest neighbours of a given molecule is z, then

$$z[2w_{ab} - w_{aa} - w_{bb}] \tag{14·12}$$

is the increase in the potential energy when one A molecule is transferred from the pure A liquid to the pure B liquid and one B molecule is transferred in the reverse direction. In §8·1 this quantity was assumed to be zero so that $\Delta U = 0$,

and this led to the laws of ideal solutions.

We now consider a rather more general model of a solution in which (14·12) is no longer taken to be zero, but the other assumptions are retained. We thus consider the case where there is preferential interaction between otherwise similar molecules in a quasi-crystalline lattice. Now the existence of this interaction can only mean that the molecules are *not* arranged entirely at random in the mixed state. However, provided that the quantity defined in equation (14·12) is not large in absolute magnitude compared to the thermal energy, kT, it may be supposed that departures from randomness are not very appreciable. We can therefore adopt the approximation

† Footnote to the Second Edition. The lattice model is now less used than formerly, but it continues to provide one of the simplest and most attractive illustrations of statistical mechanical methods, within the scope of its assumptions, and it is on these grounds that the present section has been retained for the Second Edition. For further information see Hildebrand and Scott, *Regular Solutions* (Prentice Hall, 1962); Prigogine, *The Molecular Theory of Solutions* (North Holland Publ. Co., 1957).

embodied in equations (14·7)–(14·10). In particular, we obtain for the entropy of mixing

$$\Delta_m S = k\ln\Omega_c$$

$$= k\ln\frac{(N_a+N_b)!}{N_a!\,N_b!}$$

$$= -k\left(N_a\ln\frac{N_a}{N_a+N_b}+N_b\ln\frac{N_b}{N_a+N_b}\right), \tag{14·13}$$

which is the same as for the ideal solution, as given in §8·1. It remains to obtain the energy change on mixing when (14·12) is no longer zero.

Consider two adjacent sites 1 and 2 in the mixture. In accordance with the assumption of randomness the probability of an A molecule being on site 1 is equal to the mole fraction $N_a/(N_a+N_b)$. Similarly, the probability that a B molecule is on site 2 is $N_b/(N_a+N_b)$. The probability that an A molecule is on site 1 and simultaneously a B molecule is on site 2 is therefore $N_a N_b/(N_a+N_b)^2$. This is also the probability of the reverse arrangement—A on 2 and B on 1. Therefore the probability that the particular pair of sites is occupied by an A-B pair is

$$\frac{2N_a N_b}{(N_a+N_b)^2}. \tag{14·14}$$

(Note that in this expression we have a *sum* of probabilities. This is because the events are not independent but exclude each other—either the one arrangement occurs or the other.)

Now the system contains a total of N_a+N_b molecules and, if the average number of neighbours is z, there are a total of

$$\tfrac{1}{2}z(N_a+N_b) \tag{14·15}$$

pairs of neighbouring sites in the whole mixture. (The factor $\tfrac{1}{2}$ is to prevent the counting of each pair twice.) For many liquids it seems that z is about 12 corresponding to an approximately face-centred cubic packing.

From (14·14) and (14·15) we find that the total number of A-B pairs in the mixture is

$$N_{ab} = \tfrac{1}{2}z(N_a+N_b)\times\frac{2N_a N_b}{(N_a+N_b)^2}$$

$$= z\frac{N_a N_b}{N_a+N_b}. \tag{14·16}$$

Similarly, the numbers of A-A and B-B pairs are

$$N_{aa} = \tfrac{1}{2}z(N_a + N_b) \times \frac{N_a^2}{(N_a + N_b)^2}$$
$$= \tfrac{1}{2}z \frac{N_a^2}{(N_a + N_b)}, \qquad (14\cdot17)$$

$$N_{bb} = \tfrac{1}{2}z \frac{N_b^2}{(N_a + N_b)}, \qquad (14\cdot18)$$

respectively. On the other hand, in the two pure substances before mixing there are $\tfrac{1}{2}zN_a$ A-A pairs and $\tfrac{1}{2}zN_b$ B-B pairs respectively. The increase in potential energy on mixing is readily obtained from the above relations and is

$$z\frac{N_a N_b}{(N_a + N_b)} w_{ab} + \tfrac{1}{2}z \frac{N_a^2}{(N_a + N_b)} w_{aa} + \tfrac{1}{2}z \frac{N_b^2}{(N_a + N_b)} w_{bb} - z\frac{N_a w_{aa}}{2} - z\frac{N_b w_{bb}}{2}.$$

In accordance with (14·9) this is equal to $\Delta_m U$. On simplifying we obtain

$$\Delta_m U = \frac{z}{2}\frac{N_a N_b}{(N_a + N_b)}(2w_{ab} - w_{aa} - w_{bb}), \qquad (14\cdot19)$$

and this is seen to contain the quantity expressed in (14·12). If we define

$$w \equiv w_{ab} - \frac{w_{aa}}{2} - \frac{w_{bb}}{2}, \qquad (14\cdot20)$$

then

$$\Delta_m U = zw \frac{N_a N_b}{(N_a + N_b)}. \qquad (14\cdot21)$$

This may also be put equal to $\Delta_m H$, since for a mixing process at constant pressure $\Delta_m V$ is zero or trivial by the nature of our assumptions concerning the molecules A and B. Therefore combining (14·13) and (14·21) we obtain the increase in Gibbs free energy on mixing as

$$\Delta_m G = zw \frac{N_a N_b}{(N_a + N_b)} + kT\left(N_a \ln \frac{N_a}{N_a + N_b} + N_b \ln \frac{N_b}{N_a + N_b}\right). \qquad (14\cdot22)$$

The second term on the right-hand side of this equation is equal to the free energy of mixing if the solution were ideal, as discussed previously in §8·1. Therefore the *excess* free energy is

$$G^E = zw \frac{N_a N_b}{(N_a + N_b)}$$
$$= zLw \frac{n_a n_b}{(n_a + n_b)}, \qquad (14\cdot23)$$

where L is the Avogadro constant and n_a and n_b are the mole numbers of the two species. By applying equation (9·46),

$$\frac{\partial G^E}{\partial n_i} = RT \ln \gamma_i,$$

we obtain the logarithms of the activity coefficients

$$\left. \begin{aligned} \ln \gamma_a &= z\frac{Lw}{RT}x_b^2, \\ \ln \gamma_b &= z\frac{Lw}{RT}x_a^2. \end{aligned} \right\} \tag{14·24}$$

Finally, by using equation (9·6), the partial pressures of the two components are found to be

$$\left. \begin{aligned} p_a &= p_a^* x_a\, e^{Kx_b^2/T}, \\ p_b &= p_b^* x_b\, e^{Kx_a^2/T}, \end{aligned} \right\} \tag{14·25}$$

where
$$K = zLw/R$$

This simplified account of regular solutions has been put forward mainly for the purpose of illustrating the statistical method based on the supposition that only the configurational factors are significant. For a detailed discussion on the extent to which the equations agree with experiment, the reader is referred to the literature.† It is sufficient to say, in the first place, that the class of regular solutions includes the ideal solutions as a special case, namely, when $w=0$. Secondly, it seems that the theory often gives results which are closer to experiment for the free energy than it does for either the enthalpy or entropy taken separately. That is to say the expressions (14·22) or (14·25) are frequently in better agreement with experiment than either (14·13) or (14·21).

It will be noted that the simplified theory, as given above, represents the deviations from ideality entirely in terms of an *energy* factor. The *entropy* of mixing, as given by equation (14·13), is the same as for an ideal solution.

If the ratios p_a/p_a^* and p_b/p_b^*, as given by equation (14·25), are plotted against x, the curves are symmetrical about the point $x=0.5$. Very few solutions exhibit this symmetry at all accurately. At the same time the class of regular solutions is certainly more comprehensive and closer to reality than the class of ideal solutions and is therefore a very useful working model for practical purposes.

† Hildebrand and Scott, *The Solubility of Nonelectrolytes* (New York, Reinhold, 1950); *Disc. Faraday Soc.* (1953), no. 15.

14·3. The Langmuir isotherm

This well-known equation, concerning the phenomenon of adsorption, is frequently arrived at by a kinetic method; this depends on equating the condensation and evaporation rates for adsorbed molecules at the surface. However, since it applies to a condition of equilibrium, it might be expected that the same equation could be obtained by a purely statistical method. The advantage of the statistical derivation is that it shows much more clearly the precise conditions which are necessary if the Langmuir equation is to be obeyed, and thereby it shows the causes of deviations.

The following is based on a derivation given by Everett,† with some simplifications of the notation. The necessary assumptions are (a) a gas molecule can only be adsorbed at a finite number of positions, called the 'sites', on the surface of the solid; (b) the quantum states of adsorbed gas molecules are the same for all sites and independent of the presence of neighbouring molecules. The first of these is analogous to the assumption of a quasi-crystalline lattice in the case of a solution, and its purpose is to give a 'countable' number of configurations. Its justification depends, of course, on the fact that the surface is atomic in structure and may be expected to have potential energy 'wells', where adsorption takes place most readily. The assumption will probably break down if the depth of these wells is not considerably in excess of kT, for if this condition is not satisfied the molecules will be able to adsorb with almost equal readiness anywhere on the surface, and the adsorbed layer will approach the nature of a two-dimensional gas.

The second assumption makes possible the application of equation (14·10). It may be expected to break down either if the surface is appreciably heterogeneous or if there is appreciable attractive or repulsive interaction between the adsorbed molecules themselves. A third assumption is also implicit in what follows, namely, that the gas does not dissociate on adsorption. However, this is not essential, and a form of the Langmuir isotherm can be obtained quite readily for the case where dissociation takes place.

Let there be M sites on the given surface and a total of m gas molecules adsorbed. The ratio m/M is denoted θ, the fractional coverage

$$m/M = \theta. \tag{14·26}$$

Consider any *one* of the possible arrangements, such as is shown in Fig. 46, of the m molecules on the M sites and let s be the entropy per mole of the adsorbed molecules in this configuration. Let s^0 be the entropy per mole of the gas at unit pressure. Therefore, if, from the

† Everett, *Trans. Faraday Soc.* **46** (1950), 942.

gas at unit pressure, adsorption were to take place *to give the chosen arrangement*, the entropy increase would be $s - s^0$ per mole. Since there are m/L moles of adsorbed gas (where L is the Avogadro constant), the total entropy of the adsorbed layer, relative to the gas as unit pressure, would be

$$\frac{m}{L}(s - s^0).$$

However, the above arrangement is only one out of a very large number, and at any moment the system might exist in any of them. A term for the configurational entropy, due to the randomness of mixing over the various patterns, must evidently be included.

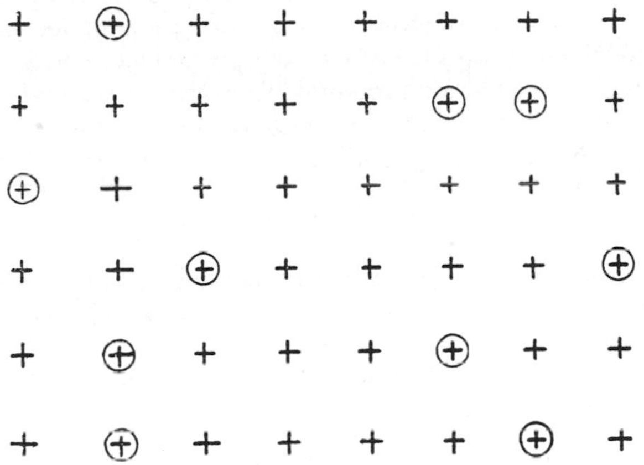

Fig. 46. + lattice sites, ◯ adsorbed molecules.

The total entropy of adsorption from the gas at unit pressure is therefore

$$\Delta S = \frac{m}{L}(s - s^0) + k \ln \Omega_c, \tag{14·27}$$

where Ω_c is the number of arrangements of m molecules on M sites.

This is

$$\Omega_c = \frac{M!}{m!(M-m)!}. \tag{14·28}$$

By applying Stirling's theorem we obtain

$$\Delta S = \frac{m}{L}(s - s^0) + k\{M \ln M - m \ln m - (M-m) \ln (M-m)\}. \tag{14·29}$$

If the number of molecules adsorbed increases by dm, the entropy of the adsorbed layer, relative to the gas at unit pressure, will increase by $(\partial \Delta S/\partial m)\, dm$. By carrying out the differentiation on (14·29), and assuming that s, the entropy in a particular configuration, is independent of the degree of coverage, we obtain

$$\frac{\partial \Delta S}{\partial m} = \frac{(s - s^0)}{L} - k \ln \frac{\theta}{1 - \theta}. \tag{14·30}$$

This is the differential entropy of adsorption per molecule, and the corresponding value per mole is obtained by multiplying by L. This will be denoted $\Delta \bar{S}$,

$$\Delta \bar{S} = s - s^0 - R \ln \frac{\theta}{1 - \theta}. \tag{14·31}$$

Let $\Delta \bar{H}$ be the corresponding differential molar heat of adsorption (and, according to the previous assumptions, this is independent of θ and also independent of gas pressure, if the gas is perfect). Let μ be the chemical potential of the adsorbed layer and let μ^0 be the chemical potential of the gas at unit pressure. Then the differential Gibbs free energy of adsorption, relative to the gas at unit pressure, is

$$\mu - \mu^0 = \Delta \bar{H} - T \Delta \bar{S}$$
$$= \Delta \bar{H} - T(s - s^0) + RT \ln \frac{\theta}{1 - \theta}. \tag{14·32}$$

This difference, $\mu - \mu^0$, is not necessarily zero because unit pressure may not give rise to a state of equilibrium at the particular coverage θ. Let p_θ be the actual gas pressure which does give rise to this state of equilibrium and let μ_{p_θ} be the corresponding chemical potential of the gas. Then the condition of equilibrium is

$$\mu = \mu_{p_\theta},$$

or, supposing that the gas is perfect,

$$\mu = \mu^0 + RT \ln p_\theta. \tag{14·33}$$

Eliminating μ and μ^0 between (14·32) and (14·33) and rearranging we finally obtain

$$p_\theta = \frac{\theta}{1 - \theta} \exp \left(\frac{\Delta \bar{H}}{RT} - \frac{s - s^0}{R} \right). \tag{14·34}$$

This may be written

$$b p_\theta = \frac{\theta}{1 - \theta}, \tag{14·35}$$

where

$$b = \exp \left(-\frac{\Delta \bar{H}}{RT} + \frac{s - s^0}{R} \right). \tag{14·36}$$

If $\Delta \bar{H}$ and s are both independent of θ, as we have assumed, then b is a constant and (14·35) is the same as the well-known Langmuir isotherm. However, it is evident that this constancy can hold only under very restrictive conditions, in particular that there are a definite number of 'sites' and that the state of the adsorbed molecules is the same on all of the sites and however many of them are occupied.

CHAPTER 15

CHEMICAL EQUILIBRIUM IN RELATION TO CHEMICAL KINETICS

15·1. Introduction

To many of the early Greek philosophers it seemed that only those things which are *changeless* could be made the subject of scientific study. How, they asked, is it possible to have any knowledge about something which is in process of becoming something else? The very notion of change seems unreal, for how can one thing cease to exist and become another?

This problem continues to give trouble, although in a less acute form, and the atomic theory provided what is at least a partial answer. According to this theory the occurrence of a natural process is regarded as being simply a change in the mutual positions of the atomic particles. The latter are thought of as being entities which are permanent and changeless, and it is this postulate which makes it possible to put forward an idea of change, i.e. in terms of changes of pattern, which is at least comprehensible.

The modern theory of rates is therefore based, in the first place, on the particular types of particles which may be assumed to remain unchanged in the process which is under discussion, e.g. the atoms in chemical reactions. However, this way of looking at things serves only to diminish the difficulties and not to eliminate them. There is no theory of rates which stands, so to speak, on its own feet; all existing theories depend, in one form or another, on ideas carried over from the study of matter at equilibrium, which is to say in an unchanging condition.

It is very difficult to think of the temporal duration of a natural process as a kind of complete entity. Instead, it is usually pictured as a sequence of instantaneous states which are spread out along a time co-ordinate. Each state consists of molecules, free radicals, etc., whose numbers change but whose properties are supposed to remain the same as if they were actually at equilibrium. In short, the fact that there is a reaction taking place is assumed not to cause any alteration in the equilibrium properties of the various entities— only of their numbers. But even in the latter respect it is often necessary to take over further conceptions from the theory of equilibrium, as will be discussed in §15·7 in connexion with the transition state theory.

15·2. Kinetic species

In a footnote to §4·16 it was remarked that the number of chemical equations which is sufficient to represent the stoichiometry aɪːʊ the equilibrium of a reaction may not be sufficient to represent its kinetics. For example, in the combination of hydrogen and oxygen the equation

$$2H_2 + O_2 = 2H_2O$$

represents the stoichiometry to a very high accuracy, but an understanding of the kinetics involves the writing of equations containing OH and other transient species.

Any chemical reaction is a rearrangement in the pattern of the atomic nuclei relative to each other, and in this process the distance between the various atoms changes over a continuous range. However, at any moment, the vast majority of the atoms which are present in the system are present as one or another of a small number of distinct chemical species, e.g. H_2, O_2, OH, H_2O, etc., and it is for this reason that the mechanism of reactions may be interpreted in terms of a finite (and usually quite small) number of species, and not an infinite number. For example, a pair of H and O atoms at a separation of, say, 1 cm, does not constitute a chemical substance in this sense, and its 'life' is far too minute for an observation of its properties.

The chemical species which are used in kinetics are therefore the more or less stable arrangements of atomic nuclei, each characterized in its ground state by a certain geometrical configuration, about which it vibrates. One species is distinct from another (which may be composed of precisely the same atoms, as in isomerism) if they are separated by a barrier *large* compared to kT.

Of course each configuration will comprise a great many quantum states of rotation, etc., which 'belong' to the particular ground state, and in principle each of these quantum states might be regarded as a distinct species. Let [OH'], [OH''], etc., be the concentrations of OH radicals in the various quantum states and let [A'], [A''] be the concentrations of some other molecule in its quantum states. Then the total reaction rate between OH and A might be written in the form

$$k_1[OH'][A'] + k_2[OH''][A'] + \ldots,$$

where k_1, k_2, etc., are velocity constants. Now all quantum states between which the energy separation is *small* compared to kT may be expected to remain in approximate equilibrium with each other. Thus $[OH''] = K''[OH']$, $[OH'''] = K'''[OH']$, etc. Using these relations, it is easily shown that the above expression reduces to the form

$$k[OH][A],$$

where [OH] and [A] are the *total* concentrations of OH and A in all quantum states between which there is approximate equilibrium. This last expression is the type which is normally used in chemical kinetics.

In short, the chemical species whose concentrations appear in the usual kinetic expressions comprise all quantum states in approximate equilibrium and which 'belong' to a particular configuration or ground state. These species are normally separated by quite high barriers, thereby causing slow interconversion through the improbable intermediate states.

15·3. Variables determining reaction rate

The question needs to be considered whether the rate of a reaction is primarily determined by the volume concentration of each reacting species, or by some alternative variable such as its mole fraction, chemical potential or thermodynamic activity.

In simple reactions involving perfect, or almost perfect, gases it is found from experiment that it is the volume concentration which is the significant variable. Consider a reaction of this type whose stoichiometry is
$$aA + bB = cC,$$
where a, b and c are stoichiometric coefficients. Let n_a, etc., be the amounts (mols) of the three species at any instant in the reaction system (or in any chosen element of *fixed mass*), and let V be the volume at the given moment. The rate of increase in the amount of C is dn_c/dt, and the rate of formation of C per unit volume is†

$$\frac{1}{V}\frac{dn_c}{dt},$$

In simple reactions of the type under discussion it is often found experimentally that this rate is proportional to small integral powers‡ of the concentrations of the reacting species. Thus

$$\frac{1}{V}\frac{dn_c}{dt} = k(n_a/V)^\alpha (n_b/V)^\beta,$$

where k is independent of the concentrations and α and β are small integers, not necessarily equal to a and b in the stoichiometric equa-

---

† It is desirable that what shall be called the *reaction rate* shall be the same for all substances taking part in the reaction. This is obtained by dividing by the corresponding stoichiometric coefficient. Thus *reaction rate* is given by
$$a^{-1}\frac{dn_a}{dt} = b^{-1}\frac{dn_b}{dt} = c^{-1}\frac{dn_c}{dt} = \frac{d\xi}{dt}$$
where ξ is the *extent of reaction* and the stoichiometric coefficients are taken as negative for substances on the left-hand side of the chemical equation. The corresponding reaction rate per unit volume is $V^{-1}\frac{d\xi}{dt}$.

‡ Simple fractional powers such as $\frac{1}{2}$ also occur.

tion. α and β define the *order* of the reaction; thus if $\alpha = 1$ the reaction is said to be first order with respect to substance A.

The above equation may also be written

$$\frac{1}{V}\frac{dn_c}{dt} = k[A]^\alpha [B]^\beta, \qquad (15\cdot1)$$

where $[A]$ and $[B]$ denote the concentrations. If the chosen system is one in which V, the volume of a fixed mass, is also constant, then the left-hand side of $(15\cdot1)$ can be written

$$\frac{1}{V}\frac{dn_c}{dt} = \frac{d(n_c/V)}{dt} = \frac{d[C]}{dt}, \qquad (15\cdot2)$$

where $[C]$ is the concentration of C. Thus

$$\frac{d[C]}{dt} = k[A]^\alpha [B]^\beta, \qquad (15\cdot3)$$

but this equation is only applicable if V is constant.†

The fact that it is the volume concentrations which determine the rate of a simple gaseous reaction (and not, for example, the mole fractions) receives a simple interpretation as soon as it is assumed that reaction takes place at the collision of the reacting molecules. According to kinetic theory the number of collisions, per unit time and volume, of the molecules A and B is proportional to the product $[A][B]$ of their concentrations. These considerations are very familiar and need not be elaborated. For the present it is sufficient to emphasize that the customary use of volume concentrations in kinetics, including the kinetics of liquids, has its origin (a) in the experimental results obtained from gas reactions‡ and (b) in the support obtained from the kinetic theory of gases. However, this question will be referred to again in § 15·5.

15·4. Forward and backward processes

In a number of reactions the point of equilibrium is not displaced predominantly in one direction or the other, and in such instances it is often found experimentally that *the observed reaction rate may be*

† This is important in connexion with the kinetics of flow systems. See, for example, Denbigh, *J. Appl. Chem.* **1** (1951), 227; Danckwerts, *Nature, Lond.*, **173** (1954), 222.

‡ See, for example, the work of Kistiakowsky (*J. Amer. Chem. Soc.* **50** (1928), 2315) on the decomposition of hydrogen iodide vapour. In a series of fourteen experiments in which the initial concentration varied between 0·02 and 0.47 mol dm^{-3}, the velocity 'constant' was actually constant, i.e. the rate was proportional to the square of the volume concentration. This was no longer the case at high concentrations, probably due to the appreciable fraction of the total volume occupied by the molecules themselves.

expressed as a difference of two terms, one of which contains only the concentrations of the reactants and the other only the concentrations of the products. For example, in the decomposition of hydrogen iodide vapour,

$$2HI = H_2 + I_2,$$

Kistiakowsky's results show that the rate may be expressed in the form

$$\frac{-d[HI]}{dt} = k[HI]^2 - k'[H_2][I_2],$$

with constant values of k and k'.

Now there is no specifically thermodynamic reason why the measured reaction rate must be expressible as a difference of two terms[†]—all that thermodynamics requires is that the rate shall be positive in the direction of free-energy decrease and shall reduce to zero at thermodynamic equilibrium. The existence of the two terms must therefore be given an interpretation which is *kinetic*, and it has become customary to regard the observed reaction rate as being equal to the difference of the rates in the forward and backward directions, these processes taking place simultaneously at the molecular level. This interpretation is entirely in harmony with a collisional picture of the mechanism; at the same time there is clearly an element of convention in identifying the two terms with the forward and backward rates. For given values of the concentrations it is only the net rate which can be measured, and the intrinsic forward and backward rates have meaning only by interpretation.

At equilibrium the above expression reduces to

$$k[HI]^2 - k'[H_2][I_2] = 0,$$

or

$$\frac{[H_2][I_2]}{[HI]^2} = \frac{k}{k'}.$$

The left-hand side of this equation has the correct form of the equilibrium constant expressed in terms of concentrations as in equation (4·34). Bodenstein[‡] confirmed by experiment that the ratio k/k' of the experimental velocity constants is equal to the measured equilibrium constant. It follows that the thermodynamic conditions mentioned above are entirely satisfied in this example.

† Consider the perfect gas reaction $aA + bB = cC$. The thermodynamic conditions would be satisfied if the rate of reaction were expressible in the form

$$\frac{d[C]}{dt} = \theta\left\{K_c - \frac{[C]^c}{[A]^a[B]^b}\right\}^n,$$

where K_c is the equilibrium constant and θ is a function of the concentrations and temperature having a positive value. Except for the case $n = 1$, the equation does *not* express the rate as a difference of two terms.

‡ Bodenstein, *Z. Phys. Chem.* **29** (1899), 295.

15·5. Thermodynamic restrictions on the form of the kinetic equations†

The restrictions on the permissible form of the kinetic equations are rather less stringent than is often supposed. The matter can best be discussed by considering a specific reaction whose stoichiometry is expressed by

$$aA + bB = cC.$$

Case (a). *Single reaction in a perfect gas mixture.* In this instance the equilibrium constant, expressed in terms of concentrations, is

$$\frac{[C]^c}{[A]^a [B]^b} = K_c. \tag{15·4}$$

Let it be supposed that, under conditions where the quantity of C in the system is very small, the measured velocity has been found to be expressible in the form

$$\frac{\mathrm{d}[C]}{\mathrm{d}t} = k[A]^\alpha [B]^\beta [C]^\gamma \tag{15·5}$$

(where γ is commonly zero and α and β are commonly unity). If we seek to express the velocity over the whole range of composition by adding to the above *only a single extra term*, chosen as a product of powers of the concentrations, this term will have a somewhat restricted form. Let this term, representing the 'backward' velocity, be denoted

$$k'[A]^{\alpha'} [B]^{\beta'} [C]^{\gamma'}.$$

The proposed complete expression for the rate is therefore

$$\frac{\mathrm{d}[C]}{\mathrm{d}t} = k[A]^\alpha [B]^\beta [C]^\gamma - k'[A]^{\alpha'} [B]^{\beta'} [C]^{\gamma'}. \tag{15·6}$$

Therefore at equilibrium

$$\frac{[A]^{\alpha'} [B]^{\beta'} [C]^{\gamma'}}{[A]^\alpha [B]^\beta [C]^\gamma} = \frac{k}{k'}, \tag{15·7}$$

where $[A]$, etc., now denote any set of concentrations, infinite in number, at which there is equilibrium. Now k and k', and therefore also the ratio k/k', are independent of composition and are functions only of the temperature. According to thermodynamics, the equilibrium constant K_c is also a function only of temperature. Therefore the ratio k/k' is a function of K_c,

$$k/k' = f(K_c).$$

† See also Manes, Hofer and Weller, *J. Chem. Phys.* **18** (1950), 1355; Hollingsworth, *J. Chem. Phys.* **20** (1952), 921, 1649; Blum and Luus, *Chem. Eng. Sci.* **19** (1964), 322.

Substituting from (15·4) and (15·7) in this relation we obtain

$$[A]^{(\alpha'-\alpha)}[B]^{(\beta'-\beta)}[C]^{(\gamma'-\gamma)}=f\{[A]^{-a}[B]^{-b}[C]^{c}\}.$$

This equality will be satisfied for all possible values of $[A]$, etc., if the function in question is a power function,

$$[A]^{(\alpha'-\alpha)}[B]^{(\beta'-\beta)}[C]^{(\gamma'-\gamma)}=\{[A]^{-a}[B]^{-b}[C]^{c}\}^{n}, \qquad (15\cdot8)$$

and if $(\alpha'-\alpha)$, $(\beta'-\beta)$ and $(\gamma'-\gamma)$ are each the same multiple of the respective stoichiometric coefficients $-a$, $-b$ and c. Thus

$$\frac{(\alpha'-\alpha)}{-a}=\frac{(\beta'-\beta)}{-b}=\frac{(\gamma'-\gamma)}{c}=n. \qquad (15\cdot9)$$

The power n can have any positive value, including fractions, but negative powers must be excluded because it can be readily shown that this would imply a positive reaction rate in the direction of free-energy increase.

In short, if α, β and γ have been determined experimentally, as in equation (15·5), the expression for the backward reaction will be consistent with thermodynamics if α', β' and γ' are chosen to comply with equation (15·9). The corresponding relation between the velocity constants and the equilibrium constant is

$$k/k'=K_{c}^{n}. \qquad (15\cdot10)$$

Suppose, for example, the reaction is

$$A+\tfrac{1}{2}B=C,$$

so that $a=1$, $b=\tfrac{1}{2}$, $c=1$. If it has been established for the forward reaction that $\alpha=1$, $\beta=1$, $\gamma=0$, then from (15·9) it is readily shown that permissible expressions for the backward reaction rate are

$$k'[A]^{\tfrac{1}{2}}[B]^{\tfrac{3}{4}}[C]^{\tfrac{1}{2}},$$

$$k'[B]^{\tfrac{1}{2}}[C],$$

$$k'[A]^{-1}[C]^{2}, \quad \text{etc.},$$

corresponding to $n=\tfrac{1}{2},1,2$, etc.†

Case (b). Single reaction in a solution. The discussion above was based, in the first place, on the fact that reaction rates in perfect gases are usually found to be proportional to the volume concentrations, or to their powers, and secondly on the necessity that the rate shall be positive in the direction of free-energy decrease and shall reduce to zero at equilibrium.

† For a practical application of this procedure see Denbigh and Prince, *J. Chem. Soc.* (1947), p. 790.

The first of these points has been very decisive in creating the generally accepted ideas on chemical kinetics. Consider, for example, the reaction
$$aA + bB = cC.$$

The second condition alone could be met by expressing the rate as

$$\frac{d[C]}{dt} = \theta(a\mu_a + b\mu_b - c\mu_c), \tag{15·11}$$

where θ is any function of the concentrations which is positive and the μ's are the chemical potentials. It is evident that the right-hand side of this expression is positive when there is reaction from left to right and it reduces to zero at equilibrium. However, it does not express the rate as proportional to the volume concentrations.†

When we turn from dilute gases to solutions the situation is much more complicated. The rates of reactions in solution, especially those involving ions, are no longer accurately proportional to the volume concentrations, or to their simple powers. In other words, if the experimental results are constrained into the form of an expression such as equation (15·6), the velocity 'constants' turn out to be no longer quite constant, but vary somewhat with the concentrations and they depend, in particular, on the ionic strength.

Now the equilibrium constants of reactions in solution may be expressed in terms of the products of molalities (or volume concentrations) together with the appropriate activity coefficients, as shown in equation (10·7). Consider the reaction
$$aA + bB = cC.$$

The equilibrium constant is

$$\frac{(m_C \gamma_C)^c}{(m_A \gamma_A)^a (m_B \gamma_B)^b} = K, \tag{15·12}$$

where the m's are molalities. Alternatively

$$\frac{[C]^c y_C^c}{[A]^a y_A^a [B]^b y_B^b} = K, \tag{15·13}$$

where $[A]$, etc., denote volume concentrations and the y's are appropriate activity coefficients (equation (9·21)).

---

† On the other hand, if the reaction is sufficiently close to equilibrium for the condition $|a\mu_a + b\mu_b - c\mu_c| \ll RT$ to be satisfied, it can be shown that (15·11) is equivalent to (15·6). This turns on the expansion of the logarithmic term in the equation which relates chemical potentials to concentrations in an ideal system. For further discussion see Prigogine, Outer and Herbo, *J. Phys. Chem.* **52** (1948), 321; Manes, Hofer and Weller, *J. Chem. Phys.* **18** (1950), 1355; Denbigh, *Trans. Faraday Soc.* **48** (1952), 389.

Because the equilibrium constant must be expressed in forms such as the above, throughout the period from 1915 to 1930 there was a strong school of thought holding the view that reaction rate must also be primarily dependent on activities (i.e. the products of concentrations with their appropriate activity coefficients), rather than on the concentrations themselves. According to this view the rate of the above reaction should be expressible in a form such as

$$\frac{d[C]}{dt} = k[A]^\alpha y_A^\alpha [B]^\beta y_B^\beta - k'[C]^{\gamma'} y_C^{\gamma'}, \qquad (15\cdot14)$$

where k and k' depend only on temperature and α, β and γ', are small powers, usually integers. When this equation is considered in relation to the equilibrium constant, (15·13), the same results as were obtained in the discussion of case (a) above may be readily derived, in particular $k/k' = K^n$.

On the other hand, there was a lack of clear evidence that an expression such as (15·14) adequately accounted for the observed velocities.[†] In order to meet this situation it was pointed out that an equation of the form of (15·14) is unnecessarily restrictive, and the thermodynamic conditions can be satisfied equally well if each term on the right-hand side is multiplied by the same factor β, which may itself be a function of the concentrations. At equilibrium this factor cancels and we obtain

$$k/k' = K^n, \qquad (15\cdot15)$$

as before.

For example, if the reaction is

$$A + B = 2C,$$

complete consistency with thermodynamics would be obtained if the rate were expressible in the form

$$\frac{d[C]}{dt} = k[A][B] y_A y_B \beta - k'[C]^2 y_C^2 \beta, \qquad (15\cdot16)$$

where β depends on the concentrations. A well-known physical interpretation of β, due to Bronsted, is that it is the reciprocal of the activity coefficient of an intermediate complex. This point will be referred to again in connexion with the transition state theory.

The treatment of kinetics in solution on the basis of equations of the type of (15·16) has led to generally satisfactory results.[‡] Roughly

† See, for example, R. P. Bell, *Acid-Base Catalysis* (Oxford, 1941); Belton, *J. Chem. Soc.* (1930), p. 116. For new evidence see Eckert and Boudart, *Chem. Eng. Sci.* **18** (1963), 144; Mason, *ibid.* **20** (1965), 1143.

‡ For detailed discussion see R. P. Bell, *Acid-Base Catalysis* (Oxford, 1941); Moelwyn-Hughes, *Kinetics of Reactions in Solution* (Oxford, 1933).

speaking the collision processes in solution may be regarded as being largely determined by the volume concentrations, but the forces between the molecules make it necessary to introduce additional factors such as the activity coefficients and the parameter β.

Case (c). More than one reaction. In the more general case where there may be several reactions in the system the restrictions due to pure thermodynamics are even less stringent than in the case of a single reaction. In particular, the supposition that every reaction must balance individually is *not* a consequence of thermodynamics. It owes its justification instead to the principle of microscopic reversibility, which was expressed by Tolman in the following form: '...under equilibrium conditions, any molecular process and the reverse of that process will be taking place on the average at the same rate....'

This point has been discussed in some detail by Onsager.† Consider three substances A, B and C (for example, three isomers) which are mutually interconvertible according to the scheme

At equilibrium the chemical potentials are equal,

$$\mu_a = \mu_b = \mu_c,$$

and, if the system is a perfect gas mixture, the equilibrium constants can be expressed as

$$\frac{[A]}{[B]} = K_1, \quad \frac{[B]}{[C]} = K_2, \quad \frac{[C]}{[A]} = K_3. \qquad (15\cdot17)$$

Let it be supposed that the velocities of the three reactions are expressible as first-order terms. Then at equilibrium we have from the kinetics

$$\left.\begin{aligned}
(k_1 + k_3')\,[A] &= k_1'[B] + k_3[C], \\
(k_2 + k_1')\,[B] &= k_2'[C] + k_1[A], \\
(k_3 + k_2')\,[C] &= k_3'[A] + k_2[B].
\end{aligned}\right\} \qquad (15\cdot18)$$

By manipulation of these three equations we obtain

$$\frac{[A]}{[B]} = \frac{k_3 k_2 + k_3 k_1' + k_1' k_2'}{k_3' k_2' + k_3 k_1 + k_1 k_2'}, \qquad (15\cdot19)$$

† Onsager, *Phys. Rev.* **37** (1931), 405. A fairly detailed discussion is also given by Denbigh, *Thermodynamics of the Steady State* (London, Methuen, 1951).

and similar expressions for $[B]/[C]$ and $[C]/[A]$. When these are compared with (15·17) it is evident that all of the conditions required by thermodynamics can be satisfied *without any necessity* for the relations

$$\frac{k_1'}{k_1}=K_1, \quad \frac{k_2'}{k_2}=K_2, \quad \frac{k_3'}{k_3}=K_3. \tag{15·20}$$

In fact, in order to obtain these relations it is necessary to use the principle of microscopic reversibility which requires that each individual reaction, when there is equilibrium in the system, shall be balanced in the forward and backward directions. That is to say

$$k_1[A]=k_1'[B], \quad k_2[B]=k_2'[C], \quad k_3[C]=k_3'[A],$$

and when these are combined with the thermodynamic equations (15·17) we obtain the relations (15·20).

15·6. The temperature coefficient in relation to thermodynamic quantities

To speak of the temperature coefficient of the reaction rate is already to assume that the notion of temperature has a clear meaning, but this is strictly true only for systems which are at equilibrium. In the case of a reaction, if it is not too fast, this difficulty seems not to be very formidable. On account of the high frequency of the ordinary molecular collisions, as compared to those leading to reaction, the distribution of the energy of the system between translational, rotational and vibrational states probably remains very close to the equilibrium distribution, despite the fact that the total energy of the system may be in process of change. In brief there is a single statistical parameter T which determines a condition of approximate equilibrium between the translational, rotational and vibrational states. (However, under extreme conditions, perhaps in flames, it may be that each of these forms of energy must be specified by a somewhat different temperature.)

Leaving aside this question, if the logarithm of a measured velocity constant is plotted against the reciprocal of the absolute temperature, it is usually found that the points fall very nearly on a straight line having a negative gradient. At any temperature T the gradient is used for the purpose of defining the *Arrhenius activation energy* E_a by the relation

$$\frac{\mathrm{d}\ln k}{\mathrm{d}(1/T)}\equiv-\frac{E_a}{R},$$

or

$$\frac{\mathrm{d}\ln k}{\mathrm{d}T}\equiv\frac{E_a}{RT^2}, \tag{15·1}$$

and, of course, if the gradient is constant E_a is constant. For the reverse reaction, whose velocity constant is k', the corresponding activation energy E'_a is given by

$$\frac{d \ln k'}{dT} \equiv \frac{E'_a}{RT^2}. \tag{15·22}$$

Consider a reaction in a perfect gas mixture. As shown previously the following relation is compatible with thermodynamics:

$$k/k' = K_c^n,$$

where K_c is the equilibrium constant expressed in terms of concentrations and n is a small positive integer or its reciprocal. Taking logarithms and differentiating with respect to temperature

$$\frac{d \ln k}{dT} - \frac{d \ln k'}{dT} = \frac{n \, d \ln K_c}{dT}.$$

Therefore by substituting from (15·21), (15·22) and also (4·37) we obtain

$$E_a - E'_a = n \Delta U, \tag{15·23}$$

where ΔU is the increase in internal energy from left to right of the reaction. The difference of the two kinetic quantities E_a and E'_a is thus equal to a thermodynamic quantity.

On the other hand, if partial pressures had been used in place of concentrations for the purpose of defining the velocity constants we should have obtained

$$k/k' = K_p^n \tag{15·24}$$

and

$$E_a - E'_a = n \Delta H. \tag{15·25}$$

It follows from (15·23) and (15·25) that the Arrhenius activation energy is either an internal energy or an enthalpy, according to whether the velocity constants are expressed in terms of concentrations or partial pressures respectively. The use of concentrations is the more usual.

15·7. Transition-state theory

The two important theories of chemical kinetics, the collision theory and the transition-state theory, both depend on essentially the same assumption. Since this is concerned with the existence of a kind of equilibrium in the system, even during the course of its reaction, it is appropriate to give a short outline of one of these theories in the present volume.

We commence with a discussion of the potential energy surface. If n is the number of atomic nuclei which are involved in each elementary reaction, their positions relative to a frame of reference could be specified by $3n$ co-ordinates. However, the translation and rotation of the centre of mass are irrelevant as far as reaction is concerned, and therefore, for the present purpose, it is sufficient to use only the $3n-6$ co-ordinates required to describe the positions of the nuclei relative to each other. The potential energy of the system could therefore be represented as a surface† in a $3n-5$-dimensional-space.

The surface can be visualized at all easily only in the simplest instances. As an example consider the reaction

$$XY + Z = X + YZ,$$

involving the free atom Z and the diatomic molecule XY. If it is assumed for simplicity that the reaction only occurs when Z approaches towards Y along the line of centres of the XY molecule, the potential energy can be represented in three-dimensional space as a function of the X-Y and Y-Z distances. On the potential energy surface there will be two deep valleys‡ which occur at short X-Y and Y-Z distances and correspond to the stable XY and YZ molecules. Surrounding these valleys are high mountain ranges but between them there is a pass or col, which is of much higher potential energy than the bottom of the valleys, but lower than that of the ranges. It is assumed that this is the 'path' of the reaction.

Thus, as Z comes near to XY the potential energy of the trio rises on account of the repulsive forces between the electronic envelopes. Y gradually stretches away from X until it is somewhere about midway between X and Z. The system XYZ is then on the top of the pass or col and is said to be in the *transition state*. Finally, the atom X breaks away, leaving the new molecule YZ and the co-ordinates of the system simultaneously move down into the YZ valley. Plotted in terms of distance along the reaction path, or *reaction co-ordinate*, the potential energy would thus appear somewhat as shown in Fig. 47.

In the general case, as mentioned above, the potential energy surface cannot be so easily visualized, but the transition state can be thought of as being the configuration at the top of the lowest pass between the reactants and the products. The term *activated complex* is also used as referring to a set of nuclei which has this configuration

---

† According to the so-called 'adiabatic hypothesis', which is an essential part of the theory, it is supposed that the electrons remain in equilibrium with the particular nuclear configuration which occurs at a given moment, due to the large speed of the electronic motions relative to those of the nuclei. This implies that there are no electronic transitions during the course of the reaction and that the whole process takes place on a single potential energy surface.

‡ On account of their vibrational energy the molecules are not to be thought of as lying along the bottom of the valleys, but rather as being in a state of rolling, as it were, from one side of a valley to the other.

and is in a state of motion such that its representative point in the $3n-5$-dimensional space can cross the pass in the one direction or the other.

Although it is freely used, the term 'complex' is not very satisfactory because it carries the implication that the reaction vessel contains molecules of reactants, molecules of products and 'complexes'. Actually the transition state is the *least populated* of any of the configurations lying along the reaction co-ordinate between the reactants in their normal states and the top of the col. This is on account of the exponential factor $e^{-E/kT}$.

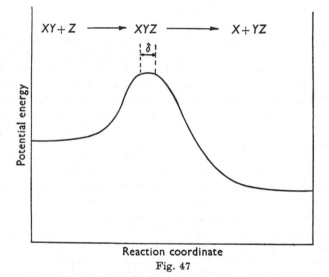

Fig. 47

Another important aspect of the activated complex is that it has no stability to small displacements in the direction of the reaction co-ordinate. A normal molecule is able to carry out vibrations without flying apart because it is surrounded on all sides by a potential energy barrier. The complex, on the other hand, has a potential energy whose shape is ⌣ in the direction perpendicular to the col, although it has the normal shape ⌣ in all directions in configuration space which are parallel to the col. In the latter directions it can carry out normal vibrations, but in the former there is no restoring force and it flies apart in the period of one 'vibration'. Provided that the top of the pass is almost flat the movement across it can be regarded as a free translatory motion of the centre of mass of the complex, and the appropriate one-dimensional partition function can be applied.

15·8. The equilibrium assumption

Let the bimolecular collision process, whose rate we are interested in, be

$$A + B = X + Y,$$

and let $\overrightarrow{[C]}$ and $\overleftarrow{[C]}$ be the concentrations of activated complexes which, at any moment, are moving across the top of the pass in the directions from left to right and from right to left respectively of this reaction.

The important assumption which is made is that the concentration $\overrightarrow{[C]}$ of complexes which are moving from left to right is in equilibrium with the reactants, and the concentration $\overleftarrow{[C]}$ of those which are moving from right to left is in equilibrium with the products. Thus, for a system which obeys the laws of a perfect gas mixture,

$$\frac{\overrightarrow{[C]}}{[A][B]} = \overrightarrow{K}_c, \tag{15·26}$$

and

$$\frac{\overleftarrow{[C]}}{[X][Y]} = \overleftarrow{K}_c. \tag{15·27}$$

The concentrations will be expressed as molecules per cm³ of the reaction space.

On the basis of this assumption the reaction rates from left to right and from right to left are calculated by dividing the concentrations $\overrightarrow{[C]}$ and $\overleftarrow{[C]}$ respectively by the time taken to go over the top of the pass. First, however, it is necessary to consider the assumption itself. It is clearly sound enough when the whole reaction system is at equilibrium, but when reaction is taking place it is very difficult to justify.†

Some degree of relief may be obtained by considering, first of all, the somewhat analogous situation concerning the effusion of a gas through a small hole. The system consists of a vessel divided into two parts by means of a plane wall through which there is a hole small compared to the mean free path. Let the gas pressures in the left- and right-hand halves of the vessel be p_l and p_r respectively, these being equal at equilibrium. According to equation (12·114) the rates of collision of gas molecules with unit area of the left- and right-hand sides of the dividing wall are $p_l(2\pi mkT)^{-\frac{1}{2}}$ and $p_r(2\pi mkT)^{-\frac{1}{2}}$ respectively. Then the analogous form of the equilibrium assumption is that these expressions also give the rates of passage from left to right

---

† The analogous postulate in the older collision theory is the maintenance of the Maxwellian distribution of velocities whilst reaction is occurring.

and from right to left through unit area of the hole. This is an example of a rate problem answered by means of a theorem of equilibrium.

The validity of this procedure clearly depends on the diameter d of the hole being small compared to the mean free path λ. The last collision before a molecule passes through the hole takes place at an average distance from the hole which is of the order of λ. The molecules may therefore be thought of as being supplied from a hemisphere of radius λ and whose centre is at the hole. If the ratio d/λ is small, the solid angle subtended by the hole at a point on the hemisphere is also small, and therefore only a very small fraction of the collisions which occur on this hemisphere actually result in a passage through the hole. In short, those molecules which would have rebounded from the wall if the hole had not been there probably make only a small contribution to the maintenance of the equilibrium distribution of velocities and concentration. The actual loss of these molecules makes little difference, and the rate of impact on the area of the hole is effectively the same as if either the hole were closed or the gas pressure were the same on the other side.

Returning to the case of the reaction

$$A + B = X + Y,$$

let it be supposed that the reactants and products are at complete equilibrium so that the relations (15·26) and (15·27) are entirely valid. From this equilibrium mixture it may be imagined that a quantity of X and Y are removed through semi-permeable membranes, thereby resulting in a state of non-equilibrium and an overall reaction from left to right. At the new values of the concentrations it is assumed that (15·26) and (15·27) continue to apply. That is to say, the energetic collisions amongst the reactants are sufficient to maintain the concentration $\overrightarrow{[C]}$ at its equilibrium level, and similarly the collisions amongst the products are sufficient to maintain $\overleftarrow{[C]}$ at *its* particular equilibrium value.

There are two rather distinct factors which might cause errors in this assumption: (1) The effect of complexes returning across the col in maintaining the equilibrium distribution. It is assumed in fact that the concentration $\overrightarrow{[C]}$ of complexes which are travelling across the col from left to right is not appreciably *diminished* by the fact that a smaller number are returning from right to left than there would be if there were complete equilibrium. Errors arising from this assumption are perhaps rather small if the activation energy per mole is large compared to RT, for in this case the collisions leading to reaction are a small fraction of the total number. (2) The effect of the heat of reaction. In exothermic reactions, for example, the products of reaction at the moment of their formation have an excess thermal energy, and this might cause an *increase* in the rate of activation of the reactants.

Various aspects of the equilibrium assumption have been discussed by Guggenheim and Weiss† and by Polanyi.‡ Fowler§ has remarked that the analogous assumption of a Maxwellian distribution of velocities in dealing with viscosity and heat conduction in gases leads to an error of 100–200 %. If the same error occurs in chemical kinetics it is probably not very large compared to other uncertainties.

More recently a number of workers ‖ have subjected the assumption to a quantitative examination. In general, for reactions in which the activation energy is large compared to RT, it seems unlikely that the assumption is seriously in error, except perhaps in highly exothermic reactions. In such instances, as Prigogine has shown, the 'hot' reaction products may cause a significant increase in the rate of activation.

15·9. The reaction rate

We shall consider the rate of the reaction from left to right, the rate of the corresponding backward reaction being calculable in an analogous manner.

Using equation (15·26), together with (12·66), the concentration $[\overrightarrow{C}]$ is obtained as the following function of the partition functions:

$$[\overrightarrow{C}] = \frac{\phi_c}{\phi_a \phi_b} e^{-E_0/RT} [A][B], \tag{15·28}$$

where ϕ_a, ϕ_b and ϕ_c are the molecular partition functions per unit volume of the reactants, A and B, and the complex, moving from left to right across the col, respectively. E_0 is the amount by which the zero energy level of the complex exceeds that of the reactants per mole.

The partition functions can be approximately factorized, in the usual way, as a product of translational, rotational and vibrational terms. These depend on the co-ordinates which specify the species in question and, as regards the complex, one of these co-ordinates can be chosen as a short length δ lying in the direction of the reaction

† Guggenheim and Weiss, *Trans. Faraday Soc.* **35** (1938), 57.

‡ Polanyi, *Trans. Faraday Soc.* **35** (1938), 75.

§ Fowler, *Trans. Faraday Soc.* **35** (1938), 75, 124.

‖ Kramers, *Physica*, **7** (1940), 284. Zwolinsky and Eyring, *J. Amer. Chem. Soc.* **69** (1947), 2702. Hirschfelder, *J. Chem. Phys.* **16** (1948), 22. Prigogine, *J. Phys. Chem.* **55** (1951), 765; Prigogine and Xhrouet, *Physica*, **15** (1949), 913; Prigogine and Mahieu, *Physica*, **16** (1950), 51.

co-ordinate at the top of the pass (Fig. 47). (It will be shown shortly that the precise value of δ is not important.) The top of the pass being almost flat, the potential energy is almost constant over the length δ and the corresponding motion can be treated as a free translation.

One of the factors in ϕ_c is therefore the translational partition function corresponding to one-dimensional motion over the distance δ. If m is the mass of the complex for motion in this direction, the one-dimensional partition function was shown in equation (12·48) to have the value

$$\frac{\delta}{h} (2\pi m kT)^{\frac{1}{2}}, \tag{15·29}$$

or rather it is just *half* this value, since the concentration $[\overrightarrow{C}]$ refers to complexes moving in the one direction over the pass and this is just half the total concentration† of complexes which would occur if there were complete equilibrium in the system.

If this factor is removed from (15·28) we obtain

$$\overrightarrow{[C]} = \frac{\delta}{2h} (2\pi m kT)^{\frac{1}{2}} \frac{\phi^{\ddagger}}{\phi_a \phi_b} e^{-E_0/RT} [A][B], \tag{15·30}$$

where $\phi^{\ddagger}$ is now a quasi-partition function which does not include the factor due to this special type of translatory motion. $\phi^{\ddagger}$ may be regarded as being the partition function of a stable molecule having one less than the normal number of vibrational modes—this missing mode corresponding, of course, to the fact that the transition complex does not have stable vibration in the direction along the reaction co-ordinate.

Let $\bar{v}$ be the average velocity of motion over the length δ. The average time taken to traverse this length is therefore $\tau = \delta/\bar{v}$. Since there are $[\overrightarrow{C}]$ complexes in unit volume of the reaction system which populate this length δ, and which are moving from left to right, the corresponding reaction rate is $[\overrightarrow{C}]/\tau$. Thus

$$\text{forward reaction rate} = [\overrightarrow{C}] \bar{v}/\delta$$
$$= \frac{\bar{v}}{2h} (2\pi m kT)^{\frac{1}{2}} \frac{\phi^{\ddagger}}{\phi_a \phi_b} e^{-E_0/RT} [A][B], \tag{15·31}$$

and from this expression δ has disappeared.

It remains to obtain an estimate of the mean velocity $\bar{v}$, and for this purpose we again assume free translatory motion of the representative point along the reaction co-ordinate. Now, according to

† It is this total concentration, given the symbol $C_{\ddagger}$ which is used in the treatment of Glasstone, Laidler and Eyring, *Theory of Rate Processes* (New York, McGraw-Hill, 1941), Chapter IV.

equation (12·99) the fraction of a set of freely moving particles which have a component of velocity in a particular direction which lies between v and $v + dv$ is

$$\left(\frac{m}{2\pi kT}\right)^{\frac{1}{2}} e^{-mv^2/2kT}\, dv.$$

The mean velocity is obtained by multiplying this fraction by v and taking the average over all fractions.† Thus

$$\bar{v} = \int_0^\infty v\, e^{-mv^2/2kT}\, dv \Big/ \int_0^\infty e^{-mv^2/2kT}\, dv$$

$$= \left(\frac{2kT}{\pi m}\right)^{\frac{1}{2}}. \tag{15·32}$$

Combining this result with (15·31) we finally obtain

$$\text{forward reaction rate} = \frac{kT}{h}\frac{\phi^\ddagger}{\phi_a \phi_b} e^{-E_0/RT}[A][B]. \tag{15·33}$$

A similar expression may be readily obtained for the backward reaction rate, based on equation (15·27), and it is easily verified, by using expressions such as (12·66), that the forward and backward rates are equal at equilibrium.

The important result (15·33) shows, in the first place, that the reaction rate from left to right of the bimolecular gas reaction

$$A + B = X + Y$$

is proportional to the product of the concentrations $[A][B]$. The theory thus provides an interpretation of the empirical law of mass action which was discussed in § 15·3. Secondly the bimolecular velocity constant k is shown to have the value

$$k = \frac{kT}{h}\frac{\phi^\ddagger}{\phi_a \phi_b} e^{-E_0/RT}. \tag{15·34}$$

The velocity would thus be calculable if sufficient were known about the complex and about the potential energy surface which determines the value of E_0.

The quantity kT/h in the above equation has the dimensions of a frequency and its value at room temperature is about 0.6×10^{13} sec.$^{-1}$. The rate constant as given by (15·34) is, of course, in terms of *molecules* and not moles.

---

† Or rather over all fractions which correspond to the complexes moving from left to right and not from right to left. For this reason the integrals are from 0 to ∞, corresponding to positive values of v only.

Equation (15·34) is often expressed in the form

$$k = \frac{kT}{h} K^{\ddagger},$$ (15·35)

where $K^{\ddagger}$ is a kind of equilibrium constant. However, it is not the same as the equilibrium constant, denoted $\overrightarrow{K_c}$ in equation (15·26), and this is because the factor (15·29) has been removed from the complete partition function of the complex. The relation between them is readily shown from the preceding equations to be

$$\overrightarrow{K_c} = K^{\ddagger} \left(\frac{\pi m kT}{2} \right)^{\frac{1}{2}} \frac{\delta}{h}.$$ (15·36)

It remains to obtain a relation between E_a, the Arrhenius activation energy, and some characteristic of the complex. By taking logarithms of (15·35) and (15·36) and differentiating with respect to temperature, we obtain

$$\frac{d \ln k}{dT} = \frac{1}{T} + \frac{d \ln K^{\ddagger}}{dT},$$ (15·37)

$$\frac{d \ln \overrightarrow{K_c}}{dT} = \frac{d \ln K^{\ddagger}}{d T} + \frac{1}{2T}.$$ (15·38)

Therefore, by eliminating $K^{\ddagger}$ between these equations,

$$\frac{d \ln k}{dT} = \frac{d \ln \overrightarrow{K_c}}{dT} + \frac{1}{2T}.$$ (15·39)

Now the term on the left-hand side defines the Arrhenius activation energy by equation (15·21). The first term on the right-hand side is related to $\overrightarrow{\Delta U}$, the increase in internal energy on forming the complex from the reactants, by the standard thermodynamic relation (4·37). Hence

$$\frac{E_a}{RT^2} = \frac{\overrightarrow{\Delta U}}{RT^2} + \frac{1}{2T},$$

or $$E_a = \overrightarrow{\Delta U} + \frac{RT}{2}.$$ (15·40)

This equation† therefore gives a theoretical interpretation to the experimental quantity E_a.

It is to be noted that neither E_a nor $\overrightarrow{\Delta U}$ are necessarily equal to E_0, which appears in (15·34), on account of the temperature-depend-

† $\overrightarrow{\Delta U}$ is the same as $\Delta E^{\ddagger}$ of Glasstone, Laidler and Eyring. However, equation (15·40) differs from their equation (167) in Chapter 4 by a factor $\frac{1}{2}RT$. These authors seem to have neglected that $K^{\ddagger}$ does not have quite the same temperature dependence as a normal concentration equilibrium constant on account of the presence of $T^{\frac{1}{2}}$ in equation (15·36) above.

ence of the partition functions. On the other hand, the difference between these three quantities is probably not very large.

In so far as the quantity denoted $K^{\ddagger}_k$ above may be regarded as an equilibrium constant (although not in the same sense as $\overrightarrow{K_c}$), it may be related to a 'standard' free energy change $\Delta G^{\ddagger}$

$$-RT \ln K^{\ddagger} = \Delta G^{\ddagger}. \tag{15·41}$$

Thus equation (15·35) becomes

$$k = \frac{kT}{h} e^{-\Delta G^{\ddagger}/RT} \tag{15·42}$$

or in terms of entropy and enthalpy factors,

$$k = \frac{kT}{h} e^{\Delta S^{\ddagger}/R} e^{-\Delta H^{\ddagger}/RT}. \tag{15·43}$$

However, in view of the remarks following equation (15·35), it will be clear that $\Delta G^{\ddagger}$, $\Delta H^{\ddagger}$ and $\Delta S^{\ddagger}$ do not refer to the same kind of standard state for the complex as in an ordinary reaction. Nevertheless, the use of these quantities, particularly $\Delta S^{\ddagger}$, has been of value in the understanding of kinetics.

The above outline of the transition state theory has been primarily concerned with bimolecular collision processes in perfect gas reactions for which (15·26) and (15·27) are correct thermodynamic expressions for the assumed equilibrium. For reactions in solution these should be replaced by

$$\frac{\overrightarrow{[C]}}{[A][B]} \frac{y_C}{y_A y_B} = \overrightarrow{K}, \tag{15·44}$$

and

$$\frac{\overleftarrow{[C]}}{[X][Y]} \frac{y_C}{y_X y_Y} = \overleftarrow{K}, \tag{15·45}$$

where the y's are appropriate activity coefficients. Provided that the other assumptions of the theory continue to hold good, the appropriate form of equation (15·33) for the forward reaction rate now contains the factor $y_A y_B/y_C$ on the right-hand side. Similarly, the backward reaction rate contains the factor $y_X y_Y/y_C$. These expressions are of the same form as those which occur in equation (15·16). The quantity β may therefore be interpreted as the reciprocal of the activity coefficient of the activated complex.

For further discussion of the transition-state theory, and in particular the introduction into the above equations of the rather problematic 'transmission coefficient', the reader is referred to the literature.†

† For example Glasstone, Laidler and Eyring, *The Theory of Rate Processes* (New York, McGraw-Hill, 1941) and Fowler and Guggenheim, *Statistical Thermodynamics* (Cambridge, 1949). For a discussion of the thermodynamics of the activated state see Guggenheim, *Trans. Faraday Soc.* 33 (1937), 607.

APPENDIX

ANSWERS TO PROBLEMS AND COMMENTS

Chapter 1

2. (a) 3060 J mol^{-1}; (b) 3100 J mol^{-1}.

5. (a) -6.3; (b) $+7.0$; (c) $+0.7$ J K^{-1}.

6. (a) for 80 °C, $+5.39$ and -4.44 J K^{-1} mol^{-1}.
   (b) for 120 °C, $+8.51$ and -7.81 J K^{-1} mol^{-1}.

8. Note that it is *not* necessary to assume that c_V or γ is constant. Along the reversible adiabatics we have

$$dU = -p\,dV = \frac{-k\theta\,dV}{V},$$

$$\therefore \frac{dU}{\theta} = \frac{-k\,dV}{V}.$$

Since U is a function of θ only, the left-hand side of this equation is also a function only of θ. Hence $\int \dfrac{dV}{V}$ is the same for any reversible adiabatic between the same pair of isothermals θ_1 and θ_2.

11. (a) -20.7 J K^{-1} mol^{-1}; (b) $+21.5$ J K^{-1} mol^{-1}.
The first result is obtained by considering the reversible path -10 °C $\rightarrow 0$ °C $\rightarrow -10$ °C for the transformation water$\rightarrow$ice. The second result is obtained by first calculating the heat given to the environment in the spontaneous freezing at -10 °C. This heat is 5656 J mol^{-1} and the corresponding entropy change of the environment is

$$5656/263.15 = 21.5 \text{ J K}^{-1} \text{ mol}^{-1}.$$

(The student should be quite confident about this question before proceeding further.)

12. 55 °C.
The steam is compressed under conditions of approximately constant entropy. Hence carry out an entropy balance on the steam (which is 'wet'), showing that its volume diminishes to 3.93ft^3 and therefore the volume of air admitted is 6.07 ft^3. Also carry out an internal energy balance on the steam and thus calculate the work done on the steam by the air. Finally, carry out an internal energy balance on the air which is admitted:

$$\Delta U = \text{(work done on admitted air by air in supply main)}$$
$$- \text{(work done on steam)}.$$

Assuming that the air is a perfect gas, $\Delta U = nc_V(t-15)$, where n is the number of mols of air in 6.07 ft^3 at t °C. Hence solve for n and finally for t. The result is rather sensitive to the accuracy with which the calculation is carried out but should be 50–60 °C.

13. For the process materials the energy balance is

$$\Delta U = q - q_0 - p_0 \Delta V, \qquad (1)$$

where q_0 is heat loss to the environment. The second law can be expressed

$$\Delta S + \Delta S_s + \Delta S_0 = \sigma,$$

where ΔS_s and ΔS_0 are the entropy changes of the steam and the environment respectively and $\sigma \geqslant 0$. Hence

$$\Delta S - \frac{q}{T_s} + \frac{q_0}{T_0} = \sigma. \qquad (2)$$

Eliminate q_0 between (1) and (2) and consider the special case where $\sigma = 0$.

CHAPTER 2

1. $\Delta S - 131$ J K^{-1} mol^{-1}; $\Delta U - 47.7$ kJ mol^{-1}; $\Delta H - 51.7$ kJ mol^{-1}.

2. The initial and final states of the process denoted by Δ must refer to the same temperature.

3. $\Delta G = -103.4$ kJ mol^{-1}; $\Delta H = -95.0$ kJ mol^{-1}; $\Delta S = 28$ J K^{-1} mol^{-1}; heat absorbed $= T\Delta S = 8.4$ kJ mol^{-1}.

4. Since U is extensive it is proportional to V and thus $U = Vf(T)$ or $u = f(T)$. For the particular system the equation (2·76)

$$\left(\frac{\partial U}{\partial V}\right)_T = T\left(\frac{\partial p}{\partial T}\right)_V - p,$$

can be rewritten

$$u = \frac{T}{3}\frac{du}{dT} - \frac{u}{3}$$

since $p = u/3$. Hence

$$4u = T\frac{du}{dT},$$

and integrating

$$u = \alpha T^4,$$

where α is an integration constant. Similarly s is obtained by integration from

$$dU = T\,dS - p\,dV.$$

5. Note that the gas is not perfect, but it can be proved that $c_p - c_V = \boldsymbol{R}$ by application of (2·91). For what molecular reason might a gas obeying the equation $p(v-b) = \boldsymbol{R}T$ be expected to behave like a perfect gas in regard to the difference of heat capacities?

6. The 'fundamental' equation (1·19) is here replaced by

$$dU = T\,dS + F\,dl. \tag{1}$$

Hence also

$$dA = d(U - TS) = -S\,dT + F\,dl, \tag{2}$$

and therefore by reciprocity

$$-\left(\frac{\partial S}{\partial l}\right)_T = \left(\frac{\partial F}{\partial T}\right)_l. \tag{3}$$

The required formula is then obtained from (1) and is analogous to (2·76).

The second equation to be proved refers to an isentropic change. From (1) by reciprocity

$$\left(\frac{\partial T}{\partial l}\right)_s = \left(\frac{\partial F}{\partial S}\right)_l = \left(\frac{\partial F}{\partial T}\frac{\partial T}{\partial S}\right)_l$$

$$= \left(\frac{\partial F}{\partial T}\right)_l \left(\frac{\partial T}{\partial S}\right)_l = \left(\frac{\partial F}{\partial T}\right)_l \frac{T}{c_l}.$$

Hence for an isentropic change

$$\int \frac{dT}{T} = \int \frac{1}{c_l}\left(\frac{\partial F}{\partial T}\right)_l dl.$$

Note that in this equation all quantities, including $(\partial F/\partial T)_l$, are experimentally measurable. This equation can therefore be tested—see, for example, Treloar, *The Physics of Rubber Elasticity*, Oxford, 1949.

7. Prove first

$$\left(\frac{\partial H}{\partial p}\right)_T = -T\left(\frac{\partial V}{\partial T}\right)_p + V. \tag{1}$$

Now

$$\left(\frac{\partial H}{\partial V}\right)_T = \left(\frac{\partial H}{\partial p}\right)_T \left(\frac{\partial p}{\partial V}\right)_T.$$

Substitute from (1) and use the relation (§ 2·10c)

$$\left(\frac{\partial p}{\partial V}\right)_T \left(\frac{\partial V}{\partial T}\right)_p = -\left(\frac{\partial p}{\partial T}\right)_V.$$

Note also that $\partial(T/V)/\partial V$ in the formula to be proved is equal to

$$\frac{1}{V}\frac{\partial T}{\partial V} - \frac{T}{V^2}.$$

In the second part of the question the condition of constant θ is the same as the condition of constant T. Hence from the formula proved in the first part, since

$$\left(\frac{\partial H}{\partial V}\right)_T \equiv \left(\frac{\partial H}{\partial V}\right)_\theta = 0,$$

we have

$$[\partial(T/V)/\partial V]_p = 0,$$

or

$$\left(\frac{\partial T}{\partial V}\right)_p = \frac{T}{V}.$$

Thus for a change at constant pressure

$$d \ln T = d \ln V,$$

and therefore by integration $\quad T = Vf(p),$ \hfill (3)

where $f(p)$ is an integration constant which is a function of pressure.

Similarly from the condition $(\partial U/\partial V)_\theta = 0$, together with equation (2·76), we obtain

$$\left(\frac{\partial p}{\partial T}\right)_V = \frac{p}{T},$$

and therefore $\quad T = pf(V).$ \hfill (4)

Hence between (3) and (4) $\quad \dfrac{f(p)}{p} = \dfrac{f(V)}{V}.$

Since the left-hand side depends only on p and the right-hand side only on V, this equation can be satisfied only if

$$f(p) = p/c,$$
$$f(V) = V/c,$$

where c is a constant. Combining with (3) or (4) we obtain $pV = cT$.

8. Similar principles to Problem 7 but easier. Note that the formula to be proved is to be applicable to *any* substance, not necessarily a gas. The only purpose of the gas, in the question as it is expressed, is to define the θ scale of temperature.

First prove for any substance

$$\left(\frac{\partial U}{\partial V}\right)_\theta \equiv \left(\frac{\partial U}{\partial V}\right)_T = T\left(\frac{\partial p}{\partial T}\right)_V - p = T\left(\frac{\partial p}{\partial \theta}\right)_V \frac{d\theta}{dT} - p, \quad (1)$$

where $d\theta/dT$ is a complete differential, since θ depends only on T.

Now use the properties of the gas to transform (1) to the θ scale of temperature. Thus

$$\left(\frac{\partial H}{\partial p}\right)_T = -T\left(\frac{\partial V}{\partial T}\right)_p + V,$$

and putting this equal to zero for the gas

$$\left(\frac{\partial V}{\partial T}\right)_p = \left(\frac{\partial V}{\partial \theta}\right)_p \frac{d\theta}{dT} = \frac{V}{T},$$

and applying the other condition, $pV = nc\theta^2$, we obtain

$$\frac{d\theta}{dT} = \frac{\theta}{2T}.$$

This can now be substituted in (1).

10. Let F, L and V be the moles of feed, of residual liquid and of vapour in unit time. Let x_F, x_L and x_V be the mole fractions of component B in these three streams respectively. By mass balance

$$F = L + V, \hfill (1)$$
$$Fx_F = Lx_L + Vx_V. \hfill (2)$$

Substituting (1) in (2) to eliminate F

$$(L + V) x_F = L x_L + V x_V,$$

and hence rearranging

$$\frac{x_F - x_V}{x_L - x_F} = \frac{V}{L}. \tag{3}$$

By enthalpy balance

$$Q + F h_F = L h_L + V h_V, \tag{4}$$

where Q is the input of heat and h_F, h_L and h_V are the enthalpies per mole of the F, L and V streams respectively. Substituting from (1) to eliminate F and rearranging

$$(L + V) h_F = L h_L + V(h_V - q),$$

where $q = Q/V$, the heat requirement per mole of vapour. Hence

$$\frac{h_F - (h_V - q)}{h_L - h_F} = \frac{L}{V}. \tag{5}$$

Equating (3) and (5)

$$\frac{x_F - x_V}{x_L - x_F} = \frac{h_F - (h_V - q)}{h_L - h_F}. \tag{6}$$

The left-hand side of this equation is equal to the ratio KM/GN on the diagram. The denominator of the right-hand side is represented by the length KN. Hence, by similar triangles,

$$PM = h_F - (h_V - q).$$

$$\therefore \quad q = PM + (h_V - h_F)$$

$$= PM + MH$$

$$= PH. \qquad \text{(Q.E.D.)}$$

Which geometrical construction would show Q/L, the input of heat per mole of the residual liquid?

CHAPTER 3

1. As piston 1 moves out slowly and reversibly the pressure exerted on it is due entirely to gas B because it is permeable to gas A. Conversely with regard to piston 2. Calculate the work of the isothermal reversible expansion of each gas to the final state, using the gas laws. Also since $\Delta U = 0$ (the gases being assumed perfect) the work done is equal to the heat taken in. Hence obtain the entropy of mixing.

2. Use equation (2·55). The atmosphere must be assumed to be a perfect mixture. The equation is also limited to heights at which g is still sensibly constant.

3. 14.5 kW.
The minimum work required is that of a reversible process. This work is equal to the heat rejected since $\Delta U = 0$. ΔS, and hence the heat, can be calculated by use of equation (3·37) (with the sign reversed since we are here concerned with *demixing*).

4. On the van der Waals isotherms the critical point is where $\left(\frac{\partial p}{\partial V}\right)_T$ and $\left(\frac{\partial^2 p}{\partial V^2}\right)_T$ are both zero. Hence applying these conditions to the van der Waals equation:

$$v_c = 3b,$$

$$T_c = \frac{8a}{27Rb},$$

$$p_c = \frac{a}{27b^2}.$$

The 'reduced' pressure is the ratio of the actual pressure of the gas to the critical pressure and is thus a dimensionless quantity. Similarly with regard to T_r and v_r.

$$p_r = p/p_c, \quad T_r = T/T_c, \quad v_r = v/v_c.$$

Using these quantities the van der Waals equation can be expressed

$$\left(p_r + \frac{3}{v_r^2}\right)(3v_r - 1) = 8T_r.$$

5. About 2.6×10^{-6} kPa^{-1}. (The observed value is 2.620×10^{-6} kPa^{-1} in close agreement.) From the van der Waals equation obtain $\partial V/\partial T$ and substitute in equation (3·44). a and b are calculated from T_c and p_c.

6. 9.64 atm, 41.7 atm.

Note that if Z is plotted against p the plot is almost linear, at least up to 50 atm. Hence from the gradient

$$Z = 1 - 0.36 \times 10^{-2} p.$$

This allows of a direct integration of equation (3·52).

7 1.29 kW; 1.17 kW

It is implied in the question that it is a steady flow process. Hence equation (3·48) gives the minimum shaft work per mol. Use the results of Problem 6.

8. 93.9 atm.

Note that the value of RT in the appropriate units is the value of the right-hand side of the given equation as $p \to 0$. This expression is taken from *International Critical Tables* (vol. III, p. 8), and much of the p-V-T data in the literature are quoted in this form. It is equivalent to the use of the *Amagat unit of volume* which is the volume of 1 mol of gas at 0 °C and 1 atm and is close to 22.4 dm^3, but varies slightly from gas to gas.

9. The equation proved in the first part of the question can be rearranged to give

$$p_2 = p_1\left(\frac{v_1}{v_2}\right)^\gamma e^{(S_2 - S_1)/C} V.$$

Let p_1, v_1 and v_2 have chosen values; then p_2 will have a certain value if the expansion or compression is conducted under reversible conditions

$(S_2 = S_1)$ and will have a *larger* value if it is conducted under irreversible conditions $(S_2 > S_1)$. Hence the *low*-pressure side of the curve $pv^\gamma = $ constant is adiabatically inaccessible. (Cf. the discussion of 'impossible' processes in §1·9.) The matter may perhaps be seen more clearly by obtaining the corresponding equation relating T_1, V_1 and S_1 with T_2, V_2 and S_2. In an irreversible compression more work is done *on* the gas than in a reversible compression and there is therefore a greater temperature rise for the same change in volume. Similarly in an irreversible expansion the gas does less work than in a reversible one and the fall in temperature is less.

10. Only the third equality is conditional on the temperature being measured on an absolute scale. The proof of the first two equalities depends only on the definitions of C_p and C_V in terms of H and U, and is *not* dependent on the second law (i.e. on the existence of entropy as a function of state). On the other hand, the proof of the third equality requires the second law.

Note that

$$\left(\frac{\partial H}{\partial V}\right)_T = \left(\frac{\partial H}{\partial p}\right)_T \left(\frac{\partial p}{\partial V}\right)_T,$$

and also

$$\left(\frac{\partial p}{\partial V}\right)_T \left(\frac{\partial T}{\partial p}\right)_V \left(\frac{\partial V}{\partial T}\right)_p = -1.$$

The required equation of state is $p(v - b) = RT$, where b is a constant. The statement that at constant volume p is proportional to T may be written

$$p = \phi T, \tag{1}$$

where ϕ is a function of volume. Hence $(\partial p / \partial T)_V = \phi = p/T$. Using this result together with the other item of information, $c_p - c_V = R$, the first equality proved in the first part of the question may be written

$$R = p\left(\frac{\partial v}{\partial T}\right)_p.$$

Integrating at constant pressure

$$RT = pv + f(p),$$

or

$$\frac{RT}{p} = v + \frac{f(p)}{p}. \tag{2}$$

Now by equation (1) RT/p is actually a function only of volume. The integration constant $f(p)$ must therefore be of the form constant $\times p$, in order that the right-hand side of (2) shall be a function only of volume. Hence $p(v - b) = RT$.

CHAPTER 4

1. Relative to the elements the free energy of the system has a value -693 cal at zero conversion (due to the free energy of mixing), a minimum value of about -1555 cal at 57·6 % conversion and a value of -1000 cal at 100 % conversion.

2. Since heat-capacity data are not given it is necessary to assume, as a very rough approximation, that ΔH is constant over the temperature range. On this basis it is readily calculated that the value of K_p at $T = 773$ K is 4×10^{-6}. Consider an initial gas consisting of 1 mole CO and 2 moles H_2. If the number of moles of CH_3OH at equilibrium is x, the corresponding number of moles of CO and H_2 are $1-x$ and $2(1-x)$ respectively. Hence

$$\frac{(3-2x)^2 x}{4(1-x)^3 P^2} = K_p = 4 \times 10^{-6}.$$

A value of x of 0.1 is feasible for an industrial process, because unchanged CO and H_2 can be recirculated. Hence $P = 260$ atm. (A more accurate calculation, allowing for the change of ΔH, would give about 400 atm for the same degree of conversion.)

3. $K_p = \dfrac{p_{NH_3}}{p_{N_2}^{\frac{1}{2}} p_{H_2}^{\frac{3}{2}}}$. Change the variables in this equation by putting

$$p_{H_2} + p_{N_2} + p_{NH_3} = p,$$

$$p_{H_2}/p_{N_2} = r,$$

thus obtaining K_p as a function of r, p and p_{NH_3}. Finally, differentiate at constant p and T and show that the condition for p_{NH_3} to be a maximum with respect to r is $r = 3$.

4. From the given value of the true equilibrium constant K_f, calculate the value of K_p. This is 8.0×10^{-3}. Then proceed as in Problem 2 and show that an initial mixture of 1 mole of N_2 and 3 moles of H_2 will give 0.86 mole of NH_3 at equilibrium. The yield is therefore 43 %, because this initial mixture could give *two* moles of NH_3 if conversion were complete.

5. It occurs at a temperature at which ΔH is zero.

6. The enhanced heat capacity is due to the heat of reaction. For simplicity assume that the mixture is perfect and let 1 g of mixture consist of w g of NO_2 and $(1-w)$ g N_2O_4. Then the enthalpy per gram of mixture is

$$wh + (1-w)H,$$

where h and H are the enthalpies per gram of NO_2 and N_2O_4 respectively. Differentiating with respect to temperature, the heat capacity per gram of the mixture is

$$c_p = w\left(\frac{\partial h}{\partial T}\right)_p + h\left(\frac{\partial w}{\partial T}\right)_p + (1-w)\left(\frac{\partial H}{\partial T}\right)_p - H\left(\frac{\partial w}{\partial T}\right)_p$$

$$= wc + (1-w)C + (h-H)\frac{\partial w}{\partial T},$$

where c and C are the heat capacities per gram of NO_2 and N_2O_4 respectively. It follows from Le Chatelier's principle, or from equation (4·21), that the last term in this equation is positive. Hence c_p is larger than would be expected if there were no reaction. (It is quoted in the literature as having the very large value of 1.65 cal/g °C at room temperature.)

The theory can be developed in more detail by expressing $\partial w/\partial T$ in terms of the equilibrium constant and the ΔH of reaction. In this way it can be shown that c_p passes through a maximum value at a certain temperature—as is intuitively obvious. At low temperature there is very little NO_2 and at high temperature there is very little N_2O_4. Hence the contribution to c_p, due to the shift in the equilibrium per unit rise in temperature, reaches a maximum at some intermediate temperature.

7. Note that the addition of steam displaces the reaction

$$H_2O + CO = H_2 + CO_2$$

to the right and produces hydrogen. However, each mol of added steam does not produce as much as 1 mol of hydrogen, because of the equilibrium. Beyond a certain excess of steam the gain in hydrogen ceases to pay for itself.

In setting up the problem mathematically assume that x mols of steam are added to each mol of producer gas, and let this give rise to y moles of H_2. The equilibrium gas is thus:

H_2O	$(x-y)$ mols	CO	$(0.22-y)$ mols
H_2	$(0.14+y)$ mols	CO_2	$(0.07+y)$ mols

Substitute in K_p.
The monetary profit P is

$$P = \text{constant} \times (ny - x),$$

where the constant is a measure of the value of a mole of steam. Hence

$$\frac{dP}{dx} = \text{constant} \times \left(n\frac{dy}{dx} - 1\right),$$

and at the optimum point $dP/dx = 0$. Hence

$$n\frac{dy}{dx} = 1.$$

Apply this condition to the K_p expression and thus obtain an equation for the optimum value of x.

8. NO_2, 4.7%; N_2O_4, 1.7%; N_2O_3, 0.06%; NO, 1.9%.

Note that each mole of N_2O_4 and of N_2O_3 counts as *two* moles as regards the alkali absorption. As regards the behaviour towards the oxidizing agent each mole of N_2O_4 behaves as *two* moles of NO_2 whilst each mole of N_2O_3 behaves as *one* mole of NO_2 and *one* mole of NO.

9. $p_{CO}/p_{O_2}^{\frac{1}{2}} = 10^9$ at 1200 K. Rather tedious but a useful exercise. Note that

$$\frac{p_{CO}}{p_{O_2}^{\frac{1}{2}}} = \frac{p_{CO}^2}{p_{CO_2}} \times \frac{p_{CO_2}p_{H_2}}{p_{CO}p_{H_2O}} \times \frac{p_{H_2O}}{p_{H_2}p_{O_2}^{\frac{1}{2}}}$$

$$= 63 \times \frac{1}{1.4} \times \frac{p_{H_2O}}{p_{H_2}p_{O_2}^{\frac{1}{2}}}.$$

Therefore it is a question of working out the equilibrium constant of the reaction

$$H_2 + \tfrac{1}{2}O_2 = H_2O$$

at 1200 K.

10. Note that, since ΔH is positive, K_p becomes more favourable with rise of temperature. On the other hand, 550 °C is about the highest temperature at which ordinary mild steel plant can be operated continuously. Hence work out K_p at 550 °C.

For this purpose it is necessary to estimate a value for Δc_p, and this may be done by the methods of §12·12, using the principle of equipartition. All molecules may be assumed to behave classically as regards their translational and rotational degrees of freedom. On the other hand, the vibrational modes may be only partially excited. The upper limit for Δc_p corresponds to complete vibrational excitation of the product molecules and no excitation of the reactant molecules. Conversely for the lower limit.† The limits can therefore be shown to be $\Delta c_p = -9$ or $+9$ cal K^{-1} mol^{-1}. The corresponding limits for K_p at 823 K are 320 and 9.5. Thus the ratio of partial pressures p_c/p_a lies between $320^{1/2}$ and $9.5^{1/2}$. The percentage conversion is $\dfrac{100 p_c}{p_a + p_c}$ and lies between 95 and 75 %.

12. This problem requires much numerical working, but it is instructive in showing the difficulty in obtaining accurate values of ΔG and ΔH at $T = 298$ K by extrapolation from typical experimental results at high temperature.

Use equation (4·31). The author finds $\Delta_f H_{298}$ to be -19 to -22 kg cal and $\Delta_f G^0_{298}$ to be -14 to -16 kg cal. The uncertainty in the latter quantity could be reduced by using additional physical information concerning the heat of combustion of CH_4. (See the treatment of Pring and Fairlies' data given by Lewis and Randall, *Thermodynamics*, ch. xi.)

13. $CH_3OH : C_2H_5OH : H_2O = 1 : 46 : 46$.
The independent reactions can be chosen as

$$CO + 2H_2 = CH_3OH,$$

$$2CO + 4H_2 = C_2H_5OH + H_2O,$$

and for each mole of C_2H_5OH one mole of H_2O is also formed.

The algebraic complexity which usually arises in solving simultaneous equilibria can be avoided in this instance. This is because the question asks merely to determine the *ratio* of the three reaction products. Now the equations above can be combined to give

$$2CH_3OH = C_2H_5OH + H_2O.$$

This is therefore the only equilibrium which needs to be considered, and the data quoted in the question on CO is actually irrelevant.

14. $-39\,980$ cal mol^{-1}.

† The possibility of excitation of electronic degrees of freedom is neglected.

15. Reactions for which there is a considerable decrease in the standard free energy are

$$2NO_2 + H_2O = HNO_3 + HNO_2,$$

$$3NO_2 + H_2O = 2HNO_3 + NO,$$

and the second of these is thermodynamically more favoured than the first. Several other reactions, such as

$$NO_2 = \tfrac{1}{2}N_2 + O_2,$$

would also result in a decrease of free energy but are known to be extremely slow. Under the conditions of an industrial nitrous gas absorption process it is the second reaction above which is of principal importance —together with the equilibrium $2NO_2 = N_2O_4$ which is rapidly established.

16. 67 % (wt.).

Calculate first the value of K_p at 20 °C, using the value of ΔH in order to allow for the appreciable effect of change in temperature. Note that K_p can be written

$$K_p = \frac{p_{NO}}{p_{NO_2}^3} \times \frac{p_{HNO_3}^2}{p_{H_2O}}.$$

Now the absorption system is counter-current and the strongest acid which can possibly be made in the first absorption tower is that which would be at equilibrium with the gas entering this tower. For this gas the term $p_{NO}/p_{NO_2}^3$ has a value of $10\,\text{atm}^{-2}$.

Convert the vapour pressure data to atmospheres and plot $p_{HNO_3}^2/p_{H_2O}$ (preferably its logarithm) against the acid concentration. Find the particular acid which satisfies the above equation.

17. $\Delta H = -52\,\text{kg cal}$; heat evolved $= 5.5\,\text{kg cal}$.

With decrease of the p.d. across the terminals, below the reversible e.m.f., the heat evolved would increase and in the limit where the terminals are 'shorted', and no work is done by the cell, the heat evolved would become equal to $-\Delta H$.

18. Calculate the equilibrium constant of the reaction

$$2Ag + \tfrac{1}{2}O_2 = Ag_2O.$$

Now the *total* gas pressure in the process is stated to be only 1 atm, and from the magnitude of the equilibrium constant it is evident that p_{O_2} would need to be very much larger than 1 atm in order that silver oxide should be formed. Hence this substance is not a stable product (at any rate not as bulk phase—the calculation does not exclude the possibility of chemisorbed oxygen).

19. The lowest temperature at which the reaction

$$BaCO_3 = BaO + CO_2$$

could be carried out at an appreciable speed in a furnace open to the atmosphere is that at which the partial pressure of the CO_2 becomes equal to 1 atm. (This temperature is analogous to a boiling-point. At temperatures at which $p_{CO_2} < 1$ atm the process would be very slow, being determined by the rate of diffusion of CO_2 out of the neck of the furnace.) Hence find the temperature at which K'_p is 1 atm.

20. 0.039% SO_3. Sulphate is not formed.
Consider the reaction $SO_2 + \frac{1}{2}O_2 = SO_3$. It is readily calculated that

$$\frac{p_{SO_3}}{p_{SO_2}p_{O_2}^{\frac{1}{2}}} = 1.66 \times 10^{-2}, \tag{1}$$

and from this equation it is evident that the concentration of SO_3 is very small, and its formation does not significantly affect the concentration of oxygen in the outgoing gas.

Assume that $ZnSO_4$ is *not* formed. Then the only significant reaction as regards the materials balance of the system is $ZnS + \frac{3}{2}O_2 = ZnO + SO_2$. Since there is 7% SO_2 in the outgoing gas it can be readily calculated, by means of an oxygen balance, that there is 11.3% O_2 in the outgoing gas. The formation of a trace of SO_3 does not significantly affect this result. The figures $p_{SO_2} = 0.07$ and $p_{O_2} = 0.113$ can now be substituted in (1) to give $p_{SO_3} = 3.9 \times 10^{-4}$ atm. (This kind of approximation, where it is valid, avoids the solution of simultaneous equations.)

It remains to be confirmed that $ZnSO_4$ is not a stable product at the given temperature. For this purpose calculate the equilibrium constant of the reaction
$$ZnO + SO_3 = ZnSO_4,$$

and show that the calculated partial pressure of SO_3 is far too low for the reaction to proceed from left to right.

If the actual content of SO_3 in the gas leaving the process was found to be *higher* than the above calculated equilibrium value, a likely explanation would be that the SO_3 is an intermediate substance in the reaction mechanism. Its concentration would pass through a maximum before falling to the equilibrium value.

21. 55%.
Note that in the reaction $A = B + C$ increased yield could be attained by *reduction* of pressure. However, since it is specified that the process must be carried out at a total pressure of 1 atm, the same effect can be achieved by introduction of an inert gas X, which has the effect of reducing the partial pressures of A, B and C. This is the 'device' asked for in the question.

On the other hand, too much of the inert gas will reduce the amount of B which can be condensed (the remainder is carried away in the gas stream). Therefore it is a question of maximizing the yield.

Let p_a, etc., be the partial pressures at the outlet of the reaction vessel. From the free-energy figure it is readily calculated that

$$p_a = 10p_b^2. \tag{1}$$

Also $\qquad p_a + p_b + p_c + p_x = 1$ atm,

and therefore, using (1), $\quad 10p_b^2 + 2p_b + p_x = 1. \tag{2}$

In a given interval of time let 1 mole of this gas enter the condenser and let x moles emerge from it. Since the vapour pressure of B is 0.01 atm and the total pressure of the gas is 1 atm, we have from the perfect gas laws:

Gas entering condenser: p_a moles of A, p_c $(=p_b)$ moles of C.

                                  p_b moles of B, p_x moles of X.

Gas leaving condenser: p_a moles of A, p_b moles of C.

                                  $0.01x$ moles of B, p_x moles of X.

Hence $p_a + 0.01x + p_b + p_x = x$

or $10p_b^2 + p_b + p_x = 0.99x,$ (3)

by substitution from (1).

The output of liquid B, in the given time interval, is $(p_b - 0.01x)$ moles, and this has been obtained from $p_a + p_b$ moles of A originally entering the reactor. Hence the yield of liquid B, expressed as a fraction, is

$$y = \frac{p_b - 0.01x}{p_a + p_b} = \frac{p_b - 0.01x}{10p_b^2 + p_b},$$ (4)

by (1).

Subtracting (3) from (2) $p_b = 1 - 0.99x,$ (5)

and eliminating x between (4) and (5)

$$y = \frac{p_b(r+1) - r}{10p_b^2 + p_b},$$ (6)

where $r = 0.01/0.99$.

Finally differentiate (6) and put $dy/dp_b = 0$ to obtain the maximum value of y. This is 0.55. If no inert gas were added it can be readily calculated that the maximum yield would be only 0.29.

Chapter 5

1. (*a*) 3; (*b*) 2; (*c*) 2, together with the restriction $p_{H_2} = 3p_{N_2}$ (this is equivalent to saying that there is only one component and no restriction).

2. There are $n + 2$ independent variables. The experimentalist is free to choose the n composition variables together with the temperature and *either* p_w or p_s. (The difference, $p_s - p_w$, is the osmotic pressure and is determined as soon as the other variables are fixed.) The proof is as follows.

The total number of variables is $n + 4$, i.e. p_w, p_s, T_w, T_s, and the n mole fractions. Between these variables there are two equalities, namely $T_w = T_s$ together with the equality of the chemical potentials of the solvent. Hence $F = n + 4 - 2 = n + 2.$

3. (*a*) $F = 0$. Note that along the length of the vapour-liquid equilibrium curve we have $F = 1$. However, the critical point is not *any* point along this curve but is the special point, which is the termination of the curve, where the two phases have equal density and become indistinguishable. Therefore there is a special restriction. Looked at from an alternative point of view, immediately above the critical point there is

only one phase together with *two* restrictions, namely, $\partial p/\partial V = 0$ and $\partial^2 p/\partial V^2 = 0$. This again results in $F = 0$.

(b) $F = 1$.

(c) There are four independent reactions. If MNO_2, MNO_3 and M_2O are immiscible there are four phases and $F = 1$, there being no stoichiometric restrictions on the composition of any phase, despite the fact that the system is prepared from nitrite only. If MNO_2, MNO_3 and M_2O are miscible the number of phases is reduced to two but there is now a rather subtle stoichiometric relationship which results in $F = 2$. This relationship† is

$$x_2(2x_4 + x_6 + x_7) = x_3(6x_4 - 4x_5 + x_6 - x_7),$$

where the x_i's refer to mole fractions in condensed or vapour phases and the numbers refer to the substances in the following sequence:

$$\begin{array}{ccccccc} MNO_2, & MNO_3, & M_2O, & N_2, & O_2, & NO, & NO_2 \\ 1 & 2 & 3 & 4 & 5 & 6 & 7 \end{array}$$
$\leftarrow$ condensed phase $\rightarrow$ $\leftarrow$ vapour phase $\rightarrow$

4. Four independent reactions, $F = 3$.

5. (a) $F = 1$.

(b) There are two independent reactions and these can be chosen as

$$FeO + CO = Fe + CO_2, \tag{A}$$

$$Fe_3O_4 + CO = 3FeO + CO_2. \tag{B}$$

The equilibrium temperature at 1 atm total pressure will be the temperature at which the value of the partial equilibrium constant, $K'_p = p_{CO_2}/p_{CO}$, is the same for both of the above reactions. Also

$$p_{CO} + p_{CO_2} = 1 \text{ atm.}$$

However, a simpler procedure is as follows. Subtract one of the above equations from the other to obtain

$$Fe_3O_4 + Fe = 4FeO. \tag{C}$$

Now these solid substances will only be at equilibrium with each other at a temperature such that ΔG is zero. This temperature is readily calculated by integrating the equation

$$\frac{\partial \Delta G/T}{\partial T} = -\frac{\Delta H}{T^2}$$

and using the given data. The result is $T = 1120$ K. The CO and CO_2 will also be at equilibrium with the solids at this temperature provided that the ratio of their partial pressures satisfies the equilibrium condition.

(c) It has been shown that $F = 1$. On the other hand, the equilibrium temperature will change only imperceptibly if the total pressure of CO and CO_2 is increased above 1 atm. This is because the free energy of the solid phases is not significantly affected by change of pressure.

(d) The calculated result is much higher than the observed equilibrium temperature which is stated to be 840 K. The calculation depends

† I am indebted to Mr R. Jackson of Edinburgh for drawing my attention to its existence.

on the difference between two large numbers, in regard to the free energy of FeO and Fe_3O_4, and this is probably the main source of the error. Also the value of c_p for FeO is an estimated rather than a measured value and the temperature coefficient of Δc_p has been neglected.

6. This problem is concerned with the properties of the triangular diagram used for representing ternary systems. Use the 'lever rule' and notice that if the residue were mixed with the distillate at any moment it would give a mixture represented by point A. Note also that the composition of the distillate is always somewhere on XY because Z is not volatile.

CHAPTER 6

1. Involves an integration of the Clausius–Clapeyron equation. Also

$$\ln p_2/p_1 \doteq \frac{p_2 - p_1}{p_1} \quad \text{when} \quad \frac{p_2 - p_1}{p_1} \ll 1.$$

2. -0.133 °C. Take care to use the correct energy units in this type of problem.

3. $dL/dT = 2.72$ J K^{-1} g^{-1}. (Use data from Problem 2.)

4. 2.25 kJ g^{-1}.

5. Considering a fixed quantity of vapour:

$$dV = \left(\frac{\partial V}{\partial T}\right)_p dT + \left(\frac{\partial V}{\partial p}\right)_T dp,$$

and therefore

$$\left(\frac{\partial V}{\partial T}\right)_{\text{sat.}} = \left(\frac{\partial V}{\partial T}\right)_p + \left(\frac{\partial V}{\partial p}\right)_T \left(\frac{\partial p}{\partial T}\right)_{\text{sat.}},$$

where the subscript sat. denotes a condition of equilibrium between liquid and vapour. Assume that the vapour is perfect and apply equation (6·8). Note that the equation to be proved shows that the coefficient of expansion is normally *negative*. Why is this?

6. Of the order of 7 g m^{-3}, but this result is only a rough approximation. The calculation depends on using the Trouton relation in order to estimate the latent heat. The Clausius–Clapeyron equation, together with the known boiling-point, can then be used to estimate the vapour pressure of the oil at 20 °C. This is about 1.5×10^{-3} atm. The calculation of the weight of oil vaporized is then obtained from the gas law together with the known molecular weight.

7. At any temperature at which the two phases are in equilibrium, the pressure has a fixed value. Hence V is a function only of T and q and an equation of the type of (2·78) can be applied:

$$\left(\frac{\partial q}{\partial T}\right)_V = -\left(\frac{\partial V}{\partial T}\right)_q \bigg/ \left(\frac{\partial V}{\partial q}\right)_T.$$

Also we have

$$V = qv'' + (1-q)v',$$

whence

$$(\partial V/\partial q)_T = v'' - v'.$$

8. Let s'' be the specific entropy of the vapour phase. Then

$$c_s'' = T \left(\frac{\partial s}{\partial T} \right)_{\text{sat.}},$$

where the subscript sat. denotes the condition of two-phase equilibrium.
Hence

$$c_s'' = T \left\{ \left(\frac{\partial S}{\partial T} \right)_p + \left(\frac{\partial S}{\partial p} \right)_T \left(\frac{\partial p}{\partial T} \right)_{\text{sat.}} \right\}$$

$$= T \left\{ \frac{c_p''}{T} - \left(\frac{\partial v''}{\partial T} \right)_p \frac{L}{T \Delta V} \right\},$$

by a Maxwell relation and (6·8).

Similarly for the second relation. To obtain the third relation apply equation (6·13).

The value of c_s'' for steam at 100 °C is *negative* and is about -4.5 J K^{-1} g^{-1}. Thus the temperature is raised by *removal* of heat from the steam, the pressure being increased at the same time to maintain saturation. (Note that the specific volume of the steam simultaneously decreases, by the result of Problem 5.)

9. 53 % condensed.

For lack of other information it must be assumed that the NH_3 behaves as a perfect gas. The application of equation (6·24) therefore gives the vapour pressure as 15 atm at 250 atm total pressure.

Consider the inlet gas to the condenser; this contains 0.12 mol NH_3 to every 0.88 mol of $N_2 + H_2$. In the outlet gas the partial pressure of NH_3 is estimated to be 15 atm and its mol fraction is therefore $15/250 = 0.06$. There are therefore 0.06 mol of NH_3 to every 0.94 mol of $N_2 + H_2$, or

$$0.06 \times \frac{0.88}{0.94} = 0.056$$

mol of NH_3 to every 0.88 mol of $N_2 + H_2$. The fraction of the NH_3 *not* condensed is therefore $0.056/0.12 = 0.47$, and the percentage condensed is 53 %. (Notice the use of the inerts as reference substances in this type of calculation.)

10. In the absence of p-V-T relations for the mixture, the Lewis and Randall rule must be used.

Chapter 7

2. It will be found best to plot the logarithms of the partial pressures. The data do not agree at all well with the Duhem–Margules equation, and it appears therefore that there is considerable experimental error.

3. It is only at the minimum point that the compositions of the solid and liquid phases in equilibrium are equal.

6. Let μ_h, μ_w and μ_s be the chemical potentials of the solid hydrate, the liquid water and the salt in solution respectively. Then for equilibrium between the solid hydrate and the solution,

$$\mu_h = 2\mu_w + \mu_s.$$

μ_h is a function of T and p. μ_w and μ_s are functions of T and p together with the mole fraction x of the water in the solution. Differentiate the above equation with respect to the independent variables and proceed as in the treatment leading up to equation (7·21).

CHAPTER 8

1. 244, 252 and 260 mmHg respectively. The corresponding mole fractions of benzene in the vapour are 0.27, 0.53 and 0.77 respectively.

2. $\Delta H = 28.4$ kJ mol^{-1}. (The directly measured value is 28.0 kJ mol^{-1}.)

3. Either hypothesis is equally applicable to the quoted results. Independent evidence indicates that the second hypothesis is more nearly correct.

4. The ideal solubility expressed as a mole fraction is 0.026. (The observed values are 0.018 in hexane and 0.028 in heptane.) Notice that both the boiling-point and the critical point lie on the vapour-pressure curve and can therefore be used for interpolating the vapour pressure at 20 °C by application of the Clausius–Clapeyron equation.

5. $\Delta H = -20.5$ kJ mol^{-1}; 42 m^3 h^{-1} of water.

The solubilities being very small, the figures quoted in the second line of the question can be taken as being proportional to the mole fractions of dissolved CO_2. Apply equation (8·35).

The least quantity of water required is that which would be just saturated by the CO_2 at its partial pressure (2 atm) entering the tower.

6. $B = \Delta H + RT = (-1.38 + 0.24) \times 10^4 = -1.14 \times 10^4$ J mol^{-1}. The quantity B is equal to ΔU in the process of solution.

Notice that s is the solubility when the partial pressure of CO_2 is 1 atm. Let m be the number of cm^3 of CO_2 per 100 cm^3 of rubber when the partial pressure is p. Then

$$m = ps \quad \text{or} \quad s = m/p \tag{1}$$

by Henry's law. Because the solution is very dilute m is proportional to the mole fraction x of dissolved CO_2 and the above equation can also be written

$$s = \text{constant} \times x/p. \tag{2}$$

The first equation in the question then follows from equation (8·35), if the temperature coefficient of the density of the rubber is neglected.

Now $pV = nRT$, where n is the number of moles of CO_2 in a gas volume V. Hence

$$c_A \equiv \frac{n}{V} = p/RT. \tag{3}$$

Also m, in equation (1), is proportional to c_B. Hence from (1) and (3)

$$\frac{c_B}{c_A} = \text{constant} \times \frac{m}{p} RT = \text{constant} \times sT.$$

Hence
$$\frac{d \ln c_B/c_A}{dT} = \frac{d \ln s}{dT} + \frac{d \ln T}{dT}$$

$$= \frac{\Delta H}{RT^2} + \frac{1}{T}$$

$$= \frac{\Delta H + RT}{RT^2}.$$

Therefore $B = \Delta H + RT = -1.14 \times 10^4$ J mol^{-1} at $T = 300$ K (approx.).
In the process of solution

$$\Delta H = \Delta U + \Delta(pV)$$
$$= \Delta U + p_2 V_2 - p_1 V_1$$
$$= \Delta U + 0 - RT,$$

because V_2, the volume in the dissolved state, is negligible and because $p_1 V_1$, referring to the gaseous state, is equal to RT per mol.

Hence $B = \Delta H + RT$ is equal to ΔU. (Cf. the temperature coefficients of a reaction equilibrium constant expressed in terms of partial pressures and in terms of concentrations in §4·7.)

7. The problem shows the existence of *apparent* deviations from Raoult's law, in an ideal solution, due to the occurrence of a reaction. In setting up the problem it is perhaps best to let x_A, x_B and x_0 be the true mole fractions of A, B and C in the solution. The condition of equilibrium in the reaction $2A = B$ may be expressed as

$$\frac{x_B}{x_A^2} = K,$$

as will be discussed in more detail in Chapter 10. If the solution is ideal over the whole range, the same relation will hold in the pure liquid (i.e. when $x_0 = 0$).

The quantity N_A which appears in the third equation of the problem is, of course, the actual number of moles of the monomer in the solution. If N_B is the corresponding number of moles of dimer then

$$N_A + 2N_B = N.$$

(This problem is based on a discussion by Rushbrooke, *Introduction to Statistical Mechanics*, Chapter XIV.)

CHAPTER 9

2. 0.871; 4.16 kJ mol^{-1}.

3. $\gamma = 1.06$. The equation in the problem is based on the integration of

$$\frac{\partial \Delta G/T}{\partial T} = -\frac{\Delta H}{T^2},$$

putting $\Delta G = 0$ at 1808 K at which there is equilibrium between solid and liquid iron. At any other temperature the equation gives the difference between the free energy of liquid and solid iron when they are not at equilibrium.

In the second part of the problem let μ be the chemical potential of the liquid iron in the solution. Then

$$\mu = \mu^* + RT \ln x\gamma,$$

where all quantities refer to the iron. But μ is equal to the free energy per mole of solid iron since there is equilibrium. Also μ^* is equal to the free energy per mole of pure liquid iron. Hence $\mu^* - \mu$ can be evaluated from the equation already obtained. Hence $\gamma = 1.06$ (Basis $\gamma \to 1$ as $x \to 1$).

4. The conditions of equilibrium $\mu_a = \mu_a'$ and $\mu_b = \mu_b'$ lead immediately to $x_a \gamma_a = x_a' \gamma_a'$ and $x_b \gamma_b = x_b' \gamma_b'$.

5. 0.8990 (in very close agreement with the directly measured value). The data for this question were taken from J. A. V. Butler, *Commentary on the Scientific Writings of Willard Gibbs*, vol. I, p. 140.

6. To each side of (9·41) add $d \ln x_b$. This gives

$$d \ln (\gamma_b x_b) = -\frac{1}{r} d \ln \gamma_a + d \ln x_b$$

$$= -\frac{1}{r} d \ln (\gamma_a x_a). \tag{1}$$

From the defining equation for h

$$dh = \frac{1}{r} d \ln (\gamma_a x_a) - \frac{\ln (\gamma_a x_a) \, dr}{r^2}$$

$$= \frac{1}{r} d \ln (\gamma_a x_a) - \frac{(h-1)}{r} \, dr. \tag{2}$$

Substituting (2) in (1)

$$d \ln (\gamma_b x_b) = -dh - \frac{(h-1)}{r} \, dr$$

$$= -dh - h \, d \ln r + d \ln r.$$

On rearranging
$$d \ln \left(\frac{\gamma_b x_b}{r} \right) = -dh - h \, d \ln r$$

and therefore
$$\ln \left(\frac{\gamma_b' x_b'}{r'} \right) = -h' - \int_0^{r'} \frac{h}{r} \, dr. \tag{3}$$

In this equation the integrand is finite at the lower limit. The equation may be applied by plotting values of h/r against r and taking the area under the curve.

7. 195.3 mmHg (observed value 194.9 mmHg).

Let G^E, etc., be the excess functions. Since $S^E = 0$ for the given solution,

$$G^E = H^E$$

$$= (n_1 + n_2)\,bx_1x_2$$

$$= \frac{bn_1n_2}{n_1 + n_2}.$$

The application of equation (9·46) gives the required expressions.

From the azeotropic data an activity coefficient can be worked out for each component at the azeotropic composition (equations (7·69) and (7·70)). When these are substituted in the given equations, two almost equal values of b are obtained. The data are thus not inconsistent with the postulates (but, of course, are quite insufficient to prove them).

The equations, together with the value of b, can finally be used to calculate the partial pressures of each component. Their addition gives the total pressure and this is 195.3 mmHg.

8. $A = 0.394$, $B = 0.375$.

At the lower temperature the estimated azeotropic composition is 60% mole of ethyl acetate (observed value 60.1%) and the calculated total pressure is 413 mmHg.

9. It is stated that the pressure p which is of interest lies between p_1 and p_2. Thus $p_2 > p > p_1$ and the pressure p is greater than the vapour pressure of component 1. This vapour will therefore be supersaturated and the compressibility factor for this component must be estimated indirectly, for example by extrapolation from the unsaturated region. (N.B. The compressibility factor referred to in the question is that defined in equation (3·51): $Z = pv/RT$.)

Consider component 2 whose mole fractions in liquid and vapour phases are x and y respectively. The condition of equilibrium for this component is

$$\mu^{\text{vap.}} = \mu^{\text{soln.}}.$$

In view of (3·68) and (8·14) this can be rewritten as

$$\mu_{gp}^* + RT \ln y = \mu_{lp}^* + RT \ln x,$$

where μ_{gp}^* stands for the chemical potential of the pure gaseous component at *the pressure p* of the solution in question, and μ_{lp}^* is the corresponding chemical potential of the pure liquid component, also at this pressure.

Now the pure component will be in equilibrium with its vapour at the pressure p_2. Let $\mu_{p_2}^*$ be the chemical potential of the liquid and vapour phases at this particular pressure. If this quantity is subtracted from both sides of the previous equation we obtain after rearranging

$$RT \ln y/x = (\mu_{lp}^* - \mu_{p_2}^*) - (\mu_{gp}^* - \mu_{p_2}^*).$$

The first bracket on the right-hand side is equal to $v_2'(p - p_2)$ by the integration of $(2 \cdot 111 b)$. The second bracket on the right-hand side is similarly equal to

$$\int_{p_2}^{p} v \, dp,$$

where v is the molar volume of the vapour of component 2.

Now the compressibility factor is

$$Z = \frac{pv}{RT} = a_2 + b_2 p.$$

Hence $v = RT'\left(\dfrac{a_2}{p} + b_2\right)$ and the above integral is obtained as

$$RT\left\{a_2 \ln \frac{p}{p_2} + b_2(p - p_2)\right\}.$$

Hence finally

$$RT \ln y/x = v_2'(p - p_2) - RT\left\{a_2 \ln \frac{p}{p_2} + b_2(p - p_2)\right\},$$

which can be rearranged to give the second equation quoted in the problem. The first equation, referring to component 1, follows at once in an analogous manner. It may be noted that the term $v_2'(p - p_2)$, in the last equation above, is equivalent to the very slight change in the vapour pressure of pure component 2 between the total pressures p and p_2. (See footnotes on p. 222 and p. 248.)

10. Per mole of the mixture the changes of G and H are

$$\Delta G = RT(x_1 \ln x_1 + x_2 \ln x_2) + \alpha x_1 x_2,$$

$$\Delta H = \alpha x_1 x_2.$$

(Cf. Problem 7.)

The second part of the problem raises important matters of principle. Using the numerical value for α, together with the equations above, it is readily worked out that the changes in G and H in the *demixing* process are $+1590$ and -104 J mol^{-1} of mixture respectively. Therefore the *least amount of work* required for separation is 1590 J mol^{-1}. This may be thought of as being supplied by means of a reversible heat engine taking heat from the hot reservoir at 373 K and rejecting heat to the cold reservoir (the cooling water) at 293 K. Hence from equation $(1 \cdot 12 b)$ the minimum heat to be taken from the hot reservoir is

$$1590 \times \frac{373}{373 - 293} = 7410 \text{ J},$$

and the corresponding amount of heat to be rejected by the engine to the cold reservoir is $7410 - 1590 = 5820$ J.

In addition, there is the heat effect on separating the two components, and this is *not* equal to the ΔH of demixing because the process involves forms of work other than volume change (see § 2·2). The heat effect is, of course, $T \Delta S$, since the process considered is reversible and isothermal. Now

$$T \Delta S = \Delta H - \Delta G$$

$$= -104 - 1590 = -1694 \text{ J mol}^{-1}.$$

Since this is negative it corresponds to heat rejected to the cooling water, and thus the total heat rejected is $1694 + 5820 = 7514$ J mol^{-1} of the mixture.

Notice the important clause in the question 'if the energy of the reservoir is used at maximum efficiency'. This requires a reversible separation process such as has been discussed above. How could such a process be approximated in practice?

An ordinary distillation is very far from being reversible and usually has an efficiency on the above basis of 5–20 %. This is due to finite temperature differences and also because the vapour entering any plate of the distilling column is not at equilibrium with the liquid which is on that plate.

11. For the present purposes the term 'regular binary solution' may be taken as meaning a solution having activity coefficients given by the expressions quoted. Dividing the first of these expressions by the second, we obtain

$$\left(\frac{1-x}{x}\right)^2 = \frac{\ln \gamma_A}{\ln \gamma_B}.$$

Now at the azeotropic composition the activity coefficients are given by equations (7·69) and (7·70):

$$\gamma_A = p/p_A^*, \quad \gamma_B = p/p_B^*,$$

where p is the azeotropic pressure and p_A^* and p_B^* are the vapour pressures at the azeotropic temperature T_z. Hence

$$\left(\frac{1-x}{x}\right)^2 = \ln \frac{p}{p_a^*} \bigg/ \ln \frac{p}{p_b^*}.$$

Now the azeotropic pressure p is 1 atm. Consider the integration of the Clausius–Clapeyron equation for pure component A between temperature T_z and the boiling-point T_A. Assuming the latent heat to be constant,

$$\ln \frac{p}{p_a^*} = -\frac{L_A}{R} \left(\frac{1}{T_A} - \frac{1}{T_Z}\right),$$

and similarly for the other component. The substitution of these expressions in the previous equation, together with the use of the Trouton relation, gives the required equation.

It may be remarked that this equation agrees quite well with experiment for a considerable number of azeotropic mixtures. This agreement does not imply, however, that these mixtures are accurately regular.

CHAPTER 10

1. $+1650$ cal mol^{-1}; -1080 cal mol^{-1}.

2. 0.42 mol; 0.55 mol at 200 °C (the observed value is 0.67).

3. (a) $-26\,360$ cal mol^{-1}; (b) $-19\,890$ cal mol^{-1}; (c) $-22\,270$ cal mol^{-1}.

It is stated that the vapour pressure is 76 mmHg and also that at this pressure the solubility in water is 0.001 mol kg^{-1}. It follows that the 0.001 mol kg^{-1} solution can be put into equilibrium with the substance as a pure liquid; i.e. under a pressure of 76 mmHg there will be two phases, one consisting almost entirely of water and the other consisting of the pure (or almost pure) substance. The vapour-pressure curves are therefore an extreme example of those shown in Fig. 32 on p. 228.

4. For Hg_2Cl_2, $\Delta_f G_{298} = -50\,400$ cal mol^{-1}.

Consider first the equilibrium of 4 mol kg^{-1} HCl solution with its vapour:

$$\mu_{H^+}^\square + \mu_{Cl^-}^\square + RT \ln m^2 \gamma_\pm^2 = \mu_{HCl}^0 + RT \ln p_{HCl}.$$

Hence

$$\mu_{H^+}^\square + \mu_{Cl^-}^\square - \mu_{HCl}^0 = -8623 \text{ cal}.$$

The standard free energy of formation of the chloride ion is therefore $-8623 - 22\,770 = -31\,393$ cal mol^{-1}.

This result is now combined with the quoted electrode potential to give the free energy of formation of mercurous chloride, as described in § 10·15a.

5. (a) Show, for example, that the measured partial pressures satisfy equation (10·84).

(b) It is necessary to use the Clausius–Clapeyron equation in order to estimate a value for the vapour pressure of HCl at 25 °C. It will be found that the measured partial pressures over the solution are minute compared to those expected on the basis of Raoult's law. It is therefore a case of very large negative deviations. (See also Fig. 29 on p. 225.)

6. $p = 7.5 \times 10^{-6}$ atm; $m\gamma_\pm = 455$.

7. The molalities are: (a) dissolved chlorine 3.07×10^{-2}; (b) HOCl, 2.45×10^{-2}; (c) Cl$^-$, 2.45×10^{-2}; (d) ClO$^-$, 6.6×10^{-10}.

Because activity coefficients are not quoted it is necessary to assume that the solution is approximately ideal. The concentration of dissolved chlorine is obtained immediately from the condition of equilibrium between itself and the gaseous chlorine:

$$\mu_{Cl_2}^\square + RT \ln m_{Cl_2} = \mu_{Cl_2}^0 + RT \ln p_{Cl_2},$$

whence

$$-RT \ln \frac{m_{Cl_2}}{0.5} = 1650,$$

and thus

$$m_{Cl_2} = 3.07 \times 10^{-2}.$$

Consider now the equilibria in solution. These can be chosen conveniently as

$$Cl_2 + H_2O = H^+ + Cl^- + HOCl, \tag{A}$$

$$HOCl = H^+ + ClO^-. \tag{B}$$

Using the free-energy data we obtain the values of the equilibrium constants

$$\frac{m_{\text{H}^+} m_{\text{Cl}^-} m_{\text{HOCl}}}{m_{\text{Cl}_2} x_{\text{H}_2\text{O}}} = 4.83 \times 10^{-4}, \tag{1}$$

$$\frac{m_{\text{H}^+} m_{\text{ClO}^-}}{m_{\text{HOCl}}} = 6.7 \times 10^{-10}. \tag{2}$$

From the second of these it is evident that the degree of ionization of the HOCl is very small indeed. Hence to a very high degree of approximation

$$m_{\text{H}^+} = m_{\text{Cl}^-} = m_{\text{HOCl}} \tag{3}$$

from the stoichiometry of reaction (A). Also $x_{\text{H}_2\text{O}} \approx 1$. Hence (1) and (2) can be written

$$\frac{m_{\text{Cl}^-}^3}{m_{\text{Cl}_2}} = 4.83 \times 10^{-4},$$

$$m_{\text{ClO}^-} = 6.7 \times 10^{-10},$$

respectively. Since m_{Cl_2} has already been calculated we can obtain m_{Cl^-} and hence also m_{HOCl} from (3).

8. The work to be done on the system is 20.1 kcal mol^{-1} of CaCl_2. The heat to be removed is 1.0 kcal mol^{-1}.

A 0.5 mol kg^{-1} solution contains 0.5 mol CaCl_2 and 55.51 mols of water. Hence to every 1 mol of CaCl_2 there are 111 mols H_2O. The total free energy of a solution containing 1 mol of CaCl_2 is therefore

$$111(\mu_w^* + RT \ln \gamma_w x_w) + \mu_{\text{Ca}^{2+}}^{\square} + 2\mu_{\text{Cl}^-}^{\square} + RT \ln \gamma_{\pm}^3 \, m_{\text{Ca}^{2+}} + m_{\text{Cl}^-}^2,$$

where the subscript w refers to the water. The free energy after separation is

$$111\mu_w^{\#} + \mu_s,$$

where μ_s refers to the solid CaCl_2. The increase in G on separation is therefore

$$\Delta G = \mu_s - \mu_{\text{Ca}^{2+}}^{\square} - 2\mu_{\text{Cl}^-}^{\square} - RT(111 \ln \gamma_w x_w + \ln \gamma_{\pm}^3 \, m_{\text{Ca}^{2+}} + m_{\text{Cl}^-}^2).$$

The numerical value of this expression is

$$\Delta G = 20.1 \text{ kcal}.$$

In this calculation it is necessary to assume that γ_w is unity, as a figure is not given. (This actually introduces very little error, the deviation of the water from ideality being comparatively small.) The mole fraction of the water is

$$x_w = \frac{55.51}{55.51 + 0.5 + 1.0} = \frac{55.51}{57.01}.$$

The ΔH of separation is $-190.0 + 209.1 = +19.1$ kcal. Hence $T\Delta S = \Delta H - \Delta G = -1.0$ kcal, and the heat to be removed is 1.0 kcal. This is clearly rather approximate as it depends on the difference of some large numbers.

9. About 10^{-14} atm.

The concentrations of all ions are very low, and it will be satisfactory to assume that the solution is approximately ideal. The equilibrium constant for the reaction

$$CO_2(g) + H_2O = 2H^+ + CO_3^{2-}$$

is first worked out from the free-energy data and is found to be

$$\frac{m_{H^+}^2 \, m_{CO_3^{2-}}}{p_{CO_2} \, x_{H_2O}} = 10^{-18}.$$

Also we have

$$m_{H^+} m_{OH^-} = 10^{-14},$$

$$m_{Ca^{2+}} + m_{CO_3^{2-}} = 0.87 \times 10^{-8}$$

$$m_{Ca^{2+}} + m_{OH^-}^2 = 0.0211 \times (2 \times 0.0211)^2$$

$$= 3.7 \times 10^{-5}.$$

Solving these equations and putting $x_{H_2O} = 1$, we obtain

$$p_{CO_2} = 2.4 \times 10^{-14} \text{ atm.}$$

Of course in any actual experiment such a minute partial pressure could not be attained on account of the slowness of absorption.

CHAPTER 11

1. See the remarks on p. 78.

CHAPTER 12

1. It has to be shown that the number of translational quantum states per molecule, whose energy ϵ_i is such that $e^{-\epsilon_i/kT}$ is not insignificant, is much larger than the number of molecules in the vessel. When this condition is satisfied, terms of high index i will make an appreciable contribution to the partition function Q of the whole system and the approximation of dividing by $N!$, as adopted in § 12·1, is justified.

Now $e^{-\epsilon_i/kT}$ will be significant if ϵ_i is of the order of magnitude kT or less. Putting $\epsilon = kT$ in (12·88), the number of translational quantum states per molecule for which the energy is kT or less is

$$\tfrac{4}{3}\pi V \frac{(2mkT)^{\frac{3}{2}}}{h^3}.$$

Consider a concentration of N particles per cm³. Hence

$$\frac{\text{number of quantum states per molecule}}{\text{number of molecules in vessel}} = \tfrac{4}{3}\pi \frac{(2mkT)^{\frac{3}{2}}}{Nh^3}.$$

Evaluate this expression numerically using (a) $N = 10^{19}$, $T = 10$ K and $m = $ mass of a hydrogen molecule; (b) $N = 10^{23}$, $T = 10$ K, and $m = $ mass of an electron.

2. See, for example, Herzfeld's article in Taylor and Glasstone's *Treatise on Physical Chemistry*, New York, Van Nostrand, 1951, vol. II, p. 49.

3. Note that relatively more fast-moving molecules pass through the plane in unit time than slow-moving ones, and also that it is the fast-moving molecules which carry the most kinetic energy through the plane.

CHAPTER 13

1. The first excited mode is that for which the vibrational quantum number is unity and its energy is

$$\epsilon_1 = (1 + \tfrac{1}{2}) h\nu_E$$

by equation (13·22). Hence from equation (13·43) the fraction of the modes which are in the first or higher excited levels is

$$\phi = e^{-h\nu_E/kT} = e^{-\theta_E/T}.$$

This fraction increases with temperature and it increases *most rapidly* when $d^2\phi/dT^2 = 0$.

With regard to c_V note the similarity of the situation to that in Problem 3 of Chapter 12.

2. See the remarks at the end of § 13·10.

4. Consider 1 mol of substance and let f be the fraction which is present as vapour and $1 - f$ the fraction present as crystal. Then the internal energy of the system is

$$u = f u_g + (1 - f) u_c,$$

where u_g and u_c are the internal energies per mole of the vapour and crystal respectively. Similarly the entropy is

$$s = f s_g + (1 - f) s_c.$$

From (12·56) and (13·45)

$$u_g = \tfrac{3}{2} RT,$$

$$u_c = E_0 + 3RT.$$

Hence
$$u = E_0 + 3RT - f(E_0 + \tfrac{3}{2}RT)$$

(and in this expression E_0 is a negative quantity). The internal energy of the system relative to the value when f is zero (as shown in Fig. 14) is therefore

$$u - E_0 - 3RT = -f(E_0 + \tfrac{3}{2}RT),$$

and is proportional to f. The expression can be worked out numerically putting, say, $E_0 = -4 \times 10^4$ J, $T = 10^3$ K.

Proceed similarly with regard to the entropy, using equation (12·64) and (13·47). The entropy of the system is

$$s = fs_g + (1-f)s_c$$
$$= fR(\ln M^{\frac{3}{2}}T^{\frac{5}{2}} - \ln p - 1·16) + (1-f)\,3R(1 + \ln kT/h\bar{\nu}),$$

where p is the pressure in atmospheres and Ω_0 has been taken as unity. In this equation p can be expressed in terms of the fraction f together with the vapour phase volume V (cm³)

$$p = fRT/V$$

where R in this expression is in cm³ atm K⁻¹. Choosing $T = 10^3$ and $V = 150\,\text{cm}^3$, then $p = 550f$ atm. Let the molecular weight M be chosen as 20 and the frequency $\bar{\nu}$ as 10^{12}. The above expression for the entropy can now be evaluated numerically.

What is the fraction present as vapour when there is equilibrium?

5. The value of $S_{\text{spec.}}$, as calculated by use of the Sackur–Tetrode equation, is 61.5 cal K⁻¹ mol⁻¹ and the value of $S_{\text{calor.}}$, as obtained by graphical integration, is about 61.2 cal K⁻¹ mol⁻¹.

INDEX